AF361722

Sustainable Building and Indoor Air Quality

Sustainable Building and Indoor Air Quality

Editor

Roberto Alonso González Lezcano

MDPI • Basel • Beijing • Wuhan • Barcelona • Belgrade • Manchester • Tokyo • Cluj • Tianjin

Editor
Roberto Alonso González Lezcano
Universidad CEU San Pablo
Spain

Editorial Office
MDPI
St. Alban-Anlage 66
4052 Basel, Switzerland

This is a reprint of articles from the Special Issue published online in the open access journal *Sustainability* (ISSN 2071-1050) (available at: https://www.mdpi.com/journal/sustainability/special_issues/Sustainable_Building_Indoor_Air_Quality).

For citation purposes, cite each article independently as indicated on the article page online and as indicated below:

LastName, A.A.; LastName, B.B.; LastName, C.C. Article Title. *Journal Name* **Year**, *Volume Number*, Page Range.

ISBN 978-3-0365-1106-1 (Hbk)
ISBN 978-3-0365-1107-8 (PDF)

Contents

About the Editor

Roberto Alonso González Lezcano Tenured Professor at the Department of Architecture and Design, area of Building Systems, within the Institute of Technology of Universidad CEU San Pablo. Coordinator of the Mechanical Systems area. Professor Accredited by ANECA in the figures of Senior Professor. One six-year research period with CNEAI (period 2003–2015). Coordinator of the post-graduate degree of Energy Efficiency and Sustainability in Buildings and Coordinator of the Laboratory of Building Systems within Universidad CEU San Pablo. Member of the PhD Program "Health Science and Technology" and the PhD Program "Composition, History and Techniques pertaining to Architecture and Urbanism". Coordinator of the training activity "Ethical use of scientific documentation". Lecturer of the Master MC2 relevant to Quality in Construction, and in the Postgraduate Degree in Overall Management of Buildings and Services at the Universidad Politécnica de Madrid. Coordinator of the Wind Energy Section within the Master's degree in Renewable Energy of the Institute of Technology (Universidad CEU, San Pablo). Lecturer of Advanced Manufacturing Processes in the Master's degree in Industrial Engineering at the Universidad Europea de Madrid. In addition, he is a lecturer of Mechanics of Continuous Media and Theory of Structures, Electromechanics and Materials in seven Spanish universities. Twenty-six books have been published, eighteen of them in the last five years. Coordinator and co-author of the collections of Abecé of Building Systems (Munilla-Lería Eds.), consisting of five volumes, and Building Systems in Building Design (Asimétricas Eds.), consisting of seven volumes. Additionally, has co-authored articles concerning areas of energy, systems, and mechanics of continuous media (44 articles in the last five years; 27 JCR and 17 SJR). Director of the PhD thesis "Efficiency of ventilation in residential environments for the promotion of health of its occupants: impact on architectural indoor design" and "Energy simulation as a prognostic tool in architecture: design of passive strategies in different climatic zones of Spain" which both obtained the grade of Cum Laude with mention of an International PhD, in June 2018 and September 2020, respectively. Director/co-director of three PhD theses currently in progress related to energy efficiency, comfort and indoor air quality: "Influence of occupants on energy consumption in multi-family dwellings in Madrid Characterization of their habits", "Building Systems in efficient office buildings" and "Energy simulation as a forecast tool in architecture: The importance of the solar reflectance index in the design stage of residential buildings in Spain".

Preface to "Sustainable Building and Indoor Air Quality"

We currently live in a global context where climate change has paved the way for the development of new initiatives to reduce carbon emissions. In light of this situation, the focus on greenhouse gas (GHG) emission management, the environmental impact of the built environment, and the optimization of environmental performance has become essential. Most energy losses are due to air renovations and infiltrations; therefore, building ventilation has become a key challenge in relation to improving energy management, because it is also closely related to human health and well-being. For this reason, it is vital that possible implementation techniques take into account the balance between indoor air quality and energy efficiency in the air renovations of buildings. Focusing on this will allow us to gain a deeper understanding of buildings' ventilation, as well as the quality of the air introduced. Original works that also deal with methodologies, numerical and experimental research, and case studies with a particular focus on the comfort conditions of the occupants and the influence of occupant habits in sustainable buildings and the effects of climate change on the built environment are welcome.

Roberto Alonso González Lezcano
Editor

Article

The Sedentary Process and the Evolution of Energy Consumption in Eight Native American Dwellings: Analyzing Sustainability in Traditional Architecture

María Jesús Montero Burgos [1,*], Hipólito Sanchiz Álvarez de Toledo [1], Roberto Alonso González Lezcano [2] and Antonio Galán de Mera [3]

[1] Facultad de Humanidades y Ciencias de la Comunicación, Campus de Moncloa, Universidad San Pablo-CEU, CEU Universities, 28040 Madrid, Spain; hsan@ceu.es

[2] Escuela Politécnica Superior, Montepríncipe Campus de Boadilla del Monte, Universidad San Pablo-CEU, CEU Universities, 28040 Madrid, Spain; rgonzalezcano@ceu.es

[3] Facultad de Farmacia, Montepríncipe Campus de Boadilla del Monte, Universidad San Pablo-CEU, CEU Universities, 28040 Madrid, Spain; agalmer@ceu.es

* Correspondence: mar.montero@ceindo.ceu.es

Received: 23 January 2020; Accepted: 25 February 2020; Published: 28 February 2020

Abstract: According to the research developed by André Leroi-Gourhan in 1964, entitled "Gesture and speech", the evolution of human beings during Prehistory was linked to the search for work efficiency. As time passed, man designed increasingly complex tools whose production implied a decreasing amount of energy. The aim of the present research was to determine if this evolution, which occurred in parallel to the sedentary process, also affected architecture, specifically if it can be detected on traditional dwellings, particularly in those built by the Native American Indians during the pre-Columbian period. Due to their great diversity, since both nomad and sedentary models can be found among them, and to the available information about their morphology and technical characteristics, these models offer a unique opportunity to study the consequences of this process for architecture. In order to achieve it, an alternative parameter that can be determined for any type of building was designed. It allows us to establish the amount of energy an envelope is equal to. The results obtained suggest that the efficiency of the dwellings decreased as this process went forward, but this pattern changed in its last step, when agriculture appeared and permanent settlements started to be built. Besides, statistical graphs were used in order to show graphically the relationship between it, the climate, the morphology of the dwellings and their technical characteristics.

Keywords: vernacular architecture; sustainability; energy efficiency; history; statistics; society

---

1. Introduction

Native American architecture offers a unique opportunity to reconstruct the dwellings used and designed by the prehistoric communities. When the European explorers arrived in America at the end of the 15th century, they found a world that was already impossible to reconstruct in Europe [1]. By means of the information contained in their chronicles, the dwellings built by the communities that inhabited those lands can be reconstructed. They also show that their lifestyles ranged between nomadism and sedentarism, comprising a great variety of systems as a result of the combination of both of them. However, sedentarism is considered the final step of this process which continues until today and which was consolidated with the construction of the first settlements.

Studying the evolution of culture during Prehistory, the anthropologist and historian Leroi-Gourhan [2,3] published research in 1964 which contained a graph about the evolution of flint tools. That graph showed that, as millennia went by, the amount of flint used to obtain each point decreased as the resulting sharp increased. This way, he demonstrated that the evolution of human beings was determined by the efficiency improvement at work. In parallel to this transformation, the sedentary process had gone forward and each community had chosen a different system to live in, as the North America of the 15th century shows [4].

This way, it is viable to understand that the sedentary process went forward according to the pattern found by Leroi-Gourhan. Proceeding on this basis, the aim of the present research consists in determining if that pattern can also be found in the evolution of the dwellings which were designed during the sedentary process that took place throughout Prehistory in North America. In other words, if the evolution from the nomad dwellings to the sedentary models pursued an improvement on energy efficiency.

In order to achieve it, eight of the most relevant dwellings built by the North American natives were analyzed. The dwellings which were chosen are the tipis, used by tribes such as the Crow or the Sioux [4–18]; the wigwams built by the Ojibwa or the Chippewa [4,19–25]; the Navaho hogans [4,22,26–31]; the Caddoan grass houses [1,4,32–35]; the earthlodges built by the Mandan, the Hidatsa and the Arikara [4,22,36–38]; the plank houses used by the Haida [4,39–46]; the Iroquois longhouses [4,20,21,47–54]; and the pueblos, specifically one of the adobe houses built in Acoma [55–60]. Each one corresponds to a different step of the sedentary process (Figure 1).

Figure 1. The dwellings that were analyzed. First row, from left to right: tipi [61], wigwam [62], hogan [31] and grass house [61]. Second row, from left to right: earthlodge [63], plank house [45], longhouse [47] and pueblos [63].

The most affordable way to determine the efficiency level of the chosen dwellings would were by means of the shape factor. Defined as the ratio between the envelope surface area and the volume of air contained under it [64], it is one of the most popular parameters used to estimate the relation between the design of a building and its energy losses due to outward exposure. Despite its undeniable utility, it simplifies the morphology of the building and does not take into account some of its characteristics, such as its orientation, the existence of any excavated surface area or its indoor compartmentalization.

In order to solve these lacks, the present research proposes to determine the capacity of an envelope to transform the outdoor conditions into the indoor ones, proposing to interpret these buildings as if they were machines. This way, it consists in analyzing the capacity of an envelope to transform the outdoor temperature and the outdoor humidity into the indoor temperature and the indoor humidity, just by means of its presence. This means that an envelope works as an air-conditioning machine and contributes an amount of energy.

Besides, a statistical method was used in order to understand the relation of this parameter with the morphological and the technical aspects of the dwellings, as well as with their corresponding weather data.

2. Materials and Methods

2.1. Methodology

2.1.1. Equivalent Energy

The calculation method is based on psychrometry. The particles contained in the air, both indoors and outdoors, move on their own at different speeds, which implies that they contain different amounts of energy. Therefore, those air masses contain an amount of energy produced by the movement of those particles. For example, as can be seen on a psychrometric chart, if the temperature rises at a constant level of humidity, the temperature of those particles rises too, as the energy contained in them does. In the same way, if temperature decreases, the amount of energy, known as enthalpy, also decreases. In addition, for constant temperature, if humidity rises, enthalpy also rises.

The state function that allows tracking the marks left by the energy variations at a constant pressure is the enthalpy [65]. It is only possible to determine its variations after a thermodynamic process; this is the reason why it is expressed as the variation of the amount of energy that is expelled to the environment or absorbed by a system during one of those processes. Therefore, its value is expressed in terms of exchanged energy [66].

This way, by means of psychrometry and characterizing these air masses by their temperature (t) and their humidity level (φ), it is possible to determine the amount of energy contained in them. Knowing the amount of energy contained in the outdoor air mass (h_e) and the amount of energy contained in the indoor air mass (h_i), the amount of energy that was contributed by the building just with its presence can be determined (Δh).

In other research proposals, the indexes which assess the energy efficiency of a construction are usually determined by the indoor and the outdoor temperature, but they do not depend on the humidity levels [67].

Contrary to the shape factor, by this method, it is possible to calculate the amount of energy that can be isolated by a building, taking into account multiple factors such as its morphology, the composition of its envelope, its orientation, the presence or absence of openings or its indoor compartmentalization. The proposed value does not consist in valuing the sustainability grade of a building, the aim of other researches [68], but on establishing the capacity of an envelope to modify the outdoor conditions provided by nature. The higher the difference between indoor conditions and outdoor conditions is, the higher this parameter is.

Virtual Modeling

Just as the outdoor enthalpy was obtained, the enthalpy value for the interior of each model was determined by calculating the indoor temperature and the indoor humidity level for each one of the corresponding ten locations.

Virtual reconstruction of each model by means of DesignBuilder v6.1.2.005 (DesignBuilder Software Ltd, Stroud, UK) (Figure 2) was carried out with the aim of obtaining its indoor conditions (temperature and humidity level) in each location. Occupancy has not been taken into account, so 0 persons per square meter is the value determined for the eight dwellings.

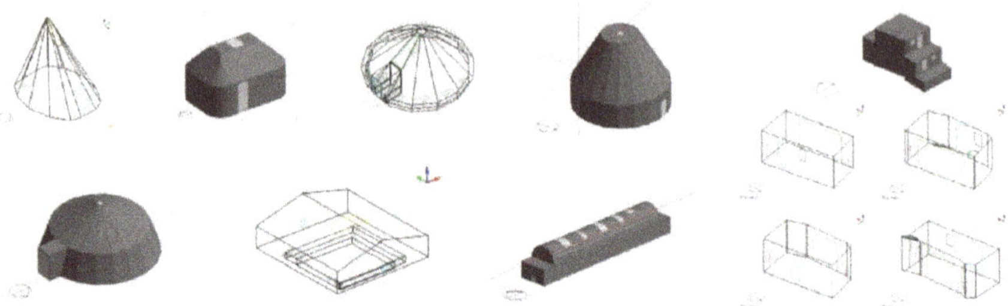

Figure 2. Virtual models developed by means of DesignBuilder v.6.0. Left group, first row, from left to right: tipi, wigwam, hogan and grass house. Left group, second row, from left to right: earthlodge, plank house and longhouse. Right group, upper part: pueblo.

Morphology and Dimensions

The modeling work of the dwellings takes as basis a previous researching work, based on the documents referenced in the Introduction.

In Table 1 the main morphological information is presented.

Table 1. Dimensions of the analyzed dwellings.

	Living Surface Area (m^2)	Volume (m^3)	Envelope Surface Area (m^2)	Openings Area (m^2)
Wigwam	17.6	45.2	65.96	3.16
Hogan	21.38	26.3	41.43	1.3
Tipi	38.54	95.68	98.15	3.16
Earthlodge	69.41	165.09	143.56	2.77
Grass house	42.3	164.78	151.05	42.3
Longhouse	338.13	2209.93	1058.58	338.13
Pueblo	49.59	105.76	136.47	49.59
Plank house	192.93	813.81	491.41	192.93

Technical Description

The materials used to build the chosen models are detailed in the section called Appendix A (Table A2). Those values were calculated according to the information gathered throughout the chronicles referenced in the Introduction.

Calculation: Outdoor and Indoor Environment

Each model was located in ten archaeological sites and one weather station was allocated to each of these sites. In each case, the nearest station was chosen, a decision that implies that some of the stations are assigned several times.

By means of using several locations for each dwelling, the results are representative. These locations are detailed in the section of Appendix A (Figure A1 and Table A1).

The climate data used in this section were obtained from https://energyplus.net/ (U.S. Department of Energy 2019) [69].

Once the outdoor conditions are known, the indoor ones can be calculated. This means that the humidity level and the temperature were determined for both environments. This way, both enthalpies can be obtained.

The enthalpy that corresponds to the pair temperature–humidity of energy in each ambient for each single hour of a year was calculated. This implies that each dwelling in each site is linked to 8760 enthalpy values and 8760 outdoor enthalpy values. These data can be seen in the Appendix A (Table A4).

Result

The difference between both enthalpies is the energy per mass unit that each envelope is able to isolate. This result receives the name of "equivalent energy" and is represented by Δh from now on. An example of the calculation process can be seen in Figure 3 and Table 2.

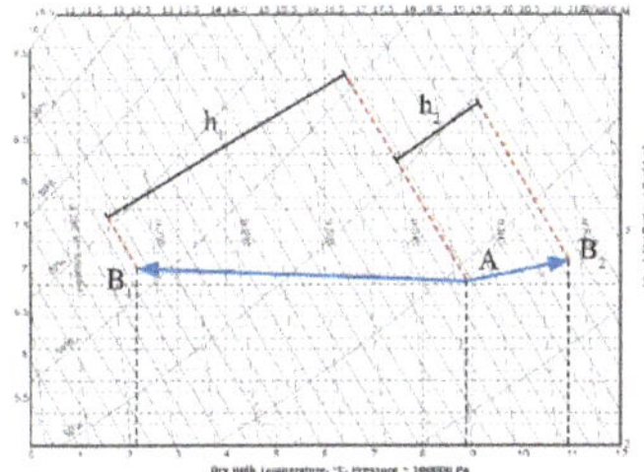

Figure 3. Psychrometric chart detail. The enthalpy values for a wigwam and for a pueblo dwelling on January 2 are marked [70].

Table 2. Environmental Conditions Corresponding to January 2 at 14:00h in the Pueblo of Taos.

	Temperature (°C)	Relative Humidity φ (%)	h (KJ/kg of dry air)	Δh (KJ/kg)
A—Outdoor conditions	8.85	39	15.9	
B1—Inside the pueblo dwelling	2.27	63.02	9.49	|15.9 − 9.49| = 6.41 KJ/kg of dry air
B2—Inside the wigwam	10.89	35.63	18.28	|15.9 − 18.28| = 2.38 KJ/kg of dry air

h: enthalpy; Δh: equivalent energy.

2.1.2. Statistical Links between the Results

Once we obtained this value, its links with the morphological aspects and the climatic circumstances of the dwellings, as well as the technical characteristics of their building materials, were analyzed, in order to better understand its functioning. By means of statistical graphs, it was possible to link morphological characteristics, that is to say qualitative data to quantitative data, such as the thermal transmittance or the wind speed. Therefore, in order to ascertain which factors influence the value of Δh, the statistics software PAST v3.25 [71–73] was used.

By means of it, a canonical correspondence analysis was carried out, that is to say a correspondence analysis based on a site/species matrix, in which values for one or more environmental variables are assigned to each site/specie.

The ordination axes of the resulting graph show the values of the combinations of those variables. This type of analysis is a direct gradient analysis, in which the gradient in environmental variables is known and the situation of the species (their presence or their absence) is the response to that gradient.

Hence, the data corresponding to each model located in each site occupy a line in a spreadsheet. The environmental variables, such as rainfall or indoor temperature, are inserted in its columns (Table 3). Last, also in columns, the information corresponding to each model about the presence or absence of the predefined species in each site, or about the presence or absence of the architectonic characteristics in each model and site, as occurs in the present research, is introduced.

Thus, both weather and environmental variables correspond to numerical data, while the morphological characteristics are the equivalent to the values of the species (Table 3). This means that the presence of one of these characteristics implies that number one is the value written in the corresponding cell, while its absence means that number zero is written in it.

The dwellings that were analyzed are represented according to the symbols shown in Table 3. The resulting table can be consulted in the Appendix A (Table A5).

The climate classification that was used is the Köppen scale, established in 1884 by Wladimir Köppen [74] and updated in 1936 [75,76]. The information about wind speed was gathered by means of https://es.windfinder.com [77], a database which presents the values obtained by more than 21,000 weather stations since 1999.

Table 3. Scatter graphs legend.

Dwellings (10 Sites per Dwelling)						Species				
W	"Wigwam"	GH	"Grass house"		Morphology		Envelope Materials		Structural Materials	
H	"Hogan"	LH	"Longhouse"	K	Entrance gallery	S	Hides	Q	Earth	
T	Tipi	P	Pueblo	L	Several levels	T	Grass	R	Wood	
E	"Earthlodge"	PH	"Plank house"	M	Domed shape	U	Turf			
				N	Vault shape	V	Bark			
				O	Conical shape	W	Mats			
				P	Expandable space	X	Earth			
						Y	Wood			

Variables		
Climate	Morphology	Technical Aspects
1 Wind speed (km/h)	9 Shape factor	10 Equivalent energy (KJ/kg)
2 Annual rainfall (mm)		11 Equivalent energy (KJ/kg m^2)
3 Average outdoor temperature (°C)		12 Effusivity (s$^{1/2}$ W/m^2 °C)
4 Average indoor temperature (°C)		13 Diffusivity (m^2/seg 10^{-6}.)
5 ΔTemperature*		14 Thermal transmittance (W/m^2 °K)
6 Average outdoor humidity (%)		
7 Average indoor humidity (%)		
8 ΔHumidity **		

* ΔTemperature = average outdoor temperature − average indoor temperature (°C); ** ΔHumidity = average outdoor humidity − average indoor humidity (%).

In conclusion, this method allows us to link three concepts: locations, species and variables. The locations are represented by black dots in the graphs. Each one is attached to a letter that indicates the dwelling, plus a number that indicates the archaeological site (Appendix A, Table A1); in total, there are ten black points per dwelling, since each dwelling was located in ten archaeological sites. The species are identified by orange dots joined to their corresponding letter (Table 3). Finally, the variables correspond to the green lines. These vectors mark the zone of the graph where the locations and the species that correspond to the higher values of that specific variable are gathered.

There are three rules that must be followed to read these graphs. First, the links between location, species and variables can be concluded by observing the distance between them. The further a location or a specie is from the vector of a variable, the smaller the influence of that variable is on that location or specie. Second, the closer a location or a specie is to the coordinate origin, the more significant is its presence in the group. Third, the length of a vector depends on the amount of information about its variable that is present in the graph. The longer a vector is, the more information about that variable is contained in the graph.

2.2. Theoretical Fundamentals

The enthalpy values were calculated by means of Equation (4), result of the substitution of Equation (2) and Equation (3) in Equation (1).

$$h = (c_{pa} \cdot t) + [W \cdot (Lo + (c_{pw} \cdot t))]) \tag{1}$$

$$W = 0.622 \cdot (p_w/(p - p_w)) \tag{2}$$

$$\varphi = p_w/p_{ws} \tag{3}$$

$$h = \left(1.004 \cdot t\left(^\circ C\right)\right) + \left[\left(\left[0.622 \cdot \frac{0.7 \cdot e^{14.2928 - \frac{5291}{t\,(^\circ K)}}}{1 - \left(\varphi \cdot e^{14.2928 - \frac{5291}{t\,(^\circ K)}}\right)}\right]\right) \cdot \left(2500.6 + t\left(^\circ C\right)\right)\right] \tag{4}$$

$$\Delta h\ (KJ/kg)_{1\ldots8760} = |h_{e,1\ldots8760} - h_{i,1\ldots8760}| \tag{5}$$

$$\Delta h\ (KJ/kg) = \underline{X}\Delta h\ (KJ/kg)_{1\text{-}8760} \tag{6}$$

where c_{pa} is specific heat of dry air (1.004 KJ/kg °K); c_{pw} is specific heat of water vapor (1.86 KJ/kg °K); t is temperature; φ is relative humidity (%); p_w is partial pressure of water vapor in the air; p_{ws} is saturation vapor pressure; p is atmospheric pressure (1 bar); h is enthalpy (KJ/kg); h_e is outdoor air enthalpy (KJ/kg); h_i is indoor air enthalpy (KJ/kg); L_0 is latent heat of vaporization of water at 0 °C (2500.6 KJ/kg); and W is absolute humidity (kg of water/kg of dry air).

Following, the difference between both values is calculated in order to determine the energy that each envelope is equal to in each hour of the year. Last, the absolute values of these results were averaged. This average is the equivalent energy, the energy that can be isolated by each envelope (Equations (5) and (6)).

Taking this as a starting point, four approaches were designed. They allow us to analyze the possible links between the equivalent energy and the shape factor, the morphology, the location and the building materials. Both the information that was used for these calculations and the results are detailed in Appendix A (Table A4).

2.3. Approaches

The following comparisons and approaches were carried out in order to achieve the aforementioned goals.

Equivalent energy—Shape factor. All the dwellings, original building materials, original locations and original morphology: The resulting values for equivalent energy were compared with the corresponding shape factors in order to check if there is any relation between them. The dwellings were situated in their original locations, and their original building materials were assigned. This way, the features that do not influence on the shape factor, but do influence on the equivalent energy, affect the results and the difference between these two values can be observed. Besides, the factors that influence these conclusions were analyzed.

Equivalent energy—Morphology. All the dwellings, same building materials, same locations, without openings: In order to establish a relation between the results of equivalent energy and the morphology of the dwellings, the models were reduced to their volumes. This means that their openings were removed, the same material was assigned to all of them and they were situated in the same ten locations (where the pueblos of New Mexico were built, that is to say, the locations the most sedentary dwelling was built). This way, all the dwellings contain the same information as the shape factor takes into account; that is to say, the surface of their envelope and the volume that is contained under it. The only characteristic which could not be eliminated was the orientation of the dwellings, not present in the shape factor and impossible to be removed from the DesignBuilder calculations. The building materials that were used in this approach are the ones that correspond to the template entitled "Timber frame-superinsulated", which appears on the database of DesignBuilder v6.1.2.005 and whose details are featured in Appendix A (Table A3).

Equivalent energy—Location. All the dwellings, both original and same locations, original building materials and original morphology: Two groups of calculations were developed for this approach in order to determine how location influence the equivalent energy. First, the equivalent energy corresponding to the original placements, their original materials and their original openings, was calculated. Second, those dwellings were moved to the locations where the New Mexican pueblos were built, and the corresponding equivalent energy was calculated as well. Therefore, the only feature that changes from one case to the other one is the location. This way, by moving the dwellings from their original locations to the ones of New Mexico, it can be determined if the values of equivalent energy for each dwelling are influenced by its location and its climatic conditions.

Equivalent energy—Building materials. Just one dwelling, original materials and same locations: By means of this approach, it was possible to determine the links between the original building materials and the equivalent energy. In order to achieve it, the morphological and environmental factors were eliminated. Thus, it is possible to establish the relation between the temperature, the humidity level, the technical characteristics of the building materials and the equivalent energy that correspond to each of them. This way, the consequences that each material has on the indoor ambient and on the equivalent energy can be determined.

Specifically, the calculations presented in this section imply to take one single dwelling and assigning it the main building materials of the other dwellings which were analyzed. Thereby, the chosen dwelling was the wigwam, and the composition of all the envelopes was assigned to it one by one: the cattail mats (its original material), the hogan envelope, the tipi hides, the multilayer envelope that covered the earthlodge, the bundles of grass typical of a grass house, the bark sheets that wrapped the longhouses, the adobe that composed the walls of the pueblos and, finally, the cedar planks that protected the interior of Haida houses. Eight versions of the same dwelling that were situated in the ten locations corresponding to the pueblos of New Mexico.

All the building materials described before were characterized by means of their diffusivity, their effusivity, their thermal transmittance and their thermal lag [78].

3. Results

3.1. Equivalent Energy—Shape Factor

In this section, the original status of the dwellings is analyzed. This way, they were assigned their original materials, were placed in their original locations and their morphology was kept.

The traditional dwellings which are built in temperate climates are those that would correspond to the highest shape factor values, whereas those from cool climates tend to be associated with lower values [64]. The orthogonal dwellings, whose presence is regular throughout the Mediterranean coasts, could be an example of the first case, whereas the snow domes built in the Arctic would represent the second group. As long as the climate is warmer and it is less necessary to modify the environmental conditions, people can extend the surface of dwellings envelopes, prioritizing other factors, such as the optimization of the available space to build on.

As can be seen (Figure 4), the highest values of the shape factor, those corresponding to the hogan and the wigwam, do not have any relation with climate, since the first one was built in a temperate desert, New Mexico, whereas the second one was typical of a zone whose humidity levels were higher and whose temperatures were lower, the vicinity of the Great Lakes.

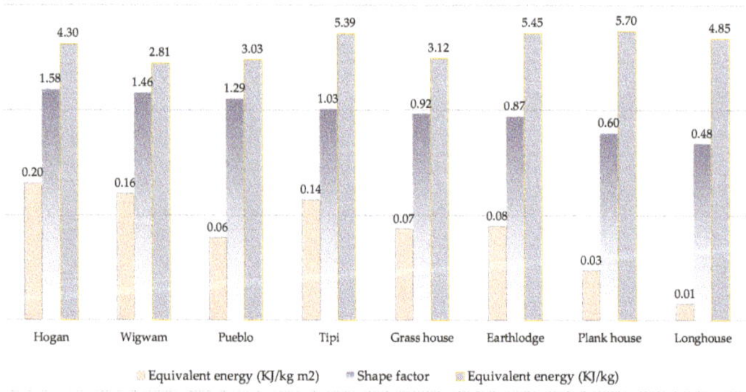

Figure 4. Equivalent energy for original locations and original materials vs. shape factor.

The results concerning the longhouse, built in the same region as the wigwam, point to the same direction. However, the first one was used by an almost sedentary community, the Iroquois community, and the second one served to a seminomadic way of life, the one developed by the Chippewa. This means that their morphologies and building systems were influenced also by their practical functioning.

By comparing the shape factor values with the equivalent energy ones, it can be seen that the order of the dwellings does not concur, unless they are calculated with respect to the living area. Taking into account the architectonic characteristics obviated by the shape factor, as the equivalent energy does, such as the building materials, the climate, the orientation or the presence or absence of openings, a more precise assessment of the way the building adapts to the environment can be obtained.

In conclusion, the designs that reach a higher difference between the indoor and outdoor conditions, those whose equivalent energy has the highest values, are those who correspond to the highest shape factor.

As explained before, the equivalent energy measures the capacity of an envelope to transform the outdoor conditions into the indoor ones. The bigger this increment or decrement is, the higher the equivalent energy is. In conclusion, the designs that reach a larger difference between the indoor and outdoor conditions, those whose equivalent energy has the biggest values, are those who correspond to the highest shape factor as a rule.

3.2. Equivalent Energy—Morphology

In order to carry out the analysis proposed in this section, the same building materials were assigned to all the dwellings, they were situated in the same locations and their openings were removed.

As can be observed in Figure 5, the highest values of equivalent energy correspond to the conical or hemispherical dwellings, such as the hogan, the wigwam and the tipi. The order of the dwellings does not change, except in the case of the pueblos, whose equivalent energy is similar to the ones of the hemispherical designs. This circumstance is possible because the pueblo design manages to reduce the surface of its envelope by overlapping its constituent volumes.

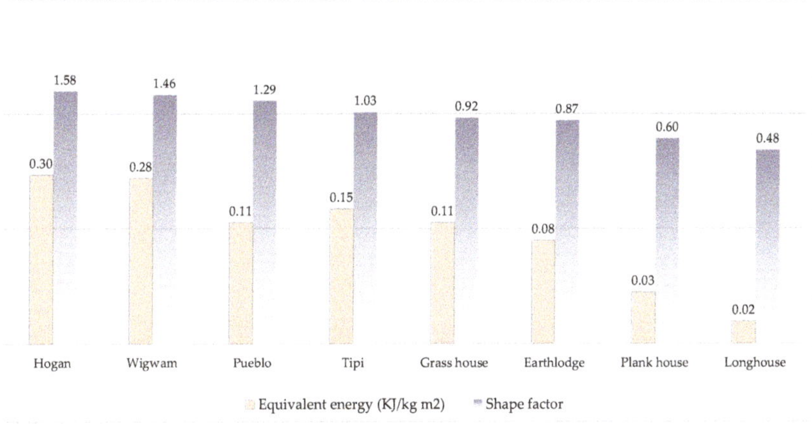

Figure 5. Timber frame-superinsulated template as the envelope of all the dwellings, situated in the New Mexican locations and without openings (KJ/kg m^2 of living surface).

This effect can be explained in the following way [64]. A cube composed by a 36-units^3 volume, has a 65-units^2 envelope, but if that same volume, those 36 units^3 are arranged horizontally, they create a 96-units^2 envelope. In the first case, the shape factor is 1.8, whereas, in the second one, it is 2.6 [64]. The opposite happens about the plank houses and longhouses, whose shape tends to be horizontal.

Even though all the models were homogenized, there is another characteristic that the shape factor cannot take into account, besides the orientation. It is the indoor compartmentalization of the buildings. As indicated before, the pueblos were composed by several volumes, whose overlapping reduces the outdoor exposure of their dwellings influencing the amount of energy isolated by these constructions, but do not influence their shape factor.

As can be seen, the rest of models appear in the same order for both values. The proportions between them are the only differences. The highest and the lowest values are more distant, whereas the ones located in the middle form a different group. This way, it can be said that the equivalent energy value qualifies the information provided by the shape factor.

Figure 6 shows the results that correspond to the original locations of the dwellings. Besides, they have also been coated with their original materials. It shows that some designing decisions, such as the entrance galleries (K), the vaulted spaces (N) or the distribution in several levels (L), are normally not related to the envelopes which achieve a high Δh (11).

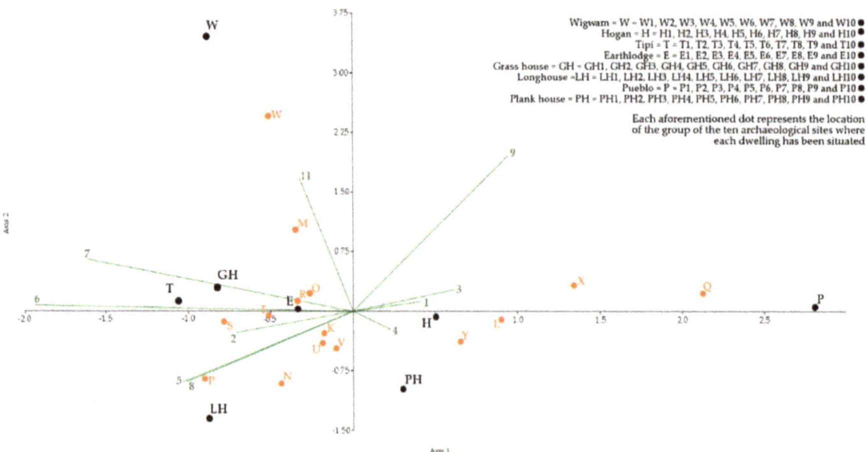

Figure 6. Scatter graph. Links between the environmental characteristics and the equivalent energy.

3.3. Equivalent Energy—Location

For this section, the dwellings were placed in both their original locations and in the same ones, their original building materials (Table 4) were assigned and their original morphology was respected.

When analyzing the dwellings in their original locations, it can be seen that the resulting order changes slightly with respect to when they were placed in the locations of New Mexico (Figure 7). The circular dwellings keep occupying the highest places. The hogan achieves the largest difference, and three dwellings increase their equivalent energy with respect to their original locations. The longhouse, the grass house and the wigwam isolate more energy in New Mexico than in their original locations. The grass house obtains the biggest difference.

However, all these differences are not significant, and even the value corresponding to some models, such as the plank house, is nearly the same. This means that the capacity of an envelope to modify the outdoor conditions, the pair temperature–humidity, does not depend on the location of the dwelling. An envelope provides a higher or a lower difference, and it is this capacity, higher or lower, that is used to adapt a building to its environment.

Table 4. Thermal lag of the building materials.

Dwelling	Envelope Layer	Thermal Lag	Dwelling	Envelope Layer	Thermal Lag
Wigwam			Earthlodge		
Wall/roof		0.3231	wall/roof		16.0424
	Cattail	0.0722		Wood	5.5892
	Air	0.0172		Wood (branches)	5.5892
	Cattail	0.0722		Grass + grass from turf	2.5518
	Cattail	0.0722		Earth from turf	2.3122
	Air	0.0172	Longhouse		
	Cattail	0.0722	wall/roof		0.5314
Tipi				Tree bark	0.5314
Wall/roof		0.1051	Pueblo		
	Hide	0.1051	wall		18.8570
Hogan				Earth	18.8570
Wall/roof		16.9469	roof		15.8050
	Wood	5.5892		Wood (branches)	5.0303
	Tree bark	0.5314		Grass	1.5311
	Earth	10.8262		Earth	9.2436
Grass House			Plank house		
Wall/roof		16.6290	wall		3.9125
	Grass	16.6290		Wooden planks	3.9125
			roof		0.5314
				Tree bark	0.5314

Dwelling	Thermal lag (h)	Dwelling	Thermal lag (h)
Tipi	0.105	Earth lodge	16.04
Wigwam	0.32	Plank house	3.91
Hogan	16.94	Longhouse	0.53
Grass house	16.63	Pueblo	18.85

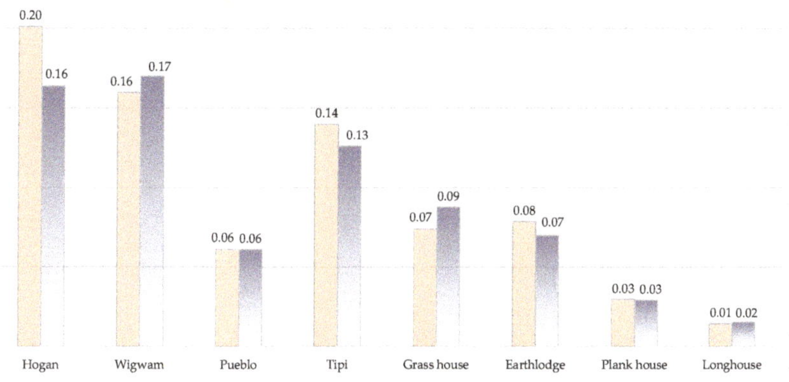

Figure 7. Equivalent energy for original building materials (KJ/kg m^2 of living surface).

Taking as an example the highest values, the hogan was built in a temperate region located in the southwest of the United States (Bsk, Bwk and Dfb zones, according to the Köppen classification), whereas the wigwam was typical of the Great Lakes region, where temperatures are cooler and the humidity level is higher (Dfa, Dfb, Cfa and Cfb). The longhouse was built in this second region too, and this is the dwelling with the lowest Δh value. The results for hogan and pueblos (Bsk, Dfb and Cfb), both built in the same region, New Mexico, point to the same direction.

By observing the scatter graph shown in Figure 6, it can be seen that the highest values of the shape factor (9) are associated with high values of average outdoor temperature (3), whereas the highest values of equivalent energy (11) are linked to the highest levels of indoor humidity (7). It can be seen that $\Delta\phi$ (8) and Δt (5) determine the value of Δh (11) equally, but they are not related to the shape factor (9).

If the data are analyzed without taking into account the living surface of the dwellings, and just the energy isolated by these specific envelopes is observed (KJ/kg) (Figure 4), it can be seen that the earthlodge achieves to duplicate the result of the wigwam. In this case, the plank house obtains the highest value of Δh.

However, if this information is analyzed from the point of view of the sedentary process, it can be seen that the dwellings that were used by the sedentary groups were also those which are equivalent to a smaller amount of energy. These three dwellings, the longhouse, the plank house and the pueblo adobe house, are equivalent to a smaller amount of energy per living square meter. All of them are orthogonal in plan, the model that is usually adopted by the sedentary communities.

Among them, the Native Americans who developed the agriculture and sedentarism the most, those groups which inhabited the zone of New Mexico and built the adobe dwellings, chose the model that was equivalent to the greatest amount of energy. However, if they wanted the model that was equivalent to the highest level of energy among the most popular designs, they should have chosen the wigwam (0.17 KJ/kg m^2), taking into account the living surface, and the plank house (5.64 KJ/kg), if comparing exactly the models presented in this research.

3.4. Equivalent Energy—Building materials

In order to develop this section, just one dwelling, the "wigwam", was considered. The original materials, which are summarized in Table 4, were assigned to it, and it was placed in the locations of New Mexico, where the pueblos were built.

This approach has made it possible to see that the elm bark sheets (from *Ulmus americana* L. or from *Ulmus rubra* Muhl.) which covered the Iroquois longhouses were the material that implied the highest amount of energy (Figures 8 and 9). The most abundant building material the natives who inhabited the forests of the Great Lakes region had at their disposal was wood. These forests, located both in Dfb and Cfa zones, according to Köppen scale, were full of coniferous trees, such as *Tsuga canadensis* (L.) Carrière or *Picea rubens* Sarg., and deciduous trees, such as *Quercus rubra* L or *Betula alleghaniensis* Britton. Besides, this region is characterized by a high ambient humidity, against which the bark tree provides a quality solution thanks to its waterproofing capacity.

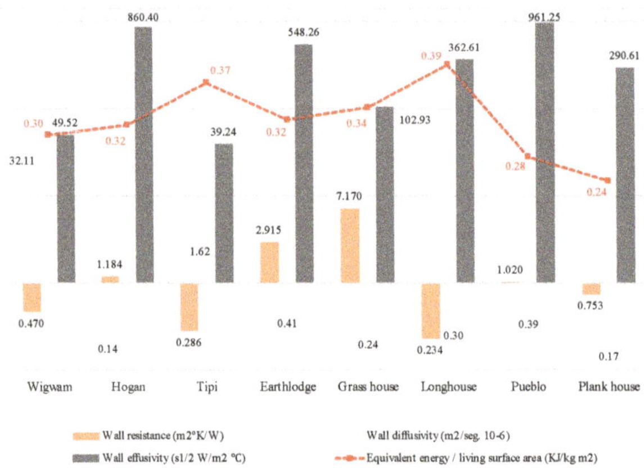

Figure 8. Thermal characteristics of the original building materials. They were assigned to the same dwelling, whose openings were removed.

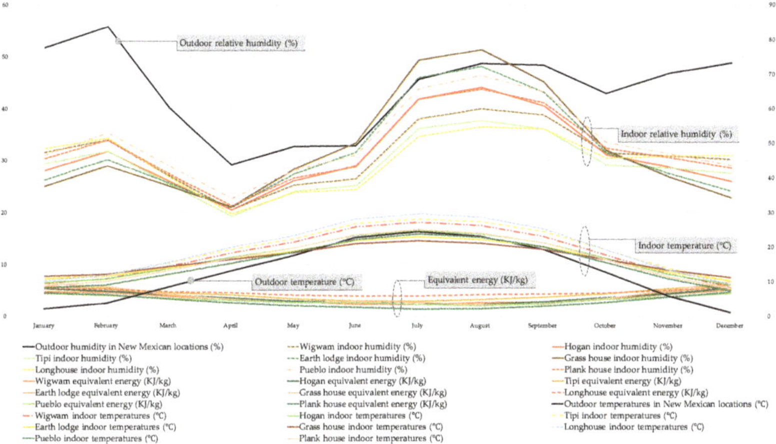

Figure 9. Indoor conditions in a dwelling whose openings were removed. All the analyzed building materials were assigned to it, and it was placed in the New Mexican locations.

Due to have the lowest thermal resistance of all of the materials that were analyzed, the bark sheets provide the most stable difference between the outdoor and the indoor conditions throughout the whole year. Unlike the other building materials, whose thermal resistance or effusivity is higher, this material neither stores heat nor offers a great resistance to its passage. Because of these reasons, it achieves a practically constant difference between the indoor and outdoor conditions throughout the year.

By observing Figures 5 and 8, it can be concluded that the amount of energy the wigwam envelope is equal to that of a timber frame superinsulated template (Figure 5) and is lower than the values obtained for the traditional materials in the same dwelling (Figure 8). The only exception is the case of wooden planks. This means that, contrary to the results obtained in previous researches about traditional architecture [79], traditional materials would have achieved better results than the present ones, if taking into account the energy they are equal to.

As can be seen in Figure 10, the transmittance (14) is the thermal characteristic more closely related to the equivalent energy (10 and 11). The second characteristic most related to it is the diffusivity (13). However, it is practically opposite to the effusivity (12). This means that the highest values of Δh (10 and 11) correspond to the highest values of transmittance (14).

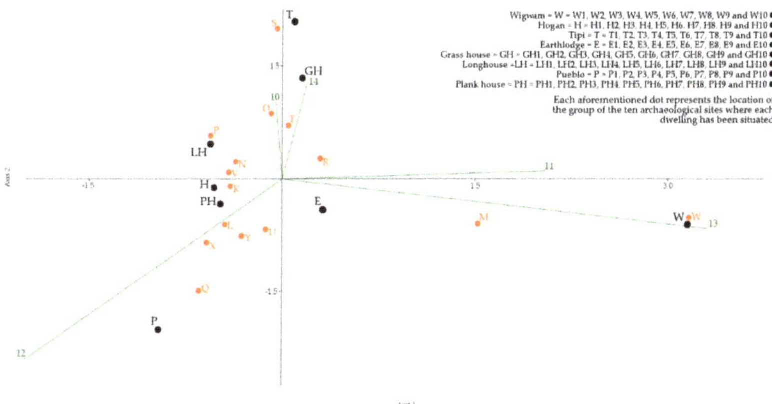

Figure 10. Scatter graph. Links between the building materials and the equivalent energy.

Moreover, in this graph, it can be seen that the tree-bark envelope (V), the one used for coating the longhouses and the roofs of the plank houses, is influenced by both the effusivity (12) and the thermal transmittance (14). However, the grass house (GH) and the tipi (T), which have the next highest values of Δh (10), are related exclusively to the transmittance (14). Its location in the graph indicates definitively that this is the factor that influences Δh the most (10 and 11).

The envelopes composed of several layers, or with a high presence of earth, offer a higher resistance to the heat transfer. The consequence of this circumstance is that indoor temperatures are lower in summer, as are their differences with respect to the outdoor temperatures. Since these differences are lower, the amount of energy these envelopes are equivalent to is usually lower during the summer too.

However, the building material that is equivalent to the lowest amount of energy is the one that covers the plank house walls, the cedar planks. Again, it is a dwelling built in a region with high humidity levels due to its proximity to the coast. This territory corresponds to a region classified as Cfb by the Köppen scale, and there are four most abundant tree species in this rainy climate, located in the northwest of the United States: *Pseudotsuga menziesii* (Mirb.) Franco, *Tsuga heterophylla* (Raf.) Sarg., *Thuja plicata* Donn ex D. Don and *Fraxinus latifolia* Benth. Specifically, it was *Thuja plicata* Donn ex D. Don, or Canadian Western red cedar, the wood used for building, since it is coated with a special type of oil that makes it resistant to water, preventing it from rotting [80].

The wood planks correspond to one of the lowest values of heating speed (diffusivity), similar to the one of the bark sheets which comprise the envelope of the longhouse. Their capacity to store heat is very similar. The biggest difference between them concerns their thermal resistance, since the value corresponding to the plank house almost triples the one of the longhouse. Their heating speed is also reflected in the thermal lag that characterizes both materials (Table 4). The dissimilarity among them provokes that the size of the difference between the indoor and the outdoor temperatures depends on the period of the year. As can be seen in Figure 11, the tree bark keeps the indoor temperature higher than the outdoor temperature during the summer, whereas the temperature achieved by the wooden planks is almost the same as it. The thermal resistance of the wooden planks, higher than the one of the tree bark, ensures that the indoor temperature takes longer to change. This means that the indoor ambient is less vulnerable to the weather changes inside a plank house and that its thermal lag reaches a higher value.

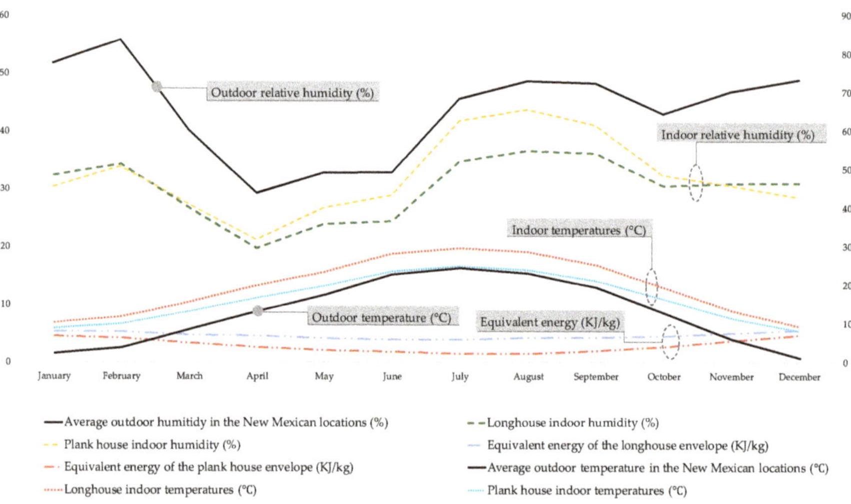

Figure 11. Comparison between the outdoor and the indoor conditions generated in the same dwelling by the envelope of a longhouse and by the envelope of a plank house.

However, the higher speed the thermal wave passes through the bark strip at ensures that the difference of temperature between the outdoors and the indoors is almost constant throughout all the year. At the same time, the humidity level changes, since it decreases when the temperatures rises and rises when the temperatures decrease.

4. Discussion of Results

As can be seen in Figure 7, the circular dwellings correspond to the highest value of equivalent energy per living square meter. This relation takes place both if the dwellings are situated in their original locations and if they are situated in the New Mexican locations, where the sedentary process had been most developed during the pre-Columbian North America. However, the models which isolate a lower amount of energy are the orthogonal ones. Besides, the former, the circular models, are related to nomad communities, whereas the latter ones were mainly designed by sedentary groups.

Contrary to the results of the research developed by Leroi-Gourhan [2], the energy the dwellings are equivalent to decreases as long as the sedentary process goes forward. The value of Δh decreased progressively, and the energy required to achieve the same indoor conditions increased as long as that process was developed. It seems that the priority in the design of these dwellings was not air-conditioning saving, neither in shape of hearths for heating nor in shape of natural ventilation.

As explained in the Introduction, the dwellings which were selected correspond to models that are built in regions where nomad lifestyle coexisted with sedentary lifestyle. This way, if these models are classified according to their provenances, it can be seen that the nomad dwelling is always equivalent to more energy than its sedentary counterpart (Table 5).

Table 5. Equivalent energy of the analyzed dwellings according to their sedentary grade.

	Nomad or Seminomad Dwelling	Equivalent Energy (KJ/kg m^2)	Sedentary Dwelling	Equivalent Energy (KJ/kg m^2)
Northeast of the United States	Wigwam	0.16	Longhouse	0.014
South of the United States	Tipi*	0.14	Pueblo	0.061
Southwest of the United States	Hogan	0.2		
Southwest of Canada	Tipi*	0.14	Grass house	0.074
North of the United States	Tipi*	0.14	Plank house	0.03
Southwest of the United States	Tipi*	0.14	Earthlodge	0.079

*Taking into account the obtained results when modifying the location of the analyzed dwellings, it can be assumed that the amount of energy a tipi is equal to is the same wherever it is placed.

This tendency was inverted in the last step of the process, the one represented by the pueblos. The longhouses represent those communities which were about to achieve the same sedentary level as the pueblos, thanks to the development of agriculture, but they are also the dwellings equivalent to the lowest amount of energy. Figure 5 shows this situation. The longhouses are the models that isolate the least amount of energy. This way and according to the evolution of the sedentary process, if the nomad dwellings are the models that isolate the highest amount of energy, the Native Americans from the pueblos, the most sedentary group, should have designed the dwelling that was equivalent to the lowest amount of energy, but that was not the case. They succeeding in designing a dwelling that isolates more energy than the longhouse and the plank house, placing it at the same level as the nomadic models. As described before, its equivalent energy was so high thanks in part to its morphology, that is to say the overlapping of volumes, the indoor compartmentalization and the reduction of the outer surface by attaching several dwellings. Regarding the building materials they were built with, the determining factor is the high value of the effusivity of earth (961.24 s$^{1/2}$ W/m^2°C). The thermal lag also stands out (18.85 h) and indicates that the changes in the outdoor ambient were not practically perceptible inside them.

These adobe dwellings were not built just in New Mexico, but they were also used in the Middle East. Several of them can be found in sites such as Çatal Hüyük or Ain Ghazal [81], linked to other agricultural and sedentary societies.

The results presented in this research may indicate that the priority when designing these dwellings changed throughout the sedentary process. According to them, the energy required to achieve the same ambient conditions inside the dwellings rose progressively until the adobe orthogonal dwellings were built for the first time. However, they did not achieve the same values as the nomadic and seminomadic dwellings, of which all of them were circular, did. The circumstances human beings lived in changed throughout this process and maybe at its end it was necessary to add the floor optimization to the resources' optimization, which had been the main objective until the rise of agriculture and fishing. This way, the pueblos design let the Native Americans have a solution for two problems in a balanced way. Its orthogonal floor plan let them dedicate as much surface as possible to agriculture, and it also isolated more energy than the other orthogonal models they could know about.

According to the results that were obtained, it can also be concluded that the dwelling designed by the Native Americans from New Mexico did not provide the most comfortable indoor environment (Figure 12). However, these adobe dwellings provided a higher level of humidity, a lack to be compensated in the desert of New Mexico.

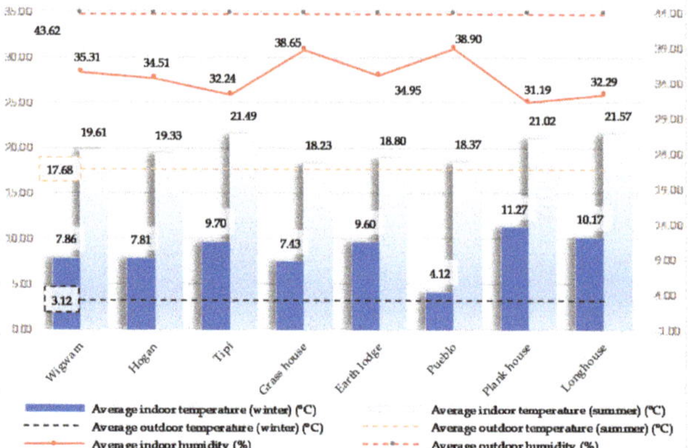

Figure 12. Environmental conditions of the analyzed dwellings. The original materials were assigned to them, and they were placed in the New Mexican locations.

4.1. The Equivalent Energy

The shape factor offers an affordable method to estimate the design quality of a building. However, it does not take into account some of the characteristics that influence the energy consumption. The equivalent energy solves these deficiencies since it depends on the outdoor and the indoor environmental conditions, with the latter being the result of the morphology of the building.

According to the obtained results, the location does not influence the amount of energy the dwellings are equivalent to in a meaningful way. The envelopes are equal to a specific amount of energy and isolate a specific amount of energy. This way, the modification of the outdoor conditions they achieve remains stable.

Taking as a basis the building materials which were analyzed, it can be concluded that those with the lower thermal resistance provide the envelopes with the highest values of equivalent energy. A high diffusivity provokes that weather changes modify the indoor conditions very fast, and thereby, the capacity of the envelope to alter the outdoor conditions remains practically stable throughout all the year. Therefore, this capability is not influenced by the placement of the building, specifically in New Mexico, where the present research is focused in and where the summer is more pronounced.

Not at all it is intended to assess that the indoor conditions achieved by low diffusivity materials reach the comfort level. The aim of this research is to determine the capacity of an envelope to modify the outdoor conditions; it is not to determine if the indoor conditions that it achieves are the most comfortable ones. However, it can be very useful for complementing other values, such as the concept of Zero Energy Buildings [82], since the higher the equivalent energy of a building is, the lower its energy consumption is.

4.2. Statistical Links between the Results

By means of the scatter graphs, it was possible to analyze the links between the technical data of the dwellings (quantitative data) and their morphological characteristics (qualitative information). These graphs show that the thermal transmittance (14) is the value that influences most in the equivalent energy (10 and 11), as the shape factor (9) and the indoor humidity (7) also do.

They have to read in terms of probability. This means that, for example, as can be seen in Figure 6, the sum of the wind speeds (1) of the plank house locations (192.61 km/h), the adobe house locations (150.01 km/h), the hogan locations (126.78 km/h) and the wigwam locations (155.57 km/h), which is equal to 624.97, is higher than the corresponding sum of the rest of the dwellings (548.19). The aforementioned dwellings are situated in the direction of the wind speed vector (1), and that is the zone of the graph where the highest sum of wind speeds is concentrated. This way, the position of the dwellings and the species on the graphs must be understood according to this system. This method is very useful for analyzing vernacular architecture in general, since it allows for the discovery of the logic of its morphological features and its links with its environmental circumstances. This would be the case of the research work developed by Varela Boydo and Moya [83]. It would allow us to identify which characteristics of the traditional windcatchers respond to cultural features and which ones are related to their adaptation to the proper circumstances of each geographical and climatic zone. The same could be determined about the Malay traditional houses analyzed by Ghaffarianhoseini, Berardi, Dahlan and Ghaffarianhoseini [84]. This method allows us to transform the morphological features of the Malay houses, such as the characteristics of their roofs, into numerical data. This way, it would be possible to establish the relation of this distinguishing element with the climate and the environmental information of each specific region.

5. Discussion

If Prehistory is understood as the pursuit of the stability provided by settling, it can be seen that the equivalent energy decreased as man approaches his objective. However, the last model, the adobe stepped dwellings, revitalized that value. It is necessary to take into account that, as long as the sedentary lifestyle went forward, the global temperatures rose progressively too, until reaching a value that made settling and agriculture viable. The greatest problem faced by the dwellings in the temperate climates, where man could live on agriculture, was that the temperatures were significantly higher in summer than in the past. That was probably the main problem to be solved, since winter could be solved, if necessary, by means of hearths. As Danny H. W. Li [85] asserted, when the temperatures began to rise, as happened 18,000 years ago, when agriculture was established, the greatest problem to be faced was the summer, and the dwellings must adapt to it. Consequently, the energy demand rises in the arid regions during these periods.

This way, of the three aforementioned sedentary dwellings, the one that isolates more energy, the model from New Mexico, is the one located in the zone that reached the highest temperatures. Facing the consequent increment of energy demand that took place during the summer, the sedentary human being designed the sedentary dwelling that isolated the highest amount of energy with respect to the known models that let him clear the largest amount of terrain for agriculture, that is to say, the orthogonal models. It would also be important to point out that the color of the envelopes would have influenced these results. As was demonstrated, the use of light colors in hot areas, such as the ones used in the pueblos, and dark colors in cold regions, such as the envelope of the earthlodge, reduce the energy consumption of the dwellings [86]. The main solution proposed in the aforementioned research [85] for this problem consists of increasing the adaptability of the dwellings built in these regions. This idea could be reinforced by analyzing the tipis, built in one of the hottest areas of North America, the Great Plains. The versatility of this shelter is one of its strengths, thanks both to its mobile envelope and to its morphology. Its smoke hole allows people to control the indoor ventilation and the indoor temperature at the same time, both at will, by means of two poles. No less important was the airtightness achieved by the envelope seams. This factor [87,88] was determinant to provide a comfortable indoor ambient. In the same way, it can be easily turned around in order to avoid strong winds [18] during a storm, since its structure is not symmetrical. This efficient design is contained in the old legend which explains the origin of tipis, since, according to it, the shape of the cottonwood leaves inspired its triangular shape. Both of them, the tipis and the leaves, use the Venturi effect to withstand wind and, in the case of tipis, improve indoor ventilation. Something similar happens in Acoma dwellings, whose shape, according to a legend, is based on the shape of the surrounding mountains. The airflow system that was used to dry the harvests on the houses' roofs works in the same way that the airflows move in the slopes of the mountains. During the day, the airflow rises from the valley, since the peak of the mountain is cooler and hot air is lighter than cool air, whereas during the night, the cycle is reversed and the air that rests in the peak turns cooler and descends to the valleys. This is the physical principle that Ralph Knowles, professor and member of the American Solar Energy Society, had already intuited and described in 1974. Thus, the mountains are not only a metaphorical reference to the design of these dwellings, but they also influence on their operation and distribution. These ideas go in the same direction as the results of the research carried out by Zahraa Saiyed and Paul D. Irwinb [89]. As they conclude, Native American legends reflect a knowledge about the environment which goes further than symbolism. These stories indicate that Native American Indians deeply knew how their surrounding environment worked, and that fact let them use the resources at their disposal in a respectful and efficient way. Moreover, as can be seen in the research developed by César J. Pérez and Carl A. Smith [90], the indigenous techniques, the so-called Indigenous Knowledge Systems (IKS), which often underlay these old stories, can be very useful for the environment protection at present.

The nomad lifestyle is more closely linked to nature than the sedentary lifestyle is. This can be seen by analyzing the Navaho culture, as the research presented by Len Necefer concludes [91]. This fact also affects their dwellings, the hogans. One of their most remarkable features is that their smoke hole cannot be closed. Unlike the tipis, whose smoke hole controls the exit of air and smoke, the hogan's can never be closed, as Thibony explains [29]: "Visitors ask what happens when it rains or snow", said a Navajo working at the visitor center. "They want to know if they cover the smoke hole. *'You let things happen'* I tell them. *'You let the rain come in. The dome represents the sky, and the floor is the earth. The earth shouldn't be covered up. It reminds you of who you are and where you came from. The hogan places you where you belong. You take your identity from it'"*. Features like this allow understanding how a culture works and the stance their members take in relation to current challenges, such as the energy consumption or the environmental resources management.

6. Conclusions

The results show that there was a decreasing progression on the energy a dwelling is equal to throughout the sedentary process. This evolution was broken in its last step by the settled agricultural communities, the pueblos, since their adobe dwellings are equal to a similar amount of energy of those used by the nomad and seminomadic groups.

This value is linked to the morphology of the analyzed building and to its building materials, but it is not related to the zone where it is set up.

Two theoretical ideas were developed to obtain this conclusion. First, the equivalent energy was the value designed to indicate the capacity of a building to transform the outdoor conditions into the indoor ones. It means that the building itself is understood as if it was a machine and its power is quantifiable. Second, a statistical method was brought from botany and archaeology to architecture. The canonical correspondence method allows us to establish links between quantitative data and qualitative information. This way, it transforms the morphological characteristics of a building into numerical information, in such a way that both quantitative and qualitative data can be related graphically.

From these bases, the present research will go on. On the one hand, the equivalent energy will be calculated and analyzed for current buildings. On the other hand, the canonical correspondence analysis will be used to determine the relation between more examples of vernacular architecture and their corresponding environments. This is the architectural field where it can be more useful, since the design of this type of dwelling derives directly from the limitations imposed by nature.

Author Contributions: Conceptualization, M.J.M.B., H.S.Á.d.T., R.A.G.L. and A.G.d.M.; methodology, R.A.G.L. and A.G.d.M.; software, M.J.M.B., R.A.G.L. and A.G.d.M.; validation, H.S.Á.d.T., R.A.G.L. and A.G.M; investigation, M.J.M.B. and H.S.Á.d.T.; writing—original draft preparation, M.J.M.B.; writing—review and editing, M.J.M.B.; supervision, H.S.Á.d.T., R.A.G.L. and A.G.d.M. All authors have read and agreed to the published version of the manuscript.

Funding: This research received no external funding.

Acknowledgments: Special thanks are due to the First Nations House of Learning and the Museum of Anthropology (University of British Columbia, Vancouver), as well as to the Library of Congress and to the Smithsonian Archives (Washington, D.C.) for the indispensable information provided for the present research. This research did not receive any specific grant from funding agencies in the public, commercial, or not-for-profit sectors.

Conflicts of Interest: The authors declare no conflicts of interest.

Appendix A

Figure A1. Location of the archaeological sites. Source: Own elaboration over Google Earth Pro 2017© cartography.

Table A1. Archaeological sites.

	Archaeological Site	Climatic Zone [74]	Nearest Weather Station
		Wigwam	
1	Skitchewaug site	Dfb	Springfield-Hartnes.State.AP.726115_TMY3
2	Site 230-3-1, Wappinger Creek	Cfa	Poughkeepsie-Dutchess.County.AP.725036_TMY3
3	Salt Pond Archaeological Site	Cfa	Groton-New.London.AP.725046_TMY3
4	Boyd's Cove	Dfb	Gander.718030_CWEC
5	Bellamy	Dfb	London.716230_CWEC
6	Pig Point	Cfa	Andrews.AFB.745940_TMY3
7	Pequot Fort	Cfa	Groton-New.London.AP.725046_TMY3
8	Figura Site	Dfb	St.Clair.County.Intl.AP.725384_TMY3
9	Localización documentada por Ezra Stiles	Cfa	Groton-New.London.AP.725046_TMY3
10	Kipp Island	Dfb	Syracuse-Hancock.Intl.AP.725190_TMY3
		Hogan	
11	Navajo Reservoir District-LA 3021	Dfb	Durango-La.Plata.County.AP.724625_TMY3
12	Navajo Reservoir District-LA 3460	BSk	Farmington-Four.Corners.Rgnl.AP.723658_TMY3
13	Bist-star BS-511 (43-2)	BSk	Farmington-Four.Corners.Rgnl.AP.723658_TMY4
14	Gobernador Canyon LA1869	BSk	Farmington-Four.Corners.Rgnl.AP.723658_TMY5
15	Chaco Canyon CM-4	BSk	Gallup-Sen.Clarke.Field.723627_TMY3
16	Kayenta	BSk	Gallup-Sen.Clarke.Field.723627_TMY3
17	Rainbow Lodge	BSk	Blanding.Muni.AP.724723_TMY3
18	Cedar Ridge	BSk	Winslow.Muni.AP.723740_TMY
19	Tuba City	BWk	Page.Muni.AWOS.723710_TMY3
20	Window Rock	BSk	Gallup-Sen.Clarke.Field.723627_TMY3
		Tipi	
21	Greasewood Creek 486	Dfb	Kalispell.727790_TMY2
22	Spring Lake 584	BSk	Cut.Bank.Muni.AP.727796_TMY
23	Souris River 32RV416	Dfb	Estevan.718620_CWEC
24	Souris River 32RV419	Dfb	Minot.727676_TMY2
25	The Cranford Site	BSk	Lethbridge.712430_CWEC
26	Indian Mountain site 5BL876	BSk	Fort.Collins.AWOS.724769_TMY3
27	Sheyenne River 32SH205	Dfb	Bismarck.Muni.AP.727640_TMY3
28	Demijohn Flats 24CB736	BSk	Cody.Muni.AWOS.726700_TMY3
29	Pinon Canyon Maneuver Site-Training Area 7	BSk	La.Junta.Muni.AP.724635_TMY3
30	Pilgrim Site 24BW675	BSk	Butte-Bert.Mooney.AP.726785_TMY3
		Earthlodge	
31	Hidatsa Village	Dfb	Bismarck.Muni.AP.727640_TMY3
32	Awatixa Village	Dfb	Bismarck.Muni.AP.727640_TMY3
33	Awatixa Xi'e Village	Dfb	Bismarck.Muni.AP.727640_TMY3
34	Rooptahee	Dfb	Bismarck.Muni.AP.727640_TMY3
35	Like-a-fishhook 32ML2	Dwb	Dickinson.Muni.AP.727645_TMY3
36	On-a-Slant Village 32MO26	Dfb	Bismarck.Muni.AP.727640_TMY3
37	Arikara Battle 1823 T20N S25 R30E	Dfa	Mobridge.Muni.AP.726685_TMY3
38	Huff site 32M011	Dfb	Bismarck.Muni.AP.727640_TMY3
39	Kansas Monument site 14RP1	Cfa	Concordia-Blosser.Muni.AP.724580_TMY3
40	Fullerton 25NC7	Dfa	Columbus.Muni.AP.725565_TMY3
		Grass house	
41	Clement site 38Mc8	Cfa	Cox.Field.722587_TMY3
42	Sanders site	Cfa	Sherman-Perrin.AFB.722541_TMY
43	Hill Farm site 41BW169 (Hatchel)	Cfa	Texarkana-Webb.Field.723418_TMY3
44	Roseborough Lake site	Cfa	Texarkana-Webb.Field.723418_TMY3
45	McLelland site 16B0236	Cfa	Shreveport.722480_TMY2
46	Caddo Indian Burial Ground Norman 3MN386	Cfa	Hot.Springs.Mem.AP.723415_TMY3
47	George C. Davis site 41CE19	Cfa	Nacogdoches.AWOS.722499_TMY3
48	Walker Creek project Pilgrim's Pride site 41CP304	Cfa	Greenville.Muni.AP.722588_TMY3
49	Pine Tree Mound 41HS15	Cfa	Longview-Gregg.County.AP.722470_TMY3
50	Vinson site 41LT1	Cfa	Waco.Rgnl.AP.722560_TMY

Table A1. *Cont.*

	Archaeological Site	Climatic Zone [74]	Nearest Weather Station
		Longhouse	
51	Mantle site (AlGt-334)	Dfb	Mount.Forest.716310_CWEC
52	Strickler site (36La3)	Cfa	Wilmington.724089_TMY2
53	Klock site	Dfb	Utica-Oneida.County.AP.725197_TMY3
54	Garoga site	Dfb	Utica-Oneida.County.AP.725197_TMY3
55	Norton site (AfHh-86)	Dfb	London.716230_CWEC
56	Lawson site (AgHh-1)	Dfb	London.716230_CWEC
57	Wiacek site (BcGw-26)	Dfb	Muskoka.716300_CWEC
58	Nodwell site (bChI-3)	Dfb	St.Clair.County.Intl.AP.725384_TMY3
59	Baumann site (BdGv-14)	Dfb	Muskoka.716300_CWEC
60	Myers Road site (AiHb-13)	Dfb	London.716230_CWEC
		Pueblo	
61	Taos	Dfb	Taos.Muni.AP.723663_TMY3
62	Isleta	BSk	Albuquerque.Intl.AP.723650_TMY3
63	Tesuque	Cfb	Santa.Fe.County.Muni.AP.723656_TMY3
64	Zia	BSk	Albuquerque.Intl.AP.723650_TMY3
65	Sandia	BSk	Albuquerque.Intl.AP.723650_TMY3
66	Acoma	BSk	Albuquerque.Intl.AP.723650_TMY3
67	Zuni	BSk	Deming.Muni.AP.722725_TMY3
68	Picuris	Cfb	Santa.Fe.County.Muni.AP.723656_TMY3
69	Jemez	BSk	Albuquerque.Intl.AP.723650_TMY3
70	San Juan	BSk	Albuquerque.Intl.AP.723650_TMY3
		Plank house	
71	Old Kasaan	Cfb	Ketchikan.Intl.AP.703950_TMY3
72	Howkan	Cfb	Hydaburg.Seaplane.Base.703884_TMY3
73	Klinkwan	Cfb	Hydaburg.Seaplane.Base.703884_TMY3
74	Kaisun	Cfb	Sandspit.711010_CWEC
75	Kiusta	Cfb	Sandspit.711010_CWEC
76	Kung	Cfb	Sandspit.711010_CWEC
77	Ninstints	Cfb	Sandspit.711010_CWEC
78	Skidegate	Cfb	Sandspit.711010_CWEC
79	Tanu	Cfb	Sandspit.711010_CWEC
80	Hiellan	Cfb	Prince.Rupert.718980_CWEC

Table A2. Building materials.

Dwelling	Layer	Thickness (m)	Specific Heat (J/kgK)	Density (kg/m^3)	Thermal Transmittance U (W/m^2 °K)
Wigwam*					
wall/roof					2.13/2.27
	Cattail	0.001	1630.00	300	
	Air	0.005	1012.00	1	
	Cattail	0.001	1630.00	300	
	Cattail	0.001	1630.00	300	
	Air	0.005	1012.00	1	
	Cattail	0.001	1630.00	300	
Tipi**					
wall/roof					3.50/3.91
	Hide	0.0058	1400	22	
Hogan					
wall/roof					0.84/0.87
	Wood	0.1	1380.00	510.00	
	Tree bark	0.0127	1364.00	482.00	
	Earth	0.15	880.00	1460.00	
Grass house					
wall/roof					0.14/0.14
	Grass	0.35	1630.00	130.00	

Table A2. *Cont.*

Dwelling	Layer	Thickness (m)	Specific Heat (J/kgK)	Density (kg/m^3)	Thermal Transmittance U (W/m^2 °K)
Earthlodge					
wall/roof					0.34/0.35
	Wood	0.1	1380.00	510.00	
	Wood (branches)	0.1	1380.00	510.00	
	Grass + grass from turf	0.05	1630.00	150.00	
	Earth from turf	0.1	880.00	1460.00	
Longhouse***					
wall/roof					4.28/4.91
	Tree bark	0.0127	1364.00	482.00	
Pueblo					
wall					0.98
	Earth	0.51	1100.00	1400.00	
roof					0.52
	Wood (branches)	0.09	1380.00	510.00	
	Grass	0.03	1630.00	150.00	
	Earth	0.25	1100.00	1400.00	
Plank house					
wall					1.33
	Wooden planks	0.07	1380	510	
roof					4.91
	Tree bark	0.0127	1364.00	482.00	

* [92]; **In order to obtain the data about tipi hides, we used the information about other nomad tents whose envelopes were also made from animal skins. First, we used the information about goat skins presented in the research carried out by Shady Attia [93]. Second, we also took the information about yurt envelopes generated by Peter Manfield [94]. *** [95].

Table A3. Details of the building materials which compose the template called "Timber frame-superinsulated" from DesignBuilder v6.1.2.005.

	Thermal Transmittance (W/m^2 °K)		Thermal Transmittance (W/m^2 °K)
Outdoor walls	0.375	Sub-surfaces	
Bellow grade walls	0.375	Walls	0.156
Flat roof	5.983	Floors	
Pitched roof	2.93	Ground floor	0.866
Semi-exposed		Internal floor	0.866
Ceilings	0.228		
Floors	0.259		

The aforementioned template contains more building materials, but only the information about those which were assigned in the present research is contained in the previous table.

Table A4. Equivalent energy for each location and for each model in the corresponding approaches.

	Original Building Materials, Original Locations, with Openings	Original Building Materials, New Mexican Locations (Locations of Pueblos), with Openings	Original Building Materials from each Model Assigned to a Wigwam in New Mexican Locations (Locations of Pueblos), without Openings	Same Materials (Timber Frame-Superinsulated), New Mexican Locations (Locations of Pueblos), without Openings
Wigwam	0.160	0.170	0.304	0.275
1	0.202	0.232	0.371	0.318
2	0.166	0.162	0.295	0.266
3	0.137	0.180	0.318	0.287
4	0.169	0.162	0.295	0.266
5	0.177	0.162	0.295	0.266
6	0.133	0.162	0.295	0.266
7	0.137	0.136	0.258	0.269
8	0.173	0.180	0.318	0.287
9	0.137	0.162	0.295	0.266
10	0.166	0.162	0.295	0.266

Table A4. *Cont.*

	Original Building Materials, Original Locations, with Openings	Original Building Materials, New Mexican Locations (Locations of Pueblos), with Openings	Original Building Materials from each Model Assigned to a Wigwam in New Mexican Locations (Locations of Pueblos), without Openings	Same Materials (Timber Frame-Superinsulated), New Mexican Locations (Locations of Pueblos), without Openings
Hogan	0.201	0.164	0.316	0.296
1	0.214	0.213	0.410	0.323
2	0.183	0.153	0.290	0.291
3	0.183	0.177	0.345	0.307
4	0.183	0.153	0.290	0.291
5	0.203	0.153	0.290	0.291
6	0.203	0.153	0.290	0.291
7	0.176	0.153	0.318	0.272
8	0.321	0.177	0.345	0.307
9	0.146	0.153	0.290	0.291
10	0.203	0.153	0.290	0.291
Tipi	0.140	0.126	0.371	0.149
1	0.145	0.160	0.456	0.157
2	0.159	0.123	0.364	0.149
3	0.132	0.129	0.372	0.149
4	0.121	0.123	0.364	0.149
5	0.129	0.123	0.364	0.149
6	0.148	0.123	0.364	0.149
7	0.128	0.106	0.323	0.141
8	0.142	0.129	0.372	0.149
9	0.118	0.123	0.364	0.149
10	0.175	0.123	0.364	0.149
Earthlodge	0.079	0.070	0.323	0.079
1	0.082	0.088	0.422	0.090
2	0.082	0.059	0.298	0.075
3	0.082	0.097	0.354	0.083
4	0.082	0.059	0.298	0.075
5	0.080	0.059	0.298	0.075
6	0.082	0.059	0.298	0.075
7	0.080	0.062	0.314	0.080
8	0.082	0.097	0.354	0.083
9	0.060	0.059	0.298	0.075
10	0.075	0.059	0.298	0.075
Grass house	0.074	0.088	0.339	0.113
1	0.057	0.098	0.441	0.139
2	0.055	0.068	0.314	0.104
3	0.061	0.147	0.368	0.121
4	0.061	0.068	0.314	0.104
5	0.057	0.068	0.314	0.104
6	0.066	0.068	0.314	0.104
7	0.066	0.072	0.326	0.121
8	0.067	0.147	0.368	0.121
9	0.074	0.068	0.314	0.104
10	0.173	0.068	0.314	0.104
Longhouse	0.014	0.015	0.394	0.016
1	0.015	0.018	0.450	0.017
2	0.012	0.015	0.389	0.015
3	0.014	0.015	0.396	0.016
4	0.014	0.015	0.389	0.015
5	0.014	0.015	0.389	0.015
6	0.014	0.015	0.389	0.015
7	0.017	0.013	0.362	0.015
8	0.014	0.015	0.396	0.016
9	0.017	0.015	0.389	0.015
10	0.014	0.015	0.389	0.015
Pueblo	0.061	0.061	0.276	0.113
1	0.077	0.077	0.353	0.112
2	0.056	0.056	0.253	0.115
3	0.062	0.062	0.298	0.108
4	0.056	0.056	0.253	0.115
5	0.056	0.056	0.253	0.115
6	0.056	0.056	0.253	0.115
7	0.075	0.075	0.292	0.118
8	0.062	0.062	0.298	0.108
9	0.056	0.056	0.253	0.115
10	0.056	0.056	0.253	0.115

Table A4. *Cont.*

	Original Building Materials, Original Locations, with Openings	Original Building Materials, New Mexican Locations (Locations of Pueblos), with Openings	Original Building Materials from each Model Assigned to a Wigwam in New Mexican Locations (Locations of Pueblos), without Openings	Same Materials (Timber Frame-Superinsulated), New Mexican Locations (Locations of Pueblos), without Openings
Plank house	0.030	0.029	0.244	0.028
1	0.033	0.037	0.327	0.031
2	0.035	0.028	0.225	0.027
3	0.035	0.031	0.270	0.029
4	0.027	0.028	0.225	0.027
5	0.027	0.028	0.225	0.027
6	0.027	0.028	0.225	0.027
7	0.027	0.023	0.229	0.026
8	0.027	0.031	0.270	0.029
9	0.027	0.028	0.225	0.027
10	0.030	0.028	0.225	0.027

Table A5. Data used in PAST v3.25.

	K	L	M	N	O	P	Q	R	S	T	U	V	W	X	Y
W1	0	0	1	0	0	0	0	1	0	0	0	0	1	0	0
W2	0	0	1	0	0	0	0	1	0	0	0	0	1	0	0
W3	0	0	1	0	0	0	0	1	0	0	0	0	1	0	0
W4	0	0	1	0	0	0	0	1	0	0	0	0	1	0	0
W5	0	0	1	0	0	0	0	1	0	0	0	0	1	0	0
W6	0	0	1	0	0	0	0	1	0	0	0	0	1	0	0
W7	0	0	1	0	0	0	0	1	0	0	0	0	1	0	0
W8	0	0	1	0	0	0	0	1	0	0	0	0	1	0	0
W9	0	0	1	0	0	0	0	1	0	0	0	0	1	0	0
W10	0	0	1	0	0	0	0	1	0	0	0	0	1	0	0
H1	1	1	0	0	1	0	0	1	0	0	0	1	0	1	0
H2	1	1	0	0	1	0	0	1	0	0	0	1	0	1	0
H3	1	1	0	0	1	0	0	1	0	0	0	1	0	1	0
H4	1	1	0	0	1	0	0	1	0	0	0	1	0	1	0
H5	1	1	0	0	1	0	0	1	0	0	0	1	0	1	0
H6	1	1	0	0	1	0	0	1	0	0	0	1	0	1	0
H7	1	1	0	0	1	0	0	1	0	0	0	1	0	1	0
H8	1	1	0	0	1	0	0	1	0	0	0	1	0	1	0
H9	1	1	0	0	1	0	0	1	0	0	0	1	0	1	0
H10	1	1	0	0	1	0	0	1	0	0	0	1	0	1	0
T1	0	0	0	0	1	0	0	1	1	1	0	0	0	0	0
T2	0	0	0	0	1	0	0	1	1	1	0	0	0	0	0
T3	0	0	0	0	1	0	0	1	1	1	0	0	0	0	0
T4	0	0	0	0	1	0	0	1	1	1	0	0	0	0	0
T5	0	0	0	0	1	0	0	1	1	1	0	0	0	0	0
T6	0	0	0	0	1	0	0	1	1	1	0	0	0	0	0
T7	0	0	0	0	1	0	0	1	1	1	0	0	0	0	0
T8	0	0	0	0	1	0	0	1	1	1	0	0	0	0	0

	K	L	M	N	O	P	Q	R	S	T	U	V	W	X	Y
GH1	0	0	0	0	1	0	0	1	0	1	0	0	0	0	0
GH2	0	0	0	0	1	0	0	1	0	1	0	0	0	0	0
GH3	0	0	0	0	1	0	0	1	0	1	0	0	0	0	0
GH4	0	0	0	0	1	0	0	1	0	1	0	0	0	0	0
GH5	0	0	0	0	1	0	0	1	0	1	0	0	0	0	0
GH6	0	0	0	0	1	0	0	1	0	1	0	0	0	0	0
GH7	0	0	0	0	1	0	0	1	0	1	0	0	0	0	0
GH8	0	0	0	0	1	0	0	1	0	1	0	0	0	0	0
GH9	0	0	0	0	1	0	0	1	0	1	0	0	0	0	0
GH10	0	0	0	0	1	0	0	1	0	1	0	0	0	0	0
LH1	1	0	0	1	0	1	0	1	0	0	0	1	0	0	0
LH2	1	0	0	1	0	1	0	1	0	0	0	1	0	0	0
LH3	1	0	0	1	0	1	0	1	0	0	0	1	0	0	0
LH4	1	0	0	1	0	1	0	1	0	0	0	1	0	0	0
LH5	1	0	0	1	0	1	0	1	0	0	0	1	0	0	0
LH6	1	0	0	1	0	1	0	1	0	0	0	1	0	0	0
LH7	1	0	0	1	0	1	0	1	0	0	0	1	0	0	0
LH8	1	0	0	1	0	1	0	1	0	0	0	1	0	0	0
LH9	1	0	0	1	0	1	0	1	0	0	0	1	0	0	0
LH10	1	0	0	1	0	1	0	1	0	0	0	1	0	0	0
P1	0	1	0	0	0	0	1	0	0	0	0	0	0	1	1
P2	0	1	0	0	0	0	1	0	0	0	0	0	0	1	1
P3	0	1	0	0	0	0	1	0	0	0	0	0	0	1	1
P4	0	1	0	0	0	0	1	0	0	0	0	0	0	1	1
P5	0	1	0	0	0	0	1	0	0	0	0	0	0	1	1
P6	0	1	0	0	0	0	1	0	0	0	0	0	0	1	1
P7	0	1	0	0	0	0	1	0	0	0	0	0	0	1	1
P8	0	1	0	0	0	0	1	0	0	0	0	0	0	1	1

Table A5. *Cont.*

	K	L	M	N	O	P	Q	R	S	T	U	V	W	X	Y		K	L	M	N	O	P	Q	R	S	T	U	V	W	X	Y
T9	0	0	0	0	1	0	0	1	1	1	0	0	0	0	0	P9	0	1	0	0	0	0	1	0	0	0	0	0	0	1	1
T10	0	0	0	0	1	0	0	1	1	1	0	0	0	0	0	P10	0	1	0	0	0	0	1	0	0	0	0	0	0	1	1
E1	1	0	1	0	0	0	0	1	0	1	1	0	0	0	1	PH1	0	1	0	1	0	0	0	1	0	0	0	1	0	0	1
E2	1	0	1	0	0	0	0	1	0	1	1	0	0	0	1	PH2	0	1	0	1	0	0	0	1	0	0	0	1	0	0	1
E3	1	0	1	0	0	0	0	1	0	1	1	0	0	0	1	PH3	0	1	0	1	0	0	0	1	0	0	0	1	0	0	1
E4	1	0	1	0	0	0	0	1	0	1	1	0	0	0	1	PH4	0	1	0	1	0	0	0	1	0	0	0	1	0	0	1
E5	1	0	1	0	0	0	0	1	0	1	1	0	0	0	1	PH5	0	1	0	1	0	0	0	1	0	0	0	1	0	0	1
E6	1	0	1	0	0	0	0	1	0	1	1	0	0	0	1	PH6	0	1	0	1	0	0	0	1	0	0	0	1	0	0	1
E7	1	0	1	0	0	0	0	1	0	1	1	0	0	0	1	PH7	0	1	0	1	0	0	0	1	0	0	0	1	0	0	1
E8	1	0	1	0	0	0	0	1	0	1	1	0	0	0	1	PH8	0	1	0	1	0	0	0	1	0	0	0	1	0	0	1
E9	1	0	1	0	0	0	0	1	0	1	1	0	0	0	1	PH9	0	1	0	1	0	0	0	1	0	0	0	1	0	0	1
E10	1	0	1	0	0	0	0	1	0	1	1	0	0	0	1	PH10	0	1	0	1	0	0	0	1	0	0	0	1	0	0	1

References

1. Bakeless, J. *America as Seen by Its First Explorers: The Eyes of Discover*; Dover Publications: New York, NY, USA, 1989.
2. Leroi-Gourhan, A. *El Gesto y la Palabra*; Universidad Central de Venezuela: Caracas, Venezuela, 1971.
3. Stringer, C.; Gamble, C.; Canals, O. *En Busca de los Neandertales: La Solución al Rompecabezas de los Orígenes Humanos*; Crítica: Barcelona, Spain, 2001.
4. Nabokov, P.; Easton, R. *Native American Architecture*; Oxford University Press: New York, NY, USA, 1989.
5. Campbell, S. The Cheyenne tipi. *Am. Anthropol.* **1915**, *17*, 685. [CrossRef]
6. Catlin, G. *Letters and Notes on the Manners, Customs, and Condition of the North American Indians*; Wiley and Putnam: New York, NY, USA, 1844.
7. Jarzombek, M. *Architecture of First Societies: A Global Perspective*; Wiley: Hoboken, NJ, USA, 2013.
8. Yue, D.; Yue, C. *The Tipi: A Center of Native American Life*; Alfred A. Knopf: New York, NY, USA, 1984.
9. Campbell, S. The tipis of the Crow Indians. *Am. Anthropol.* **1927**, *29*, 87–104. [CrossRef]
10. Hassrick, R.B. *The Sioux*; University of Oklahoma Press: Norman, OK, USA, 1964.
11. Holley, L.A. *Tipis, Tepees, Teepees: History & Design of the Cloth Teepee*; Gibbs Smith: Salt Lake City, UT, USA, 2007.
12. Lowie, R.H. *Indians of the Plains*; McGraw-Hill Book Co.: New York, NY, USA, 1954.
13. Rosoff, N.B.; Kennedy, S. *Tipi: Heritage of the Great Plains*; Brooklyn Museum en Asociación con University of Washington Press: Brooklyn, NY, USA, 2011.
14. Couchaux, D. *Habitats Nomades*; Alternatives: Paris, France, 2011.
15. Goble, P. *Tipi: Home of the Nomadic Buffalo Hunters*; World Wisdom Inc.: Bloomington, IN, USA, 2007.
16. Laubin, R.; Laubin, G. *Indian Tipi: Its History, Construction, and Use*; University of Oklahoma Press: Norman, OK, USA, 1989.
17. Tooker, E. *Lewis H. Morgan on Iroquois Material Culture*; The University of Arizona Press: Tucson, AZ, USA; London, UK, 1994.
18. Wilson, G.L. *The Horse and the Dog in Hidatsa Culture*; American Museum of Natural History: New York, NY, USA, 1924.
19. Bjorklund, K.L. *The Indians of Northeastern America*; Dodd, Mead & Co.: New York, NY, USA, 1969.
20. Morgan, L.H.; Bohannan, P.J. *Houses and House-Life of the American Aborigines*; University of Chicago Press: Chicago, IL, USA, 1965.
21. Bushnell, D.I. *Native Villages and Village Sites East of the Mississippi*; Government Printing Office: Washington, DC, USA, 1966.
22. Bushnell, D. *Villages of the Algonquian, Siouan, and Caddoan Tribes West of Mississipi*; Government Printing Office: Washington, DC, USA, 1922.
23. Densmore, F.; Marchetti, N. *Chippewa Customs*; Minnesota Historical Society Press: St. Paul, MN, USA, 1979.

24. Petersen, K.D. *Chippewa Mat-Weaving Techniques*; U.S. Government Printing Office: Washington, DC, USA, 1963.
25. Yale Indian Papers Project—New England Indian Papers Series Database. 2018. Available online: http://jake.library.yale.edu:8080/neips/search (accessed on 7 November 2018).
26. Alonso Núñez, M.P. El hogan: Es más que mi casa... es el lugar donde yo rezo. *Rev. Española Antropol. Am.* **1999**, *29*, 233–259.
27. Dobyns, H.F.; Euler, R.C. *The Navajo People*; Indian Tribal Series: Phoenix, AZ, USA, 1972.
28. Oliver, P. *Dwellings: The Vernacular House Worldwide*; Phaidon: London, UK, 2003.
29. Thybony, S. *The Hogan: The Traditional Navajo Home*; Western National Parks Association: Tucson, AZ, USA, 1998.
30. Jett, S.C.; Spencer, V.E. *Navajo Architecture: Forms, History, Distributions*; The University of Arizona Press: Tucson, AZ, USA, 2017.
31. Mindeleff, C. *Navaho Houses*; Government Printing Office: Washington, DC, USA, 1898.
32. Bolton, H. *Texas in the Middle Eighteenth Century: Studies in Spanish Colonial History and Administration*; University of California Press: Berkeley, CA, USA, 1915.
33. Harrington, M. *Certain Caddo Sites in Arkansas*; Museum of the American Indian/Heye Foundation: New York, NY, USA, 1920.
34. Curtis, E.S. *The North American Indian. The Indians of Oklahoma. The Wichita. The Southern Cheyenne. The Oto. The Comanche. The Peyote Culture*; Cambridge University Press: Cambridge, MA, USA, 1930.
35. Wilmsen, E. A suggested developmental sequence for house forms in the Caddoan area. *Bull. Tex. Archeolog. Soc.* **1959**, *30*, 35–49.
36. National Park Service—U. S. Department of the Interior 2017. Knife River Indian Villages. Available online: https://www.nps.gov/knri/index.htm (accessed on 25 February 2019).
37. Wilson, G. *The Hidatsa Earthlodge*; The American Museum of Natural History: New York, NY, USA, 1934.
38. Catlin, G. *Illustrations of the Manners, Customs & Condition of the North American Indians, with Letters and Notes Written during Eight Years of Travel and Adventure among the Wildest and Most Remarkable Tribes Now Existing*; Henry G. Bohn: London, UK, 1857.
39. Boas, F. *The Kwakiutl of Vancouver Island*; Leiden: New York, NY, USA, 1905.
40. Dawson, G.; Geological Survey of Canada. *Report of Progress for 1878–1879. Report on the Queen Charlotte Islands-1878*; Dawson Bros.: Montreal, QC, Canada, 1880.
41. Underhill, R. *Indians of the Pacific Northwest*; US Department of the Interior, Bureau of Indian Affairs, Branch of Education: Washington, DC, USA, 1944.
42. Drew, L. *Haida, their Art and Culture*; Hancock House: Surrey, UK, 1982.
43. Macdonald, G. *Haida Monumental Art: Villages of the Queen Charlotte Islands*; UBC Press: Vancouver, BC, Canada, 1995.
44. STEWART, H. *Cedar: Tree of life to the Northwest Coast Indians. Douglas & McIntyre*; University of Washington Press: Seattle, WA, USA, 1995.
45. Macdonald, G. *Ninstints, Haida World Heritage Site*; University of British Columbia Press, U.B.C. Museum of Anthropology: Vancouver, BC, Canada, 1987.
46. Vastokas, J.M. *Architecture of the Northwest Coast Indians of America*; University Microfilms International: Ann Arbor, MI, USA, 1966.
47. Boston Public Library. Norman B. Leventhal Map & Education Center at the Boston Public Library. Available online: https://collections.leventhalmap.org (accessed on 30 September 2018).
48. Chadwick, E.M. *The People of the Longhouse*; The Church of England: Toronto, ON, Canada, 1897.
49. Champlain, S.; Brébeuf, J. *Les Voyages de la Nouuelle France Occidentale, Dite Canada: Faits par le Sr. de Champlain Xainctongeois, Capitaine Pour le Roy en la Marine du Ponant, & Toutes les Descouuertes qu'il a Faites en ce Païs Depuis l'an 1603, Jusques en l'an 1629*; De l'Imprimerie de Pierre Le Mur: Paris, France, 1640.
50. Martin, P.S.; Quimby, G.I.; Collier, D. *Indians before Columbus: Twenty Thousand Years of North American History Revealed by Archeology*; The University of Chicago Press: Chicago, IL, USA, 1947.
51. Sagard, G. *Le Grand Voyage du Pays des Hurons: Situé en l'Amerique Vers la Mer Douce, és Derniers Confins de la Nouuelle France, Dite Canada*; Chez Denys Moreav: Paris, France, 1632.

52. White, J. *Graphic Sketches from Old and Authentic Works, Illustrating the Costume, Habits, and Character, of the Aborigines of America: Together with Rare and Curious Fragments Relating to the Discovery and Settlement of the Country*; J. and H.G. Langley: New York, NY, USA, 1841.

53. Morgan, L.H. *League of the Ho-Dé-No-Sau-Nee, or Iroquois*; Dodd, Mead and Company: New York, NY, USA, 1922.

54. Lafitau, J.F. *Moeurs des Sauvages Amériquains, Comparées aux Moeurs des Premiers Temps. Ouvrage Enrichi de Figures en Taille-Douce*; Saugrain: Paris, France, 1724.

55. Erdoes, R. *Native Americans: The Pueblos*; Sterling Pub. Co.: New York, NY, USA, 1983.

56. Morgan, W.N.; Swentzell, R. *Ancient Architecture of the Southwest*; University of Texas Press: Austin, TX, USA, 2013.

57. Nabokov, P.; Historic American Buildings Survey. *Architecture of Acoma Pueblo: The 1934 Historic American Buildings Survey Project*; Ancient City Press: Santa Fe, NM, USA, 1986.

58. Jackson, J.B. Pueblo. Architecture of our own. *Landsc. Mag. Hum. Geogr.* **1953**, *3*, 1953–1954.

59. Rapoport, A. The Pueblo and the Hogan. In *Shelter and Society*; Oliver, P., Frederick, A., Eds.; Praeger: New York, NY, USA, 1969.

60. Mindeleff, V. A study of Pueblo architecture: Tusayan and Cibola. In *Eighth Annual Report of the Bureau of Ethnology Government Printing Office*; Government Printing Office: Washington, DC, USA, 1891.

61. Library of Congress 2018. Available online: https://www.loc.gov (accessed on 26 December 2018).

62. Smithsonian Institution Collection Search Center. 2018. Available online: http://collections.si.edu (accessed on 15 April 2018).

63. Northwestern University-Digital Library Collections 2003. Edward S. Curtis's. The North American Indian. Available online: http://curtis.library.northwestern.edu/index.html (accessed on 19 December 2018).

64. Taylor, J. *Arquitectura Anónima: Una Visión Cultural de los Principios Prácticos del Diseño*; Editorial Stylos: Barcelona, Spain, 1984.

65. Atkins, P.; Jones, L. *Principios de Química. Los Caminos del Descubrimiento*; Editorial Médica Panamericana: Madrid, Spain, 2006.

66. Herrero García, M. *UF0902—Caracterización de Instalaciones de Climatización*; Editorial Elearning S.L.: Málaga, Spain, 2015.

67. Botsaris, P.; Prebezanos, S. A methodology for a thermal energy building audit. *Build. Environ.* **2004**, *39*, 195–199. [CrossRef]

68. El Shenawy, A.; Zmeureanu, R. Exergy-based index for assessing the building sustainability. *Build. Environ.* **2013**, *60*, 202–210. [CrossRef]

69. U.S. Department of Energy. Energy Plus. 2019. Available online: https://energyplus.net (accessed on 4 March 2019).

70. FlyCarpet Inc. Free Online Interactive Psychrometric Chart. 2019. Available online: http://www.flycarpet.net/en/PsyOnline (accessed on 9 August 2019).

71. Ryan, P.D.; Harper, D.A.T.; Whalley, J.S. *PALSTAT: User's Manual and Case Histories. Statistics for Palaeontologists and Palaeobiologists*; Chapman & Hall: London, UK, 1995.

72. Hammer, Ø.; Harper, D.A.T.; Ryan, P.D. PAST: Paleontological Statistics software package for education and data analysis. *Palaeontol. Electron.* **2001**, *4*, 9.

73. Legendre, P.; Legendre, L. *Numerical Ecology*; Elsevier: Amsterdam, The Netherlands, 1998.

74. Köppen, W. Versuch einer Klassifikation der Klimate, vorzugsweise nach ihren Beziehungen zur Pflanzenwelt. *Geogr. Z.* **1900**, *6*, 593–611.

75. Köppen, W.; Geiger, R. *Handbuch der Klimatologie. Das geographische System der Klimate*; Verlag Gebrüder Borntraeger: Berlin, Germany, 1936.

76. Kottek, M.; Rubel, F. Wold Maps of Köppen-Geiger Climate Classification. Available online: http://koeppen-geiger.vu-wien.ac.at/ (accessed on 20 January 2020).

77. Widmann, E.; Schanz, J.; Rohlfes, M.; König, O. WindFinder. 2019. Available online: https://es.windfinder.com (accessed on 12 December 2019).

78. Neila, F.J. *Arquitectura Bioclimática en un Entorno Sostenible*; Munilla-Lería: Madrid, Spain, 2004.

79. Lotfabadi, P.; Hançer, P. A comparative study of traditional and contemporary building envelope construction techniques in terms of thermal comfort and energy efficiency in hot and humid climates. *Sustainability* **2019**, *11*, 3582. [CrossRef]

80. Sobel, E.; Gahr, D.; Ames, K. *Household Archaeology on the Northwest Coast*; Berghahn Books: New York, NY, USA, 2016.
81. Roaf, M. *Cultural Atlas of Mesopotamia and the Ancient Near East*; Andromeda Oxford Limited: Oxford, UK, 1990.
82. Kylily, A.; Fokaides, P. European smart cities: The role of zero energy buildings. *Sustain. Cities Soc.* **2015**, *15*, 86–95. [CrossRef]
83. Varela-Boydo, C.; Moya, S. Inlet extensions for wind towers to improve natural ventilation in buildings. *Sustain. Cities Soc.* **2020**, *53*, 101933. [CrossRef]
84. Ghaffarianhoseini, A.H.; Berardi, U.; Dahlan, N.; Ghaffarianhoseini, A. What can we learn from Malay vernacular houses? *Sustain. Cities Soc.* **2014**, *13*, 157–170. [CrossRef]
85. Li, D.; Pan, W.; Lam, J. A comparison of global bioclimates in the 20th and 21st centuries and building energy consumption implications. *BAE Build. Environ.* **2014**, *75*, 236–249. [CrossRef]
86. Fernández-Antolín, M.; Del Río, J.; Costanzo, V.; Nocera, F.; González-Lezcano, R. Passive Design Strategies for Residential Buildings in Different Spanish Climate Zones. *Sustainability* **2019**, *11*, 4816. [CrossRef]
87. Feijó-Muñoz, J.; González-Lezcano, R.; Poza-Casado, I.; Padilla-Marcos, M.; Meiss, A. Airtightness of residential buildings in the Continental area of Spain. *Build. Environ.* **2019**, *148*, 299–308. [CrossRef]
88. Feijó-Muñoz, J.; Poza-Casado, I.; González-Lezcano, R.; Pardal, C.; Echarri, V.; Assiego, R.; Fernández-Agüera, J.; Dios-Viéitez, M.; Del Campo-Díaz, V.; Montesdeoca, M.; et al. Methodology for the Study of the Envelope Airtightness of Residential Buildings in Spain: A Case Study. *Energies* **2018**, *11*, 704. [CrossRef]
89. Saiyed, Z.; Irwin, P. Native American storytelling toward symbiosis and sustainable design. *Energy Res. Soc. Sci.* **2017**, *31*, 249–252. [CrossRef]
90. Pérez, S.; Smith, C. Indigenous Knowledge Systems and Conservation of Settled Territories in the Bolivian Amazon. *Sustainability* **2019**, *11*, 6099. [CrossRef]
91. Necefer, L.; Wong-Parodi, G.; Jaramillo, P.; Small, M. Energy development and Native Americans: Values and beliefs about energy from the Navajo Nation. *Energy Res. Soc. Sci.* **2015**, *7*, 1–11. [CrossRef]
92. Luamkanchanaphan, T.; Chotikaprakhan, S.; Jarusombati, S. A Study of Physical, Mechanical and Thermal Properties for Thermal Insulation from Narrow-leaved Cattail Fibers. *APCBEE Procedia* **2012**, *1*, 46. [CrossRef]
93. Attia, S. Assessing the Thermal Performance of Bedouin Tents in Hot Climates. In Proceedings of the 1st International Conference on Energy and Indoor Environment for Hot Climates, Doha, Qatar, 24 February 2014; p. 328.
94. Manfield, P. A Comparative Study of Temporary Shelters Used in Cold Climates. Master's Thesis, Cambridge University, Cambridge, UK, 2000.
95. Gupta, M.; Yang, J.; Roy, C. Specific heat and thermal conductivity of softwood bark and softwood char particles. *Fuel* **2003**, *82*, 919. [CrossRef]

Article

Classroom Indoor Environment Assessment through Architectural Analysis for the Design of Efficient Schools

Vicente López-Chao [1],*, Antonio Amado Lorenzo [1], Jose Luis Saorín [2], Jorge De La Torre-Cantero [2] and Dámari Melián-Díaz [2]

[1] Department of Architectural Graphics, Universidad de A Coruña, A Coruña 15008, Spain; antonio.amado@udc.es
[2] Department of Techniques and Projects in Engineering and Architecture, Universidad de La Laguna, San Cristóbal de La Laguna 38204, Spain; jlsaorin@ull.edu.es (J.L.S.); jcantero@ull.edu.es (J.D.L.T.-C.); dmeliand@ull.edu.es (D.M.-D.)
* Correspondence: v.lchao@udc.es

Received: 19 February 2020; Accepted: 3 March 2020; Published: 6 March 2020

Abstract: Optimization of environmental performance is one of the standards to be achieved towards designing sustainable buildings. Many researchers are focusing on zero emission building; however, it is essential that the indoor environment favors the performance of the building purpose. Empirical research has demonstrated the influence of architectural space variables on student performance, but they have not focused on holistic studies that compare how space influences different academic performance, such as Mathematics and Arts. This manuscript explores, under self-reported data, the relationship between learning space and the mathematics and art performance in 583 primary school students in Galicia (Spain). For this, the Indoor Physical Environment Perception scale has been adapted and validated and conducted in 27 classrooms. The results of the Exploratory Factor Analysis have evidenced that the learning space is structured in three categories: Workspace comfort, natural environment and building comfort. Multiple linear regression analyses have supported previous research and bring new findings concerning that the indoor environment variables do not influence in the same way different activities of school architecture.

Keywords: acoustics; environmental quality; learning space; occupant comfort; sustainable architecture; sustainable building; visual comfort; thermal comfort; ventilation comfort

---

1. Introduction

The term sustainable building has generally been attributed to low levels of energy consumption. However, providing quality interior environments is another goal of the so-called green buildings. On the one hand, this fact is related to the Sick Building Syndrome (SBS) that has evidenced that poor quality environments harm the health of users. On the other hand, the design of buildings must guarantee the purpose for which they were built and for which that energy cost was assumed. This effect could be measured through user performance.

Regarding education, school architecture must ensure that students learn concepts and knowledge from different disciplines. Previous literature focus on the influence of learning physical environment factors in academic outcomes. Furthermore, educational buildings are designed to last for several years, and their state remains constant for a long time without rehabilitation or diagnosis of their influence on users. It implies that some facilities do not meet the minimum quality standards that are a requirement for the new constructions, such as the case of lighting. Also, ICT (Information a Communication Technologies) implementation has promoted the use of blinds to improve the visibility of the laser projectors in classrooms.

This situation causes bad lighting conditions that can lead to poor school performance. Specifically, math performance is higher in students in classrooms with greater illumination [1]. Likewise, young children can differentiate lighting needs according to the activity performed [2], while visual comfort is a key element for arts activities, especially on drawing [3].

Schools are closed spaces that human beings occupy and in which they breathe for hours every day. Normally they do not have constant and automatic ventilation, which generates a lack of oxygen in the environment, and only schools with a mechanical supply and exhaust type of ventilation meet the recommended ventilation rate per student [4]. As well, low ventilation rates are associated with poor mathematics results [5,6], besides causing attention and concentration problems. These effects have shown to be more negative in tasks that require the use of spatial skills [7].

The thermal factor has also been correlated with academic performance, since thermal discomfort may lead to stress behaviors that influence learning [8]. Thermal alterations affect the problem-solving capacity and attention of students, which play a key role in mathematical skills competencies [9,10].

The acoustic environment correlates with attention levels in a classroom. Noise is an important factor of influence when identifying the words mentioned during classes and reducing reverberation values affect the levels of students' attention and performance [11]. Moreover, in order to solve noise problems, it is essential to understand and adapt the structure, organization and use of learning spaces in schools [12].

The environment in which students and teachers learn and teach are human-made. It is not the natural environment of a living being, and it is precisely the relationship with nature another concern in this area of research. Benfield, Rainbold, Bell and Donovan [13], studied the perceptions and behaviors of students in classrooms with landscape views. Similarly, van de Berg, Wesselius, Maas and Tanja-Dijkstra [14] conducted a controlled evaluation study on green walls as a restorative environment in the classroom. Both contributions provide a direct relationship between the inclusion of nature and performance in mathematics. Besides, views can influence the variance of reading vocabulary, language arts, and mathematics [15].

The classroom configuration has a close relationship with the teaching approach. The disposition of the space that will affect the interaction in the classroom and the choice of student seats also generates an impact on academic performance based on mathematics [16]. Besides, other studies focused on the influence of class size on the performance and behaviour of kindergarten [17–19], and satisfaction in secondary education [20]. Regarding arts, when elements, such as the aesthetics of the classroom and the furniture arrangement should allow greater interaction between students. Because it allows students to sit in a calm climate, leading to freely develop their creativity and improve their performance [21].

The literature focuses on one of the factors or dimensions mentioned above. However, some studies have developed holistic approaches to the impact of classroom spaces on learning [22–25]. The first empirical holistic model [24] included the learning space attachment factor, which were more significant than lighting for the development of mathematical performance.

Other elements of the classroom that may influence learning space attachment are the student seating location, due to issues, such as proximity to the teacher, accessibility to the halls or distance to the screen [26]. This choice of location in the classroom is influenced by the territoriality and personality of the individual [27,28]. Additionally, the learning space attachment has been associated with students' perception whose artworks were permanently exhibited [29]. This bond between students and their classroom is also related to their security and privacy feeling, which contribute to their comfort [30]. Likewise, the personalization of the space contributes to the creative development and aesthetic values of the students [31].

In a recent study, none of these factors are the answer when students were surveyed about their learning space preferences, but they mainly prefer learning spaces related to the end of their learning activities [32]. Other studies have focused on the vital relation between learning space and pedagogy [33,34], and in need to consider the perception of the student to obtain a more holistic knowledge [35]. So, it seems consistent that disciplines as different as Art and Mathematics, which respond to different teaching needs or methodologies, receive different influences from the same learning space.

Concerning the prediction of performance based on learning space, most researchers have measured mathematical performance through the Grade Point Average in different evaluation periods [24] or by national student performance evaluation programs [4]. While math outcome is considered to be accurately assessed, the evaluation of art performance is presented as a very complex task [36]. Although evidence has raised for isolated disciplines, little research has focused on how learning space factors influence the learning of different subjects [15,24,25].

In Spain, primary schools normally assign a classroom to each course group, except for Physical Education. So, students attend every subject in the same space, which seems not to be an efficient and sustainable use of the building in terms of learning. For that, this research aims to contribute to the explanation concerning how learning space influence art and mathematics performance in primary education, as well as to deepen the measurement of the learning space through the perception of the elementary school student and to investigate whether there are differences in the outcome prediction depending on the academic course.

Diagnostic studies create knowledge bases in order to support new, more efficient and sustainable classroom designs. Since it should be clarified how the learning space affects different subjects and high economic and sustainable costs must be prevented in school designs.

2. Materials and Methods

The main objective of this research is to explore the influence of learning space in mathematics and art performance in Primary Education schools in Galicia (Spain), and to explore learning space relationship with academic courses. For this, a quantitative research was designed, which relates the measurement of the perception of observable variables of the physical learning environment by students through a questionnaire with their performance in mathematics and arts.

First, the design of the questionnaire was adapted to the cognitive level of primary school students based on the *Indoor Physical Environment Perception* scale (iPEP scale) [25,37] (focused on university students), previously based on the holistic model of Barret, Davies, Zhang and Barret [38]. Next, public elementary schools are randomly chosen from the regions of A Coruña and Lugo (Galicia, Spain). Communication is established with the centers to request the possibility of conducting the questionnaires. Subsequently, the visits take place and the test data analyzed.

The psychometric properties of the instrument will be assessed through the corresponding reliability and validity analyzes to ensure that the use of the data. Then, the classrooms are described through the means and standard deviations of the variables of the scale to know in which values the data are grouped. Finally, a multiple linear regression analysis is applied to know if the space variables are predictors of mathematical performance and art education.

2.1. Sample

The sample consists of 583 Primary School (PS) students (307 boys and 276 girls), belonging to fourth, fifth and sixth grade of nine public centers. A total of 27 different classrooms in Lugo and A Coruña. The sample selection corresponds to a simple random sampling. Table 1 shows the distribution of attendants by course, center, and region.

The procedure to collect the information consisted of communicating with the centers obtained from the random selection process, of which four rejected the study proposal. The first contact was with the directors of the schools and later with the tutors of the course. The research was explained to the teaching staff, as well as the guidelines for solving the questionnaires.

Figure 1 shows the sample distribution by school and course. The fifth course of School 5 and the fourth course of School 6 stand out for a greater number of students, which is because they were centers with two small lines in both courses. Their classroom conditions were similar from a technical point of view (number and surface area of windows, orientation, sunshine, classroom size), so they have been presented together.

Table 1. Sample distribution by course, center, and region.

Region	Centre	Course			Total
		4th	5th	6th	
Lugo	School 1	24	19	14	57
	School 2	24	18	18	60
	School 3	23	24	17	64
	School 4	21	23	16	60
A Coruña	School 5	24	35	24	83
	School 6	32	20	24	76
	School 7	19	22	10	51
	School 8	21	22	20	63
	School 9	24	23	22	69
	Total	**212**	**206**	**165**	**583**

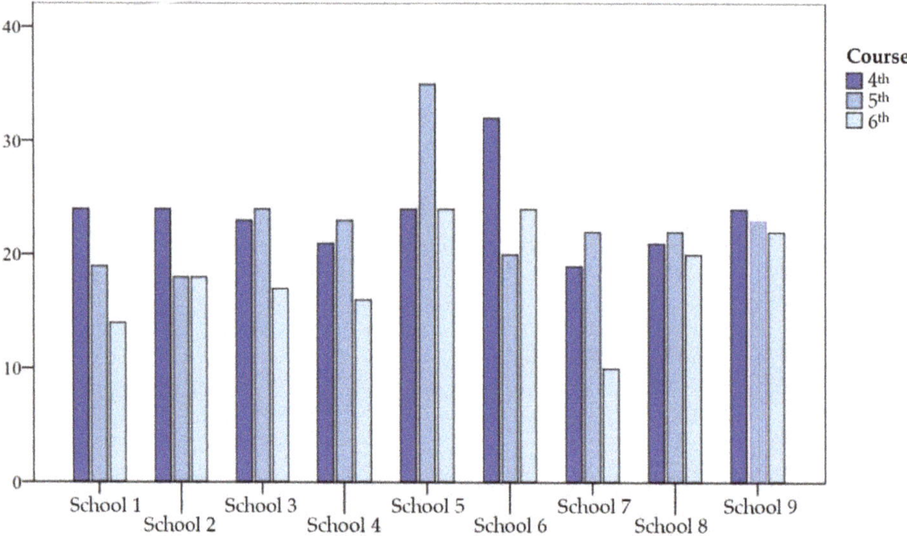

Figure 1. Student distribution by schools and courses.

2.2. Measurement Instrument

The instrument adapted is the *Indoor Physical Environment Perception* scale (iPEP scale), a 1–7 Likert scale. Students had to rate the degree of the learning space variables from 1 (low degree) to 7 (high degree). In the process of adaptation to the primary school context, the number of independent variables has been reduced to fourteen: Academic (the course), sex and age, and the physical learning space (11 variables); as well as two dependent variables: The GPA (Grade Point Average) obtained in the subjects of Mathematics and Arts. The statement of these items has been modified for students between the ages of 9 and 13. In addition, the variables School and Classroom were created in the database, based on the data obtained. The items related to the physical learning space are the following:

- Daylight quantity;
- Artificial light quantity;
- Number of times the classroom is ventilated;

- Thermal level comfort;
- Acoustic comfort;
- Room size;
- Chair comfort;
- Desk comfort;
- Connection with nature (i.e., landscapes);
- I appreciate the color of the walls;
- Learning space attachment.

3. Results

3.1. Descriptive Anaylses. Mean and Standard Deviations

Students' perception of the indoor environment of their classes ranges from values close to 4 and above 6 (see Table 2). Appreciation for the color of the walls is the worst valued (m = 3.84), followed by the chair comfort, the ventilation and the amount of natural light. While the best rated is the acoustic comfort, followed by classroom size, the learning space attachment and the connection with nature.

Table 2. iPEP descriptive values.

		Mean (1.00 to 7.00)	Standard Deviation
V1	Daylight quantity	4.95	1.469
V2	Artificial light quantity	5.18	1.618
V3	Number of times the classroom is ventilated	4.59	1.887
V4	Thermal level comfort	5.36	1.784
V5	Acoustic comfort	6.19	1.295
V6	Room size	5.82	1.313
V7	Chair comfort	4.84	1.858
V8	Desk comfort	5.17	1.739
V9	Connection with nature (i.e., landscapes)	5.43	1.804
V10	I appreciate the color of the walls	3.84	2.156
V11	Learning space attachment	5.71	1.543

Table 3 shows the descriptive analysis results in relation to the average grade of Mathematics (m_M = 7.48) and Arts (m_A = 8.24), since it has a different scale: 0–10.

Table 3. Descriptive values of mathematics and art education average grade.

Average Grade Subject	Mean (0.00–10.00)	Standard Deviation
Mathematics	7.48	1.684
Arts	8.24	1.648

Figure 2 corresponds to the analysis of the distribution of the learning space and academic variables through a boxplot diagram. In almost all cases of the iPEP variables, the lower limit is 1, while the upper limit is 7. However, the quartiles in the case of the acoustic variable are grouped in a much lower range than the rest of the variables, practically 100 per cent of the sample is located between values 5 and 7. The rest of the variables distribute the data in a range of values 1 and 7.

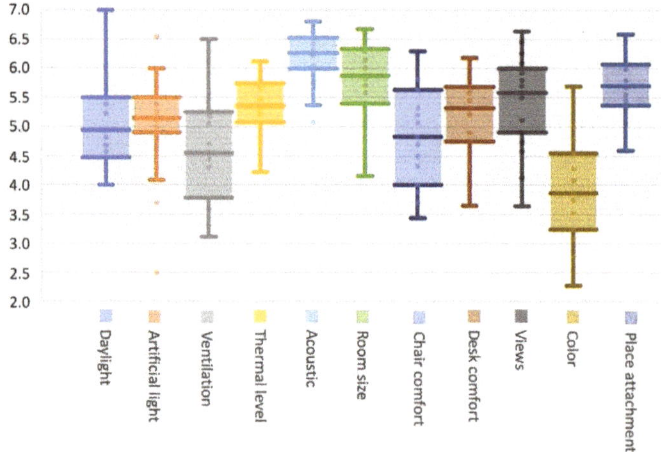

Figure 2. iPEP boxplot.

The variables Natural light, Room size, Chair comfort and Space appropriation are grouped into smaller ranges of values. While the color variable occupies a greater range between the values 2 and 6.

Subsequently, the mean variable values of the student perception of each classroom were made in order to obtain a global measurement. Figure 3 shows the dispersion of the iPEP variable mean values for each of the 27 classrooms, indicating the variability that exists in each of them.

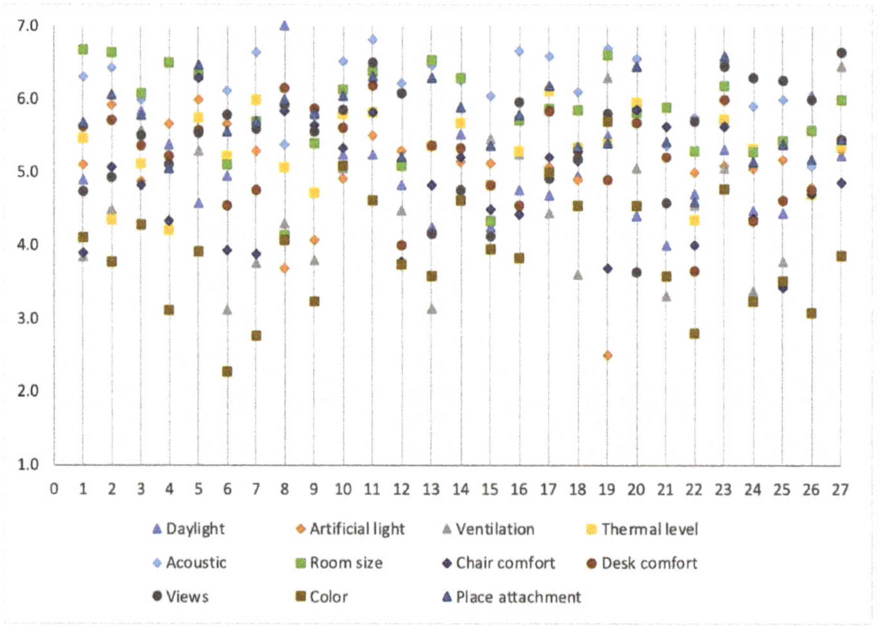

Figure 3. iPEP mean values by classroom.

Classrooms 10 and 23 have a higher concentration in the grouping of iPEP values, with values between 5–6.5; and 4.6–5.7, respectively. On the contrary, classrooms 6, 7, and 19 indicate values between 2.2 and 6.1; 2.7 and 6.6; and 2.5 and 6.7. Moreover, this figure allows us to analyze whether the sets of means are superior or inferior to the rest of classrooms. In this case, classrooms 18, 20, and 21 indicate a concentration in lower values, followed by 6 and 22. On the other hand, classrooms 5, 11, and 12 are grouped in the highest values, followed by 13 and 3.

3.2. iPEP Psychometric Roperties

The reliability of the tool has been calculated using Crombach's alpha to determine if it performs stable measurements, obtaining a result of 0.729, which indicates that the questionnaire is reliable. Subsequently, to verify the validity of the construct, the principal component method of Exploratory Factor Analysis has been applied, with varimax orthogonal rotation.

First, to verify that the factors are correlated with each other, Bartlett's Sphericity ($p < 0.001$) and the Kaiser-Meyer-Olkin test (KMO = 0.802) were calculated, in order to know the adequacy of the sample and if appropriate apply the factor analysis. The result indicates a high partial correlation coefficient, which indicates that it is possible to make a comparison between the magnitudes of correlation coefficients observed and those of partial correlation (Table 4).

Table 4. Kaiser–Meyer–Olkin (KMO) and Bartlett's Test.

Kaiser–Meyer–OlkinMeasure of Sampling Adequacy		0.802
	Aprox. Chi-square	1093.778
Bartlett's Test of Sphericity	df	55
	Sig.	< 0.001

The Exploratory Factor Analysis (EFA) results in a structure of six factors that explains 50.70 per cent of the construct. The grouping of the variables (see Table 5) is grouped into the following factors:

- Workspace comfort: Describes chair and desk comfort (V7 and V8), place attachment (V11), the appreciation for the wall color (V10), as well as the thermal level (V4).
- Natural environment: Represents the amount of daylight (V1), the connection of the classroom with the outside (V9) and the frequency of natural ventilation (V3).
- Building comfort: It is described as the amount of artificial light (V2) and acoustic comfort (V5).

Table 5. Exploratory Factorial Analysis results.

Factor	Variables	Communalities	Total Variance Explained (%)
	V8	0.816	24.39
	V7	0.806	
Workspace comfort	V11	0.627	
	V10	0.610	
	V4	0.565	
	V6	0.473	
	V1	0.748	38.53
Natural environment	V3	0.695	
	V9	0.487	
Building environment	V2	0.729	50.70
	V5	0.699	

3.3. Multiple Linear Regression Analysis

The multiple linear regression analysis is applied to investigate to what extent the iPEP scale is able to predict the performance in Mathematics and Arts. Stepwise method has been conducted.

On the one hand, the prediction of mathematical performance displays a 6-variable model (V3, V6, V7, V11, V4 and V9) that explains 7.2 per cent of the dependent variable. In addition, the value of the Durbin-Watson statistic (1.758) indicates compliance with the assumption of the independence of the residuals (Table 6). On the other hand, the prediction of performance in Arts establishes a model of 2 variables (V6 and V9) that explains 3.7 per cent of the dependent variable. The Durbin-Watson statistic is 1.408, which complies the assumption of independence of the residuals.

Table 6. Multiple linear regression results.

Model	Adjusted R^2	Std. error	Sig. F	Durbin-Watson
Mathematics performance	0.072	1.622	0.022	1.758
Arts performance	0.037	1.617	0.002	1.408

Table 6 shows the Pearson correlations. In the first model (dependent variable = Mathematics grade), the independent variables V3, V6, V11 and V9 obtain positive values reveal a direct relationship. While the variables V7 and V4, indicate an inverse relationship with mathematical performance. In the second model, both variables (V6 and V9) indicate a direct relationship with arts performance.

The validity of the models depends on the verification of the assumptions, such as the non-existence of perfect multicollinearity. The collinearity diagnosis is made through the Variance Inflation Factor (VIF), whose values are close to 1, indicating that the assumption is met (see Table 7).

Table 7. Global regression model for model 1 and 2.

Model	Variable	Beta	t	Sig.	Tolerance	VIF
1	V3	0.123	2.844	0.005	0.859	1.164
	V6	0.102	2.268	0.024	0.786	1.273
	V7	−0.182	−3.298	0.001	0.522	1.917
	V11	0.269	3.973	< 0.001	0.348	2.871
	V4	−0.176	−3.437	0.001	0.606	1.650
	V9	0.098	2.298	0.022	0.868	1.153
2	V6	0.179	4.331	< 0.001	0.970	1.031
	V9	0.130	3.141	0.002	0.970	1.031

One of the premises sought to explore whether these influences could be related to the academic level. For this purpose, multiple linear regression analyzes have been conducted for each course (see Table 8).

Table 8. Multiple linear regression results.

Predictor Variable	Course	R^2	Std. Error	Sig.	Durbin-Watson
Mathematics performance	4th	0.141	1.649	0.050	2.000
	5th	0.096	1.601	0.012	1.475
	6th	0.101	1452	< 0.001	1.290
Art performance	4th	0.135	1.543	0.026	1.739
	5th	0.321	1.437	< 0.001	1.557
	6th	0.101	1.404	0.037	1.893

Concerning mathematics performance of the fourth course, a model of four variables (V6, V4, V5 and V3) has been obtained that explain 14.1 per cent. In the fifth course, the percentage explained is 9.6 per cent for two variables (V7 and V1). In the sixth course, the V7 explains 10.1 per cent of the performance in mathematics. With regards to the performance in Arts in the fourth course, it is explained in 13.5 per cent by two variables (V6 and V4). In the fifth course, five variables explain 32.1 per cent (V9, V6, V2, V7 and V3). In the sixth course, this performance is explained in 10.1 per cent by two variables (V8 and V9).

Table 9 shows the standardized Beta coefficients that indicate whether the relationship of each variable in the model is direct or inverse towards mathematics performance. In the fourth course the V6, V5 and V3 have a direct relationship and the V4 an inverse relationship. In the fifth course V7 indicates an inverse relationship with the dependent variable and V1 a direct relationship. Moreover, in the sixth course the relation of V7 indicates a direct relation.

Table 9. Global regression model by course (predictor variable: Mathematics performance).

Course Model	Variable	Beta	t	Sig.	Tolerance	VIF
4th	V6	0.185	2.151	0.033	0.503	1.987
	V4	−0.377	−4.573	< 0.001	0.546	1.833
	V5	0.251	2.506	0.013	0.370	2.703
	V3	0.131	1.973	0.050	0.835	1.197
5th	V7	−0.346	−4.810	0.000	0.851	1.175
	V1	0.182	2.528	0.012	0.851	1.175
6th	V7	0.327	4.111	< 0.001	1.000	1.000

Table 10 shows the results regarding the multiple linear regression models regarding the art performance. In the fourth course both variables show a direct relationship with the dependent variable. In the fifth course, V9, V6 and V3 indicate a direct relationship and V2 and V7 an inverse relationship. In the sixth course, V8 indicates a direct relationship and V9 an inverse relationship.

Table 10. Global regression model by course (predictor variable: Art performance).

Course Model	Variable	Beta	t	Sig.	Tolerance	VIF
4th	V6	0.438	6.072	<0.001	0.717	1.395
	V9	0.162	2.245	0.026	0.717	1.395
5th	V9	0.165	2.013	0.045	0.496	2.018
	V6	0.287	4.479	<0.001	0.809	1.236
	V2	−0.509	−5.737	<0.001	0.420	2.380
	V7	−0.366	−5.020	<0.001	0.622	1.607
	V3	0.293	3.710	<0.001	0.533	1.878
6th	V8	0.305	3.820	<0.001	0.995	1.005
	V9	−0.168	−2.108	0.037	0.995	1.005

4. Discussion and Conclusions

The present research aimed to deepen the relationship between the physical learning space in Primary Education and the performance in Mathematics and Art, as well as at the academic course. Self-reported empirical data has confirmed the existence of direct and inverse relationships between variables of physical space and performance in mathematics and in art of Primary School students.

It has been shown that the adaptation of iPEP for the measurement of student perception of the variables of the physical learning space in Primary Education is valid and reliable. Likewise, the study has allowed us to investigate the factorial structure of the learning space construct, resulting in an organization in three factors: Workspace comfort, natural environment and building environment, which support previous results [25]. However, findings have evidenced that more question would be needed in order to improve the percentage of learning space measure in primary education level.

The distribution analysis of the learning space variables by classroom (Figure 3) has shown good levels of the indoor environment in the primary school classrooms in Galicia. In addition, this type of evaluation reports on the dispersion of responses, which have sometimes shown little variability. This has happened in the color, and acoustic variables, which may be related to their absence of significant relationships in the prediction analyzes. Therefore, it would be advisable to reconsider how to measure these variables.

The prediction analysis of mathematics performance has confirmed a direct relationship with ventilation, room size, views and place attachment; according to latest contributions of holistic approaches [24], and two inverse thermal level and chair comfort. These results support the inevitable fact of social impact in architectural spaces [33–35], and the need for their measurement through empirical and architectural analyses to understand how they work and to design more sustainable learning spaces in terms of usability. However, relations with art performance seem smaller, results that could be linked to the difficult measurement of Arts [36].

Chair comfort has evidenced an inverse relationship with mathematical performance, while it is the second variable that most explains the measurement of learning space according to the exploratory factor analysis. Despite being the second worst rated variable, the score is close to 5 on a 1–7 scale. However, anthropometry studies indicate that academic performance is reduced when furniture conditions are bad [39]. This provides evidence of an inverse relationship between a good condition of the furniture with a negative result in mathematical performance, pointing out that excess comfort could lead to a detrimental influence. A similar situation seems to happen in terms of the thermal comfort variable.

This research contributes to the literature on the existence of relationships between space and mathematical and art performance in Primary Education. In addition, an effective and easy-to-apply tool is published to diagnose situations related to the learning space in schools, in order to improve the educational purpose of the architectural space.

Concerning future lines of research, it is necessary to combine perception measurements with technical measures to estimate the best efficacy ranges of the different variables. Likewise, the influence in different areas within the study of mathematics and arts need to be studied.

Author Contributions: Conceptualization, V.L.-C. and A.A.L.; Formal analysis, V.L.-C.; Investigation, J.D.L.T.-C. and D.M.-D.; Methodology, V.L.-C., A.A.L., J.L.S. and J.D.L.T.-C.; Resources, J.D.L.T.-C.; Supervision, A.A.L., J.L.S. and J.D.L.T.-C.; Validation, V.L.-C. and J.L.S.; Visualization, D.M.-D.; Writing – original draft, V.L.-C.; Writing—review and editing, V.L.-C., A.A.L., J.L.S. and D.M.-D. All authors have read and agreed to the published version of the manuscript.

Funding: This research received no external funding.

Conflicts of Interest: The authors declare no conflict of interest.

References

1. Heschong Mahone Group. *Daylighting in Schools. An Investigation into the Relationship between Daylight and Human Performance*; Pacific Gas and Electric Company: San Francisco, CA, USA, 1999; pp. 1–31.

2. Vásquez, N.G.; Felippe, M.L.; Pereira, F.R.; Kuhnen, A. Luminous and visual preferences of young children in their classrooms: Curtain use, artificial lighting and window views. *Build. Environ.* **2019**, *152*, 59–73. [CrossRef]

3. Fernández, A.I. El trabajo por rincones en el aula de Educación Infantil. Ventajas del trabajo por rincones. Tipos de rincones. *Innovaciones y experiencias educativas* **2009**, *15*, 1–8.

4. Oluyemi, T.; Shaughnessy, R.; Turunen, M.; Putus, T.; Metsämuuronen, J.; Kurnitski, J.; Haverinen-Shaughnessy, U.; Oluyemi, T. Building characteristics, indoor environmental quality, and mathematics achievement in Finnish elementary schools. *Build. Environ.* **2016**, *104*, 114–121.

5. Bakó-Biró, Z.; Kochlar, N.; Clements-Croomel, D.; Awbi, H.; Williams, M. Ventilation Rates in Schools and Learning Performance. *Build. Environ* **2011**, *48*, 215–223. [CrossRef]

6. Haverinen-Shaughnessy, U.; Shaughnessy, R.; Cole, E.C.; Toyinbo, O.; Moschandreas, D.J. An assessment of indoor environmental quality in schools and its association with health and performance. *Build. Environ.* **2015**, *93*, 35–40. [CrossRef]

7. Bakó-Biró, Z.; Clements-Croome, D.; Kochhar, N.; Awbi, H.; Williams, M. Ventilation rates in schools and pupils' performance. *Build. Environ.* **2012**, *48*, 215–223. [CrossRef]

8. Amasuomo, T.T.; Amasuomo, J.O. Perceived Thermal Discomfort and Stress Behaviours Affecting Students' Learning in Lecture Theatres in the Humid Tropics. *Buildings* **2016**, *6*, 18. [CrossRef]

9. Huang, L.; Zhu, Y.; Ouyang, Q.; Cao, B. A study on the effects of thermal, luminous, and acoustic environments on indoor environmental comfort in offices. *Build. Environ.* **2012**, *49*, 304–309. [CrossRef]

10. Ison, M.S.; Greco, C.; Korzeniowski, C.; Morelato, G. Attentional Efficiency: A Comparative Study on Argentine Students Attending Schools from Different Socio-Cultural Contexts. *Elect. J. Res. Educ. Psychol.* **2015**, *13*, 343–368.

11. Castro-Martã-nez, J.A.; Roa, J.C.; Benítez, A.P.; González, S. Effects of classroom-acoustic change on the attention level of university students. *Interdisciplinaria: Revista de Psicología y Ciencias Afines* **2017**, *33*.

12. Longino, A. International Journal Of Environmental Research And Public Health. *Wilderness Environ. Med.* **2015**, *26*, 99. [CrossRef]

13. Benfield, J.A.; Rainbolt, G.N.; Bell, P.A.; Donovan, G. Classrooms With Nature Views. *Environ. Behav.* **2013**, *47*, 140–157. [CrossRef]

14. Berg, A.E.V.D.; Wesselius, J.E.; Maas, J.; Tanja-Dijkstra, K. Green Walls for a Restorative Classroom Environment: A Controlled Evaluation Study. *Environ. Behav.* **2016**, *49*, 791–813. [CrossRef]

15. Tanner, C.K. Effects of school design on student outcomes. *J. Educ. Adm.* **2009**, *47*, 381–399. [CrossRef]

16. Haghighi, M.M.; Jusan, M.B.M. The impact of classroom settings on students' seat-selection and academic performance. *Indoor Built Environ.* **2013**, *24*, 280–288. [CrossRef]

17. Fan, F.A. Class Size: Effects on Students' Academic Achievements and Some Remedial Measures. *Res. Educ.* **2012**, *87*, 95–98. [CrossRef]

18. Finn, J.D.; Pannozzo, G.M. Classroom Organization and Student Behavior in Kindergarten. *J. Educ. Res.* **2004**, *98*, 79–92. [CrossRef]

19. Milesi, C.; Gamoran, A. Effects of Class Size and Instruction on Kindergarten Achievement. *Educ. Evaluation Policy Anal.* **2006**, *28*, 287–313. [CrossRef]

20. Norazman, N.; Che-Ani, A.I.; Ja'Afar, N.H.; Khoiry, M.A. Standard compliance and suitability of classroom capacity in secondary school buildings. *J. Facil. Manag.* **2019**, *17*, 238–248. [CrossRef]

21. Kolb, A.Y.; Kolb, D.A. Learning Styles and Learning Spaces: Enhancing Experiential Learning in Higher Education. *Acad. Manag. Learn. Educ.* **2005**, *4*, 193–212. [CrossRef]

22. Shamaki, T.A. Influence of learning environment on students' academic achievement in mathematics: A case study of some selected secondary schools in Yobe State – Nigeria. *J. Educ. Practice* **2015**, *6*, 40–44.

23. Muñoz-Cantero, J.-M.; Mira, R.G.; Lopez-Chao, V. Influence of Physical Learning Environment in Student's Behavior and Social Relations. *Anthropologist* **2016**, *25*, 249–253.

24. Barrett, P.; Davies, F.; Zhang, Y.; Barrett, L. The Holistic Impact of Classroom Spaces on Learning in Specific Subjects. *Environ. Behav.* **2016**, *49*, 425–451. [CrossRef] [PubMed]

25. Lopez-Chao, V.; Amado, A.; Martín-Gutiérrez, J. Architectural Indoor Analysis: A Holistic Approach to Understand the Relation of Higher Education Classrooms and Academic Performance. *Sustainability* **2019**, *11*, 6558. [CrossRef]

26. Benedict, M.E.; Hoag, J. Seating Location in Large Lectures: Are Seating Preferences or Location Related to Course Performance? *J. Econ. Educ.* **2004**, *35*, 215–231. [CrossRef]

27. Marshall, P.D.; Losonczy-Marshall, M. Classroom Ecology: Relations between Seating Location, Performance, and Attendance. *Psychol. Rep.* **2010**, *107*, 567–577. [CrossRef]

28. Clément, G.; Bukley, A. Where Do You Sit in Class? A Study of Spatial Positioning During Two Courses of Different Duration. *J. Hum. Psychol.* **2017**, *1*, 1–7. [CrossRef]

29. Killeen, J.P.; Evans, G.W.; Danko, S. The Role Of Permanent Student Artwork In Students' Sense Of Ownership In An Elementary School. *Environ. Behav.* **2003**, *35*, 250–263. [CrossRef]

30. Alves, H.; Raposo, M. Conceptual Model of Student Satisfaction in Higher Education. *Total. Qual. Manag. Bus. Excel.* **2007**, *18*, 571–588. [CrossRef]

31. Pol, E. La apropiación del espacio. In Iniguez, L. & Pol, E. (coord.). In *Cognición, representación y apropiación del espacio*; Publicacions Universitat de Barcelona: Barcelona, Spain, 1996; pp. 45–62.

32. Beckers, R.; Van Der Voordt, T.; Dewulf, G. Learning space preferences of higher education students. *Build. Environ.* **2016**, *104*, 243–252. [CrossRef]

33. Daniels, H.; Leadbetter, J.; Warmington, P.; Edwards, A.; Martin, D.; Popova, A.; Apostolov, A.; Middleton, D.; Brown, S. Learning in and for multi-agency working. *Oxf. Rev. Educ.* **2007**, *33*, 521–538. [CrossRef]

34. Pérez, G.B.; Roig, A.E.; Costa, M.L. Diseño y validación de un instrumento para medir las dimensiones ambiental, pedagógica y digital del aula. *RMIE* **2019**, *24*, 1055–1075.

35. Daniels, H.; Tse, H.M.; Stables, A.; Cox, S. Design as a social practice: the experience of new-build schools. *Camb. J. Educ.* **2018**, *49*, 215–233. [CrossRef]
36. Stake, R.; Munson, A. Qualitative Assessment of Arts Education. *Arts Educ. Policy Rev.* **2008**, *109*, 13–22. [CrossRef]
37. López-Chao, V. El Impacto del Diseño del Espacio y otras Variables Socio-físicas en el Proceso de Enseñanza-Aprendizaje. Ph.D. Thesis, Universidade da Coruña, A Coruña, Spain, 17 January 2017.
38. Barrett, P.; Davies, F.; Zhang, Y.; Barrett, L. The impact of classroom design on pupils' learning: Final results of a holistic, multi-level analysis. *Build. Environ.* **2015**, *89*, 118–133. [CrossRef]
39. Castellucci, H.; Arezes, P.; Viviani, C. Mismatch between classroom furniture and anthropometric measures in Chilean schools. *Appl. Ergon.* **2010**, *41*, 563–568. [CrossRef]

 *sustainability*

Article

Volatile Organic Compound (VOC) Emissions from a Personal Care Polymer-Based Item: Simulation of the Inhalation Exposure Scenario Indoors under Actual Conditions of Use

Jolanda Palmisani *, Alessia Di Gilio *, Ezia Cisternino, Maria Tutino and Gianluigi de Gennaro

Department of Biology, University of Bari Aldo Moro, via Orabona 4, 70125 Bari, Italy;
ezia.cisternino@gmail.com (E.C.); mariatutino2015@gmail.com (M.T.); gianluigi.degennaro@uniba.it (G.d.G.)
* Correspondence: jolanda.palmisani@uniba.it (J.P.); alessia.digilio@uniba.it (A.D.G.);
  Tel.: +39-08-0544-3343 (J.P.)

Received: 19 February 2020; Accepted: 16 March 2020; Published: 25 March 2020

Abstract: Polymer-based items may release Volatile Organic Compounds (VOCs) and odors indoors, contributing to the overall VOC inhalation exposure for end users and building occupants. The main objective of the present study is the evaluation of short-term inhalation exposure to VOCs due to the use of a personal care polymer-based item, namely, one of three electric heating bags, through a strategic methodological approach and the simulation of a 'near-to-real' exposure scenario. Seventy two-hour test chamber experiments were first performed to characterize VOC emissions with the items on 'not-heating mode' and to derive related emission rates. The polyester bag was revealed to be responsible for the highest emissions both in terms of total VOC and naphthalene emissions (437 and 360 $\mu g/m^3$, respectively), compared with the other two bags under investigation. Complementary investigations on 'heating mode' and the simulation of the exposure scenario inside a 30 m^3 reference room allowed us to highlight that the use of the polyester bag in the first life-cycle period could determine a naphthalene concentration (42 $\mu g/m^3$) higher than the reference Lowest Concentration of Interest (LCI) value (10 $\mu g/m^3$) reported in European evaluation schemes. The present study proposes a strategic methodological approach highlighting the need for the simulation of a realistic scenario when potential hazards for human health need to be assessed.

Keywords: VOCs; polymer-based items; indoor air quality; test emission chamber; exposure scenario

---

1. Introduction

Volatile Organic Compound (VOC) emissions from indoor materials and consumer products have become a subject of concern among indoor air scientists [1–4]. VOCs released into indoor air from a wide range of materials and products may deteriorate Indoor Air Quality (IAQ), resulting in odor annoyance and general discomfort for building occupants as well as adverse effects on human health [5–11]. So far, the interest of indoor air scientists in this issue has been mainly motivated by the need to improve the knowledge regarding sources and their emission characteristics in both private and public settings, to investigate further the physical and chemical interactions of emitted pollutants in indoor air, and to develop innovative methodological approaches for the evaluation of emissions and their potential impacts on human health. Building and interior materials have been widely investigated in terms of VOC emissions, as highlighted by an extensive literature in the field. On the contrary, limited data are available on consumer product emissions and related inhalation exposure for end users. Recently published studies have pointed out that consumer products and polymer-based items may release VOCs and odors indoors, contributing significantly to the overall VOC exposure of consumers and building occupants [12–17]. Moreover, complaints about odor annoyance from polymer-based consumer products have enhanced the need for in-depth investigations aimed at elucidating emission patterns and characteristics. In this regard, investigations carried out on selected polymer-based items such as plastic utensils and children's toys highlighted that VOC emissions are related to the release of residual solvents and monomers from the material polymeric structure and/or the release of additives (i.e., plasticizers, inks) following surface-applied finishing processes such as coloring and printing [18–21]. VOC emissions from materials and products (i.e., building materials, furnishings, finishing products etc.) are conventionally evaluated by means of test emission chambers according to well-established procedures, standardized by the European Committee for Standardization (CEN) and by the International Organization for Standardization (ISO) [22,23]. More specifically, identification and quantification of VOCs from single or multiple sources requires emission testing inside test chambers, over a defined timescale and with selected micro-environmental parameters (i.e., temperature, relative humidity, air exchange rate) [24]. Small-scale emission testing generally fits investigation purposes when the determination of emission rates from specific materials and products is required, while large-scale experiments are more suitable for simulating realistic inhalation exposure scenarios for building occupants and consumers due to material installation and/or product use. However, some considerations regarding this point are necessary. Although testing procedures are standardized to obtain reliable data on emission characteristics, they may reveal some limitations when applied to consumer products, especially if the evaluation of emissions under actual conditions of use is required. Indeed, it is important to point out that VOC emission characteristics, in terms of the pattern of generated compounds and extent of the emission, may significantly vary during product/item use, particularly if the use involves combustion and/or heating, resulting in exposure scenarios being substantially different [25]. This typology of indoor sources is characterized by short-term emission patterns during the actual use and requires a realistic scenario to be simulated in the test emission chamber. Therefore, in these cases, the integration of conventional procedures (e.g., standardized emission testing) with complementary and innovative methodological approaches can be strategic to answer key questions on VOC emissions under effective conditions of use. The main objective of the present study is the evaluation of VOC emissions from a personal care polymer-based item, an electric heating bag, commonly used to relieve stress and reduce muscle and joint pain. The investigated item must be electrically supplied, therefore emission characteristics may significantly change under conditions of use. For this purpose, the present paper proposes a strategic methodological approach for the estimation of the inhalation exposure to VOCs emitted by the investigated items on 'heating mode' in a real setting, starting from emission data obtained through standardized methods and under controlled conditions. The experimental activity involved test emission chamber and dynamic head-space investigations on three different heating bags, commercially available at the moment of the study and also responsible for odor annoyance at ambient temperature.

Data collected were useful for the estimation of VOC emission rates (ERs) under actual conditions of use and, as a result, for the estimation of the indoor concentrations representative of short-term exposure related to the item use during the first life-cycle period in a realistic setting (e.g., reference room of EU standardized evaluation schemes), allowing the simulation of a near-to-real exposure scenario.

2. Materials and Methods

2.1. Chemicals

Authentic standards of the VOCs under consideration in the present study were included in a customized VOC standard mix in ethanol (Ultra Scientific Analytical Solutions, Italia srl). Ethanol of analytical grade was purchased by Sigma Aldrich and used as a solvent for the preparation of calibration standards.

2.2. Polymer-Based Item Description

The polymeric item under investigation is a portable electric heating bag commonly used for general comfort and/or for therapeutic purposes (i.e., warming during the winter season, reduction of muscle pain and stress). This typology of personal care item has quickly became very popular on the European (EU) market during recent years and received board consensus as it represents a low-cost and easy-to-use version of the conventional warming bag that needed to be filled with hot water to work. Electric heating bags generally appear as small bags made of polymeric material (i.e., polyvinylchloride, polyester) covering the inner bag filled with the heating liquid, provided with a plug base cover and electric cable for heating. They must be electrically supplied to provide their function and, after a few minutes of electrical charge (e.g., 5–10 min), they can be used at high temperature for at least one hour. In the present study, three different heating bags were investigated. They were all manufactured in China, distributed on the EU market and labeled with three different brands (reported as brand A, B and C). More specifically, the first one belonging to brand A was characterized by a printed and colored external coverage made of polyester (labeled in the text and tables as 'polyester-brand A'). The other two bags, belonging to brands B and C, respectively, were instead characterized by an external coverage made of polyvinylchloride (labeled in the text and tables as 'PVC-brand B' and 'PVC-brand C'). The former had an image printed onto the surface of one side, while the latter was only colored. All the investigated bags had the same shape and dimensions with an upper surface area equal to 0.04 m^2. They were part of a production lot blocked by competent authorities at port customs in a city in the South of Italy (City of Monopoli) after reporting from consumer associations. Due to end user complaints related to strong odor annoyance occurring at a greater extent during the first use events, local competent authorities formally requested the necessary investigations. At the moment of the study, the introduction on the EU market of electric heating bags manufactured in non-EU countries (e.g., China) was allowed with only the CE label ensuring electric safety and conformity in compliance with EU Directive requirements.

2.3. Experimental Design Description

The first investigation level involved 72-hour emission testing inside a small-scale emission chamber under standardized environmental conditions, according to the relevant ISO standards. As a result, emission rates (ERs) for the main detected VOCs were derived, useful for the estimation of indoor concentrations potentially determined by the item in a real setting at ambient temperature (i.e., 'not-heating' mode) and representative of a short-term exposure. This investigation level, although conventionally applied to materials and products for providing data on emission characteristics under simulated indoor conditions, was not exhaustive for the estimation of VOC inhalation exposure levels determined by the item in a real room under the actual conditions of use. In order to derive ERs under heating, non-time-consuming and cost-effective dynamic head-space experiments were performed, placing the investigated items inside customized Nalophan bags. The effect of temperature on VOC emissions from the items was evaluated, and its extent was expressed in terms of the ratio of chromatographic peak areas obtained by GC/MS analysis of the samples collected both at laboratory ambient temperature (approximately 23 °C) and under heating. This second level of investigation allowed us to estimate ERs under heating (starting from ERs derived from 72-hours data inside the chamber) and, as a result, indoor concentrations inside the reference room representative of a realistic short-term exposure scenario for end users.

2.3.1. Test Emission Chamber Experiments

Experiments were carried out inside a small-scale test emission chamber, a hermetically closed glass chemical reactor with a cylindrical shape and with the following dimensions: diameter = 29 cm, height = 61 cm and volume = 0.05 m^3. Teflon fans were installed at the top of the chamber to ensure that the air was adequately mixed. The test emission chamber was operated with controlled micro-environmental parameters, according to the relevant ISO standard. For each experiment, temperature and relative humidity were $23 \pm 2\,°C$ and $50 \pm 5\%$, respectively. Air exchange rate (AER) was $0.5 \pm 0.1\,h^{-1}$, which agrees with many European building standards for ventilation in indoor environments [26]. The chamber was supplied with ultrapure compressed air (VOC-free air). Before each experiment, the chamber was cleaned with detergent and rinsed with distilled water, and background samples were taken in order to verify VOC levels. The electric heating bag was introduced inside the chamber, and the test started when the chamber was closed ($t = 0$). Each experiment lasted 72 h and was carried out with the heating bag not electrically supplied ('not-heating mode') as the test chamber configuration was not suitable to carry out the experiments allowing electrical connections between the warming bag (inside the chamber) and outside. The test emission chamber was provided with two outlet ducts to allow chamber air to be monitored by high temporal resolution instrumentation and/or collected onto adsorbent cartridges. In the present study, Total Volatile Organic Compounds (TVOC) concentration was monitored by means of a high temporal resolution photo-ionization detector (PhoCheck® Tiger, Ion Science Ltd., UK), over the entire duration of the test, in order to verify the achievement of a steady state concentration inside the test emission chamber. VOCs were sampled in duplicate on suitable adsorbent cartridges by means of calibrated sampling pumps (Pocket pump). Adsorbent cartridges consisted of a cylindrical stainless steel net (100 mesh) containing 350 mg Carbograph 4 (35–50 mesh). Air samples were collected at 72 h after the start of the chamber experiment, with a sampling flow rate of 50 mL/min and a sampling time of 100 min, resulting in a collected air volume of 5 L.

2.3.2. Dynamic Head-Space (DHS) Experiments

Temperature-related effect on VOC emissions from the investigated items under conditions of use was evaluated by performing small-scale dynamic head-space (DHS) experiments inside customized Nalophan bags. Once the item was introduced inside the Nalophan bag, each extremity was tightly closed to avoid any kind of contamination from outside. A pressure-regulator stainless steel line, connected to an ultrapure-grade and VOC-free air cylinder, was introduced through one Nalophan bag extremity, allowing air to enter at constant air flow (50 mL/min). Through the same extremity, the electric cable connected the power outlet to the plug base cover, allowing the item to be electrically supplied. VOCs were sampled through a Teflon tube introduced at the opposite extremity of the Nalophan bag by means of a flow-controlled pump (Pocket Pump) and collected onto Carbograph 4 adsorbent cartridges (35–50 mesh). Two separate experiments were carried out for each investigated item: experiment 1 with the item at ambient temperature (i.e., 'not-heating mode'); experiment 2 with the item electrically supplied (i.e., 'heating mode'), covering the entire heating phase, the achievement of the temperature intended for use and the bag cooling. Experiment 2 involved each item at its first use. The temperature reached by the items once the heating process was completed was on average $55\,°C$. Air sampling for experiments 1 and 2 was performed with the same sampling flow rate (50 mL/min) and sampling time (100 min). Two sorbent tubes were connected in series in case of VOC breakthrough.

2.3.3. GC/MS Instrumental Analysis Setup

Adsorbent cartridges were thermally desorbed and analyzed on a thermal desorber (UNITY 1™, Markes International Ltd.) coupled to a gas chromatograph (Agilent GC-6890) and a mass selective detector (Agilent MS-5973N). For quality assurance, adsorbent cartridges were conditioned before each use at $310\,°C$ for 30 min and analyzed to verify the blank level. The chamber air background was also evaluated before each experiment. VOCs were thermally desorbed at $300\,°C$ for 10 min and refocused onto the cold trap at $-10\,°C$. The cold trap was then flash-heated at $300\,°C$, and VOCs were transferred via the heated transfer line ($180\,°C$) to the GC column. The GC column was a 30 m × 250 μm × 0.25 μm film thickness with (5%-phenyl)-methylpolysiloxane stationary phase (J&W HP5-MS, Agilent Technologies). Carrier gas (helium) flow was controlled by constant pressure mode and equal to $1.3\,mL\,min^{-1}$. The GC oven program used for optimal VOC separation was: $40\,°C$ for 1 min, ramp 1: $8\,°C\,min^{-1}$ up to $80\,°C$, ramp 2: $20\,°C\,min^{-1}$ up to $270\,°C$. A mass spectrometer was operated in electron impact (EI) ionization mode (70eV) in the mass range 25–250 m/z (SCAN acquisition mode, TIC chromatogram). Valves, transfer lines and ion source were kept at 280 °C. Single-target ions were extracted in selected ion monitoring (SIM) mode for compound identification and quantification. One quantifier ion and one qualifier ion were selected for each compound on the basis of their selectivity and abundance (Table 1). Six standard solutions with concentration levels 10, 20, 50, 100, 200 and 400 µg/mL were prepared by successive dilution in ethanol of a certified VOC standard mixture (ULTRA Scientific Italia s.r.l, Bologna, Italy). Six-point calibration curves were constructed by syringe injection of 1 µL of VOC standard solutions onto Carbograph 4 cartridges. Identification of VOCs was based on comparison of the obtained mass spectra with those included in the National Institute of Standards and Technology (NIST) library and considered positive by library search match > 800 for both forward and reverse matching. Further criteria for compound identification were: (i) the matching of relative retention times (t_R) with those of the authentic standards within the allowed deviation of ±0.05 min; (ii) the matching of ion ratios collected with those of the authentic standards within a tolerance of ± 20%. Only VOCs of particular concern due to potential adverse effects on human health and with a chamber air concentration approximately equal to or higher than 1 µg/m^3 were taken into account for further discussion. The list of VOCs (common for the three different investigated items) and related information (molecular formula, CAS number, quantifier and qualifier ions), as well as the performances of the analytical methodology in terms of Limit of Detection (LOD), Limit of Quantification (LOQ) and correlation coefficients (R^2), are reported in Table 1.

Table 1. Volatile Organic Compounds (VOCs) emitted by the heating bag: molecular formula, CAS number, retention time (t_R), quantifier and qualifier ions (m/z), Limit of Detection (LOD) (µg/m^3), Limit of Quantification (LOQ) (µg/m^3) and correlation coefficient (R^2).

Compound	Molecular Formula	CAS Number	t_R (min)	Quantifier Ion (m/z)	Qualifier Ion (m/z)	LOD (µg/m^3)	LOQ (µg/m^3)	R^2
1,2-dichloroethane	$C_2H_{14}Cl_2$	107-06-2	1.64	62	64	0.02	0.07	0.970
Benzene	C_6H_6	71-43-2	1.72	78	52	0.03	0.09	0.994
Toluene	C_7H_8	108-88-3	2.83	91	92	0.02	0.07	0.992
Tetrachloroethene	C_2Cl_4	127-18-4	3.56	166	164/129	0.05	0.15	0.970
Chlorobenzene	C_6H_5Cl	108-90-7	4.35	112	77	0.03	0.09	0.970
Ethylbenzene	C_8H_{10}	100-41-4	4.72	91	106	0.03	0.09	0.997
m-xylene	C_8H_{10}	108-38-3	4.93	91	106	0.04	0.11	0.996
p-xylene	C_8H_{10}	106-42-3	4.93	91	106	0.04	0.11	0.996
Styrene	C_8H_8	100-42-5	5.51	104	78	0.05	0.16	0.998
o-xylene	C_8H_{10}	95-47-6	5.55	91	106	0.04	0.12	0.997
3-ethyltoluene	C_9H_{12}	620-14-4	7.74	105	120	0.1	0.3	0.998
1,3,5-trimethylbenzene	C_9H_{12}	108-67-8	7.97	105	120	0.1	0.29	0.997
2-ethyltoluene	C_9H_{12}	611-14-3	8.35	105	120	0.1	0.3	0.998
1,2,4-trimethylbenzene	C_9H_{12}	95-63-6	8.83	105	120	0.1	0.29	0.997
1,2,3-trimethylbenzene	C_9H_{12}	526-73-8	9.85	105	120	0.1	0.29	0.997
4-isopropyltoluene	$C_{10}H_{14}$	99-87-6	10.02	119	134	0.03	0.08	0.939
Acetophenone	C_8H_8O	98-86-2	11.69	105	77	0.03	0.1	1.000
Naphthalene	$C_{10}H_8$	91-20-3	15.92	128	127	0.03	0.09	0.999

2.3.4. Test Chamber Experiments: Emission Rates (ERs) and Estimation of Reference Room Indoor Concentrations (Ci,ref)

The primary objective of emission testing is the determination of VOC specific emission rates (SERs), enabling description of the emission behavior of the material/product regardless of air exchange rate and loading factor. According to ISO standards and most of the existing health-related evaluation schemes at EU level, VOC specific emission rates are calculated at fixed sampling times, e.g., 3 and 28 days after the introduction of the material inside the test emission chamber [27]. The determination of specific emission rates for any individual VOC detected is addressed to model the exposure scenario and to estimate indoor air concentrations that an occupant of a real-scale room could be exposed to. More specifically, VOC specific emission rates derived at 3 days allow the estimation of indoor concentrations representative of short-term exposure. In the present study, the emission rate for compound i (ERi) emitted by the electric heating bag at 72 h in the test chamber experiment was calculated on the basis of the mass conservation Equation (1):

$$ERi = Ci \times V \times n \tag{1}$$

where Ci is the chamber concentration of compound i in the air sampled at 72 h (expressed as μg/m^3 or ng/m^3); V is the chamber volume (m^3); and n is the air exchange rate (h^{-1}). Starting from ERi, individual VOC indoor concentrations resulting from the presence or use of the investigated item inside the 30 m^3 reference room (Ci,ref) may be estimated via the following formula:

$$Ci, ref = \frac{ERi}{n \times V} \tag{2}$$

with n and V representing the air exchange rate (0.5 h^{-1}) and volume (30 m^3) of the reference room, respectively.

3. Results and Discussion

3.1. Test Chamber Experiments: Characterization of VOC Emissions and Determination of Emission Rates (ERs)

VOCs reported in Table 2 represent the pattern of gaseous pollutants identified and quantified by test chamber experiments performed under controlled conditions and with the electric bags on 'not-heating mode'. VOC emission data are expressed as emission rates (ERs) (ng/h) and chamber air concentrations (μg/m^3), with the latter reported as an average value of duplicate measurements corrected for chamber background. If a preliminary comparison among the investigated items is done, emission testing of the heating bag 'polyester-brand A' resulted in the highest total VOC chamber concentration at 72 h sampling time. More specifically, for 'polyester-brand A' the total emission expressed as the sum of concentrations of VOCs (Σ VOCs, μg/m^3) was equal to 437.0 μg/m^3, one order of magnitude higher than those obtained for 'PVC-brand B' (21.1 μg/m^3) and 'PVC-brand C' (19.6 μg/m^3). In detail, VOC chamber concentrations for 'polyester-brand A' ranged from 0.7 μg/m^3 (benzene) to 360.5 μg/m^3 (naphthalene), whilst 1,2-dichloroethane and tetrachloroethene were both below the LOQ of the applied analytical technique. As a result, the emission rates (ERs, ng/h), calculated by equation (1), ranged from 18 ng/h to 9013 ng/h. VOC chamber concentrations and ERs for 'PVC-brand B' ranged from 0.6 μg/m^3 (benzene) to 6.8 μg/m^3 (toluene) and from 15 to 171 ng/h, respectively. Finally, VOC chamber concentrations for 'PVC-brand C' were in the range from 0.7 μg/m^3 (i.e., 1,2-dichloroethane and 1,2,3-trimethylbenzene) to 4.8 μg/m^3 (toluene) and, as a result, ER values ranged from a minimum value of 16 ng/h to a maximum value of 121 ng/h. For both 'PVC-brand B' and 'PVC-brand C', 2-ethyltoluene, 3-ethyltoluene, 4-isopropyltoluene and acetophenone were below the LOQ. Additionally, 1,2-dicloroethane and benzene were below the LOQ for 'PVC-brand B' and 'PVC-brand C', respectively. Therefore, taking into account all the collected data, it is possible to observe that, whilst VOC emissions from PVC items resulted to be comparable in terms of concentration levels (for both the heating bags, chamber concentrations were generally below 10 μg/m^3), the most remarkable result was regarding the naphthalene emission from 'polyester-brand A', which resulted in a chamber concentration at 72 h equal to 360.5 μg/m^3. Previously published studies have already highlighted that materials and consumer products are responsible for naphthalene emissions and may significantly contribute to naphthalene inhalation exposure inside indoor environments [28]. Screening investigations reported by Kang et al. (2012) aimed at the identification of sources revealed that, disregarding specific sources intended to contain pure crystalline naphthalene (e.g., mothballs), interior materials as well as several consumer products may unintentionally emit naphthalene. The aforementioned study also highlighted that, across the interior materials investigated, mats consisting of PVC-coated polyester material showed the highest naphthalene emission factor, confirming that naphthalene is involved in the production and finishing of these polymeric materials. Naphthalene is, indeed, primarily used as a chemical intermediate for phthalic anhydride and naphthalene sulphonate production, both involved in the industry manufacture of plasticizers, dyes and rubber formulations. Moreover, according to the existing literature, aromatic hydrocarbon release from polymeric materials (PVC, polyester) used as covering for interior materials such as wallpapers and flooring materials as well as polymeric items such as children toys may be mainly explained by taking into account the use of specific solvents in the manufacturing process such as toluene, ethylbenzene, isopropylbenzene and potential related impurities [4,19,29].

Table 2. Test chamber concentrations (μg/m^3) at 72 h and related emission rates (ERs, ng/h) for VOCs emitted by the investigated bags on 'not-heating mode'.

Compounds	POLYESTER-Brand A		PVC-Brand B		PVC-Brand C	
	Chamber Conc. (μg/m^3)	ER (ng/h)	Chamber Conc. (μg/m^3)	ER (ng/h)	Chamber Conc. (μg/m^3)	ER (ng/h)
1,2-dichloroethane	<LOQ	/	<LOQ	/	0.7	16
Benzene	0.7	18	0.6	15	<LOQ	/
Toluene	19.4	485	6.8	171	4.8	121
Tetrachloroethene	<LOQ	/	/	/	1.3	33
Chlorobenzene	4.5	113	0.9	24	2.7	67
Ethylbenzene	1.5	38	1.2	29	1.5	38
m/p-xylene	1.9	48	1.3	32	1.9	47
Styrene	2.8	70	1.5	38	2.3	57
o-xylene	5.3	133	1.0	24	1.2	29
3-ethyltoluene	4.8	120	<LOQ	/	<LOQ	/
1,3,5-trimethylbenzene	3.3	83	0.8	19	0.8	19
2-ethyltoluene	2.6	65	<LOQ	/	<LOQ	/
1,2,4-trimethylbenzene	15.2	380	0.9	24	0.8	20
1,2,3-trimethylbenzene	8.4	210	0.8	20	0.7	18
4-isopropyltoluene	2.5	63	<LOQ	/	<LOQ	/
Acetophenone	3.6	90	<LOQ	/	<LOQ	/
Naphthalene	360.5	9013	5.3	133	1.0	10
ΣVOCs	437.0		21.1		19.6	

3.2. Estimation of Emission Rates (ERs) on 'Heating Mode'

Estimated VOC emission rates from the investigated items on 'heating mode' are listed in Table 3. They were estimated starting from the emission rate values derived for 'not-heating mode' through 72-hour test chamber experiments and taking into account the GC-MS peak area ratios obtained by dynamic head-space investigations, performed both for 'not-heating mode' and 'heating mode'. Overall, as expected, the high temperature acquired during the heating process affected VOC emissions behavior from all the investigated bags. From a lesser to greater extent, the emission process of the selected VOCs was promoted. With specific regard to the 'polyester-brand A' bag, the GC-MS peak area ratios suggest that naphthalene was the most sensitive compound to the temperature change. The estimated naphthalene emission rate when the bag was on 'heating mode', indeed, was revealed to be 70 times higher than the calculated emission rate on 'not-heating mode', with a variation from 9013 ng/h to 630.9 μg/h. For all the other VOCs, the estimated emission rates ranged from a minimum value of 0.20 μg/h for benzene to a maximum value of 6.79 μg/h for toluene. From the comparison of HS experimental data reported in Table 3, it is possible to observe that the estimated emission rates for the VOCs of concern emitted by 'PVC-brand B' and 'PVC-brand C' were generally lower compared with those estimated for 'polyester-brand A'. This evidence is not related to temperature because the effect of heating on the emission process is comparable for all the three investigated bags, with peak area ratios of the same order of magnitude. It is, instead, attributable to the starting values of emission rates for 'PVC-brand B' and 'PVC-brand C' calculated from test chamber experiments being generally lower than those for 'polyester-brand A'. More specifically, estimated emission rates for 'PVC-brand B' and 'PVC-brand C' on 'heating mode' were in the range 0.11–11.2 µg/h and 0.03–2.9 µg/h, respectively. Similarly to 'polyester-brand A', the increase in temperature significantly affected the naphthalene emission from 'PVC-brand B', resulting in an estimated emission rate on 'heating mode' 109 times higher with respect to 'not-heating mode' (with an increase from 113 ng/h to 11.2 µg/h). The remarkable effect of the high temperature on the naphthalene emissions observed for both 'polyester-brand A' and 'PVC-brand B' but not for 'PVC-brand C' may be explained by taking into account the different surface treatments. The 'polyester-brand A' and 'PVC-brand B' bags, indeed, had an external coverage characterized by an image applied onto the surface. On the contrary, the coverage of 'PVC-brand C' was only colored. The surface treatment for image application may be responsible for the higher naphthalene emission rates both on 'heating mode' and 'not-heating mode' because it is known, as highlighted above, that naphthalene is used for the production of plasticizers and dyes. Moreover, the use of naphthalene in surface treatment to preserve the items from any kind of deterioration during long-range transport cannot be excluded, i.e., naphthalene used as a repellent for undesired insects or as anti-mold. As regards all the other VOCs, taking into account the peak area ratios representing the effect of the heating process on emission behavior, it is reasonable to make the assumption that the high temperature promoted the diffusion process of compounds through the polymeric bulk and, as a result, the emission from the surface. The emitted VOCs indeed seem to be incorporated in the polymeric structure as residues, and related contaminants of the solvents used in the polymer manufacturing process, unlike naphthalene, seem to be more abundant on the surface layer.

Table 3. Estimation of VOC emission rates (ERs, µg/h) for the investigated bags on 'heating mode'.

Compounds	POLYESTER-Brand A			PVC-Brand B			PVC-Brand C		
	ER (ng/h)	Peak Area Ratio	ER (µg/h) —Heating	ER (ng/h)	Peak Area Ratio	ER (µg/h) —Heating	ER (ng/h)	Peak Area Ratio	ER (µg/h) —Heating
1,2-dichloroethane	/	/	/	/	/	/	16	11	0.18
Benzene	18	11	0.20	15	7	0.11	/	/	/
Toluene	485	14	6.79	171	15	2.57	121	9	1.09
Tetrachloroethene	/	/	/	/	/	/	33	8	0.26
Chlorobenzene	113	37	4.18	24	9	0.22	67	43	2.9
Ethylbenzene	38	11	0.42	29	10	0.29	38	25	0.95
m/p-xylene	48	9	0.43	32	9	0.29	47	15	0.71
Styrene	70	10	0.70	38	9	0.34	57	38	2.2
o-xylene	133	12	1.60	24	10	0.24	29	17	0.49
3-ethyltoluene	120	12	1.44	/	/	/	/	/	/
1,3,5-trimethylbenzene	83	18	1.5	19	11	0.21	19	17	0.32
2-ethyltoluene	65	21	1.4	/	/	/	/	/	/
1,2,4-trimethylbenzene	380	19	7.22	24	18	0.43	20	15	0.30
1,2,3-trimethylbenzene	210	30	6.30	20	20	0.40	18	14	0.25
4-isopropyltoluene	63	9	0.57	/	/	/	/	/	/
Acetophenone	90	28	2.5	/	/	/	/	/	/
Naphthalene	9013	70	630.9	133	109	11.2	10	3	0.03

3.3. Simulation of a Short-Term Exposure Scenario: Estimation of Room Reference Concentrations ($C_{i,ref}$) and Health-Related Evaluation

Once the emission rates for 'heating mode' were estimated, indoor concentrations potentially determined by each single bag under actual conditions of use were derived (Table 4). The exposure scenario taken into account is representative of a short-term inhalation exposure related to the use of the heating bag during the first period of its life-cycle. Moreover, the exposure scenario considers only one bag under heating in a 30 m^3 room resulting in VOC emissions, promoted by high temperature, diluted in the entire volume of the room. For this reason, in the present work, the estimated room reference concentration values that room occupants could be exposed to are low for all the three investigated bags, ranging overall, with the only exception of naphthalene, from 0.01 µg/m^3 (benzene for 'polyester-brand A'; benzene, chlorobenzene, and 1,3,5-trimethylbenzene for 'PVC-brand B'; 1,2-dichloroethane for 'PVC-brand C') to 0.48 µg/m^3 (1,2,4,-trimethylbenzene for 'polyester-brand A'). Attention has to be paid, instead, to the potential inhalation exposure to naphthalene occurring when the polyester bag is used, according to the selected scenario. On the basis of the obtained results, it may be predicted that, as a consequence of the heating process, the naphthalene indoor concentration determined by the polyester bag in a real setting would be equal to about 42 $\mu g/m^3$. This evidence is worthy of further discussion as human exposure to naphthalene has been recognized as a public health concern due to demonstrated harmful effects [30]. Naphthalene is indeed classified as a possible human carcinogen (group 2B) by the International Agency for Research on Cancer (IARC) and is included in EU category Carc.2 on the basis of experimental evidence in animals regarding an increased risk of contracting respiratory tract cancer [31,32]. Health-based evaluation of VOC emissions from materials is generally based on the comparison of room reference concentrations for individual compounds with guideline values. In order to assess the potential risks to health arising from inhalation exposure to individual VOCs, most of the existing health-related evaluation schemes at European level are based on the LCI (Lowest Concentration of Interest) approach. The proposed EU-LCI values are health-based reference concentrations for inhalation exposure intended as 'safe' levels where no health impairment is expected, even with a life-long exposure. EU-LCI levels, however, are usually compared to indoor concentrations representative of long-term exposure and based on emission rates derived after 28 days of chamber testing. This basic assumption would apparently limit our discussion, not allowing us to highlight if potential health risks for inhalation exposure could occur in the case of 'polyester-brand A' heating bag use. Therefore, in this regard, clarification is needed. Taking into account the most comprehensive evaluation scheme at EU level, the German AgBB scheme 'Evaluation procedure for VOC emissions from building products', chemicals with potential carcinogenic effects belonging to EU category Carc.2 are also eligible to be checked within the LCI concept at the first step of the evaluation scheme, related to 3-day chamber testing [27]. The room reference concentration estimated for naphthalene (42 $\mu g/m^3$), therefore, is eligible to be compared with the LCI value equal to 10 $\mu g/m^3$. From the comparison, it is possible to state that the actual use of the 'polyester-brand A' heating bag in the first period of its life-cycle could determine a naphthalene concentration inside a 30 m^3 room eight times higher than the reference LCI value. The inhalation exposure to naphthalene emission could represent, therefore, a risk for end users and room occupants. It is also important to point out that the item under investigation is intended to be used very close to the human upper airways, and the resulting exposure may be exacerbated. Moreover, the ventilation inside a real setting may be reduced with respect to the 'ideal' conditions (e.g., 0.5 h^{-1} air exchange rate), leading to higher VOC concentrations in the air volume in proximity with the item, and therefore close to the breathing zone, compared with the rest of the room [33]. Finally, although the most remarkable result has to be attributed to the naphthalene emission from the 'polyester-brand A' bag, it should be noted that both 'polyester-brand A' and 'PVC-brand B' released benzene, recognized as a carcinogen in humans for which no safe level of inhalation exposure can be recommended. In this regard, the World Health Organization pronounced a suggestion to reduce or eliminate the use of materials that are able to release benzene [30].

Table 4. Simulation of a short-term exposure scenario: estimation of room reference concentrations (Ci,ref, $\mu g/m^3$) determined by bags on 'heating mode'.

Compounds	POLYESTER-Brand A		PVC-Brand B		PVC-Brand C	
	ER Heating ($\mu g/h$)	Ci,ref ($\mu g/m^3$)	ER Heating ($\mu g/h$)	Ci,ref ($\mu g/m^3$)	E_R Heating ($\mu g/h$)	Ci,ref ($\mu g/m^3$)
1,2-dichloroethane	/	/	/	/	0.18	0.01
Benzene	0.20	0.01	0.11	0.01	/	/
Toluene	6.79	0.45	2.57	0.17	1.09	0.07
Tetrachloroethene	/	/	/	/	0.26	0.02
Chlorobenzene	4.18	0.28	0.22	0.01	2.9	0.19
Ethylbenzene	0.42	0.03	0.29	0.02	0.95	0.06
m/p-xylene	0.43	0.03	0.29	0.02	0.71	0.05
Styrene	0.70	0.05	0.34	0.02	2.2	0.14
o-xylene	1.60	0.11	0.24	0.02	0.49	0.03
3-ethyltoluene	1.44	0.10	/	/	/	/
1,3,5-trimethylbenzene	1.5	0.10	0.21	0.01	0.32	0.02
2-ethyltoluene	1.4	0.09	/	/	/	/
1,2,4-trimethylbenzene	7.22	0.48	0.43	0.03	0.30	0.02
1,2,3-trimethylbenzene	6.30	0.42	0.40	0.03	0.25	0.02
4-isopropyltoluene	0.57	0.04	/	/	/	/
Acetophenone	2.5	0.17	/	/	/	/
Naphthalene	630.9	42.06	11.2	0.75	0.03	/
ΣVOCs		44.41		1.09		0.64

3.4. Limitations of the Study

In the present study, replicated test chamber experiments under controlled conditions for each investigated bag on 'not-heating mode' were not carried out. In addition, investigations after 3 days in order to define VOC emission rate profiles were not performed. The aforementioned lack of data could represent a limitation of the study. However, the authors specified that the main purpose of the study was to evaluate the short-term exposure to VOC emissions resulting from the use of the heating bags during the first life-cycle time, therefore during the first use events. For this purpose, emission data from 3-day test chamber experiments were considered adequate.

4. Conclusions

The present study proposes a methodological approach for the evaluation of short-term inhalation exposure for end users handling three different personal care polymeric items, i.e., electric heating bags. A near-to-real exposure scenario was simulated for each investigated item, taking into account the actual conditions of use ('heating mode') during the first period of life-cycle (first use events). Test emission chamber experiments were performed according to the relevant ISO standards, allowing us to derive 72-hour chamber concentrations and emission rates (ERs) for the main identified VOCs. Collected chamber emission data revealed that, under controlled environmental conditions, the item 'polyester-brand A' was characterized by the highest VOC emission (expressed as the sum of VOC concentrations) equal to 437 $\mu g/m^3$, one order of magnitude higher than those of the other two bags, labeled as 'PVC-brand B' and 'PVC-brand C' (21.1 and 19.6 $\mu g/m^3$, respectively). A remarkable result was the naphthalene emission from 'polyester-brand A', with a chamber concentration equal to 360.5 $\mu g/m^3$ and an emission rate of about 9 $\mu g/h$. This investigation level, although conventionally applied for the evaluation of short-term exposure for materials and products, was not exhaustive for the estimation of VOC inhalation exposure levels determined by each investigated item on 'heating mode' in a real setting. For this purpose, the effect of the temperature on emission characteristics was evaluated through dynamic head-space experiments and, as a result, VOC emission rates for the 'heating mode' were estimated. Indoor concentrations inside a 30 m^3 reference room, representative of short-term exposure related to the item use in a realistic setting, were estimated as well. The simulation of the exposure scenario allowed us to highlight that the use of the 'polyester-brand A' heating bag in the first period of its life-cycle could determine a concentration inside a 30 m^3 room equal to 42 µg/m^3, eight times higher than the reference value for health effects (LCI value equal to 10 µg/m^3) reported in EU evaluation schemes. The inhalation exposure to naphthalene emission from 'polyester-brand A' could represent, therefore, a risk for end users and room occupants. Also of concern is the release of benzene, recognized as a carcinogen in humans, for which no safe level of inhalation exposure can be recommended.

Author Contributions: Conceptualization: J.P., G.d.G., M.T.; methodology: J.P., G.d.G., M.T.; investigation: J.P., E.C.; data curation: J.P., E.C., A.D.G.; supervision: G.d.G.; writing—original draft preparation: J.P., A.D.G. All authors have read and agreed to the published version of the manuscript.

Funding: This research received no external funding.

Acknowledgments: The authors acknowledge the Operational Section 'Pronto Impiego' of Italian Finance Police placed in the City of Monopoli (Italy) and coordinated by the Company Commander, Luigi Mario Paone. The established collaboration was fruitful and fundamental for the development of the present study.

Conflicts of Interest: The authors declare no conflict of interest.

References

1. Salthammer, T. Release of organic compounds and particulate matter from products, materials and electrical devices in the indoor environments. Indoor air Pollution 1–35. In *Indoor Air Pollution*; Springer: Berlin/Heidelberg, Germany, 2014; Volume 64, pp. 1–35. [CrossRef]

2. De Gennaro, G.; Demarinis Loiotile, A.; Fracchiolla, R.; Palmisani, J.; Saracino, M.R.; Tutino, M. Temporal variation of VOC emission from solvent and water based wood stains. *Atmos. Environ.* **2015**, *115*, 53–61. [CrossRef]

3. Katsoyiannis, A.; Leva, P.; Barrero-Moreno, J.; Kotzias, D. Building materials. VOC emissions, diffusion behavior and implications from their use. *Environ. Pollut.* **2012**, *169*, 230–234. [CrossRef]

4. Wilke, O.; Jann, O.; Brödner, D. VOC- and SVOC-emissions from adhesives, floor coverings and complete floor structures. *Indoor Air* **2004**, *14*, 98–107. [CrossRef] [PubMed]

5. Peng, Z.; Deng, W.; Tenorio, R. Investigation of Indoor Air Quality and the Identification of Influential Factors at Primary Schools in the North of China. *Sustainability* **2017**, *9*, 1180. [CrossRef]

6. Śmiełowska, M.; Marć, M.; Zabiegala, B. Indoor air quality in public environments—A review. *Environ. Sci. Pollut. Res.* **2017**, *24*, 11166–11176. [CrossRef]

7. De Gennaro, G.; Dambruoso, P.R.; Demarinis Loiotile, A.; Di Gilio, A.; Giungato, P.; Tutino, M.; Marzocca, A.; Mazzone, A.; Palmisani, J.; Porcelli, F. Indoor air quality in schools. *Environ. Chem. Lett.* **2014**, *12*, 467–482. [CrossRef]

8. Campagnolo, D.; Saraga, D.E.; Cattaneo, A.; Spinazzè, A.; Mandin, C.; Mabilia, R.; Perreca, E.; Sakellaris, I.; Canha, N.; Mihucz, V.G.; et al. VOCs and aldehydes source identification in European office buildings—The OFFICAIR study. *Build. Environ.* **2017**, *115*, 18–24. [CrossRef]

9. Mendell, M.J. Indoor residential chemical emissions as risk factors for respiratory and allergic effects in children: A review. *Indoor Air* **2007**, *17*, 259–277. [CrossRef] [PubMed]

10. Wolkoff, P. Indoor air pollutants in office environments: Assessment of comfort, health and performance. *Int. J. Hyg. Environ. Health* **2013**, *216*, 371–394. [CrossRef]

11. Kotzias, D. Indoor air and human exposure assessment—Needs and approaches. *Exp. Toxicol. Pathol.* **2005**, *57*, 5–7. [CrossRef]

12. Bartzis, J.; Wolkoff, P.; Stranger, M.; Efthimiou, G.; Tolis, E.I.; Maes, F.; Nørgaard, A.W.; Ventura, G.; Kalimeri, K.K.; Goelen, E.; et al. On organic emissions testing from indoor consumer products' use. *J. Hazard. Mater.* **2015**, *285*, 37–45. [CrossRef] [PubMed]

13. Palmisani, J.; Nørgaard, A.W.; Kofoed-Sørensen, V.; Clausen, P.A.; de Gennaro, G.; Wolkoff, P. Formation of ozone-initiated VOCs and secondary organic aerosol following application of a carpet deodorizer. *Atmos. Environ.* **2020**. [CrossRef]

14. Kwon, K.-D.; Jo, W.-K.; Lim, H.-J.; Jeong, W.-S. Characterization of emissions composition for selected household products available in Korea. *J. Hazard. Mater.* **2007**, *148*, 192–198. [CrossRef] [PubMed]

15. Brattoli, M.; Cisternino, E.; de Gennaro, G.; Giungato, P.; Mazzone, A.; Palmisani, J.; Tutino, M. Gaschromatography analysis with olfactometric detection (GC-O): An innovative approach for chemical characterization of odor active Volatile Organic Compounds (VOCs). *Chem. Eng. Trans.* **2014**, *40*, 121–126. [CrossRef]

16. Even, M.; Hutzler, C.; Wilke, O.; Luch, A. Emissions of volatile organic compounds from polymer-based consumer products: Comparison of three emission chamber sizes. *Indoor Air* **2020**, *30*, 40–48. [CrossRef] [PubMed]

17. Luca, F.-A.; Ciobanu, C.-I.; Andrei, A.G.; Horodnic, A.V. Raising Awareness on Health Impact of the Chemicals Used in Consumer Products: Empirical Evidence from East-Central Europe. *Sustainability* **2018**, *10*, 209. [CrossRef]

18. Marć, M.; Formela, K.; Klein, M.; Namieśnik, J.; Zabiegala, B. The emissions of monoaromatic hydrocarbons from small polymeric toys placed in chocolate food products. *Sci. Total Environ.* **2015**, *530–531*, 290–296. [CrossRef]

19. Wiedmer, C.; Velasco-Schön, C.; Buettner, A. Characterization of off-odours and potentially harmful substances in a fancy dress accessory handbag for children. *Sci. Rep.* **2017**, *7*, 1807. [CrossRef]

20. Abe, Y.; Yamaguchi, M.; Mutsuga, M.; Kawamura, Y.; Akiyama, H. Survey of volatile substances in kitchen utensils made from acrylonitrile–butadiene–styrene and acrylonitrile–styrene resin in Japan. *Food Sci. Nutr.* **2014**, *2*, 236–243. [CrossRef]

21. Even, M.; Girard, M.; Rich, A.; Hutzler, C.; Luch, A. Emissions of VOCs from polymer-based consumer products: From emission data of real samples to the assessment of inhalation exposure. *Front. Public Health* **2019**, *7*, 202. [CrossRef]

22. ISO 16000-9. *Indoor Air–Part 9: Determination of the Emission of Volatile Organic Compounds from Building Products and Furnishing—Emission Test Chamber Method*; ISO: Geneva, Switzerland, 2006.

23. ISO 16000-6. *Indoor Air–Part 6: Determination of Volatile Organic Compounds in Indoor and Test Chamber Air by Active Sampling on Tenax TA Sorbent, Thermal Desorption and Gas Chromatography Using MS/FID*; ISO: Geneva, Switzerland, 2011.

24. Kozicki, M.; Piasecki, M.; Goljan, A.; Deptuła, H.; Niesłochowski, A. Emission of Volatile Organic Compounds (VOCs) from Dispersion and Cementitious Waterproofing Products. *Sustainability* **2018**, *10*, 2178. [CrossRef]

25. Ahn, J.-H.; Kim, K.-H.; Kim, Y.-H.; Kim, B.-W. Characterization of hazardous and odorous volatiles emitted from scented candles before lighting and when lit. *J. Hazard. Mater.* **2015**, *286*, 242–251. [CrossRef] [PubMed]

26. Dimitroulopoulou, C. Ventilation in European dwellings: A review. *Build. Environ.* **2012**, *47*, 109–125. [CrossRef]

27. AgBB. Requirements for the Indoor Air Quality in Buildings: Health-Related Evaluation Procedure for Emissions of Volatile Organic Compounds Emissions (VVOC, VOC and SVOC) from Building Products. 2018. Available online: https://www.umweltbundesamt.de/sites/default/files/medien/360/dokumente/agbb_evaluation_scheme_2018.pdf (accessed on 27 January 2020).

28. Kang, D.H.; Choi, D.H.; Won, D.; Yang, W.; Schleibinger, H.; David, J. Household materials as emission sources of Naphthalene in Canadian homes and their contribution to indoor air. *Atmos. Environ.* **2012**, *50*, 79–87. [CrossRef]

29. Lim, J.; Kim, S.; Kim, A.; Lee, W.; Han, J.; Cha, J.-S. Behavior of VOCs and carbonyl compounds emission from different types of wallpapers in Korea. *Int. J. Environ. Res. Public Health* **2014**, *11*, 4326–43339. [CrossRef]

30. World Health Organization. Selected Pollutants. In *Indoor Air Quality Guidelines*; WHO Regional Office for Europe: Copenhagen, Denmark, 2010.

31. International Agency for Research on Cancer (IARC). Naphthalene, IARC Monograph 1, 367–435. 2001. Available online: http://monographs.iarc.fr/ENG/Monographs/vol82/mono82-8.pdf (accessed on 27 January 2020).

32. Regulation (EC) No 1272/2008 on Classification, Labeling and Packaging of Substances and Mixtures, Amending and Repealing Directives 67/548/ECC and 1999/45/ECC and Amending Regulation (EC) No 1907/2006. Available online: https://eur-lex.europa.eu/legal-content/IT/ALL/?uri=uriserv:OJ.L_.2008.353.01.0001.01.ITA (accessed on 27 January 2020).

33. Masuck, I.; Hutzler, C.; Jann, O.; Luch, A. Inhalation exposure of children to fragrances present in scented toys. *Indoor Air* **2011**, *21*, 501–511. [CrossRef]

 *sustainability*

Article

Modeling In-Vehicle VOCs Distribution from Cabin Interior Surfaces under Solar Radiation

Zheming Tong [1,2,*] **and Hao Liu** [2]

[1] State Key Laboratory of Fluid Power and Mechatronic Systems, Zhejiang University, Hangzhou 310027, China

[2] School of Mechanical Engineering, Zhejiang University, Hangzhou 310027, China; 11825065@zju.edu.cn

* Correspondence: tzm@zju.edu.cn

Received: 7 June 2020; Accepted: 5 July 2020; Published: 8 July 2020

Abstract: In-vehicle air pollution has become a public health priority worldwide, especially for volatile organic compounds (VOCs) emitted from the vehicle interiors. Although existing literature shows VOCs emission is temperature-dependent, the impact of solar radiation on VOCs distribution in enclosed cabin space is not well understood. Here we made an early effort to investigate the VOCs levels in vehicle microenvironments using numerical modeling. We evaluated the model performance using a number of turbulence and radiation model combinations to predict heat transfer coupled with natural convection, heat conduction and radiation with a laboratory airship. The Shear–Stress Transport (SST) k-ω model, Surface-to-surface (S2S) model and solar load model were employed to investigate the thermal environment of a closed automobile cabin under solar radiation in the summer. A VOCs emission model was employed to simulate the spatial distribution of VOCs. Our finding shows that solar radiation plays a critical role in determining the temperature distribution in the cabin, which can increase by 30 °C for directly exposed cabin surfaces and 10 °C for shaded ones, respectively. Ignoring the thermal radiation reduced the accuracy of temperature and airflow prediction. Due to the strong temperature dependence, the hotter interiors such as the dashboard and rear board released more VOCs per unit time and area. A VOC plume rose from the interior sources as a result of the thermal buoyancy flow. A total of 19 mg of VOCs was released from the interiors within two simulated hours from 10:00 am to noon. The findings, such as modeled spatial distributions of VOCs, provide a key reference to automakers, who are paying increasing attention to cabin environment and the health of drivers and passengers.

Keywords: in-vehicle air quality; pollution model; thermal environment; solar radiation; VOCs exposure; CFD; environmental health

---

1. Introduction

Over the past decades, China has been experiencing the world's fastest growth in vehicle population. As a result, commuters inevitably spend a substantial amount of time in vehicle cabins due to increased traffic congestions and vehicle population especially in major cities [1]. Epidemiological studies show that long-time exposure to air pollutants is associated with increased risks of morbidity and mortality [2–4], especially high volatile organic compounds (VOCs) concentrations emitted from cabin interiors [5–7] that could lead to respiratory irritation and cancer [8]. Developing advanced methods to identify in-cabin emission sources [9] has become a public health priority for consumers, car producers and government. Some attempts have been made to provide indoor air quality (IAQ) guidelines for passenger cars. For example, the Ministry of Environmental Protection of the People's Republic of China has promulgated national standard the HJ/T 400 "Determination of Volatile Organic Compounds and Carbonyl Compounds in Cabin of Vehicles" and GB/T 27,630 "Guideline for air quality assessment of passenger car". The Japan Automobile Manufacturers Association (JAMA) has introduced a voluntary approach for reducing the concentration levels of VOCs in the vehicle cabins. The World Health Organization (WHO) provides a guideline of 0.1 mg/m^3 for the protection of public health from risks due to a number of chemicals commonly found in indoor air [10,11].

The factors controlling in-vehicle VOCs levels have been identified as a combined impact of interior materials, vehicle age [12,13], microenvironment in the cabin such as temperature, relative humidity and ventilation mode [14,15], and pollutants outside the vehicle like exhaust gases [16] and fuel leakage [9]. However, for a certain parked vehicle with the engine and ventilation off, the cabin becomes a completely closed space ignoring the air leakage, and it can be reasonably inferred that temperature becomes the main factor. Some researchers have addressed that VOCs diffusing from building materials are strongly associated with temperature [17–19]. Similar studies also were carried out in vehicle with an increased focus on cabin air quality. Many evidences have proved that the VOCs concentrations significantly increased with the increase in surface and ambient temperature. Yoshida et al. [20] indicated that the total volatile organic compounds (TVOC) concentrations in summer exceeded the indoor guideline value of 300 μg/m^3; the interior temperature was the main factor affecting the interior concentrations of most compounds. Geiss et al. [21] found the VOC concentrations in the hot cabin with 70 °C were 40% higher through measuring in 23 old private cars in both summer and winter. Faber et al. [22] presented that chemical composition in vehicle air strongly depends on temperature. Chen et al. [23] investigated the VOCs in taxi cabins and found vehicle age is the most important factor, followed by interior temperature. Xiong et al. [24] derived a theoretical correlation between the steady state concentration and temperature for VOC emission from materials performing on three cars at different temperatures. Xu et al. [13] found toluene, styrene, ethylbenzene, and xylene were the most sensitive VOCs to temperature, which increased by 513.6%, 544.8%, 767.0%, and 597.7%, respectively, as the temperature increased from 11 °C to 25 °C. Huang et al. [25] found that the TVOC emission rate exponentially increased with the increase in in-cabin temperature.

As mentioned above, the VOCs emission is significantly associated with the in-cabin thermal environment, which is currently a hot topic. To evaluate the thermal comfort in cabin and further optimize the heating, ventilation, and air conditioning (HVAC) system [26–28], both experimental and numerical simulation studies have investigated the unsteady temperature and airflow profiles in buildings and passenger compartments [29–36]. Some studies have documented the dangerously high temperature in passenger compartments during exposure to the sun when parked outdoors [37]. Solar radiation is identified as the main heat source for a static vehicle in summer. More than 40% of solar heat flux enters the vehicle via the windshield. The vehicle exposed to direct solar radiation performs comparably to a greenhouse with severe thermal accumulation [38]. The cabin tends to be overheated quickly during the thermal soak period, the terminal temperature of the air and interior can reach about 60 °C and 80 °C, respectively [39,40].

This brief review shows that VOCs released from cabin interiors are proven to be temperature-dependent, and thermal simulation of the cabin environment has been conducted before. However, they were not linked together to discuss the in-cabin air quality. How VOC concentrations in the cabin vary with temperature under parked conditions is still a problem that has not been investigated quantitatively before. Gas sensing in real conditions with the use of cost-acceptable sensors is not trivial task [41], therefore, the numerical approximation of VOC distribution modeling in the car will be better justified. This research was aimed at bridging this knowledge gap and providing an effort to quantify the impact of the solar radiation on the cabin temperature and VOCs emission. This paper is organized as follows. We first evaluate the model performance by comparing it to the experiment in Section 2. Then, in-vehicle VOCs distribution from interior surfaces under solar radiation is modelled in Section 3. Section 4 presents the results and discussions. Concluding remarks are provided in Section 5.

2. Numerical Development and Verification

Computational fluid dynamics (CFD) has proved itself to be a powerful numerical tool in modeling the in-cabin environment. Figure 1 describes our analytical procedure based on the CFD approach. The solar calculator was imposed to compute the solar irradiation. The solar load model was used to calculate radiation effects from the sun's rays that enter the cabin domain. Turbulence and radiation models were applied to capture the airflow and temperature. The pollutant emission model linked the temperature and emission rate (ER) of the cabin interiors. Then, the data of hot-soak temperature, airflow distribution and pollutant concentration were obtained.

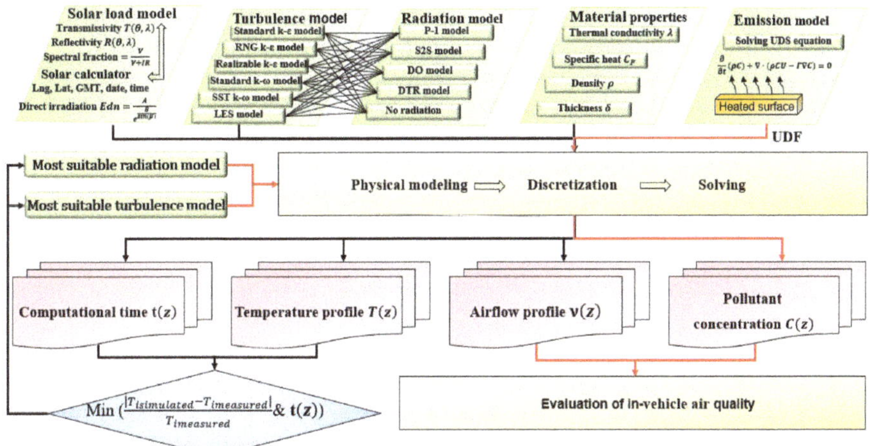

Figure 1. The framework of our methodology.

2.1. The Turbulence Models

Turbulence modeling is a critical process for the numerical investigation of thermal environments. There have been some available studies to evaluate the performance of various turbulence models. Zhai et al. [42] compared eight turbulence models for predicting airflow and turbulence in enclosed environments and found that the Re-Normalization Group (RNG) k-ε model performed best among the Reynolds-averaged Navier–Stokes (RANS) models. Hussain et al. [43] used six RANS turbulence models to simulate the thermal environment in an atrium, and found the Shear–Stress Transport (SST) k-ω model provided comparatively better results. Li et al. [44] presented a numerical evaluation of the eddy viscosity turbulence models in terms of CFD modeling of convection-radiation coupled heat transfer in the indoor environment, and demonstrated a great performance of k-ω group models. In this study, five RANS turbulence models (including the standard k-ε model, the RNG k-ε model, the realizable k-ε model, the standard k-ω model, the SST k-ω model) and the Large Eddy Simulation (LES) model were selected to evaluate the prediction of the airflow and temperature distributions in the vehicle cabin.

2.2. The Radiation Models

Thermal radiation makes the temperature distribution more uniform in an enclosed space by transferring thermal energy from a hot surface to a cold one. Generally, thermal radiation accounts for 30–70% of the total heat transfer rate [45]. In this study, radiation heat transfer was taken into consideration for modeling the cabin thermal environment.

The P-1 radiation model is the simplest case of the P-N model, which is based on the expansion of the radiation intensity into an orthogonal series of spherical harmonics functions. The directional dependence in radiative transfer equation (RTE) is integrated out, resulting in a diffusion equation for incident radiation. Equation (1) is obtained for the radiation flux q_r, only considering scattering and absorption when modeling gray radiation.

$$q_r = -\frac{1}{3(a + \sigma_s) - C\sigma_s} \nabla G \tag{1}$$

An advection-diffusion equation is solved to determine the local radiation intensity G in the P-1 model.

$$-\nabla \cdot q_r = aG - 4\sigma T^4 \tag{2}$$

where q_r is the radiation heat flux, a is the absorption coefficient, σ_s is the scattering coefficient, G is the incident radiation, C is the linear-anisotropic phase function coefficient, σ is the Stefan-Boltzmann constant.

The Surface to Surface (S2S) radiation model is applicable for modeling radiation in situations where there are no participating media. All surfaces involved in radiation are assumed to be gray and diffuse, ignoring absorption, emission and scattering and preserving only "surface-to-surface" radiation. The energy flux leaving a given surface is composed of directly emitted and reflected energy. The reflected energy flux is dependent on the incident energy flux from the surroundings, which then can be expressed in terms of the energy flux leaving all other surfaces. The energy leaving from the kth adiabatic surface can be expressed as Equation (3).

$$q_{out,k} = \varepsilon_k \sigma T_k^4 + \rho_k q_{in,k} \tag{3}$$

where $q_{out,k}$ is the energy flux leaving the surface, ε_k is the emissivity, ρ_k is the reflectivity of surface k, $q_{in,k}$ is the energy flux incident on the surface from the surroundings.

The Discrete Ordinates (DO) radiation model is regarded as the most comprehensive radiation model, which solves the RTE for a discrete number of finite solid angles, as shown in Equation (4). Each associated with a vector direction $\vec{s}$ is fixed in the global Cartesian system (x, y, z). Accuracy can be increased by using a better discretization, while it may be CPU-intensive with many ordinates.

$$\nabla \cdot \left(I\left(\vec{r}, \vec{s}\right)\vec{s} \right) + (a + \sigma_s)I\left(\vec{r}, \vec{s}\right) = an^2 \frac{\sigma T^4}{4\pi} + \frac{\sigma_s}{4\pi} \int_0^{4\pi} I\left(\vec{r}, \vec{s}\right) \Phi\left(\vec{s} \cdot \vec{s'}\right) d\Omega' \tag{4}$$

where $\vec{r}$ is the position vector, $\vec{s}$ is the direction vector, $\vec{s'}$ is the scattering direction vector, α is the absorption coefficient, n is the refractive index, I is the radiation intensity, which depends on the position $\left(\vec{r}\right)$ and direction $\left(\vec{s}\right)$, T is the local temperature, Φ is the phase function, Ω' is the solid angle.

For the Discrete Transfer Radiation Model (DTRM), the main assumption is that radiation leaving a surface element within a specified range of solid angles can be approximated by a single ray. The energy source in the fluid due to radiation is computed by summing the change in intensity dI along the path of each ray ds that is traced through the fluid control volume.

$$\frac{dI}{ds} + aI = \frac{a\sigma T^4}{\pi} \tag{5}$$

2.3. The Solar Load Model (SLM)

Thermal energy due to incident solar rays is a very common but important phenomenon. In the SLM, the solar beam direction and irradiation were calculated according to the provided geographical place and the given time based on the solar load model's ray tracing algorithm. A two-band spectral model was used for direct solar illumination and accounted for separate material properties in the visible and infrared bands. A single-band hemispherical-averaged spectral model was used for diffuse radiation. The ground was considered to be dry bare land with a reflectivity of 0.2 [40]. Solar scattering was set to the default value of 1.

2.4. Contaminants Emission Model

For the VOCs emission of building materials, the quasi-steady-state ER can be described by the following equation [46]:

$$E(t) = 2.1\frac{D_m C_0}{\delta}exp\left(-2.36\frac{D_m t}{\delta^2}\right) \tag{6}$$

where E is the emission rate factor, t is the emission time, δ is the material thickness, C_0 is the initial emittable concentration inside the material, D_m is the diffusion coefficient. The correlation between C_0 and T can be described by the following equation [47]. The correlation between D_m and T can be described by the following equation [48]. Equation (9) is obtained by substituting Equations (7) and (8) into Equation (6) and then taking the logarithm on both sides.

$$C_0 = \frac{C_1}{T^{0.5}}exp\left(-\frac{C_2}{T}\right) \tag{7}$$

$$D_m = D_1 T^{1.25}exp\left(-\frac{D_2}{T}\right) \tag{8}$$

$$In\frac{E(t)}{T^{0.75}} = A - \frac{B}{T} - 2.36F0_m \tag{9}$$

where T is the temperature in K. C_1, C_2, D_1 and D_2 are all constants determined only by the physical and chemical properties of pollutants

$$A = In\frac{2.1C_1 D_1}{\delta}, \; B = C_2 + D_2, \; F0_m = \frac{D_m t}{\delta^2}.$$

To quantify the pollutants released by materials in a certain period of time, Equation (10) is included.

$$M = \int_{t_s}^{t_e} AE(t)dt \tag{10}$$

where M is the total quality of the released contaminants, t_s is the start time, t_e is the end time, $E(t)$ is the emission rate, A is the surface area.

2.5. Model Validation and Discussion

The temperature and flow field in the airship is a result of the interplay among multiple factors including solar radiation, earth reflection, infrared radiation, external forced convection and internal natural convection, which is similar to the thermal environment of a parked car. Therefore, the experimental results of Li et al. [49] were selected for model validation. In their study, the transient thermal behaviors of an airship under different solar radiation were revealed. An airship model with a spherical tank type was built in a closed laboratory. The body was covered with 0.1 mm polyimide film, and shaped by thin metal sheets. Solar irradiation was supplied by a TRM-PD solar simulator and measured by a XLP12-1S-H2 heat flow meter. Eighteen T-type thermocouples were arranged in 18 different locations to obtain the hull and inner gas temperatures. Eight points are located on plane 1 including from point 1 to point 6 and point 8. For plane 2, there are also eight points, from point 9 to point 14 and point 16. Point 18 and point 19 are distributed on plane 3. Plane 1 and plane 2 are symmetrical about plane 3 with 150 mm axial distance. In Figure 2, a full-size computational domain was modeled with the same physical dimensions as the experimental airship. The computational domain was discretized using the structured mesh with five refined inflation layers applied close to the solid surfaces. The grid independence was achieved at 0.2 million mesh elements, as shown in Table 1. To evaluate the CFD approach to model convection-radiation coupled heat transfer in the airship, the hull and inner gas temperatures were compared with experimental data in terms of the accuracy and computational cost. Six turbulence models and four radiation models were included. In each configuration, one turbulence model was combined with or without a radiation model. Finally, thirty CFD cases were obtained totally.

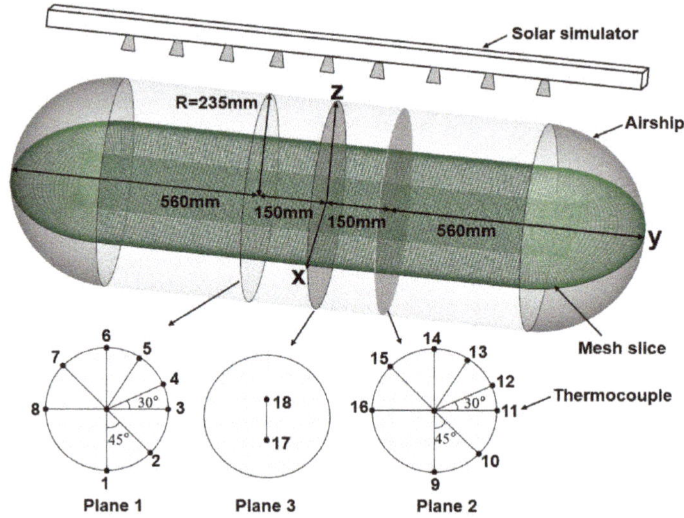

Figure 2. The airship computational domain according to [49].

Table 1. The grid independence test of the temperature profile at point 18.

Type	Cell Number	The Temperature of Point 18 (*K*)
Grid 1	5640	378.39
Grid 2	9728	378.62
Grid 3	35685	381.75
Grid 4	108547	382.31
Grid 5	207100	382.34

The hull material is polyimide film with thermal conductivity $\lambda = 0.32 \text{w}/(\text{m·K})$. The total solar absorptivity of the external surface is 0.45 with a 0.81 absorptivity in the infrared spectrum. The coupled solver with pseudo-transient relaxation was applied for the solution of the momentum, energy, and turbulence equations. The total calculation adopted 600 s according to the experimental sampling time.

The results of 30 designed test trips with different combinations of radiation and turbulence models are obtained. Except for the air temperature measured at point 17 and 18, the rest are measured at the airship surface. Since the solar simulator is installed above the airship and keeps both axes parallel, the temperature distributions on plane 1 and 2 are basically the same. The temperature of six locations including point 1, point 3, point 6, point 17 and point 18 are shown in Figure 3. The measured data at each point is selected as the benchmark. The temperature of the measuring point closer to the light source is higher. Among them, the temperature at point 6 is the highest, reaching more than 90 °C, from where the temperature drops along with the body surface. Due to the shelter of the airship, point 1 is not directly exposed to the sunlight, resulting in a similar temperature as that of the surroundings with 16.7 °C.

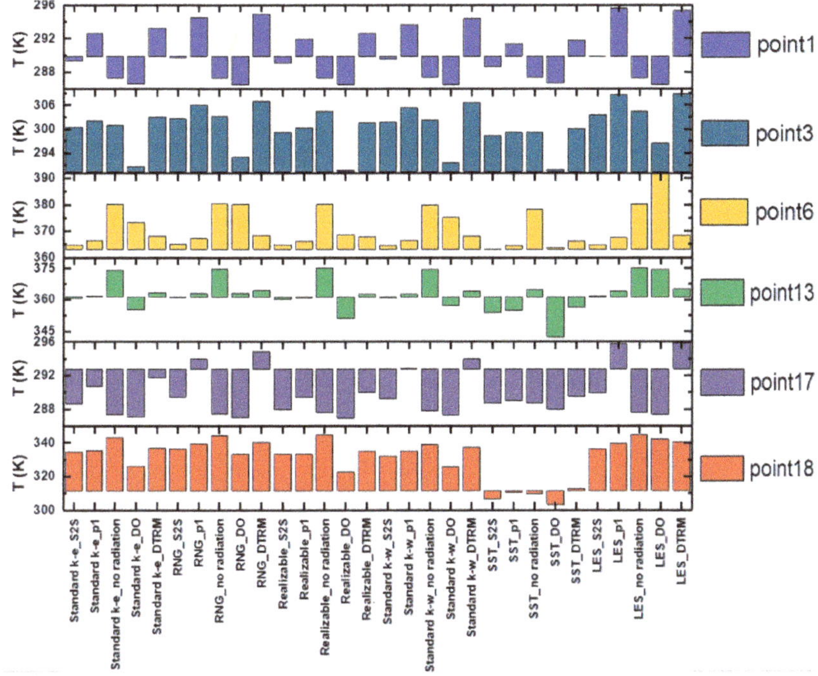

Figure 3. Comparisons of the experimental and numerical temperature $T(K)$ in the closed airship. Simulation data floats up and down based on experimental data from [49].

To evaluate the performance of model combinations, a method called the total temperature error is proposed to assess the accuracy of numerical simulations. The total temperature error is calculated according to Equation (11). In addition, the computational cost is obtained according to the CPU calculation time when using a computer with an AMD Ryzen 7 1700X, 3.40 GHz CPU, 32 GB of RAM and 8 compute nodes.

$$T_{Error}(\%) = \sum_{i=1}^{18} \frac{|T_{isimulated} - T_{imeasured}|}{T_{imeasured}} \times 100\% \tag{11}$$

where $T_{isimulated}$ is the simulated temperature of the i_{th} point, $T_{imeasured}$ is the measured temperature of the i_{th} point. In the experiment, 18 points were measured totally.

In Figure 4, comparing the temperature errors with and without radiation, the predicted air and solid surface temperature profiles agree better with the experimental results when the effects of thermal radiation are accounted for in the numerical investigation. On the contrary, the total error when ignoring the surface radiation may be twice that when including the surface radiation. When the turbulence model is fixed and combined with different radiation models, it is found that the S2S and DO models perform best, while the DTRM model has the lowest accuracy. Among all turbulence models, the standard k-ω and SST k-ω models have very close accuracies to predict the temperature profiles. The SST k-ω model has a clear advantage in predicting accuracy compared to the LES model; it works the best for the high Rayleigh number buoyancy-driven flow [50]. Figure 4 reveals an M-shape trend of the calculation time according to the order of the combined model. The radiation model has a more significant impact on computation time than the turbulence model. The S2S and P1 radiation models require much less computation time than the DO model, although some small disparities exist but still are comparable. The DTRM is not compatible with parallel processing, so it will consume more time to model radiative heat transfer. The above model validation and comparison suggest that the SST k-ω and S2S model are the best choices to predict the in-vehicle thermal environment under solar radiation. Some existing studies [51] also show that the SST k-ω model performs slightly better than the RNG k-ε model when simulating convection-radiation coupled heat transfer.

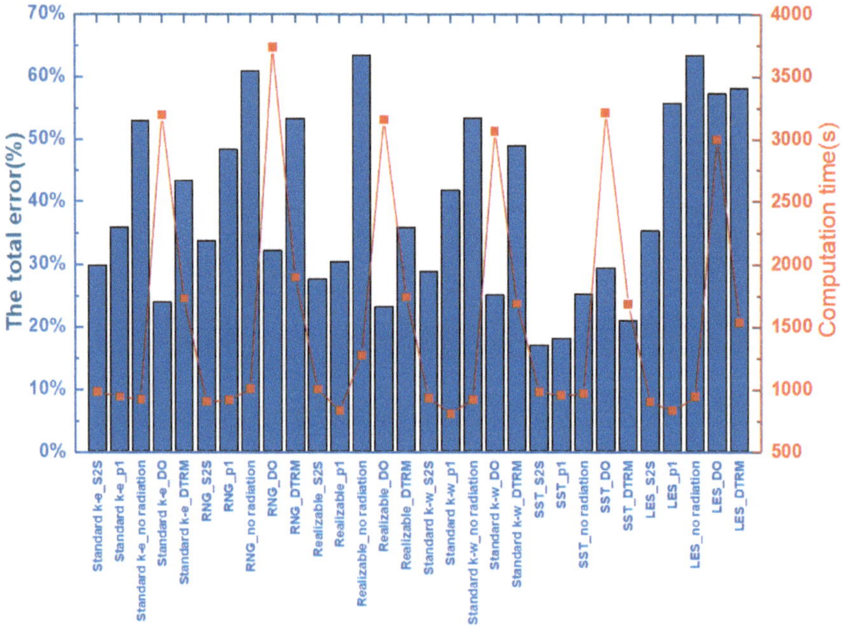

Figure 4. The total temperature error and computation time of 30 cases.

To validate the theoretical correlation between the ER and surface temperature, some experimental results in the previous literature were used. Table 2 displays the measured emission rates of five pollutants from the car mat under three varied temperatures and total volatile organic compounds (TVOC) from PBS-C at five different temperatures. The detailed experimental description can be found in references [52,53].

Table 2. The emission rate (ER) of pollutants from in-vehicle materials under varied temperature conditions.

	Type	ER at 25 °C	ER at 35 °C	ER at 50 °C
	Benzene	1.049	1.517	2.472
	Toluene	2.472	0.595	1.376
VOCs	P-xylene	0.423	0.517	0.659
	Ethylbenzene	0.235	0.315	0.459
	Styrene	0.447	0.517	0.635

	ER at 35°C	ER at 50 °C	ER at 70 °C	ER at 80 °C	ER at 90 °C
TVOC	0.393	1.836	4.459	6.033	9.311

The unit of ER for the volatile organic compounds (VOCs) and total volatile organic compounds (TVOC) is $\mu g \cdot m^{-2} \cdot h^{-1}$ and $mg \cdot m^{-2} \cdot h^{-1}$, respectively.

The linear curve fittings are illustrated in Figure 5. All R^2 are greater than 0.95, indicating a satisfactory correlation between the VOCs emission rate and temperature proposed in Equation (9).

In the follow-up study, the little impact of existing in-air VOCs on the emission rate is ignored during the emission period [54]. Moreover, the materials maintain their original appearance and properties without any bake out treatment, which indicates abundant VOCs to be volatilized in a quasi-steady state. Besides, the effect of relative humidity on the emission factor is ignored because of the constant relative humidity in that environment [35].

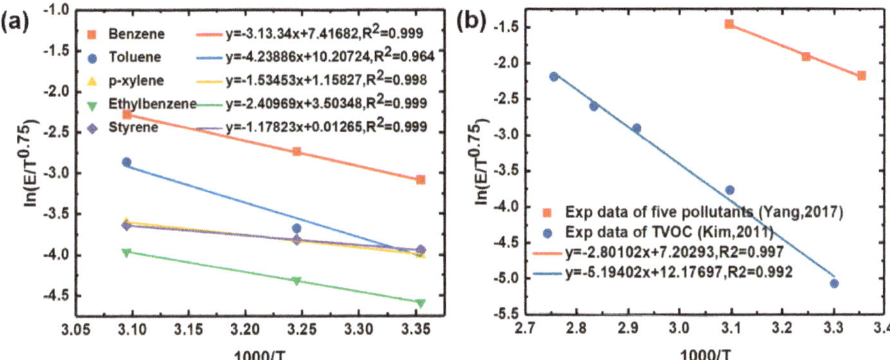

Figure 5. (a) The relationship between five VOCs emissions and temperature according to the experimental data [52]; (b) The relationship between TVOC emission and temperature according to the experimental data [53].

3. In-Vehicle VOCs Distribution Considering Solar Radiation

3.1. Vehicle Computational Model

Figure 6 presents the vehicle model established referring to the original data of a hatchback, in which the engine and luggage compartment were ignored. In addition, some interior components including safety belts, electrical equipment and console details were neglected due to the fewer effects on airflow. The details of the dimension are shown in Table 3. The passenger compartment was designed with eight air conditioning inlets and two outlets. Four inlets are located on the center console. Among them, No.1 and No.2 are, respectively, located on the middle, facing the gap between the two bucket seats. No.3 and No.4 are located on the sides. No.5 and No.6 are, respectively, arranged in the feet space of the driver and passenger. No.7 and No.8 are located on the armrest box, facing the bench seat. The air outlets No.1 and No.2 are located on the rear board. No additional airflow inlet and outlet are included anymore under the well-sealed assumption. The vehicle computational domain was discretized using unstructured tetrahedron grids [55]. Inflation layers were used at the interfaces between the air volume and the solids. The case study adopted 3.8 million elements after converting the domain to the polyhedral meshes.

Table 3. The dimensions of the vehicle model.

Parts	Dimension	Unit
body	3100*1600*1230	(L*W*H)(mm)
windshield	0.864	m^2
rear window	0.787	m^2
side window 1 and 3	0.224	m^2
side window 2 and 4	0.266	m^2
cabin volume	2.716	m^3

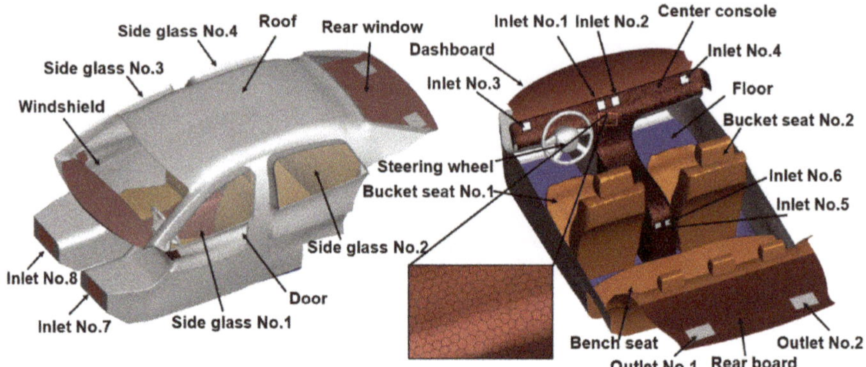

Figure 6. The computational model of the passenger compartment.

3.2. Boundary Conditions

3.2.1. The Thermal Environment Analysis

Figure 7 shows the thermal energy transfer for the cabin. External heat enters the cabin through three ways including heat conduction, heat convection and thermal radiation. When the vehicle is parked under the sunlight for a soaking period, some solar radiation enters the passenger compartment passing through the windows, some is reflected by the solid envelope, the rest is absorbed. Solar radiation leads to a considerable thermal load through heating the envelope and interiors. In addition, the cabin exchanges the heat with the external environment through the coupled convection and radiation. Due to the uniform temperature distribution, the airflow cycles are driven by buoyant force, creating the natural convection in the cabin. The scorching air is trapped inside the cabin due to the lack of openings, resulting in the greenhouse effect [56]. In this study, the simulation was based on the city of Hangzhou (118°21′–120°30′ E, 29°11′–30°33′ N), the capital of Zhejiang Province located along Southeast coast of China, characterized by long and hot summers. The ambient conditions were chosen on June 21 (summer solstice). The windshield orientation was to the south.

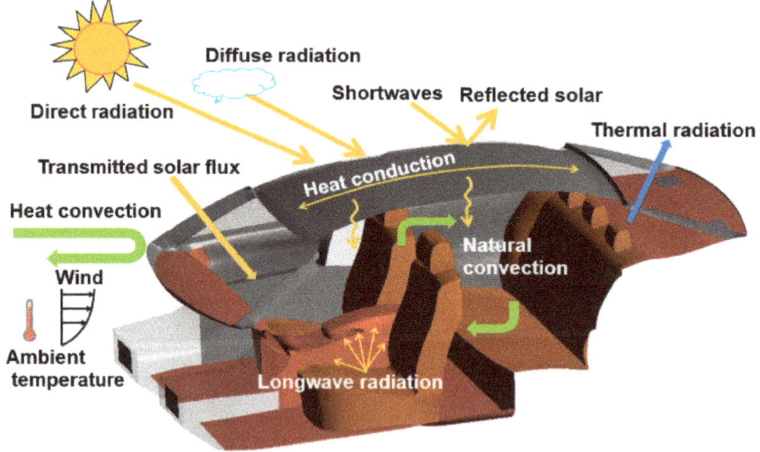

Figure 7. Heat transfer between the surface and its surroundings in the cabin.

3.2.2. The Thermal Setup

The cabin enclosure is composed of body structure, door, floor and glass, exposed to outdoor climatic conditions with sun and wind directly; the mixed wall thermal boundary was uniformly applied to the shell. According to the local summer weather condition, the ambient temperature of 38 °C was set as free steam and external radiation temperature for the whole cabin body, except for the driver's foot space adjacent to the engine cooling water. The appropriate thermal resistance across the wall thickness was imposed according to the wall thickness and material properties. The details are presented in Table 4. Besides, the shell conduction approach was utilized to model conduction in the planar direction of steel body with good thermal conductivity. All solid surfaces were considered stationary walls with No-slip conditions. The convective heat transfer coefficient was calculated based on the empirical formula in Equation (12) [29]. The inlets and outlets of the HVAC system were treated as the wall when the vehicle kept ventilation off.

$$h = 1.163\left(4 + 12v^{0.5}\right) \tag{12}$$

where v is the wind speed relative to the parked vehicle with 0.2 m/s, $h = 10.89$ W/m^{2}K.

Table 4. The material parameters and optical properties of the main parts.

Parts	Material	Density $\rho/(\mathrm{kg \cdot m^{-3}})$	Thermal Conductivity $\lambda/(\mathrm{W \cdot m^{-1} \cdot K^{-1}})$	Specific Heat $c_p/(\mathrm{J \cdot kg^{-1} \cdot K^{-1}})$	Thickness δ/mm
vehicle body	steel	8030.00	100.04	448.83	5.00
Dashboard rear board					5.00
center console	PBS	1260.00	2.70	1480.60	1.50
seats					15.00
floor	carpet	1601.85	0.29	1485.38	5.00
windows	glass	2529.58	1.17	754.04	5.00

3.2.3. The Radiation Setup

The windshield, side window, and rear window were treated optically as semi-transparent walls; all other surfaces were considered opaque. All surfaces participated in radiation heat transfer. The emissivity of interior surfaces was assumed to be 0.95 [30], and 0.88 for the windows [57]. The optical properties of the cabin surfaces are listed in Table 5.

Table 5. The optical properties of the surface.

Surfaces	Absorptivity		Transmissivity		Diffuse hemispherical	
	Visible	Infra-red	Visible	Infra-red	Absorptivity	Transmissivity
vehicle body	0.7	0.1	—	—	—	—
dashboard	0.7	0.1	—	—	—	—
rear board	0.7	0.1	—	—	—	—
center console	0.7	0.1	—	—	—	—
door	0.7	0.1	—	—	—	—
roof	0.7	0.1	—	—	—	—
seats	0.8	0.1	—	—	—	—
floor	0.8	0.2	—	—	—	—
windows	0.14	0.65	0.76	0.25	0.1	0.5

3.2.4. The Emission Model Setup

Due to the common HVAC operation during driving and brief natural ventilation when getting off, it is reasonable to expect a very low concentration of contaminants left in the cabin. Moreover, the existence of contaminants has no impact on the airflow and concentration dispersion. Under such reasonable assumptions, defining the user-defined source (UDS) equation is computationally less expensive compared to the multi-component Eulerian approach when modeling the gas transport [51]. Therefore, the unsteady variation, convection, diffusion, and generation in the domain were calculated using a UDS coupled with the user-defined functions (UDFs) based on the existing flow parameters.

$$\frac{\partial}{\partial t}(\rho S) + \nabla \cdot (\rho S U - \Gamma \nabla S) = 0 \tag{13}$$

where ρ is the density of air, S is a scalar representing the contaminant concentration, Γ is the molecular diffusivity of S.

4. Results and Discussion

Chemical mass balance results demonstrated that carpet and seats are the most important VOCs source inside a new vehicle [58], so the bucket, bench seats and floor were chosen as the conventional VOCs sources in the cabin. Besides, the contaminant emission behavior of the dashboard and rear board were additionally investigated due to their prominent representation of high solar exposure.

Solar flux and average temperature variations and distributions with respect to soaking process are shown in Figures 8 and 9, respectively. The transmitted solar flux shows a decrease tendency on window No.3 and No.4, whereas with a sustained growth for the other envelope throughout the soaking period. Most direct solar irradiation (over 350 W) enters the cabin through the windshield and rear glass, falling on the dashboard and rear board. Only a small portion of the sun's rays (below 40 W) passes through the side windows. At noon, the sun moves directly above the vehicle with about 180° solar incidence angle, causing a similar solar load on both side body. In Figure 8b, the surface temperature rises steadily with the increase of solar intensity. The highest average temperature with more than 60 °C occurs at the dashboard and rear board, while the floor has the lowest temperature with about 44 °C. The average temperature of bucket seat No.1 is slightly higher than that of No.2. This is due to the reason that the former receives more solar load from both the windshield and left glass (Figure 9a). The temperature on the two bucket seats tends to be the same at noon because of the comparable sun exposure (Figure 9b). The heat is difficult to transfer around the interior surface by conduction due to low thermal conductivity, exacerbating the thermal imbalance and local overheating [59].

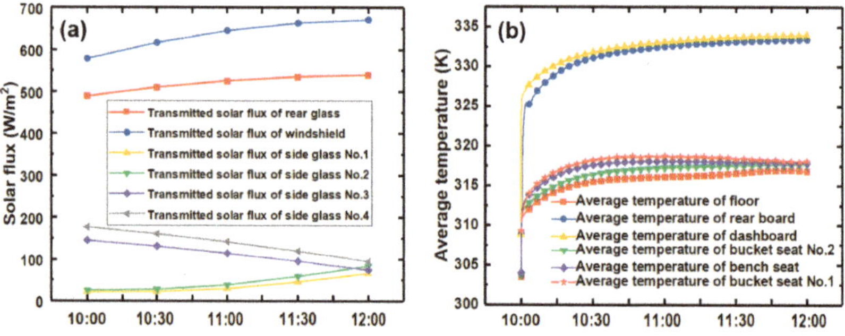

Figure 8. The solar flux (**a**) and temperature (**b**) variations from 10:00 am to noon.

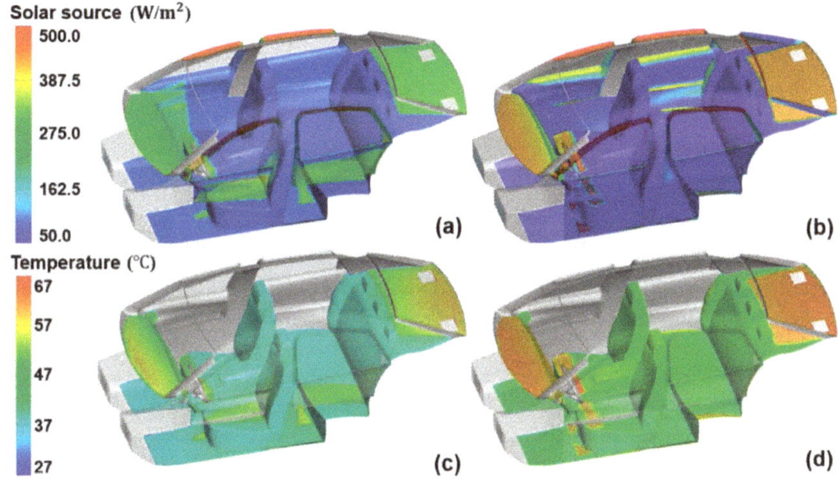

Figure 9. The solar heat distributions at 10:00 am (**a**) and noon (**b**); The temperature distributions at 10:00 am (**c**) and noon (**d**).

Figure 10 displays the temperature distributions of airflows in the driver plane. The air temperature near the hot interior surface is substantially higher than that which is far away from it; this is because of where the thermal boundary layers exist. In addition, the upper air owns a higher temperature because of the hot air rising and more heat sources. The hotter air gradually develops towards the floor, forming obvious temperature stratification phenomenon. The temperature in the driver's head position reaches 59 °C. At 10:00 am, the airflow crosses the bucket to the rear compartment, while it turns into a flow recirculation in the located temperature layer at 11:00 am and noon. Our analysis shows that the temperature distribution is primarily affected by solar radiation and airflow itself. Due to the heat exchange and direct solar heating, the temperature rises as much as 30 °C for direct exposed cabin surfaces and 10 °C for shaded ones, respectively, which supports the claim that solar radiation is a necessity in the thermal environment simulation.

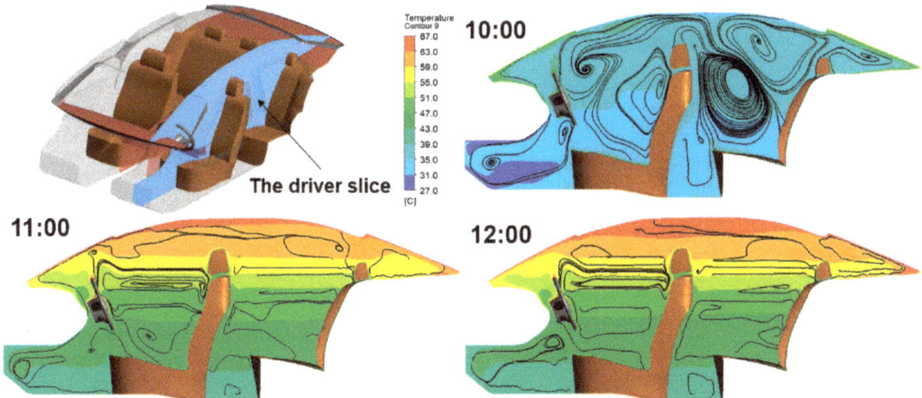

Figure 10. Temperature distributions of airflows with respect to the three time cases at driver plane.

Figure 11 shows the VOCs distributions in the whole cabin at 10:00 am and noon, respectively. And the driver plane was selected in the compartment to analyze the VOCs distributions under solar radiation in Figure 12. It is clearly noticed that the VOCs concentration at the hotter contaminant source is remarkably larger due to the strong dependence between the VOCs emission and the surface temperature. The dashboard and rear board are exposed to the strongest sunlight uniformly, therefore releasing more TVOC at the per unit area. The VOCs emission from the higher temperature area at the carpet increases approximately five-fold compared to the unexposed region. From the view of time, the concentration is about 3–4 times higher at noon than 10:00 am. Besides, the paths of concentration distribution and dispersion are significantly different in the four cases. A remarkable pollutant plume above the hotter surfaces is observed in Figure 12, which confirms that the near-wall thermal buoyancy flows are captured by the adopted turbulence and radiation models. The higher VOCs concentration distributes below the driver's knees, which is difficult to diffuse above the dashboard and rear board by natural convection (Figure 11a,b). Most TVOC is concentrated on the driver's head and above. As the concentration increases, a small amount of TVOC moves towards the floor under the driving of concentration difference. On the roof, the pollutants released from the dashboard and rear board tend to form a bridge of high concentration contaminants (Figure 11c,d). All pollutants gather near the surface sources and diffuse throughout the surroundings without ventilation. It demonstrates that thermal buoyancy and natural convection play an important role in dissipating pollutants from the contaminant source to the adjacent air in the enclosed environment.

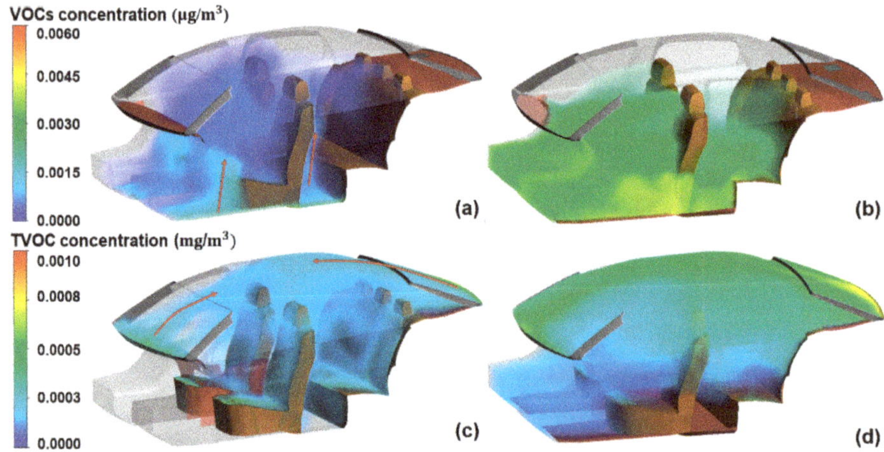

Figure 11. VOCs concentration emitted from carpet at 10:00 am (**a**) and noon (**b**); TVOC released from dashboard, rear board and seats at 10:00 am (**c**) and noon (**d**).

Table 6 lists the amount of pollutants released by materials from 10:00 am to noon, which was calculated based on the Equation (10). There is a total of 35.08 µg VOCs emitted from carpet and 19 mg TVOC released from other interiors, which far exceeds the national standards of many countries. The seats become the largest source of pollutants due to the larger surface area. Moreover, as illustrated in Figure 10, above the seats are the places where pollutants are most likely to accumulate. It will pose a great threat to health if the level of pollutant exposure cannot be mitigated effectively by natural ventilation or HVAC systems when drivers and passengers re-enter the car.

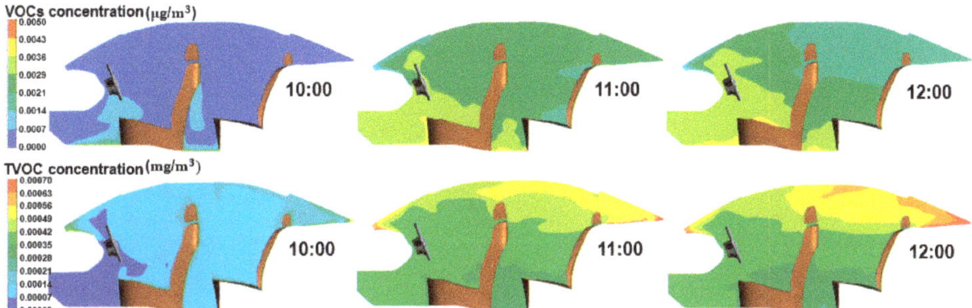

Figure 12. The pollutant distributions on the driver plane from 10:00 am to noon.

Table 6. The total pollutants mass (VOCs and TVOC) released from interior surfaces.

Parts	Area (m^2)	Pollutants Mass
floor	1.23	35.08 µg
dashboard	0.39	1.94 mg
rear board	0.68	3.40 mg
bucket seat 1	1.69	3.98 mg
bucket seat 2	1.69	3.80 mg
bench seat	2.50	5.86 mg

5. Conclusions

The in-vehicle VOCs exposure is a public health concern worldwide. This study explored the thermal environment of an in-vehicle cabin under solar radiation and provided an early effort to quantify the spatial distribution of VOCs released from interior surfaces. In this study, the in-cabin thermal environment was simulated by considering in-cabin natural convection-conduction coupled with solar radiation. The performance of different combinations of turbulence and radiation models was validated with measured temperature data in a reduced-scale airplane cabin. Combining the SST k-ω turbulence model with a surface-to-surface radiation model (S2S) performed best in terms of computational accuracy among 30 different combinations, including turbulence models such as the standard k-ε, RNG k-ε, realizable k-ε, standard k-ω, SST k-ω and LES model and radiation models such as the P-1, S2S, DO, DTRM model. Our findings suggest that solar radiation plays a critical role in determining the temperature distribution in the cabin, which can increase as much as 30 °C for direct exposed cabin surfaces and 10 °C for shaded ones. The maximum average temperature was observed over 60 °C on the dashboard and rear board that were exposed the direct sunlight. The lowest temperature was found on the cabin floor at 44 °C. Such a high cabin temperature profile considerably lowers the thermal comfort and promotes VOCs emissions due to a strong temperature dependence. The dispersion of VOCs strongly depended on the local emission rate and airflows. The dashboard and rear board were shown to have a larger emission rate than that of the seats and floor because of the higher surface temperature. The VOC plume from the seats rose upward towards the ceiling, whereas the VOCs from the floor stayed below the seats. From the 2-h simulation period from 10:00 am to noon, there was a total of 35.08 µg VOCs and 19.02 mg TVOC accumulated throughout the cabin. With increasing attention to the cabin environment and the health of drivers and passengers, the findings, such as modeled spatial distributions of VOCs, provide automakers with an important design guide that could improve the current ventilation design for the summer time.

Author Contributions: H.L.: conceptualization, methodology, and writing—original draft preparation. Z.T.: methodology, project administration and writing—review and editing. All authors have read and agreed to the published version of the manuscript.

Funding: This work was supported by the National Natural Science Foundation of China (51708493); Zhejiang Provincial Natural Science Foundation (LR19E050002); the Zhejiang Province Key Science and Technology Project (2020C01111);

Conflicts of Interest: The authors declare no conflict of interest.

Nomenclature

q_r	Radiation heat flux
a	Absorption coefficient
σ_s	Scattering coefficient
C	Linear-anisotropic phase function coefficient
G	Incident radiation
σ	Stefan–Boltzmann constant, $5.672 \times 10^{-8} \mathrm{W/m^2K^4}$
$q_{out,k}$	Energy flux leaving the surface
$q_{in,k}$	Incoming energy flux on the surface
ε_k	Surface emissivity
ρ_k	Surface reflectivity
T	Gas local temperature, K
T_k	Surface temperature, K
I	Radiation intensity
$\vec{r}$	Position vector
$\vec{s}$	Direction vector
$\vec{s'}$	Scattering direction vector
n	Refractive index
Φ	Phase function
Ω'	Solid angle
$E(t)$	Emission rate
D_m	Diffusion coefficient
C_0	Initial emittable concentration
δ	Material thickness, mm
t	Emission time, s
M	Contaminant quality
A	Surface area, $\mathrm{m^2}$
t_s	Release start time
t_e	Release end time
T_{Error}	Temperature error
$T_{isimulated}$	Simulated temperature of the i_{th} point, K
$T_{imeasured}$	Measured temperature of the i_{th} point, K
λ	Thermal conductivity, $\mathrm{W \cdot m^{-1} \cdot K^{-1}}$
c_p	Specific heat, $\mathrm{J \cdot kg^{-1} \cdot K^{-1}}$
ρ	Material Density, $\mathrm{kg \cdot m^{-3}}$
h	Convective heat transfer coefficient, $\mathrm{W/m^2K}$
v	Wind speed, m/s
ρ_{air}	Air density, $\mathrm{kg \cdot m^{-3}}$
S	Scalar
Γ	Molecular diffusivity, $\mathrm{m^2/s}$

References

1. Tong, Z.; Li, Y.; Westerdahl, D.; Adamkiewicz, G.; Spengler, J.D. Exploring the effects of ventilation practices in mitigating in-vehicle exposure to traffic-related air pollutants in China. *Environ. Int.* **2019**, *127*, 773–784. [CrossRef] [PubMed]
2. Tong, Z.; Li, Y.; Westerdahl, D.; Freeman, R.B. The impact of air filtration units on primary school students' indoor exposure to particulate matter in China. *Environ. Pollut.* **2020**, 115107. [CrossRef]

3. Tong, Z.; Whitlow, T.H.; MacRae, P.F.; Landers, A.J.; Harada, Y. Quantifying the effect of vegetation on near-road air quality using brief campaigns. *Environ Pollut* **2015**, *201*, 141–149. [CrossRef]

4. Tong, Z.; Chen, Y.; Malkawi, A.; Adamkiewicz, G.; Spengler, J.D. Quantifying the impact of traffic-related air pollution on the indoor air quality of a naturally ventilated building. *Environ. Int.* **2016**, *89*, 138–146. [CrossRef]

5. Zhang, G.; Li, T.; Luo, M.; Liu, J.; Liu, Z.; Bai, Y. Air pollution in the microenvironment of parked new cars. *Build. Environ.* **2008**, *43*, 315–319. [CrossRef]

6. Brodzik, K.; Faber, J.; Łomankiewicz, D.; Gołda-Kopek, A. In-vehicle VOCs composition of unconditioned, newly produced cars. *J. Environ. Sci.* **2014**, *26*, 1052–1061. [CrossRef]

7. You, K.; Ge, Y.; Hu, B.; Ning, Z.; Zhao, S.; Zhang, Y.; Xie, P. Measurement of in-vehicle volatile organic compounds under static conditions. *J. Environ. Sci.* **2007**, *19*, 1208–1213. [CrossRef]

8. Khanchi, A.; Hebbern, C.A.; Zhu, J.; Cakmak, S. Exposure to volatile organic compounds and associated health risks in windsor, Canada. *Atmos. Environ.* **2015**, *120*, 152–159. [CrossRef]

9. Borecki, M.; Prus, P.; Korwin-Pawlowski, M.L. Capillary Sensor with Disposable Optrode for Diesel Fuel Quality Testing. *Sensors* **2019**, *19*, 1980. [CrossRef]

10. Zulauf, N.; Dröge, J.; Klingelhöfer, D.; Braun, M.; Oremek, G.M.; Groneberg, D.A. Indoor Air Pollution in Cars: An Update on Novel Insights. *Int. J. Environ. Res. Public Health* **2019**, *16*, 2441. [CrossRef]

11. World Health Organization. WHO Guidelines for Indoor Air Quality: Selected Pollutants. Available online: https://www.euro.who.int/_data/assets/pdf_file/0009/128169/e94535.pdf (accessed on 30 June 2020).

12. Xu, B.; Wu, Y.; Yu, G.; Wu, S.; Wu, X.; Zhu, S.; Tao, L. Investigation of volatile organic compounds exposure inside vehicle cabins in China. *Atmos. Pollut. Res.* **2016**, *7*, 215–220. [CrossRef]

13. Faber, J.; Brodzik, K.; Da-Kopek, A.G.; Lomankiewicz, D. Benzene, toluene and xylenes levels in new and used vehicles of the same model. *J. Environ. Sci.* **2013**, *25*, 2324–2330. [CrossRef]

14. Yang, X.; Srebric, J.; Li, X.; He, G. Performance of three air distribution systems in VOC removal from an area source. *Build. Environ.* **2004**, *39*, 1289–1299. [CrossRef]

15. Kim, K.-H.; Szulejko, J.E.; Jo, H.-J.; Lee, M.-H.; Kim, Y.-H.; Kwon, E.; Ma, C.-J.; Kumar, P. Measurements of major VOCs released into the closed cabin environment of different automobiles under various engine and ventilation scenarios. *Environ. Pollut.* **2016**, *215*, 340–346. [CrossRef] [PubMed]

16. Cao, X.; Yao, Z.; Shen, X.; Yu, Y.; Xi, J. On-road emission characteristics of VOCs from light-duty gasoline vehicles in Beijing, China. *Atmos. Environ.* **2015**, *124*, 146–155. [CrossRef]

17. Wiglusz, R.; Sitko, E.; Nikel, G.; Jarnuszkiewicz, I.; Igielska, B. The effect of temperature on the emission of formaldehyde and volatile organic compounds (VOCs) from laminate flooring—Case study. *Build. Environ.* **2002**, *37*, 41–44. [CrossRef]

18. Xiong, J.; Zhang, Y. Impact of temperature on the initial emittable concentration of formaldehyde in building materials: Experimental observation. *Indoor Air* **2010**, *20*, 523–529. [CrossRef] [PubMed]

19. Xiong, J.; Zhang, P.; Huang, S.; Zhang, Y. Comprehensive influence of environmental factors on the emission rate of formaldehyde and VOCs in building materials: Correlation development and exposure assessment. *Environ. Res.* **2016**, *151*, 734–741. [CrossRef]

20. Yoshida, T.; Matsunaga, I. A case study on identification of airborne organic compounds and time courses of their concentrations in the cabin of a new car for private use. *Environ. Int.* **2006**, *32*, 58–79. [CrossRef]

21. Geiss, O.; Tirendi, S.; Barrero-Moreno, J.; Kotzias, D. Investigation of volatile organic compounds and phthalates present in the cabin air of used private cars. *Environ. Int.* **2009**, *35*, 1188–1195. [CrossRef]

22. Faber, J.; Brodzik, K.; Łomankiewicz, D.; Gołda-Kopek, A.; Nowak, J.; Świątek, A. Temperature influence on air quality inside cabin of conditioned car. *Siln. Spalinowe* **2012**, *51*, 49–56.

23. Chen, X.; Feng, L.; Luo, H.; Cheng, H. Analyses on influencing factors of airborne VOC S pollution in taxi cabins. *Environ. Sci. Pollut. Res.* **2014**, *21*, 12868–12882. [CrossRef] [PubMed]

24. Xiong, J.; Yang, T.; Tan, J.; Li, L.; Ge, Y. Characterization of VOC emission from materials in vehicular environment at varied temperatures: Correlation development and validation. *PLoS ONE* **2015**, *10*, e0140081. [CrossRef] [PubMed]

25. Huang, W.; Lv, M.; Yang, X. Long-term volatile organic compound emission rates in a new electric vehicle: Influence of temperature and vehicle age. *Build. Environ.* **2020**, *168*, 106465. [CrossRef]

26. Tong, Z.; Liu, H.; Tong, S.; Xu, J. Inverse Design for Thermal Environment and Energy Consumption of Vehicular Cabins with PSO–CFD Method. In *International Conference on Man-Machine-Environment System Engineering*; Springer Singapore: Singapore, 2020.

27. Tong, Z.; Li, Y. Real-Time Reconstruction of Contaminant Dispersion from Sparse Sensor Observations with Gappy POD Method. *Energies* **2020**, *13*, 1956. [CrossRef]

28. Chen, Y.; Tong, Z.; Zheng, Y.; Samuelson, H.; Norford, L. Transfer learning with deep neural networks for model predictive control of HVAC and natural ventilation in smart buildings. *J. Clean Prod.* **2020**, *254*, 119866. [CrossRef]

29. Wu, J.; Jiang, F.; Song, H.; Liu, C.; Lu, B. Analysis and validation of transient thermal model for automobile cabin. *Appl. Therm. Eng.* **2017**, *122*, 91–102. [CrossRef]

30. Lee, J.W.; Jang, E.Y.; Lee, S.H.; Ryou, H.S.; Choi, S.; Kim, Y.; Brodzik, K. Influence of the spectral solar radiation on the air flow and temperature distributions in a passenger compartment. *Int. J. Therm. Sci.* **2014**, *75*, 36–44. [CrossRef]

31. Mao, Y.; Wang, J.; Li, J. Experimental and numerical study of air flow and temperature variations in an electric vehicle cabin during cooling and heating. *Appl. Therm. Eng.* **2018**, *137*, 356–367. [CrossRef]

32. Torregrosa-Jaime, B.; Bjurling, F.; Corberán, J.M.; Di Sciullo, F.; Payá, J. Transient thermal model of a vehicle's cabin validated under variable ambient conditions. *Appl. Therm. Eng.* **2015**, *75*, 45–53. [CrossRef]

33. Zhang, H.; Dai, L.; Xu, G.; Li, Y.; Chen, W.; Tao, W.-Q. Studies of air-flow and temperature fields inside a passenger compartment for improving thermal comfort and saving energy. Part I: Test/numerical model and validation. *Appl. Therm. Eng.* **2009**, *29*, 2022–2027. [CrossRef]

34. Moon, J.H.; Jin, W.L.; Chan, H.J.; Lee, S.H. Thermal comfort analysis in a passenger compartment considering the solar radiation effect. *Int. J. Therm. Sci.* **2016**, *107*, 77–88. [CrossRef]

35. Zhou, X.; Lai, D.; Chen, Q. Experimental investigation of thermal comfort in a passenger car under driving conditions. *Build. Environ.* **2019**, *149*, 109–119. [CrossRef]

36. Marcos, D.; Pino, F.J.; Bordons, C.; Guerra, J.J. The development and validation of a thermal model for the cabin of a vehicle. *Appl. Therm. Eng.* **2014**, *66*, 646–656. [CrossRef]

37. Grundstein, A.; Meentemeyer, V.; Dowd, J. Maximum vehicle cabin temperatures under different meteorological conditions. *Int. J. Biometeorol.* **2009**, *53*, 255–261. [CrossRef] [PubMed]

38. Pan, H.; Qi, L.; Zhang, X.; Zhang, Z.; Salman, W.; Yuan, Y.; Wang, C. A portable renewable solar energy-powered cooling system based on wireless power transfer for a vehicle cabin. *Appl. Energy* **2017**, *195*, 334–343. [CrossRef]

39. Soulios, V.; Loonen, R.; Metavitsiadis, V.; Hensen, J. Computational performance analysis of overheating mitigation measures in parked vehicles. *Appl. Energy* **2018**, *231*, 635–644. [CrossRef]

40. Al-Kayiem, H.H.; Sidik, M.F.B.M.; Munusammy, Y.R.A.L. Study on the Thermal Accumulation and Distribution Inside a Parked Car Cabin. *Am. J. Appl. Sci.* **2010**, *7*, 784–789. [CrossRef]

41. Borecki, M.; Gęca, M.; Duk, M.; Korwin-Pawlowski, M.L. Miniature gas sensors heads and gas sensing devices for environmental working conditions—A review. *J. Electron. Commun. Eng. Res.* **2017**, *1*, 1–11.

42. Zhai, Z.J.; Zhang, Z.; Zhang, W.; Chen, Q.Y. Evaluation of various turbulence models in predicting airflow and turbulence in enclosed environments by CFD: Part 1–Summary of prevalent turbulence models. *Hvac&R Res.* **2007**, *13*, 853–870.

43. Hussain, S.; Oosthuizen, P.H.; Kalendar, A. Evaluation of various turbulence models for the prediction of the airflow and temperature distributions in atria. *Energy Build.* **2011**, *43*, 18–28. [CrossRef]

44. Li, X.; Tu, J. Evaluation of the eddy viscosity turbulence models for the simulation of convection–radiation coupled heat transfer in indoor environment. *Energy Build.* **2019**, *184*, 8–18. [CrossRef]

45. ASHRAE Handbook-Fundamentals (SI Edition). American society of Heating, Refrigerating and Air-Conditioning Engineers: Atlanta, GA, USA, 2009. Available online: http://shop.iccsafe.org/media/wysiwyg/material/8950P217-toc.pdf (accessed on 7 July 2020).

46. Xiong, J.; Wei, W.; Huang, S.; Zhang, Y. Association between the emission rate and temperature for chemical pollutants in building materials: General correlation and understanding. *Environ. Sci. Technol.* **2013**, *47*, 8540–8547. [CrossRef]

47. Huang, S.; Xiong, J.; Zhang, Y. Impact of temperature on the ratio of initial emittable concentration to total concentration for formaldehyde in building materials: Theoretical correlation and validation. *Environ. Sci. Technol.* **2015**, *49*, 1537–1544. [CrossRef] [PubMed]

48. Deng, Q.; Yang, X.; Zhang, J. Study on a new correlation between diffusion coefficient and temperature in porous building materials. *Atmos. Environ.* **2009**, *43*, 2080–2083. [CrossRef]

49. Li, D.F.; Xia, X.L.; Sun, C. Experimental investigation of transient thermal behavior of an airship under different solar radiation and airflow conditions. *Adv. Space Res.* **2014**, *53*, 862–869. [CrossRef]

50. Zhang, Z.; Zhang, W.; Zhai, Z.J.; Chen, Q.Y. Evaluation of various turbulence models in predicting airflow and turbulence in enclosed environments by CFD: Part 2—Comparison with experimental data from literature. *Hvac& R Res.* **2007**, *13*, 871–886.

51. Li, X.; Yan, Y.; Tu, J. Effects of surface radiation on gaseous contaminants emission and dispersion in indoor environment—A numerical study. *Int. J. Heat Mass Transf.* **2019**, *131*, 854–862. [CrossRef]

52. Yang, T.; Zhang, P.; Xu, B.; Xiong, J. Predicting VOC emissions from materials in vehicle cabins: Determination of the key parameters and the influence of environmental factors. *Int. J. Heat Mass Transf.* **2017**, *110*, 671–679. [CrossRef]

53. Kim, K.W.; Lee, B.H.; Kim, S.; Kim, H.J.; Yun, J.H.; Yoo, S.E.; Sohn, J.R. Reduction of VOC emission from natural flours filled biodegradable bio-composites for automobile interior. *J. Hazard Mater.* **2011**, *187*, 37–43. [CrossRef]

54. Qian, K.; Zhang, Y.; Little, J.C.; Wang, X. Dimensionless correlations to predict VOC emissions from dry building materials. *Atmos. Environ.* **2007**, *41*, 352–359. [CrossRef]

55. ANSYS®. *Academic Research Release 17.2, Help System, Coupled Field Analysis Guide*; ANSYS, Inc.: Canonsburg, PA, USA, 2017.

56. Lahimer, A.; Alghoul, M.; Sopian, K.; Khrit, N. Potential of solar reflective cover on regulating the car cabin conditions and fuel consumption. *Appl. Therm. Eng.* **2018**, *143*, 59–71. [CrossRef]

57. Mezrhab, A.; Bouzidi, M. Computation of thermal comfort inside a passenger car compartment. *Appl. Therm. Eng.* **2006**, *26*, 1697–1704. [CrossRef]

58. Liang, B.; Yu, X.; Mi, H.; Liu, D.; Huang, Q.; Tian, M. Health risk assessment and source apportionment of VOCs inside new vehicle cabins: A case study from Chongqing, China. *Atmos. Pollut. Res.* **2019**, *10*, 1677–1684. [CrossRef]

59. Wu, W.; Yoon, N.; Tong, Z.; Chen, Y.; Lv, Y.; Ærenlund, T.; Benner, J. Diffuse ceiling ventilation for buildings: A review of fundamental theories and research methodologies. *J. Clean Prod.* **2019**, *211*, 1600–1619. [CrossRef]

Article

Experimental Validation of Water Flow Glazing: Transient Response in Real Test Rooms

Belen Moreno Santamaria [1], Fernando del Ama Gonzalo [2,*], Benito Lauret Aguirregabiria [1] and Juan A. Hernandez Ramos [3]

[1] Department of Construction and Architectural Technology, Technical School of Architecture of Madrid, Technical University of Madrid (UPM), Av. Juan de Herrera, 4, 28040 Madrid, Spain; belen.moreno@upm.es (B.M.S.); benito.lauret@upm.es (B.L.A.)

[2] Department of Sustainable Product Design and Architecture, Keene State College, 229 Main St, Keene, NH 03435, USA

[3] Department of Applied Mathematics, School of Aeronautical and Space Engineering, Technical University of Madrid (UPM), Plaza Cardenal Cisneros 3, 28040 Madrid, Spain; juanantonio.hernandez@upm.es

* Correspondence: fernando.delama@keene.edu

Received: 16 June 2020; Accepted: 14 July 2020; Published: 16 July 2020

Abstract: The extensive use of glass in modern architecture has increased the heating and cooling loads in buildings. Recent studies have presented water flow glazing (WFG) envelopes as an alternative building energy management system to reduce energy consumption and improve thermal comfort in buildings. Currently, commercial software for thermal simulation does not include WFG as a façade material. This article aims to validate a new building simulation tool developed by the authors. Simulation results were compared with real data from a scale prototype composed of two twin cabins with different glazing envelopes: a Reference double glazing with solar-control coating and a triple water flow glazing. The results showed a good agreement between the simulation and the real data from the prototype. The mean percentage error of the indoor temperature cabin was lower than 5.5% and 3.2% in the WFG cabin and in the Reference glazing one, respectively. The indoor air temperature of the WFG cabin was 5 °C lower than the Reference one in a free-floating temperature regime when the outdoor air temperature was 35 °C and the maximum value of solar radiation was above 700 W/m^2. WFG has energy-saving potential and is worthy of further research into the standardization of its manufacturing process and its ability to increase building occupants' comfort.

Keywords: building energy simulation; water flow glazing; experimental validation

1. Introduction

Energy consumption in buildings shows a growing trend worldwide and is of primary concern for the world population [1]. Over the last decade, nearly 60% of total net electricity consumption in Organization for Economic Co-operation and Development (OECD) economies, was in the building sector, both residential and commercial [2]. The residential building sector is responsible for more than half of the electricity consumption in developing countries [3].

In the frame of the Paris agreement in 2015, 195 countries adopted 17 sustainable development goals (SDGs) as the outcome of the UN's inclusive and comprehensive negotiations in the frame of the 2030 agenda [4]. The seventh goal states that using clean and sustainable energy sources is an opportunity to transform economies and lives, especially in developing countries.

Annual power consumption depends on the use of the building, construction year, number of floors, building structure, and building location [5]. When it comes to heating and cooling consumption, the heating, ventilation, and air conditioning (HVAC) system, exterior walls, and glazing are the essential elements [6,7]. Building energy management systems (BEMS) and energy-saving measures are aimed at reducing buildings' energy requirements for heating and cooling [8–10].

In countries with a hot, humid climate, the excessive use of inefficient cooling systems leads to an increase in electricity consumption and causes pollution [11–13]. The energy performance of a building also depends on the solar radiation and the correlation between cooling/heating loads and the colors of surfaces [14].

In hot climate areas, the glazing solar heat gain coefficient (SHGC) must be low, and it is more relevant than the U-value because solar radiation causes the most significant part of the cooling load [15]. In cold climate areas, the goal is to reduce the need for heating energy, making the most of solar radiation [16,17]. Heating, ventilation, and air conditioning (HVAC) systems have to be efficient in providing users with a healthy environment. When fossil fuels and oil resources run out, solar energy and other renewable sources are alternatives to overcome the clean-energy demand growth [18,19]. The annual solar irradiation ranges between 100–200 W/m^2 as an average in Mediterranean countries, so the potential of solar energy is more than enough to provide as much energy as the building consumes [20].

Solar energy is aligned with the concept of "Regenerative Design." It implies a proactive attitude of the building beyond the traditional sustainable design practice. Regenerative buildings reduce their energy consumption to zero, and can recollect, generate, and distribute renewable resources [21]. Glass is a fundamental element in the design of regenerative buildings. Still, its extensive use has increased the heating and cooling loads. Using transparent materials requires understanding their spectral properties and developing systems to solve some of the issues regarding heat gain, heat loss, and daylight [22,23]. Accurate prediction of the performance of glazing facades has to include a thorough analysis of thermal and spectral properties that depend on the glass, spacers, coatings, and gas fillings. Solar control layers reflect and filter solar radiation, and low emissivity coatings reduce the emissivity of the glass and retain the heat charge inside [24]. Acting in the chamber can improve the insulation capacity of the double-glazed windows [25]. The chamber can also be filled with inert gas, or vacuumed, to reduce the transmittance in large glazed surfaces [26]. Thermochromic and electrochromic glazing vary in color and transparency as a reaction to light and heat excess [27]. Double-pane windows, in which the exterior photovoltaic pane produces electricity, can be designed and manufactured today [28,29]. Double-pane windows can also be developed with circulating water through the chamber, instead of inert gas, allowing the water to absorb the heat of direct and diffused solar radiation [30].

The use of the building, the orientation of the facade, and the location of the project are relevant inputs to determine the glazing composition [31]. The Fourier model does not predict variations in thermal properties as a function of time [32]. Water flow glazing (WFG) facades are considered dynamic envelopes able to react or adapt to the building's external and internal conditions. Most of the simulation engines do not include dynamic properties, so developing new tools to calculate the impact of WFG has become a goal of researchers [33,34]. Water is opaque to the near-infrared (NIR) spectrum of light, while its visible transmittance is very high [35].

Complicated simulation engines provide the designer with multiple options, and sometimes they are not useful at an early design stage because decisions have not been made yet. Architects might find better support in simple energy simulation tools than in complicated ones [35,36]. Building information modeling (BIM) has the potential to achieve performance improvements and high-quality construction, and architecture, engineering, and construction (AEC) industries have applied BIM in construction projects over the last few decades [37]. One of the features of BIM is the energy analysis of buildings. It makes the most of a friendly interface that has been tested over decades of experience by many users. However, users have identified gaps between the expected building energy consumption and the actual measured performance [38–40]. The causes of these gaps are diverse, including behavioral habits of occupants and construction flaws [41]. The evaluation of the actual thermal properties of the building stock from monitored data is widely considered advantageous compared to tabulated data to improve the overall quality of the building process by feeding back the measured data [42].

The steady-state model is not a reliable means to analyze dynamic forms of heat transfer. Temperature, solar radiation, occupancy, and HVAC systems affect the transient state of the building envelopes. Those parameters are time-dependent and non-linear. Remote sensing systems have become indispensable in comparing the actual energy performance with simulation models and understanding the dynamic heating and cooling loads [43–45]. Cooling has represented a small share of the final energy use in buildings, but demand has been rising over the last decade [46,47]. This article considers the best available technologies (BAT) for cooling, which are innovative and economically viable [48]. The energy efficiency ratio (EER) is the parameter that measures the efficiency of cooling systems. Hydronic technologies, such as water-to-water heat pumps, are compatible with WFG and radiant floors and walls. WFG can improve the performance coefficient of cooling systems by increasing the indoor comfort temperature and the inlet temperature of the fluid through the glazing [49]. The technology of WFG has been studied in previous scientific articles. Some papers have studied the physical structure and energy performance of WFG in cooling-demand climates through numerical computation [50,51]. Recent research studied the performance of WFG compared with conventional double glazing with low-emissivity coatings. Dynamic simulation has been used to evaluate different options of glazing, and the presented simulation results concluded that improving SHGC is more efficient for thermal performance than improving the U-value [52]. Other papers have validated the numerical simulations using test prototypes. The dimensions of the tested devices varied depending on the goals of the research. Cubic boxes measuring $60 \times 60 \times 60$ cm, with one side open, have been used to place different glass panes [53]. If the goal was to validate the performance of WFG, the prototype was designed as an adiabatic box, with high thermal insulation in the opaque walls with U values below 0.1. Other tests focused on analyzing the influence of coatings applied to the indoor surface and the heat gains by measuring the water flow rate and the inlet/outlet temperatures of WFG. These test facilities were slightly bigger (the length was 1.55 m, the width, 0.9 m, and the height, 0.9 m). In this case, the insulation of opaque walls was not relevant, and the indoor air temperature was set to 24 °C by a direct expansion cooling coil with an electrical heater [54]. The authors of the present paper have developed a set of equations to take into account the influence of multiple diffuse reflections, direct reflections between glazing and parallel surfaces, indirect reflections between the glazing, parallel surfaces, and perpendicular surfaces. These equations have been included in the simulation tool tested in the present article [55]. The simulation of the indoor air temperature and the water absorption in a transient state affected by changes in temperature and solar radiation was relevant when the test facility was bigger. In these cases, validating simulation tools with real data was essential in predicting thermal behavior and the fluctuations of indoor air temperature [56,57]. This paper aims to investigate the dynamic thermal parameters of WFG. The influence of WFG as a means of energy management was tested by comparing the indoor temperatures of two prototypes. The empirical tests under variable weather conditions and have been carried out over two years. There are three objectives in the analysis of the prototype. First, it allows for the comparison of the indoor temperatures of the WFG cabin and the Reference cabin. Second, the simulation tool based on the mathematical model to predict the performance of WFG was validated using real data. Finally, it aims to study the improvement of a water-to-water heat pump's performance by reducing the temperature gap between the water and the indoor air. Two cases have been tested. In the first case, the water was flowing without controlling its temperature. In the second case, there was a source of energy that provided the desired boundary conditions. The thermal performances of the WFG cabin and the Reference cabin have been recorded using a proper monitoring system. Different boundary conditions based on real data are given to the mathematical models to carry out the simulation.

2. Materials and Methods

This section aims to provide a simplified model that helps designers to analyze the energy strategy of a dynamic envelope. Commercial BES tools do not include WFG as an option, so it is necessary to validate the hypothesis with data from real prototypes. The first subsection set the criteria to select the spectral and thermal parameters of the WFG. The second subsection described geometry, energy management, and the materials used in the prototype.

2.1. Simplified Model of Triple WFG

Water flow glazing (WFG) allows the flow of water between two glass panes. Water captures the solar NIR and increases its temperature through the window. The flow of the water enables the homogenization of the building envelope temperature so that designers can apply energy-saving strategies, such as energy storage or solar energy rejection. Table 1 shows the list of symbols that have been used in equations.

Table 1. List of symbols.

Symbol	Meaning
A_j	Absorptance of glass layers.
A_w	Absorptance of water.
A_v	Total absorptance of water flow glazing.
h_i	Interior heat transfer coefficient (W/m^{2}K)
h_w	Water heat transfer coefficient (W/m^{2}K)
h_g	Air chamber heat transfer coefficient (W/m^{2}K)
h_e	Exterior heat transfer coefficient (W/m^{2}K)
q_j	Heat fluxes through the different layers of the glazing
i_0	Normal incident solar irradiance (W/m^2)
g^{OFF}	Solar heat gain coefficient without flow rate.
g^{ON}	Solar heat gain coefficient with high flow rate.
θ_i	Interior temperature (K)
θ_e	Exterior temperature (K)
θ_j	Temperature of the glass layer (K)
θ_{IN}	Inlet temperature of the water chamber (K)
θ_{OUT}	Outlet temperature of the water chamber (K)
θ_w	Temperature of the water (K)
$\theta_{STAGNATION}$	Temperature of the water when $\dot{m} = 0$ (K)
U	Thermal transmittance (W/m^{2}K)
U_i	Interior thermal transmittance (W/m^{2}K)
U_e	Exterior thermal transmittance (W/m^{2}K)
U_w	Thermal transmittance (water chamber–interior) (K)
T	Transmittance of the glazing
R	Reflectance of the glazing
$\dot{m}$	Mass flow rate (kg/s m^2)
c	Specific heat of the water (J/Kg K)
P	Heat gain in the water chamber (W)

Figure 1 shows the heat flux and the temperature distribution when the outdoor temperature (θ_e) is higher than the indoor (θ_i) through a triple WFG. The thermal transmittance, U, is the parameter that explains the heat transfer. The European Standard determines its value and a calculation method [58,59].

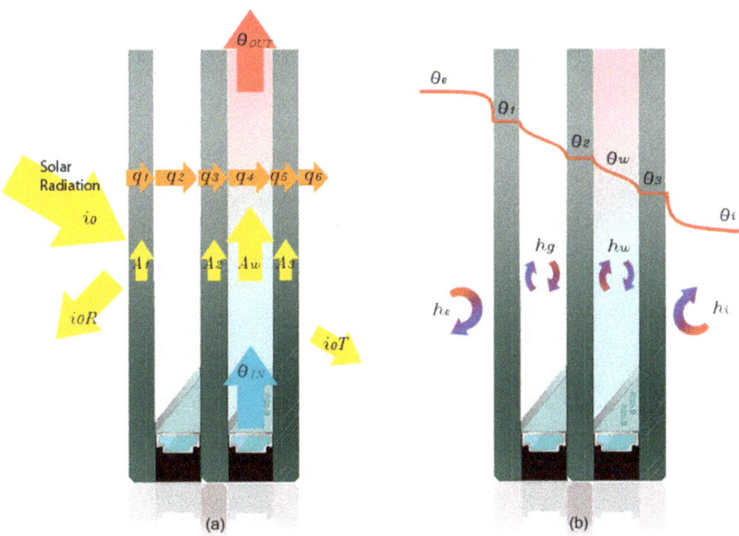

Figure 1. (**a**) Heat fluxes through a triple water flow glazing (WFG). Solar radiation and absorptance of layers, A_1, A_2, A_w, and A_3. (**b**) Temperature distribution in a triple WFG at a specific height, with heat transfer coefficients, h_i, h_w, h_g, and h_e.

Previous studies explained the thermal and spectral properties of WFG and its behavior [60,61]. This research used a simplified set of equations from those studies, along with data from commercial software, to assess the performance of the prototype defined in Section 2.2. Equations (1) to (10) show the heat fluxes through the different layers of the glazing. They consider the energy balance at each layer and the Newton's definition of heat flux.

$$q_1 = h_e(\theta_e - \theta_1), \tag{1}$$

$$q_2 = q_1 + A_1 i_0, \tag{2}$$

$$q_2 = h_g(\theta_1 - \theta_2), \tag{3}$$

$$q_3 = q_2, \tag{4}$$

$$q_4 = h_w(\theta_2 - \theta_w), \tag{5}$$

$$q_4 = q_3 + A_2 i_0, \tag{6}$$

$$q_5 = h_w(\theta_w - \theta_3), \tag{7}$$

$$q_5 = q_4 + A_w i_0 + \dot{m}c(\theta_{IN} - \theta_w), \tag{8}$$

$$q_6 = h_i(\theta_3 - \theta_i), \tag{9}$$

$$q_6 = q_5 + A_3 i_0. \tag{10}$$

The U values depend on the interior heat transfer coefficient, h_i, the exterior heat transfer coefficient, h_e, the air chamber heat transfer coefficient, h_g, and the water heat transfer coefficient.

$$\frac{1}{U_e} = \frac{1}{h_e} + \frac{1}{h_g} + \frac{1}{h_w}, \tag{11}$$

$$\frac{1}{U_i} = \frac{1}{h_i} + \frac{1}{h_w}. \tag{12}$$

A_1 is the absorptance of the exterior glass pane, A_2, is the absorptance of the middle glass pane, and A_3 is the absorptance of the interior one. A_w is the absorptance of the water chamber. The absorptance, A_v, depends on the energy absorbed by the glass panes and by the water:

$$A_v = A_1\left(\frac{U_e}{h_e}\right) + A_2\left(\frac{1}{h_g} + \frac{1}{h_e}\right)U_e + A_3\left(\frac{U_e}{h_i}\right) + A_w. \tag{13}$$

Solving the Equations (1) to (10) and using the values of Equations (11) to (13):

$$\theta_{OUT} = \frac{i_0 A_v + U_i\theta_i + U_e\theta_e + \dot{m}c\theta_{IN}}{\dot{m}c + U_e + U_i}. \tag{14}$$

Water heat gain is a power magnitude, and it is measured in watts (W). Equation (2) shows the analytical expression.

$$P = \dot{m}c(\theta_{OUT} - \theta_{IN}), \tag{15}$$

where P is the power absorbed by the water, θ_{OUT} and θ_{IN} the temperature of water leaving and entering the glazing, respectively, $\dot{m}$ is the mass flow rate, and c is the specific heat of the water. The mass flow rate is a measurement of the amount of mass passing by a point over time. The goal of absorbing the same power, P, can be achieved with a high $\dot{m}$, which results in a low-temperature increase or a low $\dot{m}$, which results in a high-temperature difference between the inlet and outlet. Equation (16) results by combining Equations (14) and (15).

$$P = \frac{\dot{m}c}{\dot{m}c + U_e + U_i}(i_0 A_v + U_i(\theta_i - \theta_{IN}) + U_e(\theta_e - \theta_{IN})), \tag{16}$$

where θ_e and θ_i are, respectively, the temperature outdoors and indoors of the room. Analyzing the Equation (16), the power absorbed by the water varies with θ_{IN}. Considering the rest of the equation as a constant, P_0, P linearly decreases with slope ($U_i + U_e$). Figure 2 shows that at a specific value of $\dot{m}$, there is a maximum absorbed power, P, depending on θ_{IN}.

$$P = P_0 - \theta_{IN}(U_e + U_i). \tag{17}$$

If boundary conditions do not change with time, the solution becomes constant when the system reaches a steady state. In this set of test cases, sunrays are perpendicular to the glazing. This assumption eliminates uncertainties associated with the dependence of each layer's absorptance with the angle of incidence. Two case studies are considered: (a) and (b). Table 2 defines the outdoor and indoor boundary conditions. Case (a) studies the influence of θ_{IN} with a fixed absorptance, and case (b) considers the impact of the absorptance when θ_{IN} is fixed.

Table 2. Outdoor and indoor boundary conditions for WFG steady thermal performance.

Glazing	i_0 (W/m^2)	U_e (W/m^{2}K)	U_i (W/m^{2}K)	c (J/Kg °C)	θ_i (°C)	θ_e (°C)	θ_{IN} (°C)	A_v
	800	1.08	6.89	3600	25	30	15	0.5
Case (a)	800	1.08	6.89	3600	25	30	20	0.5
	800	1.08	6.89	3600	25	30	25	0.5
	800	1.08	6.89	3600	25	30	20	0.4
Case (b)	800	1.08	6.89	3600	25	30	20	0.5
	800	1.08	6.89	3600	25	30	20	0.6

Figure 2 illustrates the power variations with water flow rate, $\dot{m}$. There is a maximum water heat gain when $\dot{m}c \gg U_e + U_i$. The maximum power absorption occurs when A_v is high and θ_{IN} is low.

If the goal is to reject the solar energy, A_v must be as low as possible, with solar control coatings in the outermost glass panes. In this case ($\dot{m}c >> Ue + Ui$), the absorbed power (P) is the maximum power (P_{max})

$$P_{max} = i_0 A_v + U_i(\theta_i - \theta_{IN}) + U_e(\theta_e - \theta_{IN}),\tag{18}$$

$$P_{max} = i_0 A_v + U_i\theta_i + U_e\theta_e - \theta_{IN}(U_i + U_e).\tag{19}$$

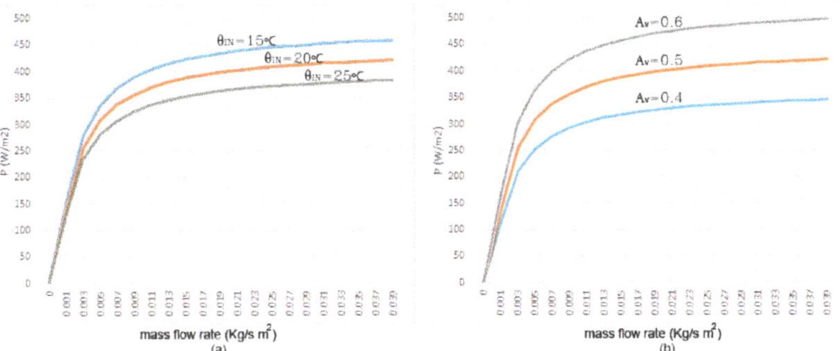

Figure 2. Power absorbed by the water chamber of the WFG: (**a**) constant A_v at different inlet temperatures (θ_{IN}); (**b**) constant inlet temperature (θ_{IN}) with different A_v.

Combining Equations (14) and (19), the value of θ_{OUT} is:

$$\theta_{OUT} = \theta_{IN} + \frac{P_{max}}{\dot{m}c + U_e + U_i}.\tag{20}$$

$\theta_{STAGNATION}$ is the temperature of the water when the mass flow rate is stopped. Figure 3 illustrates the relationship between θ_{OUT} and $\dot{m}$ in two cases: (a) and (b). Case (a) shows that $\theta_{STAGNATION}$ is the same if the water absorption A_v does not change; case (b) shows the variations of $\theta_{STAGNATION}$ with different conditions of A_v. When $\dot{m}$ is close to zero, then θ_{OUT} gets the maximum value, which is $\theta_{STAGNATION}$. When $\dot{m}$ is very high, then $\theta_{OUT} = \theta_{IN}$.

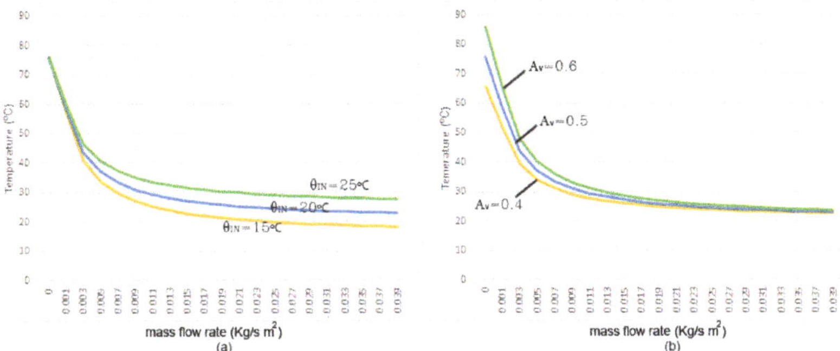

Figure 3. Outlet temperature (θ_{OUT}) of the WFG: (**a**) constant A_v at different inlet temperatures; (**b**) constant inlet temperature (θ_{IN}) with different A_v.

Equation (21) shows the total heat flux, q:

$$q = gi_0 + U(\theta_e - \theta_i) - U_w(\theta_i - \theta_{IN}),\tag{21}$$

where g describes the proportion of solar energy transmitted indoors, $U(\theta_e-\theta_i)$ expresses the heat exchange between the room and outdoors, and U_w $(\theta_i-\theta_{IN})$ represents the heat exchange between the water chamber and indoors. As per Equations (34) and (35) in [50]:

$$U_w = \frac{U_i \dot{m}c}{\dot{m}c + U_e + U_i}, \tag{22}$$

$$U = \frac{U_e U_i}{\dot{m}c + U_e + U_i}. \tag{23}$$

Since the WFG is a dynamic envelope, the g factor depends on the flow rate. When the water flows, the g factor decreases, and when the water flow stops, the solar energy enters the building, and the g factor increases. The thermal transmittance, U, is almost zero when the flow rate is the design flow rate because the water chamber isolates the building from outdoor conditions. When the water flow stops, the thermal transmittance depends mainly on the components that make up an insulated glass unit: the glass panes, coatings, spacer, sealing, and the gas filling the sealed space.

2.2. Description of the Prototype

A prototype has been built to compare the thermal performance of WFG with a double-glazing solution. It is located near Madrid, Spain (latitude 41°22′6″ N, longitude 3°29′57″ W, altitude 1037 MAMSL). Conceptually, this mock-up is a mobile and autonomous prototype consisting of two cabins named WFG and Reference. The prototype design allows both the primary and secondary circuits to be housed within the demonstrator structure, although in independent sectors. Besides, the mock-up has four wheels at the bottom of the structure for easy transport and orientation.

The prototype has two levels: the lower level of 500 mm high and the upper level of 1000 mm high. The lower level holds the primary circuit, which is composed of a "Peltier" unit, and a lithium battery that feeds it. The upper level includes the two cabins. Finally, a photovoltaic panel has been installed on the top roof allowing to store electric energy in the battery for the operation of the cooling device. A circulating device is made up of a 10 W solar pump. The primary circuit connects the circulating system with a "Peltier" device. The secondary circuit goes from the circulating system to the WFG, with a design flow rate of 0.5 L/min. Figure 4 illustrates a schematic of the prototype with its main components, and Figure 5 shows the layout of the prototype. It consists of two semi-detached insulated cabins, one with WFG facing South and the other with a Reference glazing in the same orientation. The dimensions of the windows are 1 m × 0.5 m.

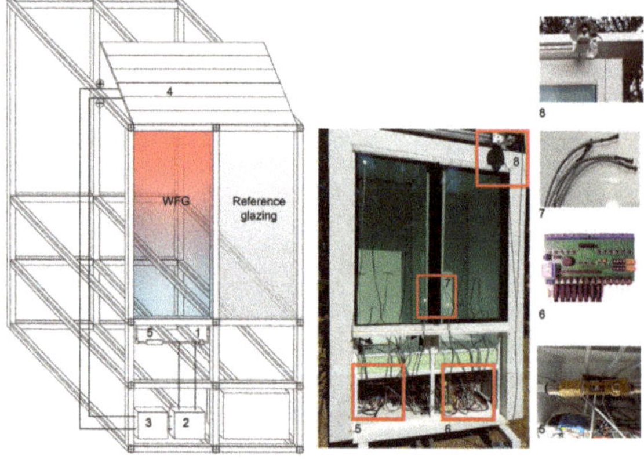

Figure 4. Schematics of the prototype. 1. Solar Pump 10 W, 2. "Peltier" Thermo-Electric Liquid Cooler 184 W, 24 V, 3. Battery 12 V (2 units serial connection), 4. PV Panel Polycrystalline cells. 236 W 24 V, 5. Flow meter, 6. Control unit, 7. Temperature sensors, 8. Pyranometer.

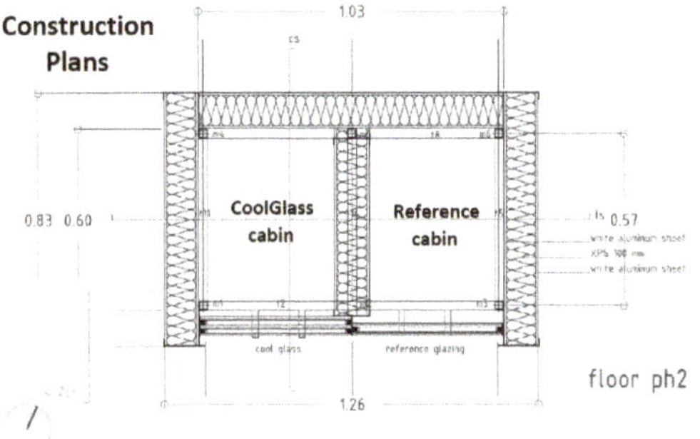

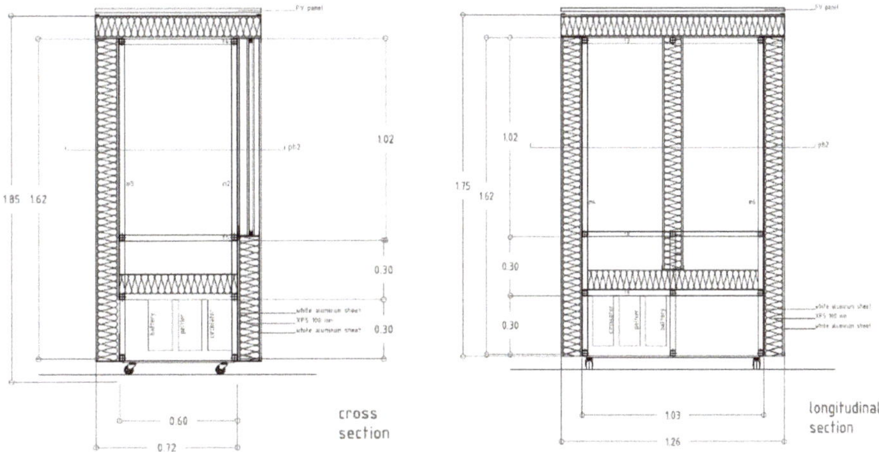

Figure 5. Construction plans of the prototype: Plan, cross, and long sections.

Regarding the construction materials, the prototype is made up of a steel structure formed by welded tubular profiles. Furthermore, for the cladding of the opaque parts, a white aluminum sandwich panel with 100 mm of XPS has been selected. Hence, for the glazed facade, a high selective double glazing has been chosen for the Reference cabin and a highly selective triple glazing with a water chamber towards the inside, for the WFG cabin. Figure 6 shows the configuration of both glazing facades.

The prototype has undergone an exhaustive commissioning process from the design stage to the manufacturing, factory assembly, on-site assembly, and commissioning of all the systems involved. Hence, high reflective WFG is chosen to reject sun energy and use the water chamber of the glazing facing indoors to eliminate heat loads when required. Since the energy absorption should be minimized, a high reflective coating (Xtreme 60.28) is positioned close to the outer glass pane (face 2). This coating allows for the reduction of the U-value because it can be considered a low emissivity coating at the same time (Planitherm XN). The maximum g value (g^{OFF}) and the minimum g value (g^{ON}) are almost the same and around 0.2. Table 3 shows the spectral and thermal properties of the glazing defined in Figure 6. WFG presents different values of U and g, depending on the mass flow rate. U^{ON} and U^{OFF} have been calculated with the equation (23) using $\dot{m} = 1$ L/min m^2 and $\dot{m} = 0$ L/min m^2, respectively. The steady values of the reference glazing have been placed on the columns U^{OFF} and g^{OFF}.

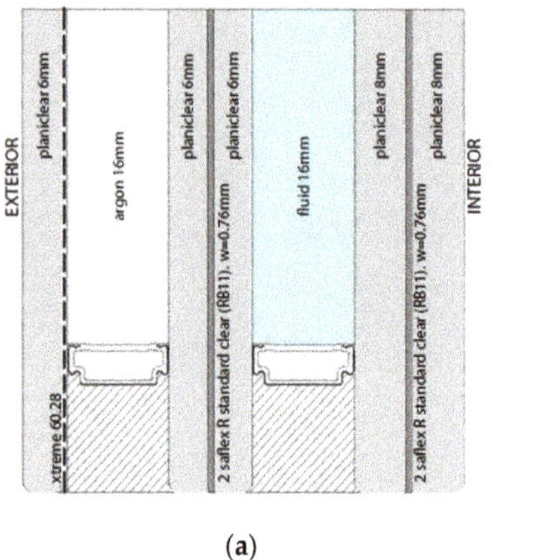
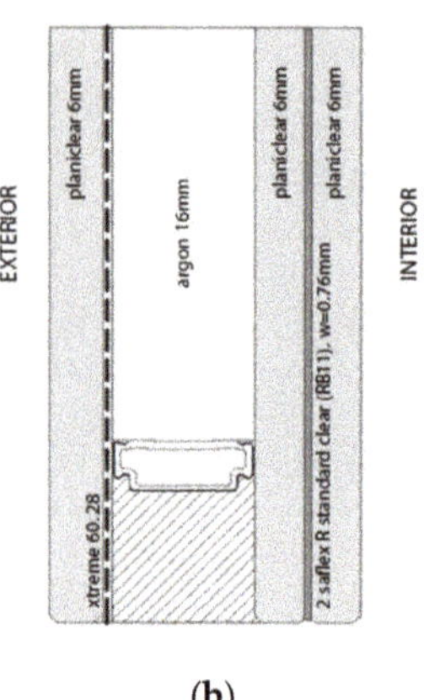

(a)   (b)

Figure 6. Glass configuration for the Spanish mock-up. (**a**) Water Flow Glazing (WFG), (**b**) Reference glazing.

Table 3. Thermal and spectral properties of the reference glazing and the WFG.

	R	T	U^{ON} (W/m²K)	U^{OFF} (W/m²K)	g_{ON}	g_{OFF}
Reference	0.433	0.252	-	1.017	-	0.27
WFG	0.433	0.215	0.061 [1]	0.977 [2]	0.22	0.26

[1] Equation (23) with $\dot{m}$ = 1 L/min m²; [2] Equation (23) with $\dot{m}$ = 0 L/min m².

The WFG was selected to eliminate internal heat loads by circulating cool water through the water chamber facing indoors. This cool isothermal envelope allows insulating the building from external boundary conditions. Furthermore, the "Peltier" device connected to a buffer tank produces cool water for the glazing. To minimize the electrical consumption of the "Peltier" device, the temperature of the buffer tank should be close to the cool water of the glazing. When the outdoor conditions made it possible, evaporative cooling and cooling by night were considered to cool down the buffer tank. Besides, the prototype allows the understanding of the real issues of glass facades, analyzing deep concepts such as overheating and thermal mass.

Figure 7 shows the prototype and the position of sensors. A short description of the monitoring devices is listed below:

- Pyranometer: The Delta Ohm LP PYRA 03 Pyranometer measures the irradiance on a horizontal surface (W/m²). The measured irradiance is the sum of the direct irradiance of the sun and the diffuse irradiance.
- One wire temperature probe: The DS18B20-PAR uses Dallas' exclusive 1-Wire bus protocol that implements bus communication using one control signal.
- Flow meter: TacoSetter Inline 100 Potermic: Balancing valve made of brass with female thread 3/4" × 1/2". Measuring range 0.3–1.5 L/min), Kvs 0.25 (m³/h).

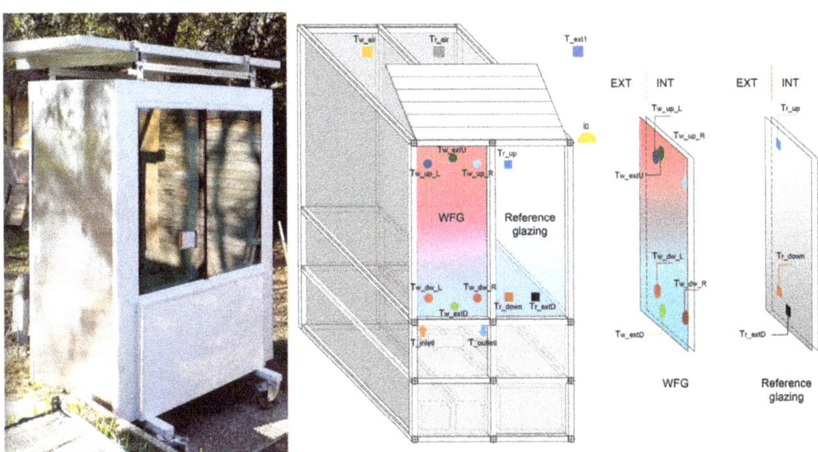

Figure 7. Set up of the Spanish mock-up and schematic of sensors distribution of the cabins. Sensors nomenclature and position for the WFG cabin and the Reference cabin.

3. Results

The empirical tests, conducted by the experimental setup, were run over the years 2018 and 2019 to collect data in all the possible weather conditions. Some logged parameters were used as input data to the simulation tool, and other measurements compared the performance of the selected glazing and its importance on the indoor temperature. Continuous experimental data were taken to reflect the influence of variable weather conditions. The experimental data logging time step was set at 1 min.

3.1. Analysis of the Prototype. Free-Floating Temperature Regime

In 2019, the "Peltier" device was switched off, in a free-floating temperature regime. Figure 8 presents the interior temperature of both cabins and the exterior temperature over the last week of April, May, June, July, September, and October of 2019. In all the cases, the interior temperature of the WFG cabin (T_{w_air}) was 5 °C below the internal temperature of the Reference cabin (T_{r_air}) and reached similar or slightly lower values compared to the outside temperature ($T_{\_ext1}$). When analyzing 27/04/2019, T_{w_air} almost reached 20 °C, while T_{r_air} was 25 °C. Likewise, during the night on that day, the minimum T_{w_air} remained at 7 °C, while T_{r_air} dropped to 3 °C when the outdoor temperature reached 0 °C. Furthermore, in July, the maximum values for T_{w_air} were below T_{r_air} and $T_{\_ext1}$. The minimum temperature inside the WFG cabin did not drop as much as that inside the Reference cabin. On 24/07/2019 the graph shows that the maximum T_{w_air} was 34 °C, while the T_{r_air} reached 40 °C, and $T_{\_ext1}$ remained at 38 °C. During the night, T_{w_air} reached 25 °C, while T_{r_air} was 21 °C, when $T_{\_ext1}$ dropped to 18 °C.

Besides, Figure 8 shows the curve of the indoor temperature of the WFG cabin as a damped wave shape compared to the interior temperature curve of the Reference cabin or the outside temperature. The temperature difference between day and night inside the WFG cabin did not exceed 10 °C in most cases. However, the temperature difference between day and night inside the Reference cabin was around 20 °C. Analyzing the same parameters on 24 July 2019, it can be observed that the difference between the maximum (34 °C) and the minimum (25 °C) temperature inside the WFG cabin was 9 °C, while the difference between the maximum (40 °C) and the minimum (21 °C) temperature inside the Reference cabin was 19 °C. However, the difference between the maximum (38 °C) and the minimum (18 °C) outside temperature was 20 °C. The interior temperature curve of the WFG cabin is flatter than the reference glazing one.

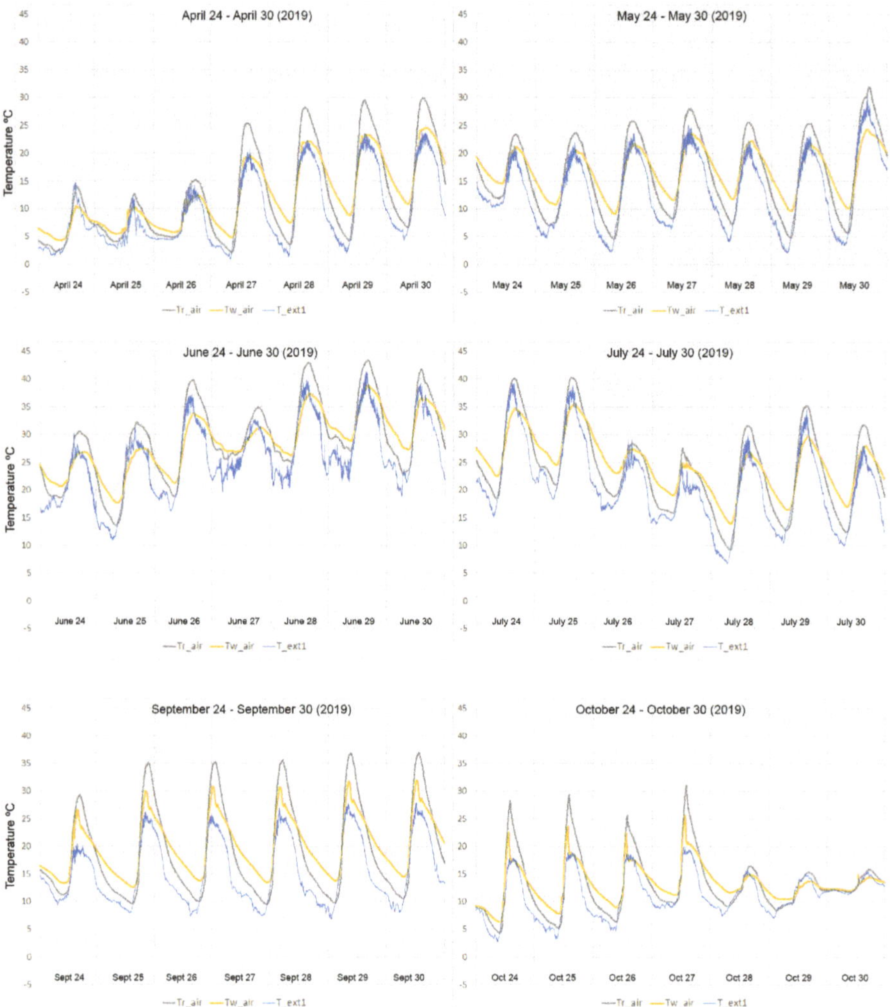

Figure 8. Comparison between indoor temperature of WFG cabin (T_{w_air}), indoor temperature of Reference cabin (T_{r_air}), and outdoor temperature ($T_{\_ext1}$). Free floating temperature regime.

Figure 9 presents the indoor temperature of both cabins (T_{r_air}, T_{w_air}), the outdoor temperature ($T_{\_ext1}$), and solar radiation from 24 July 2019 to 30 July 2019. The maximum value of solar radiation was above 720 W/m^2, almost every day. Despite these high values of solar radiation, the interior temperature of the WFG cabin remained under 35 °C on 24 July 2019, while the Reference cabin temperature was above 40 °C.

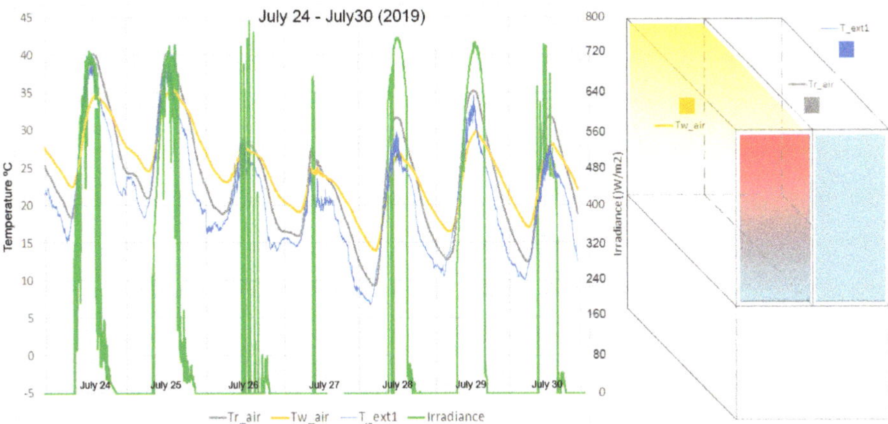

Figure 9. Indoor temperature of both cabins (T_{r_air}, T_{w_air}), the outdoor temperature ($T_{\_ext1}$), and solar radiation from 24 July 2019 to 30 July 2019.

The next step was to analyze a typical summer day in both cabins. Figures 10 and 11 show the glazing temperatures, measured in different points. There is a little difference between the surface temperatures of the outer glass pane in the WFG cabin (T_{w_extU} and T_{w_extD}) with the Reference cabin (T_{r_extD}). The effect of the water flowing through the glazing affected the indoor surface temperatures in both cabins. Figure 10 illustrates that the measured indoor glass surface temperatures in the Reference cabin (T_{r_up}, T_{r_down}) are remarkably variable during the daily hours, according to the relevant variations of the outdoor thermal conditions and the low thermal inertia of the glazing. However, there is no difference between T_{r_up} and T_{r_down}. The indoor surface temperatures in the WFG was measured at the bottom and the top of the indoor glass pane. The water heats up as it moves through the glazing, and the figure explains the influence of the water flow in the temperature distribution.

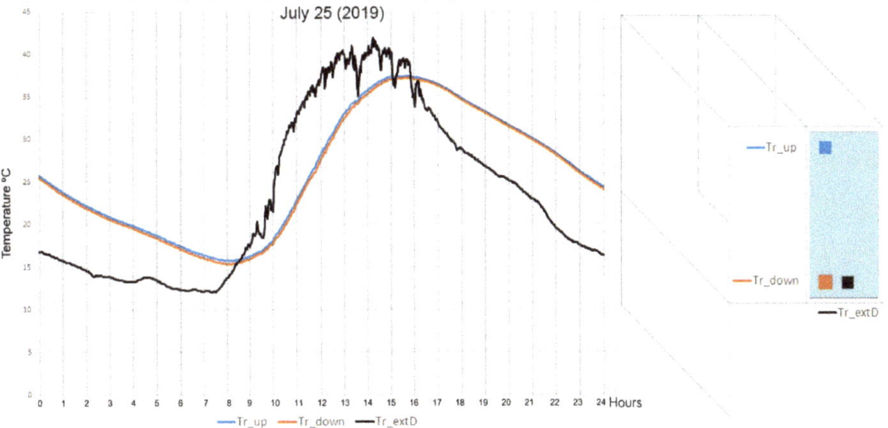

Figure 10. Temperatures of the Reference glazing, both on the outside (T_{r_extD}) and the inside face of the glazing (T_{r_up}, T_{r_down}). Sample day 25 July 2019.

Figure 11 shows the surface temperatures of the WFG facade, both on the outside and the inside face of the glazing. The external pane reached temperatures above 41 °C, while the maximum surface temperature on the inner pane was 34 °C. There was no significant difference between $T_{w\_ext\_U}$ and

$T_{w\_ext\_D}$. When it comes to the interior pane, the water absorbed part of the solar radiation as it flowed. There were two sensors in the upper part ($T_{w\_up\_L}$, $T_{w\_up\_R}$), and two sensors in the lower part ($T_{w\_down\_L}$, $T_{w\_down\_R}$). A 2 °C difference was observed between the lower and upper probe of the inside surface of the glazing. The lower part of the inner glass reaches a maximum temperature of 32 °C, while the upper part reaches 34 °C.

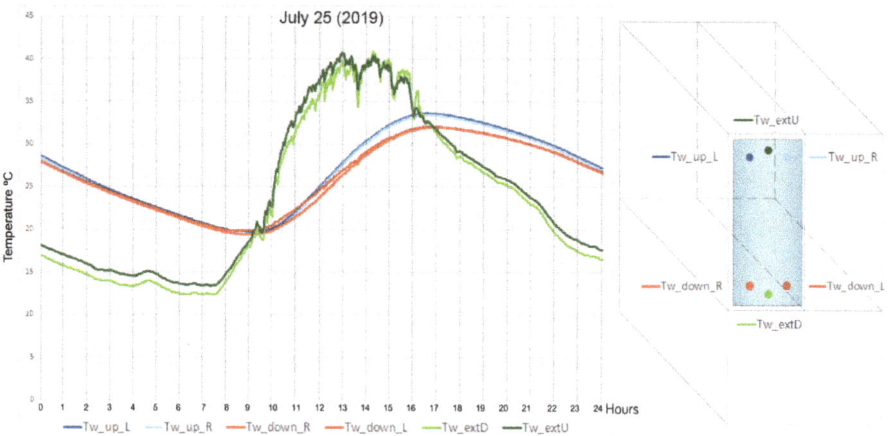

Figure 11. Temperatures of the WFG facade, both on the outside ($T_{w\_ext\_U}$, $T_{w\_ext\_D}$) and the inside face of the glazing ($T_{w\_up\_L}$, $T_{w\_up\_R}$ for the upper probes, $T_{w\_down\_L}$, $T_{w\_down\_R}$ for the lower ones). Sample day 25 July 2019.

3.2. Analysis of the Prototype. "Peltier" Device ON

Figure 12 shows the interior temperature of both cabins and the exterior temperature from 22 July 2018 to 27 July 2018. The "Peltier" cell was in operation according to a simple control logic based on the interior temperature programmed in the monitoring control unit. It was set to operate every summer day from 12:30 to 20:00. The goal was to keep the inlet temperature between 15 and 17 °C to test the indoor air conditions inside the WFG cabin. The mean maximum temperature reached inside the WFG cabin (T_{w_air}) was 26.5 °C, with a mean maximum solar radiation of 720 W/m². However, the mean maximum temperature of the Reference cabin (T_{r_air}) exceeds 37 °C, when the mean maximum outdoor temperature ($T_{\_ext1}$) is 34.5 °C. Therefore, there were more than 10 °C of difference inside both cabins. The Reference glazing replicated the thermal oscillations of the outside temperature, generating discomfort inside the building and contributing to overheating. The same behavior was observed when analyzing the minimum temperatures. The minimum value of $T_{\_ext1}$ was 10 °C, while the Reference cabin remained at 12 °C, and the WFG cabin reached 15 °C. Therefore, the thermal inertia that characterizes the WFG facade managed to dampen the oscillation of the interior temperature. The temperature oscillations of the Reference cabin were similar to the outside temperature. Hence, the WFG cabin allowed for the bringing of the maximum and minimum values close to the comfort temperature. Specifically, analyzing 23 July 2018, the mean value between the maximum T_{w_air} daytime temperature (26.5 °C) and the mean minimum value of night time temperature (16.5 °C) was 21.5 °C, which was very close to the comfort temperature.

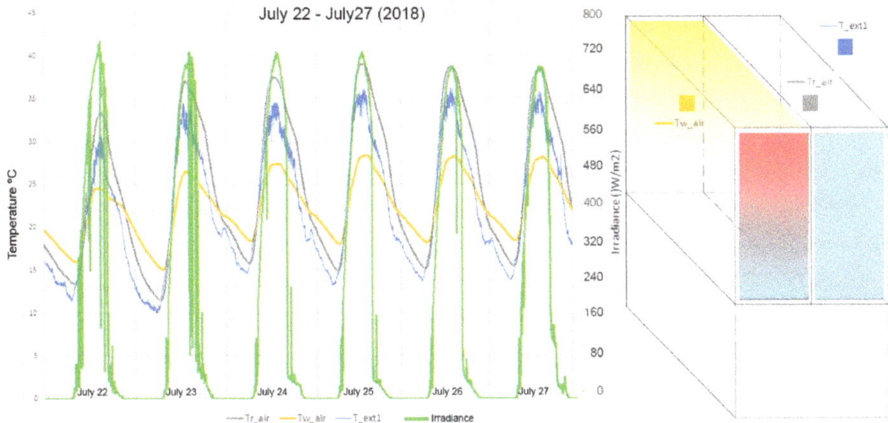

Figure 12. Indoor temperature of both cabins (T_{r_air}, T_{w_air}), the outdoor temperature (T_{ext1}), and solar radiation from 22 July 2018 to 27 July 2018.

4. Discussion

When the glazing is part of an insulated room, the thermal problem of the glazing is coupled with the thermal problem of the room, and the indoor temperature should be determined. The indoor boundary condition disappears to be part of the solution to the thermal problem. The prototype was a rectangular room with glazing facing south. The dimensions of the room, the near and far-infrared absorption (α_{NIR}, α_{FIR}), and the thermal transmittance of the opaque envelope are defined in Table 4.

Table 4. Dimensions of the cabins with thermal and spectral properties of opaque walls.

	Height (m)	Length (m)	Width (m)	U_{wall} (W/m^2 K)	α_{NIR}	α_{FIR}
Reference cabin	1	0.6	0.5	0.3	0.4	0.9
WFG cabin	1	0.6	0.5	0.3	0.4	0.9

In these following test cases, outdoor temperature and solar irradiance varied during the day, and thermal performances depended on time. The indoor temperature was unknown, and it should have been obtained at the same time as the glazing temperature profile. Regarding the water flow glazing, the flow rate and the inlet temperature were constant values given by Table 5.

Table 5. Parameters of WFG.

Glazing	$\dot{m}$ (l/min m^2)	θ_{IN} (°C)	h_g (W/m^{2}K)	h_w (W/m^{2}K)	c (J/Kg K)
WFG	1	18–22	1.16	50	3600

Transient behavior occurred when the outdoor temperature and solar irradiance varied during the day. Besides, each wall had a different temperature due to the luminance of the direct beam solar radiation. Since the WFG was facing south, the north indoor wall absorbed solar radiation. The rest of this energy was diffusely reflected and created the indoor diffuse irradiance. Later, this irradiance was absorbed in each indoor surface. Hence, in this test case, the water flow glazing absorbed extra energy from the indoor irradiance. Figure 13 shows the validation of the Software Tool using real data from both cabins. The figure illustrates six days, from 22 July 2018 to 27 July 2018, in which we can see how the measured curves replicate the simulation curves in all the cases.

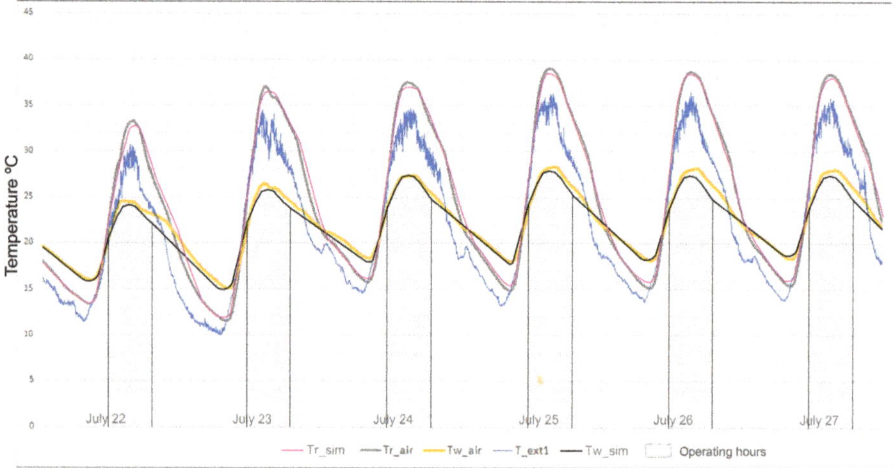

Figure 13. Indoor air temperature. Real results and simulation of the Reference cabin and the WFG cabin. Sample summer days 22 July 2018 to 24 July 2018.

The software tool tested in this article is an open software code written in modern Fortran, with a graphic user interface. The functionalities are grouped visually and logically into thematic units. There are libraries of spectral and absorption properties with different glasses and coatings. These libraries can allow developers to integrate WFG in existing building energy simulators. A thermal simulator of zones with glass and opaque envelopes includes properties such as thermal mass and reflections inside the zone. Some papers on the functionality of this tool have been published to present the design approach [62].

4.1. Analysis of the Reference Cabin

The reference glazing is considered a high-performance transparent envelope. It is made up of glass panes with coatings and gas cavities. Figure 14 illustrates real data and simulated indoor temperatures of the Reference cabin. In both measured and simulated curves, the interior peak temperature is slightly above 40 °C between 03:00 and 10:00 on 23 July 2018. Equation (24) represents the mean error (ME), which is the difference between the measured value and simulation results. Equation (25) represents the mean percentage error (MPE), with a total number of measurements of $n = 1440$.

$$ME = \frac{1}{n} \sum_{i=1}^{n} |T_{Si} - T_{Ri}|, \tag{24}$$

$$MPE = \frac{1}{n} \sum_{i=1}^{n} \left| \frac{T_{Si} - T_{Ri}}{T_{Ri}} \right| 100. \tag{25}$$

where T_{Si} is the simulated value, and T_{Ri} is the measured value. By computing ME and MPE, the sample summer day, 23 July 2018, MEs and MPEs of the indoor temperature of the Reference cabin, T_{r_air}, were lower than 0.6 °C and 3.2%, respectively. Predictions of T_{r_air} were more accurate because the boundary conditions were more suitable to predict.

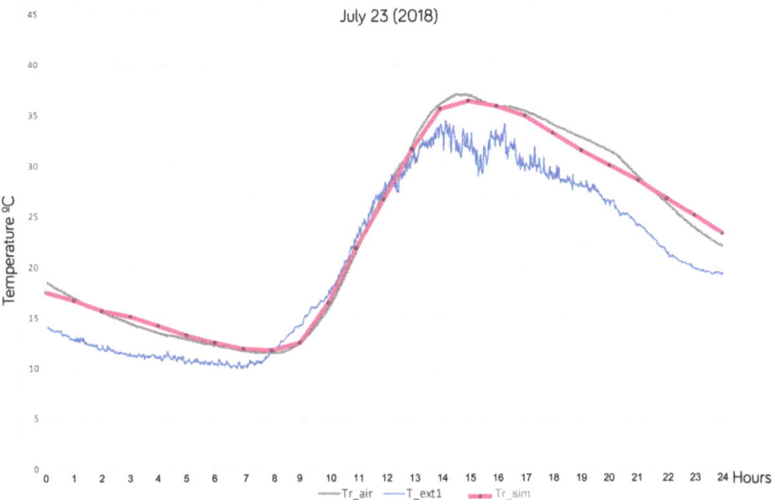

Figure 14. Indoor air temperature. Real results and simulation of the Reference cabin. Sample summer day 23 July 2018.

4.2. Analysis of the WFG Cabin

Two different mechanisms appeared to modify the thermal performances of the WFG. The first one was the value of the indoor temperature, which can be very high, depending on the wall insulation. The second one was the absorbed back irradiance. Depending on the insulation of the walls and the near-infrared absorptances, the water heat gain can differ when compared to the water heat gain of isolated glazing. Figure 15 shows the comparison between real data and simulation over the same day, 23 July 2018. MEs and MPEs of the indoor temperature of the WFG cabin were lower than 1.2 °C and 5.5%, respectively.

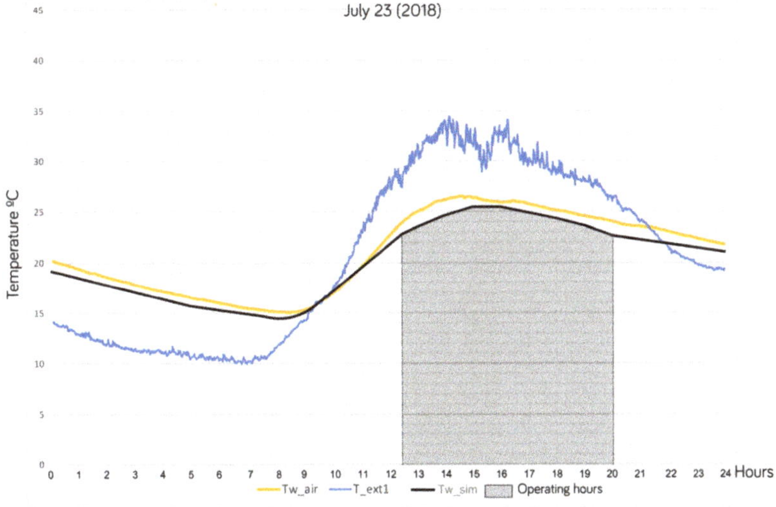

Figure 15. Indoor air temperature. Real results and simulation of the WFG cabin. Sample summer day 23 July 2018.

Figure 16 illustrates the time histories of indoor and outdoor air temperatures and the solar irradiance in W/m^2 on the outdoor horizontal roof surface on a sample summer day (23 July 2018). The indoor air temperature of the Reference cabin (T_{r_air}) varied according to the solar irradiance and the outdoor temperature ($T_{\_ext1}$). The measured inlet temperature ($T_{\_inlet1}$) was input in the simulation tool as boundary conditions. By considering the activation of a solar fed "Peltier" device, the inlet temperature was set between 18 and 22 °C. The operation schedule for the cooling system activation was from 12:30 to 20:00. Figure 16 reports the results of the daily analysis. The thermal effect of the "Peltier" device kept the indoor temperature below 27 °C over the hottest hours of the day.

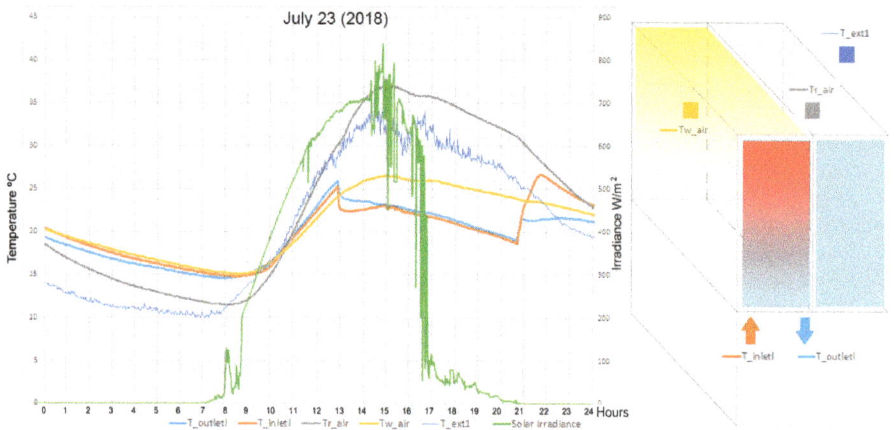

Figure 16. WFG cabin with the "Peltier" cell on from 12:30 to 20:00, 23 July 2018.

The daily cooling demands are shown in Figure 17. The daily energy that the flow of water can absorb was calculated using Equation (15). WFG prevented this energy from entering the cabin. The "Peltier" device provided the cooling power that explains the difference in temperatures between the WFG cabin and the Reference cabin. The energy absorbed by the WFG was 117.5 Wh in 0.5 m^2, so the ratio of energy per area was 0.235 kWh/m^2.

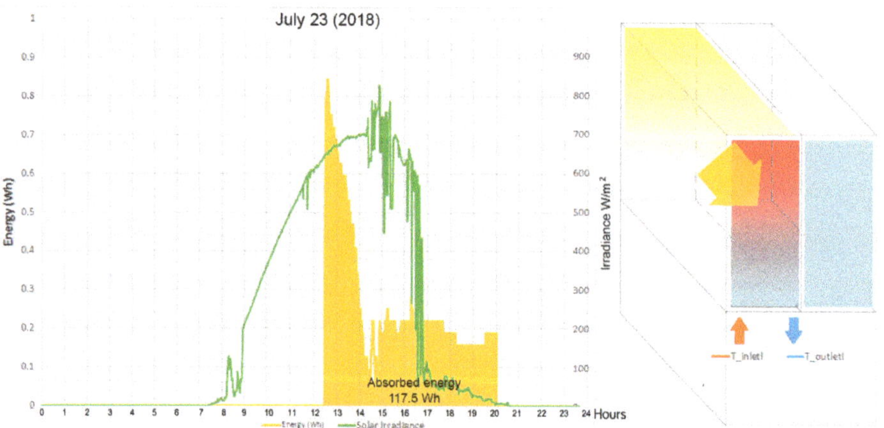

Figure 17. WFG cabin with the "Peltier" cell on from 12:30 to 20:00, 23 July 2018. Absorbed power $\dot{m}c(\theta_{IN}-\theta_{OUT})$ in W/m^2.

4.3. Final Energy and Cost Considerations

This section has considered best available technologies (BAT), which are innovative and economically viable [48]. Hydronic technologies are compatible with radiant systems, such as WFG and radiant floors and walls. Air-to-air heat pumps are used to compare the final energy consumption and cost of different cooling systems. WFG can be a part of hydronic HVAC systems, and it is compatible with water-to-water heat pumps. Tables 6 and 7 illustrate the final energy consumption with different energy generators. It takes into account the effect of the operative temperature of the system [63]. The performance of water-to-water heat pumps (WWHP) depends on the inlet temperature of the WFG (θ_{IN}), and the source inlet temperature ($T_{s,i}$) in the heat pump. A typical value of $T_{s,i}$ ranges from 20 °C in ground source heat pumps (GSHP) to 35 °C in other WWHPs. The source and load sides are relevant when it comes to calculating the performance of the cooling device. Air-to-air heat pumps (AAHP) for heat recovery on ventilation are also analyzed. The parameters that influence air-to-air heat pumps' performance are the dry bulb exterior air temperature (T_{e_db}) and the dry bulb interior return air temperature (T_{ri_db}).

Table 6. Final energy analysis. Water-to-water heat pump.

	Water-to-Water Heat Pump			
$\theta_{IN}(°C)$	7 °C		18 °C	
$T_{s,I}$ (°C)	20	35	20	35
Energy consumption (kWh/m^2)	0.235	0.235	0.235	0.235
EER [1]	5.93	3.27	7.11	4.22
FE consumption (kWh/m^2)	0.040	0.072	0.033	0.056
NRFE consumption (kWh/m^2)	0.077	0.140	0.065	0.109
CO$_2$ emissions (KgCO$_2$)	0.013	0.024	0.011	0.018

[1] Energy efficiency ratio (ERR) values are taken from Appendix A in [63].

Table 7. Final energy analysis. Air-to-air heat pump.

	Air-to-Air Heat Pump			
T_{ri_db} (°C)	22 °C		26 °C	
T_{e_db} (°C)	35	40	35	40
Energy consumption (kWh/m^2)	0.235	0.235	0.235	0.235
EER [1]	3.22	2.94	2.60	2.90
FE consumption (kWh/m^2)	0.073	0.080	0.090	0.081
NRFE consumption (kWh/m^2)	0.143	0.156	0.177	0.158
CO$_2$ emissions (KgCO$_2$)	0.024	0.026	0.030	0.027

[1] ERR values are taken from Appendix A in [63].

PEF stands for the primary energy factor from final energy (FE) to non-renewable final energy (NRFE). The Spanish code for thermal systems in buildings (RITE) recommends a value of 1.954. The factor of CO$_2$ emissions for electricity was 0.331 [64]. The RITE aims to establish primary energy factors and CO$_2$ emission factors, for each final energy consumed by buildings in Spain and for each geographic area with a different electricity generation source.

Table 6 shows that the best performance belongs to WWHP when the inlet temperature (θ_{IN}) is close to 18 °C, and the source inlet temperature ($T_{s,i}$) in the heat pump is 20 °C. Figure 16 shows that it is possible to keep the cabin's indoor temperature within the comfort range by setting θ_{IN} between 18 °C and 22 °C. Other systems, such as fan-coils, need lower operating temperatures. The lower the difference between $T_{s,i}$ and θ_{IN} the higher the EER coefficient. The NRFE consumption of GSHP is 0.065 kWh per m^2 of WFG, and the average NRFE use of air-to-air heat pumps is 0.155 kWh per m^2 of

WFG. When it comes to CO_2 emissions, air-to-air heat pumps generate an average of 0.027 $KgCO_2$, twice as much as the CO_2 emitted by GSHP.

5. Conclusions

This paper has studied a model to assess innovative building envelopes' energy performance, such as water flow glazing (WFG), the system equations for load calculation, and the relationships with steady and transient parameters. Some of the magnitudes can be measured accurately in the prototype. The presented tool has been developed and tested by the authors. Details of the prototype and the on-site measures have been used to validate the tool. The analysis included a free-floating temperature regime and a cooling system with simple logic to keep the prototype within a comfort range. Results included simulated indoor air and glazing temperatures along with the potential final energy savings.

1. When the WFG cabin's interior temperature was below the exterior temperature, the WFG facade cooled down the room. As the "Peltier" device was not in operation over 2019, it can be concluded that WFG working on a free-floating temperature regime, without auxiliary energy systems, results in smaller indoor temperature fluctuations.

2. Circulating water increased the temperature gap between external and internal glass panes. The outer pane reached temperatures above 41 °C, while the maximum surface temperature on the inner pane was 34 °C. The reference glazing showed a smaller gap between outer pane (41 °C) and inner pane (38 °C). The reduction of radiant temperature of indoor envelopes can improve the occupant's comfort.

3. The damping effect on the WFG cabin's temperature is shown in Figure 12. The WFG system provided the facade with thermal inertia, and the cabin did not suffer large thermal oscillations between day and night. This effect can increase the thermal comfort inside the building and reduce energy consumption.

4. The WFG increased the thermal inertia of the facade. Once the maximum temperature was reached, the interior of the WFG cabin cooled down more slowly than the Reference cabin did.

There was a good agreement between the simulation and the real data from the prototype. MEs and MPEs of the indoor temperature in the WFG cabin, were lower than 1.2 °C and 5.5%, respectively. The simulation results of the Reference cabin were more accurate because the boundary conditions were more suitable to predict.

The weather and indoor conditions impact the efficiency coefficients of heat pumps. The EER values of the cooling systems were evaluated for different combinations of indoor and outdoor conditions.

Ground source heat pumps (GCHP) coupled with borehole heat exchangers in a closed loop are very effective and make the most of the near-constant ground temperature over the year.

Finally, if the electricity is supplied from solar cells, it is possible to use a renewable and CO_2 free energy source to provide thermal comfort.

Author Contributions: Conceptualization, B.M.S., F.d.A.G., J.A.H.R.; methodology, B.M.S., F.d.A.G.; software, J.A.H.R.; formal analysis, B.M.S., F.d.A.G.; data curation, J.A.H.R.; writing—original draft preparation, J.A.H.R.; writing—review and editing, F.d.A.G., B.L.A.; visualization, B.M.S., F.d.A.G., B.L.A.; supervision, J.A.H.R., B.L.A.; project administration, B.M.S.; funding acquisition, F.d.A.G. All authors have read and agreed to the published version of the manuscript.

Funding: This article has been funded by KSC Faculty Development Grant (Keene State College, New Hamshire, USA).

Acknowledgments: This work was supported by program Horizon 2020-EU.3.3.1: Reducing energy consumption and carbon footprint by smart and sustainable use, project Ref. 680441 InDeWaG: Industrialized Development of Water Flow Glazing Systems.

Conflicts of Interest: The authors declare that they have no conflict of interest.

References

1. Santamouris, M. *Energy and Climate in the Urban Built Environment*; James and James Science Publishers: London, UK, 2001.
2. The Department of Energy's Renewable Energy Efforts. 2012. Available online: https://www.energy.gov/sites/prod/files/OAS-M-12-04_0.pdf (accessed on 3 June 2020).
3. Organisation for Economic Cooperation and Development, Energy Statistics of OECD Countries. Available online: http://www.oecd.org/ (accessed on 3 June 2020).
4. García-Feijoo, M.; Eizaguirre, A.; Rica-Aspiunza, A. Systematic Review of Sustainable-Development-Goal Deployment in Business Schools. *Sustainability* **2020**, *12*, 440. [CrossRef]
5. Kim, M.-K. An Estimation Model of Residential Building Electricity Consumption in Seoul. *Seoul Stud.* **2013**, *14*, 179–192.
6. Eum, M.-R.; Hong, W.-H.; Lee, J.-A. Deriving Factors Affecting Energy Usage for Improving Apartment Energy Consumption Evaluation. *J. Archit. Inst. Korea Struct. Constr.* **2018**, *34*, 27–34.
7. Lee, K.-H.; Chae, C.-U. Estimation Model of the Energy Consumption under the Building Exterior Conditions in the Apartment Housing—Focused on the Maintenance Stage. *J. Archit. Inst. Korea Plan. Des.* **2008**, *24*, 85–92.
8. Synnefa, A.; Santamouris, M.; Akbari, H. Estimating the effect of using cool coatings on energy loads and thermal comfort in residential buildings in various climatic condition. *Energy Build.* **2007**, *39*, 1167–1174. [CrossRef]
9. Del Ama Gonzalo, F.; Ferrandiz, J.; Fonseca, D.; Hernandez, J.A. Non-intrusive electric power monitoring system in multi-purpose educational buildings. *Int. J. Power Electron. Drive Syst.* **2019**, *9*, 51–62. [CrossRef]
10. Kim, M.; Jung, S.; Kang, J. Artificial Neural Network-Based Residential Energy Consumption Prediction Models Considering Residential Building Information and User Features in South Korea. *Sustainability* **2020**, *12*, 109. [CrossRef]
11. Katili, A.R.; Boukhanouf, R.; Wilson, R. Space cooling in buildings in hot and humid climate—A review of the effect of humidity on the applicability of existing cooling techniques. In Proceedings of the International Conference on Sustainable Energy Technologies—SET 2015, Nottingham, UK, 25–27 August 2015.
12. Dhariwal, J.; Banerjee, R. An approach for building design optimization using design of experiments. *Build. Simul.* **2017**, *10*, 323–336. [CrossRef]
13. Lai, C.M.; Wang, Y.H. Energy-saving potential of building envelope design in residential house in Taiwan. *Energies* **2011**, *4*, 2061–2076. [CrossRef]
14. González Lezcano, R.A.; Ros Garcia, J.M.; Del Ama Gonzalo, F. Influence of Envelope Colour on Energy Saving in Different Climate Zones in an Emergency House. *Contemp. Eng. Sci.* **2020**, *13*, 69–87. [CrossRef]
15. Sudhakar, K.; Winderl, M.; Shanmuga Priya, S. Net-zero building designs in hot and humid climates: A state-of-art. *Case Stud. Therm. Eng.* **2019**, *13*, 100400. [CrossRef]
16. Huw, H. *101 Rules of Thumb for Low Energy Architecture*; RIBA Publishing: London, UK, 2012; pp. 180–189.
17. Omar, O. Near Zero-Energy Buildings in Lebanon: The Use of Emerging Technologies and Passive Architecture. *Sustainability* **2020**, *12*, 2267. [CrossRef]
18. Hoel, M.; Kverndokk, S. Depletion of fossil fuels and the impacts of global warming. *Resour. Energy Econ.* **1996**, *18*, 115–136. [CrossRef]
19. Shafiee, S.; Topal, E. When will fossil fuel reserves be diminished? *Energy Policy* **2009**, *37*, 181–189. [CrossRef]
20. Shaahid, S.; Elhadidy, M. Economic analysis of hybrid photovoltaic–diesel–battery power systems for residential loads in hot regions—A step to clean future. *Renew. Sustain. Energy Rev.* **2008**, *12*, 488–503. [CrossRef]
21. Aksamija, A. Regenerative Design of Existing Buildings for Net-Zero Energy Use. *Procedia Eng.* **2015**, *118*, 72–80. [CrossRef]
22. Li, D.H.; Wong, S.L.; Tsang, C.L.; Cheung, G.H. A study of the daylighting performance and energy use in heavily obstructed residential buildings via computer simulation techniques. *Energy Build.* **2006**, *38*, 1343–1348. [CrossRef]
23. Gueymard, C.; duPont, W. Spectral effects on the transmittance, solar heat gain, and performance rating of glazing systems. *Solar Energy* **2009**, *83*, 940–953. [CrossRef]

24. Hermanns, M.; del Ama, F.; Hernández, J.A. Analytical solution to the one-dimensional non-uniform absorption of solar radiation in uncoated and coated single glass panes. *Energy Build.* **2012**, *47*, 561–571. [CrossRef]

25. Fang, Y.; Eames, P.C.; Norton, B. Effect of glass thickness on the thermal performance of evacuated glazing. *Solar Energy* **2007**, *81*, 395–404. [CrossRef]

26. Ismail, K.A.R. Non-gray radiative convective conductive modeling of a double glass window with a cavity filled with a mixture of absorbing gases. *Int. J. Heat Mass Transf.* **2006**, *47*, 2972–2983. [CrossRef]

27. Baetens, R.; Jelle, B.P.; Gustavsen, A. Properties, requirements and possibilities of smart windows for dynamic daylight and solar energy control in buildings: A state-of-the-art review. *Solar Energy Mater. Solar Cells* **2010**, *94*, 87–105. [CrossRef]

28. Attia, H.A.; Del Ama Gonzalo, F. Stand-alone PV system with MPPT function based on fuzzy logic control for remote building applications. *Int. J. Power Electron. Drive Syst.* **2019**, *10*, 842–851. [CrossRef]

29. Grantham, A.; Pudney, P.; Ward, L.A.; Whaley, D.; Boland, J. The viability of electrical energy storage for low energy households. *Sol. Energy* **2017**, *155*, 1216–1224. [CrossRef]

30. Gil-Lopez, T.; Gimenez-Molina, C. Environmental, economic and energy analysis of double glazing with a circulating water chamber in residential buildings. *Appl. Energy* **2013**, *101*, 572–581. [CrossRef]

31. Del Ama Gonzalo, F.; Moreno, B.; Hernandez, J.A. Dynamic Solar Energy Transmittance for Water Flow Glazing in Residential Buildings. *Int. J. Appl. Eng. Res.* **2018**, *13*, 9188–9193.

32. Chow, T.T.; Li, C.; Lin, Z. Thermal characteristics of water-flow double-pane window. *Int. J. Therm. Sci.* **2010**, *50*, 140–148. [CrossRef]

33. Bambardekar, S.; Poerschke, U. The architect as performer of energy simulation in the early design stage. In Proceedings of the IBPSA 2009—International Building Performance Simulation Association 2009, Glasgow, Scotland, 27–30 July 2009; pp. 1306–1313.

34. Zevenhoven, R.; Fält, M.; Gomes, L.P. Thermal radiation heat transfer: Including wavelength dependence into modelling. *Int. J. Therm. Sci.* **2014**, *86*, 189–197. [CrossRef]

35. Fernandez-Antolin, M.; del-Rio, J.M.; del Ama Gonzalo, F.; Gonzalez-Lezcano, R. The Relationship between the Use of Building Performance Simulation Tools by Recent Graduate Architects and the Deficiencies in Architectural Education. *Energies* **2020**, *13*, 1134. [CrossRef]

36. Stein, J.R.; Meier, A. Accuracy of home energy rating systems. *Energy* **2000**, *25*, 339–354. [CrossRef]

37. Zero Carbon Hub. *Closing the Gap between Design & As-Built Performance*; End of term report; Tech. Rep.: 2014; Zero Carbon Hub: London, UK, 2014.

38. Delghust, M.; Roelens, W.; Tanghe, T.; Weerdt YDJanssens, A. Regulatory energy calculations versus real energy use in high-performance houses. *Build. Res. Inf.* **2015**, *43*, 675–690. [CrossRef]

39. Van Dronkelaar, C.; Dowson, M.; Spataru, C.; Mumovic, D. A review of theregulatory energy performance gap and its underlying causes innon-domestic buildings. *Indoor Environ.* **2016**, *1*, 17. [CrossRef]

40. Rouchier, S.; Woloszyn, M.; Kedowide, Y.; Béjat, T. Identification of the hygrothermal properties of a building envelope material by the covariance matrix adaptation evolution strategy. *J. Build. Perform. Simul.* **2016**, *9*, 101–114. [CrossRef]

41. Soudari, M.; Srinivasan, S.; Subathra, B. Adaptive Disturbance Observers for Building Thermal Model. In Proceedings of the International Conference on Artificial Intelligence, Smart Grid and Smart City Applications (AISGSC), Coimbatore, India, 3–5 January 2019. [CrossRef]

42. Soudari, M.; Kaparin, V.; Srinivasan, S.; Seshadhri, S.; Kotta, U. Predictive smart thermostat controller for heating, ventilation, and air-conditioning systems. *Proc. Est. Acad. Sci.* **2018**, *67*, 291–299. [CrossRef]

43. Echarri, V.; Espinosa, A.; Rizo, C. Thermal Transmission through Existing Building Enclosures: Destructive Monitoring in Intermediate Layers versus Non-Destructive Monitoring with Sensors on Surfaces. *Sensors* **2017**, *17*, 2848. [CrossRef]

44. Pires, I.M.; Garcia, N.M.; Pombo, N.; Flórez-Revuelta, F.; Rodríguez, N.D. Validation Techniques for Sensor Data in Mobile Health Applications. *J. Sens.* **2016**, *2016*, 2839372. [CrossRef]

45. Li, Z.; Luan, X.; Liu, T.; Jin, B.; Zhang, Y. Room Cooling Load Calculation Based on Soft Sensing. In Proceedings of the Computational Intelligence, Networked Systems and Their Applications (ICSEE 2014, LSMS 2014), Shanghai, China, 20–23 September 2014.

46. Mastrucci, A.; Byers, E.; Pachauri, S.; Rao, N.D. Improving the SDG energy poverty targets: Residential cooling needs in the Global South. *Energy Build.* **2019**, *186*, 405–415. [CrossRef]

47. Prieto, A.; Knaack, U.; Klein, T. Auer. 25 Years of cooling research in office buildings: Review for the integration of cooling strategies into the building façade (1990–2014). *Renew. Sustain. Energy Rev.* **2017**, *71*, 89–102. [CrossRef]
48. Lun, V.; Tung, D. *Heat Pumps for Sustainable Heating and Cooling*; Springer International Publishing: Cham, Switzerland; Springer Nature: Cham, Switzerland, 2020. [CrossRef]
49. Torregrosa-Jaime, B.; Martínez, P.J.; González, B.; Payá-Ballester, G. Modelling of a variable refrigerant flow system in EnergyPlus for building energy simulation in an Open Building Information modelling environment. *Energies* **2019**, *12*, 22. [CrossRef]
50. Chow, T.; Li, C.; Lin, Z. Innovative solar windows for cooling-demand climate February. *Solar Energy Mater. Solar Cells* **2010**, *94*, 212–220. [CrossRef]
51. Chow, T.; Li, C.; Lin, Z. The function of solar absorbing window as water-heating device. *Build. Environ.* **2011**, *46*, 955–960. [CrossRef]
52. Gutai, M.; Kheybari, A.G. Energy consumption of water-filled glass (WFG) hybrid building envelope. *Energy Build.* **2020**, *218*, 110050. [CrossRef]
53. Gil-Lopez, T.; Gimenez-Molina, C. Influence of double glazing with a circulating water chamber on the thermal energy savings in buildings. *Energy Build.* **2013**, *56*, 56–65. [CrossRef]
54. Chow, T.; Chunying, L.; Clarke, J.A. Numerical prediction of water-flow glazing performance with reflective coating. In Proceedings of the Building Simulation 2011: 12th Conference of International Building Performance Simulation Association, Sydney, Australia, 14–16 November 2011.
55. Moreno BHernández, J.A. Analytical solutions to evaluate solar radiation overheating in simplified glazed rooms. *Build. Environ.* **2018**, *140*, 162–172. [CrossRef]
56. Moreno Santamaria, B.; del Ama Gonzalo, F.; Pinette, D.; Gonzalez-Lezcano, R.-A.; Lauret Aguirregabiria, B.; Hernandez Ramos, J.A. Application and Validation of a Dynamic Energy Simulation Tool: A Case Study with Water Flow Glazing Envelope. *Energies* **2020**, *13*, 3203. [CrossRef]
57. Liu, Y.L.; Chow, T.T.; Wang, J.L. Numerical prediction of thermal performance of liquid-flow window in different climates with anti-freeze. *Energy* **2018**, *157*, 412–423. [CrossRef]
58. EN 673. *Glass in Building—Determination of Thermal Transmittance (U Value)—Calculation Method*; German Institute for Standardization: Berlin, Germany, 2011.
59. EN 410. *Glass in Building—Determination of Luminous and Solar Characteristics of Glazing*; German Institute for Standardization: Berlin, Germany, 2011.
60. Sierra, P.; Hernandez, J.A. Solar heat gain coefficient of water flow glazing. *Energy Build.* **2017**, *139*, 133–145. [CrossRef]
61. Chow, T.T.; Li, C. Liquid-filled solar glazing design for buoyant water-flow. *Build. Environ.* **2013**, *60*, 45–55. [CrossRef]
62. Moreno, B.; Hernandez, J.A. Software tool for the design of water flow glazing envelopes. In Proceedings of the 7th European Conference on Renewable Energy Systems, Madrid, Spain, 10–12 June 2019.
63. Priarone, A.; Silenzi, F.; Fossa, M. Modelling Heat Pumps with Variable EER and COP in EnergyPlus: A Case Study Applied to Ground Source and Heat Recovery Heat Pump Systems. *Energies* **2020**, *13*, 794. [CrossRef]
64. Spanish Regulation of Thermal Installations in Buildings (RITE). *Factores de Emisión de CO2 y Coeficientes de Paso a Energía Primaria de Diferentes Fuentes de Energía Final Consumidas en el Sector de Edificios en España*; Instituto para la Diversificación y Ahorro de la Energía (IDAE): Madrid, Spain, 2016.

Article

Overheating in Schools: Factors Determining Children's Perceptions of Overall Comfort Indoors

Samuel Domínguez-Amarillo [1,*], Jesica Fernández-Agüera [1,*], Maella Minaksi González [2] and Teresa Cuerdo-Vilches [3]

[1] Instituto Universitario de Arquitectura y Ciencias de la Construcción, Escuela Técnica Superior de Arquitectura, Universidad de Sevilla, 41015 Sevilla, Spain
[2] Facultad de Arquitectura de la Universidad Autónoma de Yucatán, Merida 97000, Mexico; maella.gonzalez@correo.uady.mx
[3] Instituto de ciencias de la construcción Eduardo Torroja (IETcc), Consejo Superior de Investigaciones Científicas (CSIC), 28033 Madrid, Spain; teresacuerdo@ietcc.csic.es
*   Correspondence: sdomin@us.es (S.D.-A.); jfernandezaguera@us.es (J.F.-A.)

Received: 15 June 2020; Accepted: 13 July 2020; Published: 17 July 2020

Abstract: Climate change is raising the length and intensity of the warm season in the academic year, with a very significant impact on indoor classroom conditions. Increasingly frequent episodes of extreme heat are having an adverse effect on school activities, whose duration may have to be shortened or pace slackened. Fitting facilities with air conditioning does not always solve the problem and may even contribute to discomfort or worsen health conditions, often as a result of insufficient ventilation. Users have traditionally adopted measures to adapt to these situations, particularly in warm climates where mechanical refrigeration is absent or unavailable. Implementation of such measures or of natural ventilation is not always possible or their efficacy is limited in school environments, however. Such constraints, especially in a context where reasonable energy use and operating costs are a primary concern, inform the need to identify the factors that contribute to users' perceptions of comfort. This study deploys a post-occupancy strategy combined with participatory action to empower occupants as agents actively engaging in their own comfort. It addresses user-identified classroom comfort parameters potentially applicable in the design and layout of thermally suitable spaces meriting occupant acceptance.

Keywords: schools; heat perception; user's perception; thermal comfort; qualitative technique; POE

---

1. Introduction

Human beings depend on energy for almost all of their daily activities. Energy is not only required to cover basal needs, but also those which allow them to remain comfortable to face climate dynamic variations outdoors [1], even more for vulnerable populations, as children. These variations have been altered by anthropogenic activity, boosting extreme weather conditions related to climate change, or more complex effects, such as urban heat islands [2].

This effect also impacts on indoor air quality, resulting in discomfort and even affecting health [3–5], especially in risk groups, such as the elderly [6,7], children [8] and births [9], but also with a significant incidence in the active population [10,11].

Among the most studied buildings in the field of indoor comfort, schools represent a relevant group. One of the main reasons is the exposure of children to spending a long time under indoor environmental conditions. These children are considered a risk population, and there are also other considerations, such as social or vulnerability aspects, which can influence, so investigating in this regard has become a global priority, as a development objective sustainable by 2030 [12].

Comfort studies in schools have progressed in the last fifty years [13–16] and recently on indoor environments [17]. Recently, many of the studies in educational buildings include more innovative methods: student performance, cognitive processes or disruptive and engaging techniques, such as storytelling [18], gamification [19] or adaptation, and comparison among more traditional ones, as Post-Occupancy Evaluation (POE) [20–22]. Qualitative techniques are sometimes included as part of mixed methods, such as open-ended interviews [23]. Techniques to evaluate subjective aspects of users' perception on comfort have been developed [24], as well as on emotional state [25,26], but often are misnamed as "qualitative" [27–29].

According to the ASHRAE 55 standard [30] and the subsequent ISO 7730 [31], the thermal sensation experienced by human beings is mainly related to the global thermal balance in their body. It depends on physical activity, clothing and on environmental parameters, such as air temperature, average radiant temperature, air velocity and air relative humidity, whose values can be measured or estimated, to calculate the Predicted Mean Vote (PMV) comfort index. However, "nonthermal environmental parameters are not considered, such as air quality, acoustics and illumination or other physical, chemical or biological space contaminants that may affect comfort and health" [30]. Since thermal comfort is considered only a part of environmental comfort, and other environmental factors affect the thermal sensation, productivity, concentration and health of occupants indoors, this research proposes an overall comfort perception study, by a mixed method that allows to deepen the global satisfaction related to environmental comfort of users at class.

Questionnaires are a common practice either in internal environmental comfort research [32], or just in thermal comfort. These surveys are commonly based on ISO 7730 thermal comfort parameters and may include those related to potential local thermal discomfort (by unwanted air flows, temperature asymmetries, etc.). However, in recent years, research shows an increasing trend to complete the perception analysis also including nonthermal environmental factors that affect overall satisfaction and comfort perception, such as illumination, noises, odors, ventilation frequencies, spatial design elements of finishing and so on [33].

The inclusion of physiological indexes was not an objective of this study, due to several reasons: as considered in [32], physiological measurement (skin temperature, blood flow, core temperature, heart rate, etc.) can be correlated with thermal comfort/discomfort, but it consists of an invasive (contact) method that implies to have certain devices and a deep knowledge of the correct measuring method, for avoiding potential measuring mistakes, adding uncertainties to results. Since users' surveys are considered a traditional (contact) measure, the author did not want to interrupt the daily tasks at the classroom. Otherwise, the proposed technique, with its limitations, is understood as more effective and user-friendly, since it is not as invasive as physiological measures and is easier to carry out, with results more engaging and fun for kids, and it deals with behavior-change by debate sessions.

In the Mediterranean climate, thermal comfort studies in schools are scarce compared with tropical climates, however [34], there are many comfort studies in schools that develop research focused on the cold season, since it occupies most of the school period, sometimes also including midseason [15,35]. However, climate change is effectively lengthening the warm season, especially in areas of Southern Europe, where episodes of overheating affect the performance of daily tasks and health inside these buildings [36]. With the slow but constant rise in temperatures, the hot season extends over months that until a few years ago were months that were considered merely warm. This has forced the coincidence of the hot season with the start of classes in schools. In Seville, where "heat" implies exceeding 35 °C, the classrooms become spaces with an increasing lack of thermal comfort.

Despite some research, the literature has not taken into account children to express how they feel or perceive emotions when they interact with their built environment [37], or they are considered passive agents subject to the teacher's preferences [38]; recent studies have been demonstrated that the non-adult population is able to offer interesting insights to researchers, and they are in a position to express their thermal sensation and make adjustments to improve the thermal acceptability in the classroom [34]. Jindal's study carried out in India reported that students demonstrated adaptability to indoor temperature variations if they were able to adjust windows and switch off/on fans [39].

Fostering participative research with engaging techniques allows children motivation and broad-mindedness. The power of these techniques in sustainability, indoor environmental quality and users' global comfort in schools may not only make students aware, but also move their knowledge to homes and share it with their families [40].

These new approaches recently discussed in schools in the Netherlands [41–44] demonstrate that internal comfort in schools results in a wide-spread research topic. However, there are still some gaps that could be studied in greater depth and solved.

The main objective in this study consists of knowing the perception of overall satisfaction and environmental comfort perception in classrooms by secondary school children (12–16 years), using participatory techniques. It had been established a participatory, mixed method, where subjective and objective aspects can contribute to build a coherent discourse on the perception of comfort in the classroom, taking into account the active participation of the users.

We consider occupant perception to be a fundamental aspect of the process of architectural creation and engineering of environmental control. The purpose is therefore to provide mechanisms which can improve the response to the specific requirements of individuals. An improved adjustment to these needs can contribute to the design of more suitable (mechanical and constructive) systems, as well as to better energy use, improving both the practical energy efficiency of buildings and user experience.

2. Materials and Methods

This study presents a participative research on the overall environmental satisfaction and comfort perception at class, using the emotional design as a driver through a mixed approach to classroom users (students). The multifactorial character makes it necessary to collate and identify the (conscious or subconscious) information which users can provide as active "sensors" of their ambient environment to be able to feed predictive models for management of the indoor thermal environment.

The qualitative-exploratory and participatory part of the study is carried out using two techniques: emotional drawings, and group debate around them. Through the drawings, students visually express their understanding of indoor comfort in the school [43]. This technique also allows them to graphically communicate which elements provide comfort, and which contribute to a lack thereof.

Then, they develop their group discourse at the level of global environmental comfort perception inside the classroom. The drawings previously selected can act as triggers to elicit deliberation. Students build a consensus on what they understand by thermal and environmental comfort in the classroom, which aspects affect it positively and negatively and which solutions will be needed to achieve it. Finally, two questionnaires complete the whole students' evaluation, one based on the user perception of how they feel indoors (aligned with ISO 7730 [31]), and the second one about classroom features that may affect comfort and health in the classroom, under their point of view [44].

In parallel, indoor environment parameters (air temperature and relative humidity) were monitored with portable sensors during working sessions with children. Measurements of relative humidity and outdoor temperature conditions were continuously monitored during measurements. However, the humidity and the indoor temperature of the classrooms were only punctual measures during the development of the different sessions, to know the average temperature values during the different sessions for each class. During the research, the environmental conditions were obtained for both indoor and outdoor spaces to ensure similar boundary conditions for all users, and to be able to evaluate the different responses and contributions in these conditions.

Data monitoring allows assessing the adaptation degree of the method to the specific characteristics of the proposed study case.

2.1. Location and Climate

The city of Seville, in the south of Spain, has a typical Mediterranean climate with mild winters between December and March, very hot summers between June and September and short periods of variable temperatures between seasons in April–May and October–November, although each time with higher temperatures. Relative average humidity ranges from 44% to 80%, with variations inversely proportional to the daytime temperature due to air heating. The predominant wind direction is southwest with low speeds. There is a differentiation between the prevailing winds in winter, coming from the northeast, and those in summer, coming from the southwest. Seville also presents a large number of sunny days, with little cloudy or clear skies [45].

Figure 1 shows the school building, the study rooms are oriented south-southwest. Figure 2 shows the classroom typology.

Figure 1. Location and orientation of the case studies.

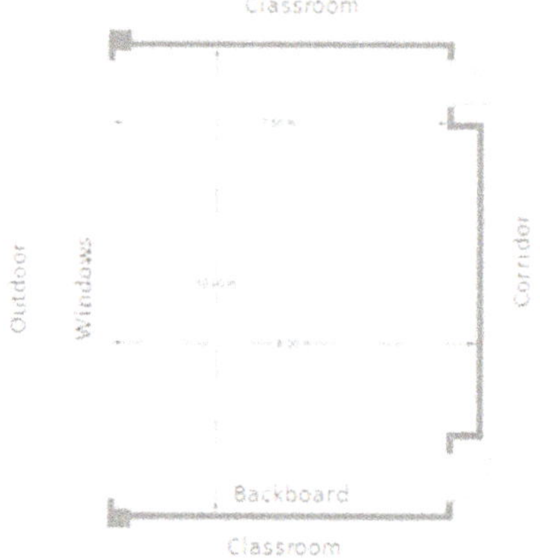

Figure 2. Classroom typology.

2.2. Case Study

The evaluation of this methodology was carried out in a secondary school in Seville (Spain). A total of 99 students were part of this study, aged between 12 and 14. According to the literature, this range of age is more appropriate to express descriptions and relationships between students and the built environment [37].

The study period was between 2 and 11 October 2019. The monitoring of indoor environmental parameters (temperature, relative humidity) was carried out simultaneously with the reflection-by-drawing exercise and the completion of questionnaires by the students. For the field measurements, a portable device was arranged in each class during the whole class time for every session and classroom. Despite the date, and as an evidence of the climate change ravages, the range of outdoor temperatures corresponds to the studied period, the temperature variation exceeds 15 °C, recording outdoor maximum temperature of 34.3 °C and minimum temperature of 19.4 °C. This difference has a great influence on the variation in the thermal sensation of the students throughout the day. The stablished schedule for students' evaluations was set between 08:00 and 14:30 hours, since children complete their activities during the class time. The indoor environmental conditions in classrooms in terms of support for heating, ventilation and air conditioning (HVAC) equipment were freely evolving.

To carry out the exercise in each classroom, four sessions were needed, taking three consecutive classes in the same week and lasting as much as 1 hour per session. The facilitator-researcher led these sessions. During this time, the tasks to complete were the following:

1. Shallow presentation of the activity, emotional drawing by the students, coding of the drawings;
2. First group debate after drawing selection by the facilitator;
3. Keywords categorization, jotted down by the facilitator from the students' consensus and questionnaires;
4. Thinking about possible solutions to settle what they considered negative for comfort.

2.3. Qualitative Technique: Drawings

Currently, drawing is a tool with unquestionable scientific rigor. It is used in medical tests, in pediatrics and psychology for instance, and its validation is based on evidence, recognized as an emotional driver to know and evaluate children behavior, even for the detection and monitoring of emotional, cognitive or behavioral disorders [46]. Despite that images in general are considered qualitative data [27], there are quantitative tests able to assess children drawings [47].

Drawings are also used as part of mixed methods, as well as triangulated with other evaluation techniques that support or refute the results, often applied in the child population. Well-known and validated drawing tests are Draw-a-man Test (DAMT), Family Drawing, linked to Attachment Theory [48], or Kinetic Family Drawing [49], validated with others, such as problem behavior tests, to predict and mediate internal childhood behavior problems [49]. Researchers often compare with questionnaires and objective visual indicators and given insights on familiar relationships, following the art-based phenomenological analytic approach [50]. Other research applies creative techniques as writing and storytelling to work out psychological disorders [51].

2.4. Questionnaires

Two questionnaires were given to students, with scaled answers for a better understanding and evaluation on their part, as well as to collect aggregated data.

Firstly, they were asked to fill in the questionnaire on user comfort perception, aligned with ISO 7730, with questions about comfort perception, which included environmental issues such as classroom location, lighting perception [52], indoor air quality and possible inconveniences linked to the pupil's position in the classroom. The survey was used to establish the boundary conditions of the classrooms, but its results are not discussed in this article in depth. However, the results of those surveys and the internal measurements of humidity and temperature and the quantitative indices such as PMV and

Predicted Percentage of Dissatisfied (PPD) in classrooms for the same typology have been discussed in [14]. The CO2 concentration and symptomatology of the students in [15] have also been analyzed for the same climate and typology. A version of this questionnaire adapted to schools has been adapted in [53].

Secondly, the facilitator distributed to the students the questionnaire on what conditions influence their comfort in the classroom. It also contained scaled-answers and included a wide variety of issues that potentially could affect to the environmental conditions of the classroom, as well as to their hypothetical improvement according to the students' own perspectives. In this questionnaire, the main questions use language aimed at obtaining an impulsive response, without prior reflection or necessary observation. The objective of this method is to extract the unconscious information that influences the state of thermal comfort. The surveys distributed to the students in their original version in Spanish are found in Appendix A.

2.5. Group Debate

The facilitator selected among all the drawings collected those that could represent most of the pupils, or which provoked more interesting questions, so that the debate flowed easily and effectively. The facilitator asked the creators to explain what they have drawn in order to better understand what they want to reflect, after a discussion started. The following questions were asked:

- What do you see here?
- What is really happening here?
- How does this relate to our lives?
- Why does this concern, situation or strength exist?
- How can we become empowered through our new understanding?
- What can we do?

One of the questions that was most affected during the debate was "empowerment", and it was interesting to obtain a group feedback through their answers, to know the perspective of how they understood it and what solutions could be put to such problems.

3. Results

In Seville, temperatures are highly variable between 8:00 and 14:30 hours. This implies that the adaptability of people must be able to assimilate all the conditions included in that hour range. When the temperature of a space deviates towards warmer or colder, it is altering the thermal comfort of individuals. In the short time of a morning, that adaptability is very difficult. Therefore, it is very likely that there will be situations in which one easily moves away from the comfort zone. To avoid extreme discomfort, people turn to conscious and unconscious systems of adaptability. The immediate system is the readjustment of clothing, putting on and taking off clothes throughout the morning.

Other factors of special interest are:

- The overcrowding of classrooms (there are 34 students in each class);
- The distribution of the students in the classrooms;
- The students' activity, it is not the same when coming from physical education or technology classes as coming from theoretical classes;
- Classrooms' orientation, three case studies have a south orientation.

3.1. Boundary Conditions

The school year begins in mid-September. This study was developed one month after the beginning of classes, with the autumn just begun. Autumn 2019 presented itself as a time of slightly higher temperatures than usual. The average temperature of this season tends to be around 20 °C, but in this case, it has been recorded at an average of 1 °C more. No rainfall was recorded in the days leading up to the survey.

On the day of the survey, the outdoor temperature variation exceeded 15 °C. In the morning at the beginning of classes, 18 °C was measured and at the exit, 34 °C was reached. This difference has a great influence on the variation in the thermal sensation of the students throughout the morning (Figure 3).

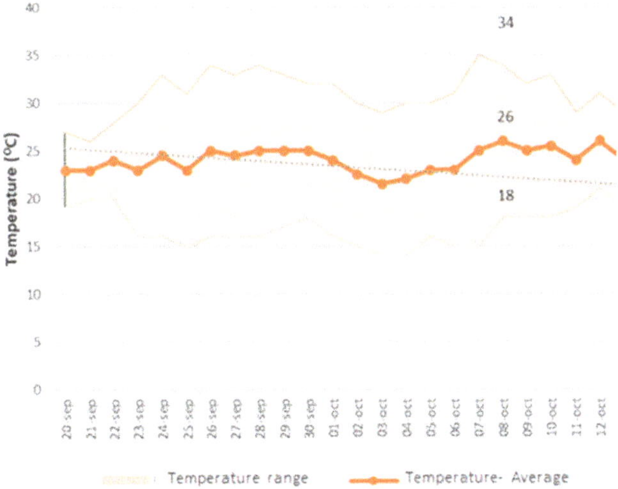

Figure 3. Outdoor temperature evolution in Seville during the study period.

Figure 4 shows a slight deviation from the CS1 and CS2 class towards heat sensations with respect to the CS3 class, who were more comfortable thermally because their class schedule was 10 to 11 hours and the average temperature measured was 26.2 °C. Meanwhile, for the CS1 and CS2 groups, the time frame in which they did the questionnaires and the discussion was 11:30–12:30 and 12:30–13:30, respectively, and the average temperature measured was 27.4 °C.

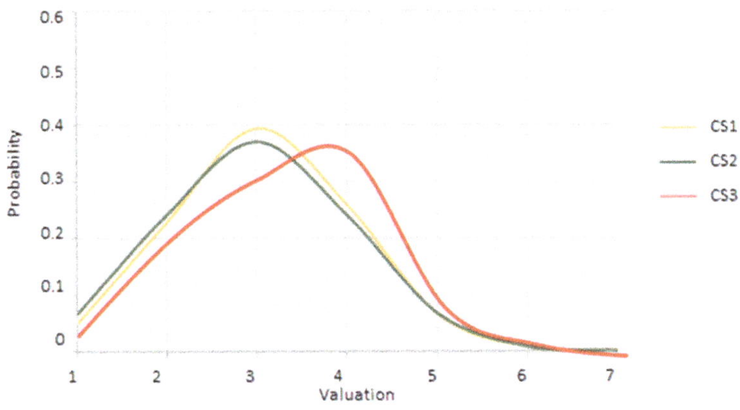

Figure 4. Thermal sensation reported by students.

Despite this, when assessing overall comfort (Figure 4), CS1 gets a rating somewhat worse than CS3. This is indicative that there are factors that are affecting comfort beyond temperature.

Figure 5 shows the comfort sensation reported by students. The valuations were affected by several overall comfort parameters, as well as the schedules and time-frames for the assessments, resulting in different distribution of rating among CS1, CS2 and CS3.

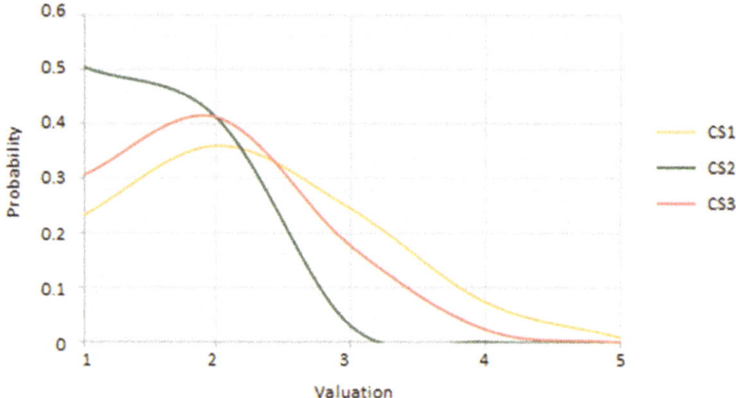

Figure 5. Comfort sensation reported by students.

Another value that can influence the results is the clothing of the occupants, as it has been evaluated and established that the conditions in this aspect are very similar, since the variation is minimal (Figure 6).

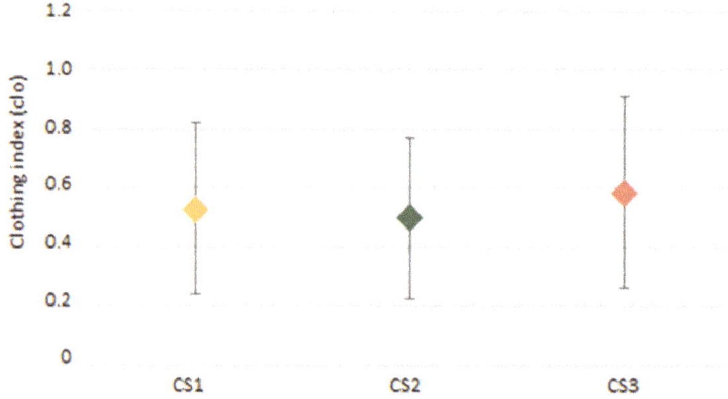

Figure 6. Clothing index (clo) in the classrooms.

3.2. Drawings

In the first session, students had to make a drawing that showed elements or factors that represented the discomfort in the classrooms (Figure 7). This explanation was just indicative, trying not to skew their own perceptions about the existing problems and their way of showing them with images. A code instead of the name was provided to each survey, in order so that they could be expressed more freely. For this study, spontaneous or freely expressed answers could give much more valuable data than pre-defined answers, even if it is harder to quantify.

Figure 7. Emotional drawings performed by students.

In this case study, 20% of the students did not know what to draw about the classroom, and 5% wrote a small opinion instead of drawing. A total of 50% of the drawings included an air conditioner and the student's position in the classroom. A total of 45% of the students drew windows, being one of the most significant elements because the sun's rays enter through them, and some even drew the views in front of the windows. Many of the drawings included the blackboard, although in the debate it was said that it was a component to be placed in the classroom. Other drawings included water, doors, airflow, clothes and hairstyles.

3.3. Questionnaires

The surveys distributed included three fundamental questions. Scale/rank questions (from 0 to 5) were used to ask respondents to rate items in order of importance or preference, where 0 meant that the element had no influence and 5 that it was very influential.

Figure 8 shows the mean values of factors that influence comfort in the classroom. A total of 50% of students voted that the most relevant elements influencing classroom comfort were noise, heat, smell and the number of students in the classroom. The elements voted as less important were wall color, roof color, floor material and table distribution. Regarding the trend in the different classes, there is a small divergence of less than 1 point in almost all the factors.

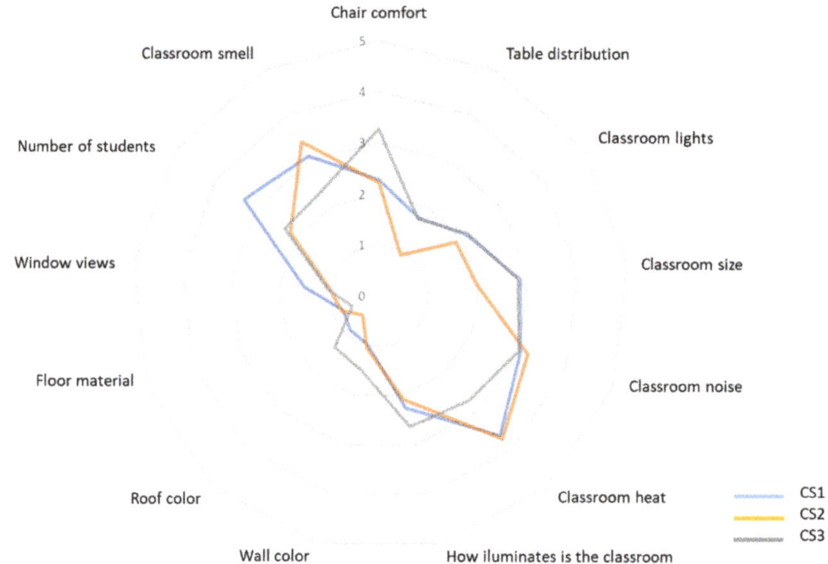

Figure 8. Response rates for factors that influence comfort in the classroom.

Figure 9 shows the mean values of factors that influence the temperature in the classroom. The most voted elements were to have air conditioning and that the air conditioning was on. Other factors voted as important were the number of students and the movement of the air in the classroom. Other elements were considered of medium importance, including elements that were of little importance in comfort such as wall color, roof color and the activity. Regarding the trend in the different classes, there is a small divergence in the importance of air conditioning, air movement and classroom size.

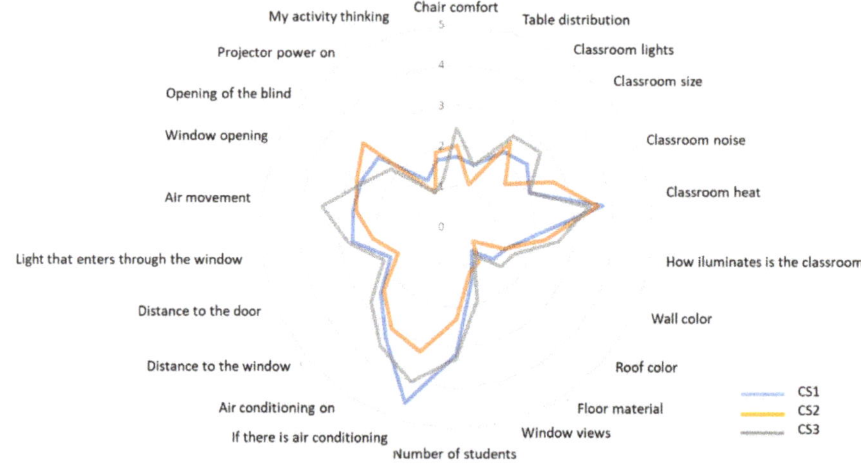

Figure 9. Response rates for factors that influence the temperature in the classroom.

Figure 10 shows the mean values for factors that should be changed to improve the temperature in the classroom. The answers were similar to question 2, given that the most important elements were related to air conditioning. Other additional factors were the distance to the window and the existence of air flow/breeze.

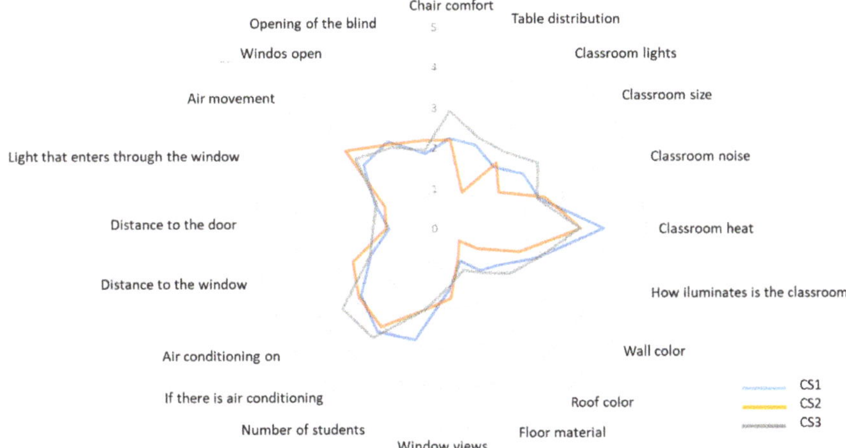

Figure 10. Response rates for factors that should be changed to improve the temperature in the classroom.

3.4. Group Debate

In session 2, the facilitator asked the creators to explain what they had drawn in order to better understand what they wanted to show, after other students expressed their opinions about those drawings. We conducted a thorough content analysis of conversation transcripts and other meaningful information collected, encompassing the researchers' notes, interviews and children's drawings. Table 1 describes the main themes identified.

Question "What do you see here?". The most repeated response was the entry of sun through the window. They also said that people, lights, walls and digital equipment were hot.

Question "What is really happening here?". It is very hot in the class, there are many students inside the class, the air does not move inside the class, the windows are broken and some reported bad odor and humidity.

Question "How does it relates to our lives?". It is mandatory to study until the age of 16 in person, though in spring and summer, the temperatures are very high so it is more difficult to study, with some even saying that their parents force them to go to school when they do not want to.

Question "Why does this concern exist?". They believe the problem is that they do not put air conditioning on when the temperatures are very high in Seville in summer. Therefore, each year they pass, there is more heat and the heat is more durable, with some suggesting that this is due to climate change.

Question "How can we become empowered through our new understanding?". They think that to become empowered you have to put air conditioning on in all classrooms, allow them to drink water in class and adapt their clothes. They also think that if they create problems, do not go to class on hotter days and complain to the authorities, someone will listen to them and solve the problem. Further, they think that if someone measures the temperatures throughout the day it will be seen that they are not suitable for teaching.

In session 4, it was proposed to think about possible solutions to improve thermal comfort. The windows were proposed to be changed, using improvements such as corbels, sunshades or curtains. Larger trees could be planted in the playground to provide shade and a cooler environment. It was also necessary to include some element that moved the air, or open the windows when there were windy conditions outside to obtain ventilation.

Table 1. Frequency of Words in the Group Debate.

Letter	Question	Concept More Repeated	Number
S	What do you **See** here? SUN WINDOW PEOPLE WALL DIGITAL BLACKBOARD LIGHT	Sun	60
		Window	56
		Wall	38
		People	29
		Light	25
		Digital Blackboard	16
H	What is really **Happening** here? STUDENTS CLASSROOM ODOUR HEAT AIR WINDOW ENTER HUMIDITY MOVEMENT	Classroom	40
		Heat	40
		Students	35
		Air	28
		Window	26
		Enter	24
		Humidity	8
		Odour	7
		Movement	6
O	How does it relate to **Our** lives? STUDY TEMPERATURE SCHOOL MANDATORY SPRING SUMMER PARENTS	Temperature	52
		Study	30
		Mandatory	25
		School	25
		Summer	18
		Spring	17
		Parents	12
W	Why does this concern exist? NO COOLING TEMPERATURE LONG SEVILLE SPAIN CLIMATE CHANGE SUMMER	No Cooling	35
		Temperature	30
		Sevilla	20
		Spain	10
		Summer	10
		Climate change	8
		Long	7
E	How can we become **Empowered** through our new understanding? AIR CONDITIONING CLOTHES TROUBLE WATER WHITE SHUTTERS NO ASSIST BLIND REPORT AUTHORITY HAIRSTYLE MONITORING	Air conditioning	28
		Trouble	25
		Water	20
		Clothes	20
		White	15
		Hairstyle	14
		Shutters	12
		Blind	10
		No assist	8
		Report authority	3
		Monitoring	2
D	What can we **Do**? AIR CONDITIONING FAN DRINK CLOTHES SHORT WATER HAIRSTYLE IMPROVE WHITE HAND FAN TREES TIMETABLE OPEN WINDOWS PAINT	Air conditioning	38
		Clothes	25
		Fan	25
		Drink	17
		White	16
		Ventilation	16
		Short	12
		Hand fan	10
		Improve	8
		Open Windows	7
		Water	8
		Hairstyle	7
		Timetable	5
		Trees	5
		Paint	5

Many students said that it was important the kind of clothes they wore, although sometimes this measure was insufficient to achieve well-being conditions indoors. Other measures were to turn off the lights, install fans or hand-fans, drink water or even change the hairstyle.

In addition, students performed claims such as "overheating due to excessive solar radiation through the windows" or that "with such heat cannot think", even that "they get very distracted during the day to try to find solutions to have less heat".

Having to engineer "homemade" mechanisms to acclimatize or seek comfort often resulted in wasted time, according to the testimonies of the participants, such as using the notebooks as fans, or they got wet in the playground so as not to pass heat when they entered the classroom.

Some of them even thought that small actions such as painting the classrooms white or better thinking about the hours in which the different subjects are carried out can help to improve comfort in the classroom.

4. Discussion

The students proposed diagnoses and possible solutions to the problems of the indoor environment of the classroom. The teaching–learning indoor environment must have a global environmental comfort assessment (including aspects of thermal comfort [54], but not exclusively), and students should be considered as a fundamental part in this assessment, as well as the costs for its operation and adequacy [55].

The factors that in this study were considered the most influential with 50% of the students' votes were noise, heat and odor. In the study carried out by Zhang et al., where students were divided into different and independent clusters, noise was considered the most annoying followed by odor [43], while in the Sense Lab's experience room, students reported that the problems that affected them most were noise (58%) followed by temperature (53%) [43].

In the content analysis of the debate group a clear tendency towards student discomfort was found caused by overheating. Zhang et al.'s study in the Netherlands coincides, due to more than 70% of teachers from different types of schools indicating that the most requested students' modification was the adjustment of windows, with the argument of thermal discomfort [41].

The modifications proposed by the students in this studio range from modifications to clothing, window adjustments, equipping the classroom with air conditioners and fans to planting trees around the classrooms. In the study carried out by Bluyssen et al., students proposed similar solutions [43]. The students of both studies agreed on the need to implement a device that regulates ventilation in the indoor environment, adapting to changes in the outdoor environment automatically.

The differences in the perception of aspects that cause discomfort in the classroom between this study and the contrasted studies [41–44] may be due to the fact that the variation in the thermal environment not only affects thermal comfort but also has an impact on the perception of other factors related to the indoor environment [56].

5. Conclusions

Currently, the perceptions of overall comfort in general, and at schools in particular, are incomplete or biased, often leading to failure when carrying out HVAC interventions, or in prior installed air-conditioned classrooms, where students still declare that they are not comfortable. In educational buildings, the approach that engages students in POE provides researchers with highly contextualized information about which elements are most influential in global (thermal and environmental) comfort, helping the analysis to be done more accurately and thinking about the factors that maximize the performance solutions.

The adapted POE methods motivated the students: feeling involved, they seemed excited to share their opinion and propose improvements. When feeling empowered, they create real opportunities for change, even discussing it with their families and close friends to explore new possibilities. Leveraging this motivation offers a real opportunity for change.

On the other hand, the application of techniques or methods that promote the active participation of users (belonging to the Participatory Action Research (PAR)) implies reflection on a certain topic or issue, common for a collective (pupils in class or school). This reflection leads to deepen the problems' roots, so environmental comfort misinformation, ignorance or lack of autonomy in comfort-related device control could be detected. Group debates contributed to build a common and aware discourse that tried to fix those gaps, that also may be communicated to decision-makers (teachers, school managers and directors, etc.) as part of a potential intervention strategy.

On the other hand, using the qualitative participatory technique based on drawings and group debates, insights independent of this kind of research are unveiled, not before considered from researchers, that could result in solutions that not only conduct possibly energy-saving but also cheaper and faster environmental comfort-related interventions, since real needs and lacks are soon-detected, revealed and analyzed from the users' experiences, to the researchers and intervenors.

The solutions that students gave, which covered a wide range of issues, also served to address the heat adaptation concern and even to raise awareness about energy savings.

Taking into account starting points that do not cost money, such as the position of the tables in the classroom or the study of the hours in which the different subjects must be taught, including access to the operation of the windows, blinds and lights can contribute to improving comfort in spaces. In addition, there are other adaptive solutions like changing your hairstyle or clothes. There are even improvement proposals that cost less and can greatly influence the microclimate, such as planting trees. Classroom ventilation was determined as an important element of indoor air quality and thermal comfort.

This research has provided information to the design community that is not generally obtained through Post-Occupational Evaluation (POE) but is essential in addressing design quality, since the approach gives importance to the people who really use the school buildings.

Author Contributions: Conceptualization, S.D.-A.; J.F.-A. and T.C.-V.; data curation, S.D.-A. and J.F.-A.; formal analysis, S.D.-A., J.F.-A., M.M.G. and T.C.-V.; funding acquisition, S.D.-A. and J.F.-A.; investigation, S.D.-A., J.F.-A. and T.C.-V.; methodology, S.D.-A., J.F.-A. and T.C.-V.; project administration, S.D.-A.; resources, S.D.-A. and J.F.-A.; supervision, S.D.-A.; Validation, S.D.-A., J.F.-A. and T.C.-V.; visualization, J.F.-A.; writing—original draft, S.D.-A., J.F.-A., M.M.G. and T.C.-V. All authors have read and agreed to the published version of the manuscript.

Funding: This research was funded by Data collection and study for the development of an Energy Efficiency Assessment Prototype of Active Participation Pilot Centers, grant number 3620/0451.

Conflicts of Interest: The authors declare no conflict of interest.

Appendix A

ALIAS:		CLASE:

¿Eres?	Chica ☐	Chico ☐	Edad:	
Estoy sentada/o	Cerca de la ventana ☐ En mitad ☐	Lejos de la ventana ☐	Cerca de la pizarra ☐	Lejos de la pizarra ☐

En términos generales me parece un *Aula* **CONFORTABLE**

☐ Totalmente en desacuerdo	☐ En desacuerdo	☐ Indiferente	☐ De acuerdo	☐ Totalmente de Acuerdo

De la siguiente tabla, ¿Qué elementos te han influido para establecer dicha valoración?

	Elemento	influye		En caso afirmativo, ¿Cuánto te influye?				
		SI	**NO**	Muy poco	poco	regular	bastante	mucho
1	Comodidad de las sillas			1	2	3	4	5
2	Distribución de las mesas			1	2	3	4	5
3	Luces del aula			1	2	3	4	5
4	Tamaño del aula			1	2	3	4	5
5	Ruido del aula			1	2	3	4	5
6	Calor en el aula			1	2	3	4	5
7	Como de iluminada está el aula			1	2	3	4	5
8	Color de la pared			1	2	3	4	5
9	Color del techo			1	2	3	4	5
10	El tipo de suelo			1	2	3	4	5
11	Las vistas de la ventana			1	2	3	4	5
12	El número de compañeros en el aula			1	2	3	4	5
13	Olor en el aula			1	2	3	4	5

Figure A1. Original Questionnaire.

ALIAS:	CLASE:

¿Eres?	Chica ☐		Chico ☐	Edad:	
Estoy sentada/o	Cerca de la ventana ☐	En mitad ☐	Lejos de la ventana ☐	Cerca de la pizarra ☐	Lejos de la pizarra ☐

En términos generales me parece un **Aula CON BUENA TEMPERATURA**

☐ Totalmente en desacuerdo	☐ En desacuerdo	☐ Indiferente	☐ De acuerdo	☐ Totalmente de Acuerdo

De la siguiente tabla, ¿Qué elementos te han influido para establecer dicha valoración?

	Elemento	Influye		En caso afirmativo, ¿Cuánto te influye?				
		SI	NO	Muy poco	poco	regular	bastante	mucho
1	Comodidad de las sillas			1	2	3	4	5
2	Distribución de las mesas			1	2	3	4	5
3	Luces del aula			1	2	3	4	5
4	Tamaño del aula			1	2	3	4	5
5	Ruido del aula			1	2	3	4	5
6	Calor en el aula			1	2	3	4	5
7	Como de iluminada está el aula			1	2	3	4	5
8	Color de la pared			1	2	3	4	5
9	Color del techo			1	2	3	4	5
10	El tipo de suelo			1	2	3	4	5
11	Las vistas de la ventana			1	2	3	4	5
12	El número de compañeros en el aula			1	2	3	4	5
13	Que haya aire acondicionado			1	2	3	4	5
14	Que el aire acondicionado esté encendido			1	2	3	4	5
15	La distancia a la ventana			1	2	3	4	5
16	La distancia a la puerta			1	2	3	4	5
17	La luz que entra por la ventana			1	2	3	4	5
18	La corriente de aire que me llega			1	2	3	4	5
19	La apertura de la ventana			1	2	3	4	5
20	La apertura de la persiana			1	2	3	4	5
21	El encendido del proyector			1	2	3	4	5
22	Mi actividad: pensando			1	2	3	4	5

Figure A2. Original Questionnaire.

ALIAS:					CLASE:	
¿Eres?	Chica ☐		Chico ☐		**Edad:**	
Estoy sentada/o	Cerca de la ventana ☐	En mitad ☐	Lejos de la ventana ☐	Cerca de la pizarra ☐	Lejos de la pizarra ☐	

De la siguiente tabla, ¿Qué querrías cambiar para que la sensación de la temperatura fuera más acorde a tus necesidades?

	Elemento	influye		En caso afirmativo, ¿Cuánto te influye?				
		SI	NO	Muy poco	poco	regular	bastante	mucho
1	Comodidad de las sillas			1	2	3	4	5
2	Distribución de las mesas			1	2	3	4	5
3	Luces del aula			1	2	3	4	5
4	Tamaño del aula			1	2	3	4	5
5	Ruido del aula			1	2	3	4	5
6	Calor en el aula			1	2	3	4	5
7	Como de iluminada está el aula			1	2	3	4	5
8	Color de la pared			1	2	3	4	5
9	Color del techo			1	2	3	4	5
10	El tipo de suelo			1	2	3	4	5
11	Las vistas de la ventana			1	2	3	4	5
12	El número de compañeros en el aula			1	2	3	4	5
13	Que haya aire acondicionado			1	2	3	4	5
14	Que el aire acondicionado esté encendido			1	2	3	4	5
15	La distancia a la ventana			1	2	3	4	5
16	La distancia a la puerta			1	2	3	4	5
17	La luz que entra por la ventana			1	2	3	4	5
18	La corriente de aire que me llega			1	2	3	4	5
19	Que la ventana esté abierta			1	2	3	4	5
20	Que la persiana esté bajada			1	2	3	4	5

Figure A3. Original Questionnaire.

References

1. Lusinga, S.; de Groot, J. Energy consumption behaviours of children in low-income communities: A case study of Khayelitsha, South Africa. *Energy Res. Soc. Sci.* **2019**, *54*, 199–210. [CrossRef]
2. Sanchez-Guevara, C.; Peiró, M.N.; Taylor, J.; Mavrogianni, A.; González, J.N. Assessing population vulnerability towards summer energy poverty: Case studies of Madrid and London. *Energy Build.* **2019**, *190*, 132–143. [CrossRef]
3. Ballester, F.; Díaz, J.; Moreno, J.M. Cambio climático y salud pública: Escenarios después de la entrada en vigor del Protocolo de Kioto. *Gac. Sanit.* **2006**, *20*, 160–174. [CrossRef]
4. Linares, C.; Jiménez, J.D. Impact of heat waves on daily mortality in distinct age groups. *Gac. Sanit.* **2008**, *22*, 115–119. [CrossRef] [PubMed]
5. Ortiz, C.; Linares, C.; Carmona, R.; Díaz, J. Evaluation of short-term mortality attributable to particulate matter pollution in Spain. *Environ. Pollut.* **2017**, *224*, 541–551. [CrossRef] [PubMed]

6.   Jimenez, E.; Linares, C.; Martinez, D.; Diaz, J. Particulate air pollution and short-term mortality due to specific causes among the elderly in Madrid (Spain): Seasonal differences. *Int. J. Environ. Health Res.* **2011**, *21*, 372–390. [CrossRef] [PubMed]

7.   Williams, A.; Spengler, J.D.; Catano, P.; Allen, J.; Laurent, J.C.C. Building Vulnerability a Chamging Climate: Indoor Temperature Exposures and Health Outcomes in Older Adults Living in Pubic Housing during an Extreme Heat Event in Cambridge, MA. *Int. J. Environ. Res. Public Health* **2019**, *16*, 2373. [CrossRef]

8.   Linares, C.; Díaz, J. Efecto de las partículas de diámetro inferior a 2,5 micras (PM2,5) sobre los ingresos hospitalarios en niños menores de 10 años en Madrid. *Gac. Sanit.* **2009**, *23*, 192–197. [CrossRef]

9.   Arroyo, V.; Díaz, J.; Carmona, R.; Ortiz, C.; Linares, C. Impact of air pollution and temperature on adverse birth outcomes: Madrid, 2001–2009. *Environ. Pollut.* **2016**, *218*, 1154–1161. [CrossRef]

10. Díaz, J.; Linares, C.; Tobías, A. Impact of extreme temperatures on daily mortality in Madrid (Spain) among the 45-64 age-group. *Int. J. Biometeorol.* **2006**, *50*, 342–348. [CrossRef]

11. Laurent, J.C.C.; Williams, A.; Oulhlote, Y.; Zanobetti, A.; Allen, J.; Spengler, J.D. Reduced cognitive function during a heat weave among residents of non-air-conditioned buildings: An observational study of young adults in the summer of 2016. *PLoS Med.* **2018**, *15*, 1–20.

12. UNICEF. *Progress for Every Child in the SDG Era*; UNICEF: New York, NY, USA, 2019.

13. Singh, M.K.; Ooka, R.; Rijal, H.B.; Kumar, S.; Kumar, A.; Mahapatra, S. Progress in thermal comfort studies in classrooms over last 50 years and way forward. *Energy Build.* **2019**, *188*, 149–174. [CrossRef]

14. Campano, M.Á.; Domínguez-Amarillo, S.; Fernández-Agüera, J.; Sendra, J.J. Thermal perception in mild climate: Adaptive thermal models for schools. *Sustainability* **2019**, *11*, 3948. [CrossRef]

15. Fernández-Agüera, J.; Campano, M.Á.; Domínguez-Amarillo, S.; Acosta, I.; Sendra, J.J. CO_2 Concentration and Occupants' Symptoms in Naturally Ventilated Schools in Mediterranean Climate. *Buildings* **2019**, *9*, 197. [CrossRef]

16. Schweiker, M.; André, M.; Al-Atrash, F.; Al-Khatri, H.; Alprianti, R.R.; Alsaad, H.; Amin, R.; Ampatzi, E.; Arsano, A.Y.; Azar, E.; et al. Evaluating assumptions of scales for subjective assessment of thermal environments—Do laypersons perceive them the way, we researchers believe? *Energy Build.* **2020**, *211*, 109761. [CrossRef]

17. López-Chao, V.; Lorenzo, A.A.; Saorín, J.L.; de la Torre-Cantero, J.; Melián-Díaz, D. Classroom Indoor Environment Assessment through Architectural Analysis for the Design of Efficient Schools. *Sustainability* **2020**, *12*, 2020. [CrossRef]

18. Ebersbach, M.; Brandenburger, I. Reading a short story changes children's sustainable behavior in a resource dilemma. *J. Exp. Child. Psychol.* **2020**, *191*, 104743. [CrossRef]

19. Konis, K.; Blessenohl, S.; Kedia, N.; Rahane, V. TrojanSense, a participatory sensing framework for occupant-aware management of thermal comfort in campus buildings. *Build. Environ.* **2020**, *169*, 106588. [CrossRef]

20. Martinez-Molina, A.; Boarin, P.; Tort-Ausina, I.; Vivancos, J.L. Post-occupancy evaluation of a historic primary school in Spain: Comparing PMV, TSV and PD for teachers' and pupils' thermal comfort. *Build. Environ.* **2017**, *117*, 248–259. [CrossRef]

21. Merabtine, A.; Maalouf, C.; Hawila, A.A.; Martaj, N.; Polidori, G. Building energy audit, thermal comfort, and IAQ assessment of a school building: A case study. *Build. Environ.* **2018**, *145*, 62–76. [CrossRef]

22. Rodriguez, C.M.; Coronado, M.C.; Medina, J.M. Classroom-comfort-data: A method to collect comprehensive information on thermal comfort in school classrooms. *MethodsX* **2019**, *6*, 2698–2719. [CrossRef] [PubMed]

23. Nakanishi, H. Children's travel behavior and implication to transport energy consumption of household: A case study of three Australian cities. In *Transport and Energy Research*; Elsevier: Amsterdam, The Netherlands, 2020; pp. 129–154.

24. Wang, Z.; Wang, J.; He, Y.; Liu, Y.; Lin, B.; Hong, T. Dimension analysis of subjective thermal comfort metrics based on ASHRAE Global Thermal Comfort Database using machine learning. *J. Build. Eng.* **2019**, *29*, 101120. [CrossRef]

25. Moyano, D.B.; Fernández, M.S.J.; Lezcano, R.A.G. Towards a Sustainable Indoor Lighting Design: Effects of Artificial Light on the Emotional State of Adolescents in the Classroom. *Sustainability* **2020**, *12*, 4263. [CrossRef]

26. Villaplana, A.; Yamanaka, T. Effect of smell in Space Perception. *Int. J. Affect. Eng.* **2015**, *14*, 175–182. [CrossRef]

27. Cuerdo Vilches, M.T. *User Participation in Energy Management of Buildings: Application of Photovoice Method in Workplaces (Spanish)*; University of Seville: Seville, Spain, 2017.

28. Papazoglou, E.; Moustris, K.P.; Nikas, K.S.P.; Nastos, P.T.; Statharas, J.C. Assessment of human thermal comfort perception in a non-air-conditioned school building in Athens, Greece. *Energy Procedia* **2019**, *157*, 1343–1352. [CrossRef]

29. Shahzad, S.; Calautit, J.K.; Hughes, B.R.; Satish, B.K.; Rijal, H.B. Visual thermal landscaping (VTL) model: A qualitative thermal comfort approach based on the context to balance energy and comfort. *Energy Procedia* **2019**, *158*, 3119–3124. [CrossRef]

30. ASHRAE 55. *Thermal Environmental Conditions for Human Occupancy*; ANSI/ASHRAE Standard 55-2017; American National Standards Institute: Washington, DC, USA, 2017.

31. ISO 7730:2005. *Ergonomics of the Thermal Environment—Analytical Determination and Interpretation of Thermal Comfort Using Calculation of the PMV and PPD Indices and Local Thermal Comfort Criteria*; ISO: Geneva, Switzerland, 2005.

32. Yang, B.; Li, X.; Hou, Y.; Meier, A.; Cheng, X.; Choi, J.-H.; Wang, F.; Wagner, A.; Yan, D.; Li, A.; et al. Non-invasive (noncontact) measurements of human thermalphysiology signals and thermal comfort/discomfort poses—A review. *Energy Build.* **2020**, *224*, 110261. [CrossRef]

33. Díaz, E.; Siegel, J. Indoor envorinmental quality in social housing: A literature review. *Energy Procedia* **2018**, *131*, 231–241.

34. Singh, M.; Ooka, R.; Rijal, H. Thermal comfort in Classrooms: A critical review. In Proceedings of the 10th Windsor Conference, Windsor, UK, 12–15 April 2018; pp. 649–668.

35. Campano-Laborda, M.Á.; Domínguez-Amarillo, S.; Fernández-Agüera, J.; Acosta, I. Indoor Comfort and Symptomatology in Non-University Educational Buildings: Occupants' Perception. *Atmosphere* **2020**, *11*, 357. [CrossRef]

36. Krüger, E.; Zannin, P. Acoustic, thermal and luminous comfort in classrooms. *Build. Environ.* **2004**, *39*, 1055–1063. [CrossRef]

37. Fusco, C.; Moola, F.; Faulkner, G.; Buliung, R.; Richichi, V. Toward an understanding of children's perceptions of their transport geographies: (Non)active school travel and visual representations of the built environment. *J. Transp. Geogr.* **2012**, *20*, 62–70. [CrossRef]

38. De Guili, V.; da Pos, O.; de Carli, M. Indor environmental quality and pupil perception in italian primary schools. *Build. Eviron.* **2012**, *49*, 129–140.

39. Jindal, A. Investigation and analysis of thermal comfort in naturally ventiladed secondary schools classrooms in the composite climate of India. *Archit. Sci. Rev.* **2019**, *62*, 466–484. [CrossRef]

40. Tucker, R.; Izadpanahi, P. Live green, think green: Sustainable school architecture and children's environmental attitudes and behaviors. *J. Environ. Psychol.* **2017**, *51*, 209–216. [CrossRef]

41. Zhang, D.; Bluyssen, P. Actions of primary school teachers to improve the indoor environmental quality of classrooms in the Netherlands. *Intell. Build. Int.* **2019**, 1–14. [CrossRef]

42. Zhang, D.; Ortiz, M.; Bluyssen, P. Clustering of Dutch school children based on their preferences and needs of the IEQ in classrooms. *Build. Environ.* **2019**, *147*, 258–266. [CrossRef]

43. Bluyssen, P.M.; Kim, D.H.; Eijkelenboom, A.; Ortiz-Sanchez, M. Workshop with 335 primary school children in The Netherlands: What is needed to improve the IEQ in their classrooms? *Build. Environ.* **2020**, *168*, 106486. [CrossRef]

44. Bluyssen, P.M.; Zhang, D.; Kurvers, S.; Overtoom, M.; Ortiz-Sanchez, M. Self-reported health and comfort of school children in 54 classrooms of 21 Dutch school buildings. *Build. Environ.* **2018**, *138*, 106–123. [CrossRef]

45. Kottek, M.; Grieser, J.; Beck, C.; Rudolf, B.; Rubet, F. World map of the Köpen-Geiger climate classification updete. *Meteorologische Zeischrift* **2006**, *15*, 259–263. [CrossRef]

46. Chollat, C.; Joly, A.; Houivet, E.; Bénichou, J.; Marret, S. School-age human figure drawings by very preterm infants: Validity of the Draw-a-Man test to detect behavioral and cognitive disorders. *Arch. Pediatr.* **2019**, *26*, 220–225. [CrossRef]

47. Galli, M.; Cimolin, V.; Stella, G.; de Pandis, M.F.; Ancillao, A.; Condoluci, C. Quantitative assessment of drawing tests in children with dyslexia and dysgraphia. *Hum. Mov. Sci.* **2019**, *65*, 51–59. [CrossRef] [PubMed]

48. Pace, C.S.; Guerriero, V.; Zavattini, G.C. Children's attachment representations: A pilot study comparing family drawing with narrative and behavioral assessments in adopted and community children. *Arts Psychother.* **2020**, *67*, 101612. [CrossRef]
49. Kim, J.K.; Suh, J.H. Children's kinetic family drawings and their internalizing problem behaviors. *Arts Psychother.* **2013**, *40*, 206–215. [CrossRef]
50. Zaidman-Zait, A.; Yechezkiely, M.; Regev, D. The quality of the relationship between typically developing children and their siblings with and without intellectual disability: Insights from children's drawings. *Res. Dev. Disabil.* **2020**, *96*, 103537. [CrossRef]
51. Altay, N.; Kilicarslan-Toruner, E.; Sari, Ç. The effect of drawing and writing technique on the anxiety level of children undergoing cancer treatment. *Eur. J. Oncol. Nurs.* **2017**, *28*, 1–6. [CrossRef]
52. Lourenço, P.; Pinheiro, M.D.; Heitor, T. Light use patterns in Portuguese school buildings: User comfort perception, behaviour and impacts on energy consumption. *J. Clean. Prod.* **2019**, *228*, 990–1010. [CrossRef]
53. Schweiker, M.; Abdul-Zahra, A.; André, M.; Farah, A.; Alkhatri, H.; Alprianti, R.R.; Alsaad, H.; Amin, R.; Ampatzi, E.; Arsano, A.Y.; et al. The Scales Project: A cross-national dataset on the interpretation of thermal perception scales. *Sci. Data* **2019**, *6*, 1–10. [CrossRef]
54. Amasuomo, T.; Amasuomo, J. Perceived Thermal Discomfort and Stress Behaviors Affecting Students' Learning in Lecture Theathers in Humid Tropics. *Buildings* **2016**, *6*, 18. [CrossRef]
55. Katafygiotou, M.; Serghides, D. Thermal comfort of a typical secundary school buiding in Cyprus. *Build. Eviron.* **2014**, *13*, 303–312.
56. Geng, Y.; Ji, W.; Lin, B.; Zhu, Y. The impact of thermal environment on occupant IEQ perception and productivity. *Build. Eviron.* **2017**, *121*, 158–167. [CrossRef]

 *sustainability*

Article

Impact Assessment for Building Energy Models Using Observed vs. Third-Party Weather Data Sets

Eva Lucas Segarra [1,*,†], Germán Ramos Ruiz [1,†], Vicente Gutiérrez González [1], Antonis Peppas [2], Carlos Fernández Bandera [1]

1 School of Architecture, University of Navarra, 31009 Pamplona, Spain; gramrui@unav.es (G.R.R.); vgutierrez@unav.es (V.G.G.); cfbandera@unav.es (C.F.B.)
2 School of Mining and Metallurgical Engineering, National Technical University of Athens (NTUA), 15780 Athens, Greece; peppas@metal.ntua.gr
* Correspondence: elucas@unav.es; Tel.: +34-948-425-600 (ext. 800000)
† These authors contributed equally to this work.

Received: 20 July 2020; Accepted: 17 August 2020; Published: 21 August 2020

Abstract: The use of building energy models (BEMs) is becoming increasingly widespread for assessing the suitability of energy strategies in building environments. The accuracy of the results depends not only on the fit of the energy model used, but also on the required external files, and the weather file is one of the most important. One of the sources for obtaining meteorological data for a certain period of time is through an on-site weather station; however, this is not always available due to the high costs and maintenance. This paper shows a methodology to analyze the impact on the simulation results when using an on-site weather station and the weather data calculated by a third-party provider with the purpose of studying if the data provided by the third-party can be used instead of the measured weather data. The methodology consists of three comparison analyses: weather data, energy demand, and indoor temperature. It is applied to four actual test sites located in three different locations. The energy study is analyzed at six different temporal resolutions in order to quantify how the variation in the energy demand increases as the time resolution decreases. The results showed differences up to 38% between annual and hourly time resolutions. Thanks to a sensitivity analysis, the influence of each weather parameter on the energy demand is studied, and which sensors are worth installing in an on-site weather station are determined. In these test sites, the wind speed and outdoor temperature were the most influential weather parameters.

Keywords: weather file management; weather datasets; weather stations; building energy simulation; sensitivity analysis of weather parameters; thermal zone temperature

---

1. Introduction

The sustainable development goals report of 2019 highlighted the concern of the United Nations toward a more sustainable world where people can live peacefully on a healthy planet. One of the most important areas for the protection of our planet is the actions to mitigate climate change. *"If we do not cut record-high greenhouse gas emissions now, global warming is projected to reach 1.5 °C in the coming decades"* [1]. This concern was also endorsed by 186 parties in the Paris agreement on climate change in 2015 [2]. One of the strategies for tackling climate change is to reduce energy consumption (by increasing the system efficiency) and increase the use of clean energy so that greenhouse gas emissions are reduced. In this process of decarbonization, the buildings and the construction sector are critical elements, since as the Global Status Report for Buildings and Construction highlighted, they are responsible *"for 36% of final energy use and 39% of energy and process-related carbon dioxide (CO_2) emissions in 2018, 11% of which resulted from manufacturing building materials and products such as steel, cement and glass."* [3].

For this reason, building energy models (BEMs) play an important role in the understanding of how to reduce the energy consumed by buildings (lighting, equipment, heating, ventilation, and air-conditioning (HVAC) systems, etc.) and how to optimize their use. As Nguyen et al. highlighted, there is a huge variety of building performance simulation tools [4], and EnergyPlus is one of the most used [5]. In any simulation program, to obtain reliable results, the model must not only be adjusted to the behavior of the building, but all the files on which it depends must be reliable and suitable for the use that will be given to the model. Of all these files, the weather file is, perhaps, one of the most important [6].

Bhandari et al. highlighted that there are three kinds of weather data files, *typical*, *actual*, and *future*, which should be selected according to the use of the energy model [6]. The first corresponds to a representation of the weather pattern of a specific place taking into account a set of years (commonly 20–30 years). For each month, the data are selected from the year that was considered most "typical" for that month so that it represents the most moderate weather conditions, excludes weather extremities, and reflects long-term average conditions for a location. In general, they are used to design and study the behavior of the building under standard conditions, to obtain Energy Performance Certificates, to study the feasibility of a building's refurbishment, etc. The typical files are not recommended in extreme conditions, such as designing HVAC systems for the worst case scenario.

There are two main sources of typical weather files: those developed by the National Renewable Energy Laboratory (NREL), which come from stations in the United States and its territories, where the different versions (typical meteorological year (TMY), TMY2, and TMY3) take into account different numbers of stations, time periods, solar radiation considerations, etc. [7,8]; and those developed by the American Society of Heating, Refrigerating and Air-Conditioning Engineers (ASHRAE) as a result of the ASHRAE Research Project 1015 [9,10]: the International Weather for Energy Calculations (IWEC), which takes into account weather stations outside the United States and Canada. The latest version (IWEC2) covers 3012 worldwide locations [11] taking into account a 25 year period (1984–2008) from the Integrated Surface Hourly (ISH) weather database.

The second kind of weather data file, *actual*, corresponds to a specific location and time period. It could be obtained from an on-site weather station or by processing data from several nearby stations. The latter option is commonly used by external data providers. This type of file is usually used to carry out building energy model calibrations, calculate energy bills and utility costs, study the specific behavior of HVAC systems, etc. As can be seen, these files take into account extraordinary situations, such as heat waves, adverse or extraordinary atmospheric phenomena, etc. [12].

Finally, the last kind of weather data file, *future*, is mainly used to simulate how it is possible to adapt the building energy demand to an external energy requirement (demand response [13]) or to obtain better use strategies of HVAC systems in certain situations (model predictive control [14,15]). As Lazos et al. highlighted, there are three common groups of forecasting techniques: statistical, machine learning, and physical and numerical methods [16].

Therefore, each kind of weather data file has a certain purpose; therefore, the reliability of the results will largely depend on the accuracy and suitability of the file selected [17,18].

Many studies highlighted the importance of the weather file when performing building energy analysis, as its accuracy is generally assumed, although it is out of the control of the person in charge of the simulation, its difference being significant. The following are some examples of studies that use *typical* weather files for different applications where the weather data are relevant. It is meaningful in retrofitting scenarios [19], or when quantifying renewable energy as in the cases when sizing photovoltaic solar panels [20], or when used to obtain ground temperatures [21], even when studying climate change; the accuracy of the weather files is also very important, as they are the baseline files in the process of morphing the data [22–24]. Therefore, when using such files, it is important to try to use the most recent [25].

There are other studies that analyze the building energy performance using *actual* weather files, either from commercial vendors [26], or developed using nearby weather stations, not located in the building [6,27–29], or the lesser ones, from weather stations placed in the building or in its surroundings [30]. Finally, regarding the *future* weather files, the uncertainty of the forecast files is closely related to the accuracy of the files on which they are based [31–33]. There are many studies that highlight the importance of the weather files, measuring, for example, their impact on passive buildings [34], on micro-grids [35,36], etc., calculating the loads of district energy systems [37], the electricity consumption with demand response strategies [13], or evaluating the effect on comfort conditions [38]. Some analyzed the effect that certain parameters of the weather file have, emphasizing the temperature as the most sensitive value for load forecasting [39,40].

In terms of temporal resolutions, most research focused on annual results when comparing different sets of weather data. Wang et al. analyzed the uncertainties in annual energy consumption due to weather variations and operation parameters for a reference office prototype, concluding that uncertainties caused by operation parameters were much more significant than weather variations [26]. Crawley et al. analyzed the energy results using measured weather data for 30 years and several weather datasets for a set of five locations in the USA, and the variation in annual energy consumption was on average ±5% [17]. Seo et al. conducted a similar study, also analyzing the peak electrical demand with similar results: a maximum difference of 5% [41].

In terms of research that focused on monthly criteria, Bhandari et al. found that, when using different weather datasets, the annual energy consumption could vary around ±7%, but up to ±40% when monthly analysis was performed [6]. Radhi compared the building's energy performance of using past and recent (up-to-date) weather data with annual and monthly criteria. This showed a difference of 14.5% between the annual electricity consumption simulated with past data and actual consumption, while this difference grew up to 21% for one month when monthly criteria were used [42]. Finally, there were a few studies where the temporal resolution was less than one month. With weekly criteria, Silvero et al. compared five different weather data sources with the observed meteorological data, showing that, for the annual criteria, the results were similar, but for the hottest and coldest weeks of the year, the inaccuracies increased [28].

The aim of this work is to show how to evaluate the impact of using two different *actual* weather datasets on the building energy model simulations, measuring both their effect on energy demand and indoor temperature. The purpose is to analyze if the data provided by the *on-site* weather stations (with a high economic cost and maintenance) could be replaced by the *actual* data provided by a *third-party*. For this study, four test sites based on real buildings were used to compare the existing variations when weather files with data obtained from real stations and external provider were used in the simulations. These test sites are part of an EU funded H2020 research and innovation project SABINA (SmArt BI-directional multi eNergy gAteway) [43], which aims to develop new technology and financial models to connect, control, and actively manage the generation and storage of assets to exploit synergies between electrical flexibility and the thermal inertia of buildings. The energy demand variation analysis was measured by grouping it into different periods (annual, seasonal, monthly, weekly, daily, and hourly) since as explained by ASHRAE [44], "…*the aggregated data will have a reduced scatter and associated CV(RMSE), favoring a model with less granular data.*" The objective of the paper is to highlight these differences in the results when using different granular criteria since, depending on the use of the building energy model, their influence can be significant, for example for calibration purposes, where the monthly or hourly criteria are required. A sensitivity analysis was also performed to evaluate the influence of each weather parameter on the energy demand variation when using the two different *actual* weather datasets.

The main contributions of this research are: (1) four real test sites, with different uses and architectural characteristics, located in three different climates were employed in the study; (2) while most of studies that analyze weather data influence in building energy simulation use the typical meteorological year (TMY) [19–24], this study performed a comparison of two *actual* datasets: *on-site* and *third-party* weather data; (3) when the studies used *actual* weather data, they usually lacked a local weather station due to its expensive installation [6,27–29]; instead, the observed weather data from this study were provided by three weather stations installed on the building roofs or in their surroundings, providing on-site measurements; and (4) the energy results are shown with different temporal resolutions (from annual to hourly) in order to highlight the differences in the variations when the data are accumulated.

The paper is structured as follows. In Section 2, we describe the methodology used to analyze both the differences between the *on-site* and *third-party* weather datasets and the variation produced by these weather files in the results of the simulations in terms of the energy demand and in terms of the indoor temperature. In Section 3, we show the results obtained from the different approaches: the weather datasets comparison (Section 3.1), energy demand (Section 3.2), and indoor temperature (Section 3.3). Finally, in Section 4, we discuss the results obtained in the study, and in Section 5, the conclusions are presented.

2. Methodology

Figure 1 shows the diagram of the methodology used in this study and the three analyses that were performed: (1) the weather data comparison between the data provided by the *on-site* weather stations and the *third-party* data; and through energy model simulations using these two weather datasets; (2) the energy demand comparison; and (3) the indoor temperature comparison.

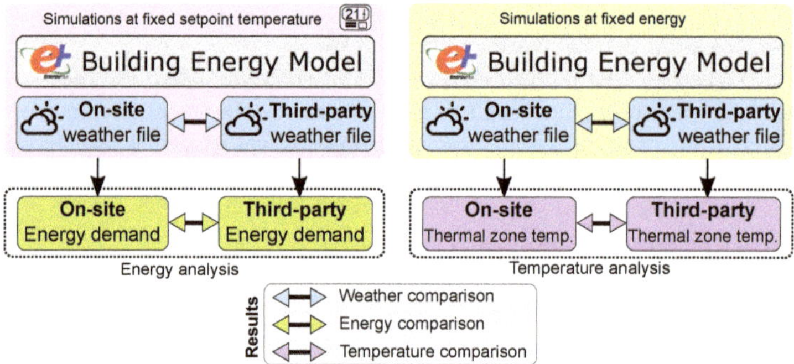

Figure 1. Diagram of the methodology used to check the effect of the use of different *actual* weather data in the building energy simulations. Three analyses are performed in the study: weather data comparison (methodology explained on Section 2.1), energy comparison (Section 2.2), and indoor temperature comparison (Section 2.3).

2.1. Weather Data Comparison Methodology

Once the weather data from the *on-site* weather station and the *third-party* are gathered, the first analysis is the comparison using a Taylor Diagram [45,46]. This provides a simple way of visually showing how closely a pattern matches an observation, and it is a useful tool to easily compare different parameters at a glance using the same plot. This type of comparison is widely used when weather parameters are analyzed [28,47–51]. Developed by Taylor in 2001, this diagram shows the correspondence between two patterns (in this case, *third-party* weather data as the test field (f) and *on-site* weather data as the reference field (r)) using three statistical metrics: the correlation R, the centered root-mean-squared difference RMS_{diff}, and the standard deviation σ of the test and reference field.

The correlation R (3) is used to show how strongly the two fields are related, and it ranges from -1 to 1. The centered root-mean-squared difference RMS_{diff} (4) measures the degree of adjustment in amplitude. The closer to 0, the more similar the patterns are. Both indexes provide complementary information quantifying the correspondence between the two fields, but to have a more complete characterization, their variances are also necessary, which are represented by their standard deviations σ_f (1) and σ_r (2) [46]. To allow the comparison between different weather parameters and to show them in the same plot, RMS_{diff} and σ_f are normalized by dividing both by the standard deviation of the observations (σ_r). Thus, the normalized reference data have the following values: $\sigma_r = 1$, $RMS_{diff} = 0$, and $R = 1$.

Figure 2 shows the Taylor diagram baseline plot and how it is constructed. The reference point appears in the x-axis as a grey point. The azimuthal positions show the correlation coefficient R between the two fields. The standard deviation for the test field σ_f is proportional to the radial distance from the origin, with the solid dashed arc as the reference standard deviation σ_r. Finally, the centered root-mean-squared difference RMS_{diff} between the test and reference patterns is proportional to the distance to the reference point, and the arcs indicate its value. The diagram allows us to determine the ranking of the *test* fields by comparing the distance to the reference point.

In Figure 2, two test fields are represented as an example: Example 2 has a correlation ±0.99, a RMS_{diff} ±0.24, and a σ_f ±1.15. Example 1 has a correlation ±0.48, a RMS_{diff} ±1.40, and a σ_f ±1.55. Example 2 performs better than Example 1 since all the metrics are better. The diagram shows this in a visual way as Example 2 is closer to the reference point.

$$\sigma_f^2 = \frac{1}{N} \sum_{n=1}^{N} (f_n - \bar{f})^2 \tag{1}$$

$$\sigma_r^2 = \frac{1}{N} \sum_{n=1}^{N} (r_n - \bar{r})^2 \tag{2}$$

$$R = \frac{\frac{1}{N} \sum_{n=1}^{N} (f_n - \bar{f})(r_n - \bar{r})}{\sigma_f \sigma_r} \tag{3}$$

$$RMS_{diff} = \left[\frac{1}{N} \sum_{n=1}^{N} [(f_n - \bar{f}) - (r_n - \bar{r})]^2 \right]^{1/2} \tag{4}$$

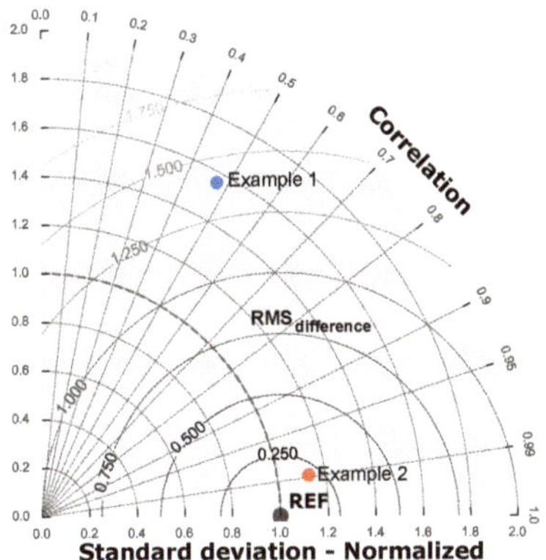

Figure 2. A Taylor diagram example.

2.2. Energy Analysis Methodology

The study of the impact on the energy demand due to the use of two different *actual* weather datasets is based on the energy simulation, and therefore, building energy models (BEM) are needed. For this study, detailed BEMs are employed using the EnergyPlus engine [52,53]. BEMs require weather files in the *EPW* format, which are created using the weather data from both sources (*on-site* and *third-party* weather data) and employing the Weather Converter tool [54] provided as an auxiliary program by EnergyPlus.

As shown in Figure 1, the energy analysis studies the impact on the energy demand when using both *on-site* and *third-party* weather files. BEM provides the energy demand that it requires each weather file to accomplish with the defined requirements (the temperature setpoint of each space).

The energy results using the *on-site* weather file are established as the reference as they corresponded with the weather data measured in the building's surroundings. Comparing the variation between the energy demand results using both weather files allows us to analyze the impact of using weather data from *third-party* sources with respect to the reference.

In order to perform a deeper study, a sensitivity analysis is performed to analyze the effects on the energy demand generated by each weather parameter. The methodology consists of replacing variables (one variable at a time) from the *on-site* weather file with data from the *third-party* weather file and generating specific weather files for each parameter. For example, when the dry bulb temperature is analyzed, a weather file is prepared that has the dry bulb temperature data from the *third-party*, but the rest of the weather parameters are maintained the same as in the *on-site* weather file. This way, the impact on energy demand when using only dry bulb temperature data from a *third-party* can be studied. This procedure is done for the weather parameters analyzed in the weather comparison: the dry bulb temperature (*Temp*), relative humidity (*RH*), direct normal irradiation (*DNI*), diffuse horizontal irradiation (*DHI*), wind speed (*WS*), and wind direction (*WD*). These weather parameters are selected for the study as they are all used by EnergyPlus in the simulations, unlike other parameters, such as the global horizontal irradiation [54].

In the case of the other weather parameters provided by the weather stations, such as the atmospheric pressure and precipitation, they are not presented in this study as their impact in the BEMs is low. The process to perform the energy analysis is the same as before: BEM is simulated with the generated weather file with one parameter changed, obtaining an energy demand that is compared to the reference (the energy demand obtained using the *on-site* weather data).

As was explained in the Introduction, most of the studies that analyzed the effect of using different weather datasets in the building energy simulations used only the annual energy results, and only some of them used smaller temporal resolutions (monthly or weekly). This study presents the analysis according to different temporal resolutions and discusses the differences in the results. The time granularity levels proposed are annual, seasonal, monthly, weekly, daily, and hourly. Thus, the uncertainty metrics calculated for the energy results are related to the accumulated energy demand provided by the model in year, season, month, week, day, and hour periods.

For the statistical analysis of the results, three metrics are used in the study: the mean absolute deviation percent (*MADP*) (5), the coefficient of variation of the root-mean-squared error (*CV*(*RMSE*)) (6), and the coefficient of determination (R^2) (7). The equations of these statistical indexes are shown as:

$$MADP = \frac{\sum_{i=1}^{n} |y_i - \hat{y}_i|}{\sum_{i=1}^{n} |y_i|} \tag{5}$$

$$CV(RMSE) = \frac{1}{\bar{y}_i} \sqrt{\frac{\sum_{i=1}^{n} (y_i - \hat{y})^2}{n - p}} \tag{6}$$

$$R^2 = \left(\frac{n \cdot \sum_{i=1}^{n} y_i \cdot \hat{y} - \sum_{i=1}^{n} y_i \cdot \sum_{i=1}^{n} \hat{y}}{\sqrt{(n \cdot \sum_{i=1}^{n} y_i^2 - (\sum_{i=1}^{n} y_i)^2) \cdot (n \cdot \sum_{i=1}^{n} \hat{y}^2 - (\sum_{i=1}^{n} \hat{y})^2)}} \right)^2 . \tag{7}$$

In the equations, n is the number of observations, y_i the *on-site* measured data at moment i, and $\hat{y}_i$ the *third-party* value at that moment.

MADP and *CV*(*RMSE*) are both quantitative indexes that show the results in percentage terms. They allow the comparison between different test sites, weather parameters, and time resolutions. *MADP*, which is also called the MAD/mean in some studies [55], has advantages that overcome some shortcomings of other metrics. It is not infinite when the actual values are zero, is very large when actual values are close to zero, and does not take extreme values when managing low-volume data [55–57]. *CV*(*RMSE*), which gives a relatively high weight to large variations, is the other percentage metric selected for this study because it is a common metric in energy analysis. Indeed, the ASHRAE (American Society of Heating, Refrigerating and Air-Conditioning Engineers)

Guidelines [44], FEMP (Federal Energy Management Program) [58], and IPMVP (International Performance Measurement and Verification Protocol) [59] use it to verify the accuracy of the models.

The coefficient of determination (R^2) allows us to measure the linear relationship of the two patterns [60]. It ranges between 0.00 and 1.00, and higher values are better. It should be noted that uncertainty cannot be assessed using only this metric as the linear relationship may be strong, but with a substantial bias.

In the study, the $MADP$ and $CV(RMSE)$ metrics are shown for all the temporal resolutions, from annual to *hour*. However, R^2 is only analyzed for the hourly time grain as the study of the linear relationship of larger time grains variations, which has few points, is meaningless.

2.3. Indoor Temperature Analysis Methodology

The third comparison analysis studies the impact of using the two weather datasets (*on-site* and *third-party* weather files) for the building's indoor temperature. In order to allow the temperature comparison, the energy used by each model is fixed. In other words, both simulations with *on-site* and *third-party* weather data use the exact same energy; however, due to the differences in the weather parameters, the indoor temperature is different. The methodology consists of performing the first simulation with the *on-site* weather file to obtain the baseline energy demand for each thermal zone of the model. This baseline energy demand is then injected into the model using an EnergyPlus script for an HVAC machine that distributes that energy in each thermal zone. Then, the model is simulated for both the *on-site* and *third-party* weather files. The results of the building temperature—unifying thermal zone temperature, weighing it with respect to its volume—of these two last simulations are compared to analyze the impact on the indoor temperature conditions.

In this case, two quantitative indexes (mean absolute error MAE (8) and root-mean-squared error $RMSE$ (9)) and a qualitative index (R^2 (7)) are used to quantify the variation in the shape of the temperature curves.

$$MAE = \frac{1}{n} \sum_{i=1}^{n} |y_i - \hat{y}_i| \tag{8}$$

$$RMSE = [\frac{1}{n} \sum_{i=1}^{n} (y_i - \hat{y}_i)^2]^{\frac{1}{2}}. \tag{9}$$

The MAE and $RMSE$ indexes are used to determine the average variation of the indoor temperature when using the different weather files in the simulations [61]. Both measure the average magnitude of the variation in the units of the variable of interest and are indifferent to the direction of the differences, overcoming cancellation errors. However, $RMSE$ gives a relatively high weight to large deviations [62–64]. $RMSE$ will always be greater than or equal to MAE (due to its quadratic nature); thus, the greater the difference between MAE and $RMSE$, the greater the variance between the individual dispersions on the sample. In this case, the three metrics are calculated for the hourly criteria.

3. Results

The methodology described in the previous section was applied to four buildings located in three different real test sites for a period of one year (2019). The test sites were an office building at the University of Navarre in Pamplona (Spain), a public school in Gedved (Denmark), and two buildings in the Lavrion Technological and Cultural Park (LTCP) in Lavrion, Greece: H2SusBuild and an administrative building. As shown in Figure 1, three different analyses were performed in the study: weather dataset comparison, energy comparison, and indoor temperature comparison. The following three subsections develop the three analyses.

3.1. Weather Data Comparison

For the analysis and comparison of the weather data, the first step was the data gathering from the *on-site* and *third-party* sources for the three locations for the period of study, which is the whole year 2019. *On-site* weather data were provided by weather stations installed in the buildings' surroundings. In Pamplona and Gedved, the weather station was installed on the buildings' roofs. In the case of Lavrion, the weather station was placed in the Technological and Cultural Park where the two test sites were located, near H2SusBuild. Table 1 shows the range, resolution, and accuracy of the sensors that formed part of each weather station. In general, the sensors of Pamplona's weather station had the best accuracy. In the case of Lavrion, the diffuse solar radiation had a manual shadow bar that required readjustment every two days.

The time period of the measured data gathered from the three weather stations is the whole year 2019. Despite the good quality of the measured weather data, the raw data contained small gaps, usually a few hours, so interpolation was performed in order to fill in the missing data. On the other hand, the Weather Converter tool, which is used to generate the weather files, allows undertaking a complementary validation of the data since it produces a warning if data out of the range are used in the weather files' generation process.

The *third-party actual* weather data for the year 2019 and for three locations were provided by meteoblue [65]. They are simulated historic data for a specific place and time calculated with models based on the NMM (nonhydrostatic meso-scale modeling) or NEMS (NOAA Environment Monitoring System) technology, which enables the inclusion of the detailed topography, ground cover, and surface cover. Further information about the computation of the weather data provided by meteoblue is available in [66].

The results of the weather parameter comparison between data from *on-site* weather stations and *third-party* (meteoblue) are shown using the Taylor diagram described in Section 2.1. Figure 3 summarizes all the results: showing the three statistics (R, $RMS_{difference}$, and σ) for the whole period of study (2019) calculated on an hourly basis, for the six weather parameters analyzed (*Temp*, *RH*, *DNI*, *DHI*, *WD*, and *WS*), and for the three locations (each one represented in a different color).

For Pamplona weather (represented in blue), the diagram shows that *Temp* provided the weather parameter for this location that better agreed with the *on-site* observations as it had the highest correlation R of around 0.95, the smallest RMS_{diff} (± 0.3), and a very close σ_f (standard deviation) to the reference (± 0.95). *RH*, *DNI*, and *DHI* provided similar results with a correlation around 0.7–0.8, an RMS_{diff} between 0.5–0.6, and a good standard deviation. The parameters that correlated worse with the observed values were the wind parameters, especially *WD* ($R = 0.09$).

For Gedved weather (red color), the Taylor diagram shows that *Temp* was the *third-party* weather parameter that agreed best with the *on-site* observations, with an R of around 0.97. As in Pamplona, the wind parameters delivered the results furthest from the reference point. *WS* had an acceptable correlation, but a very high standard deviation, and *WD* performed better for σ_f, but had a low correlation (less than 0.5). For the third location, Lavrion (green color), *Temp* also had a good correlation R (higher than 0.95). *RH*, *DHI*, and *WS* had a medium R for the observed data (around 0.8), but they presented differences in the other metrics. *RH* had a better standard deviation than the other two, and *RH* and *DHI* had a lower RMS_{diff} than *WS*. In this location, the parameter that provided the worst results was *WD*, which had the worst R (-0.22) and the highest RMS_{diff} (1.5).

Table 1. Technical specifications of the sensors of the weather stations installed in the office building in Pamplona (Spain), in Gedved School (Denmark), and in the Technological and Cultural Park in Lavrion (Greece).

Sensor	Pamplona (Spain)			Gedved (Denmark)			Lavrion (Greece)		
	Range	Resolution	Accuracy	Range	Resolution	Accuracy	Range	Resolution	Accuracy
Temperature (°C)	−50 to +60	0.1	±0.2	−40 to +60	0.1	±0.3	−40 to +65	0.1	±0.5
Relative Humidity (%)	0 to 100	0.1	±2	0 to 100	1	±2.5	0 to 100	1	±3 (0–90%) ±4 (90–100%)
Atmospheric Pressure (mbar)	300 to 1200	0.1	±0.5	150 to 1150	0.1	±1.5	880 to 1080	±0.1	±1
Precipitation (mm)	0.3 to 5.0	0.01	-	-	0.2	±2%	-	0.2	±4%/0.25 (<50 mm/h) ±5%/0.25 (>50 mm/h)
Wind Direction (°)	0 to 359.9	0.1	<3	1 to 360	1	1%	1 to 360	1	±4%
Wind Speed (m/s)	0 to 75	0.1	±3% (0–35) ±5% (>35)	1 to 96	1	0.1 (5–25)	1 to 67	0.44	±1/±5%
Global Solar Radiation (W/m^2)	0 to ±1300	1	<±10%	0 to ±1300	1	<±10%	0 to1500	1	<10
Diffuse Solar Radiation (W/m^2)	0 to ±1300	1	<±10%	0 to ±1300	1	<±10%	0 to 1500	1	<20

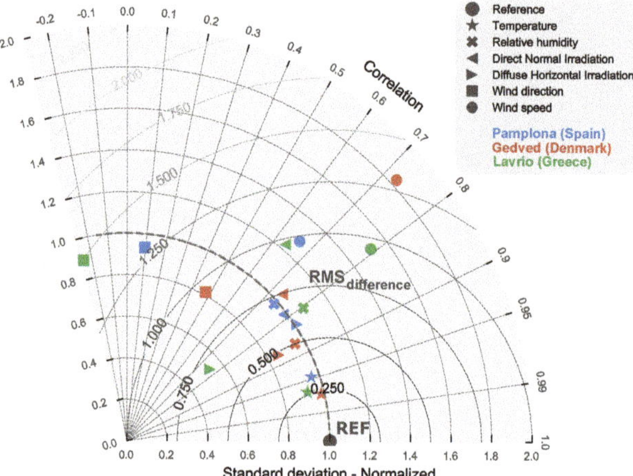

Figure 3. The normalized Taylor diagram for the three weather locations (Pamplona (Spain), Gedved (Denmark), and Lavrio (Greece)) compared on an hourly basis for *on-site* and *third-party* weather data for the year 2019. This shows the dry bulb temperature, relative humidity, direct normal irradiation, diffuse horizontal irradiation, wind speed, and wind direction.

Comparing the statistical results for the three locations, the performance of data provided by the *third-party* varied for each location. Gedved provided the best results for four of the six weather parameters (*Temp*, *RH*, *DHI*, and *WD*). In the three cases, *Temp* was the parameter that best matched the reference (R around 0.95, RMS_{diff} lower than 0.4, and σ_f near one), and *WD* was the worst (correlations lower than 0.5 and RMS_{diff} higher than 0.9). *WS* also provided poor correspondence with the observations, especially for the Gedved location. The rest of the parameters were scattered in the medium part of the diagram.

In Appendix A, a deeper analysis is shown where the statistical indexes for the monthly and seasonal data are represented in order to analyze their homogeneity. Figures A1–A3 show that the *Temp*, *RH*, *DNI*, *DHI*, and *WD* parameters for the three weathers were quite homogeneous, with the seasonal and monthly indexes quite concentrated, providing similar R, RMS_{diff}, and σ_f. There are some exceptions, such as *DNI* for November in Pamplona and January in Gedved, which agreed worse with the observations than for the rest of the months. On the other hand, *WS* had more heterogeneous monthly and seasonal results since more scattered points were seen in the diagrams. In general, the winter and autumn months correlated the worst with the observed data.

Since the wind parameters produced higher variations when comparing *on-site* and *third-party* weather datasets, a deeper comparison analysis was performed using wind rose diagrams (see Figure 4). This diagram shows the distribution of the wind speed and wind direction. The rays point to the direction the wind is coming from, and their length indicates the frequency in percentage. The color depends on the wind speed, growing from blue to red colors. Pamplona's wind rose shows that *WS* from the *third-party* data was much higher than observed, and although the prevailing direction was north in both cases, there were important differences in the frequency percentages. For Gedved, the *third-party* data provided much higher wind speed (yellow to red colors in the wind rose) than the observed data (blue colors) and a different prevailing wind direction. Finally, the wind roses for Lavrion show differences in the prevailing wind direction and very different wind speeds, being higher in the *third-party* wind rose.

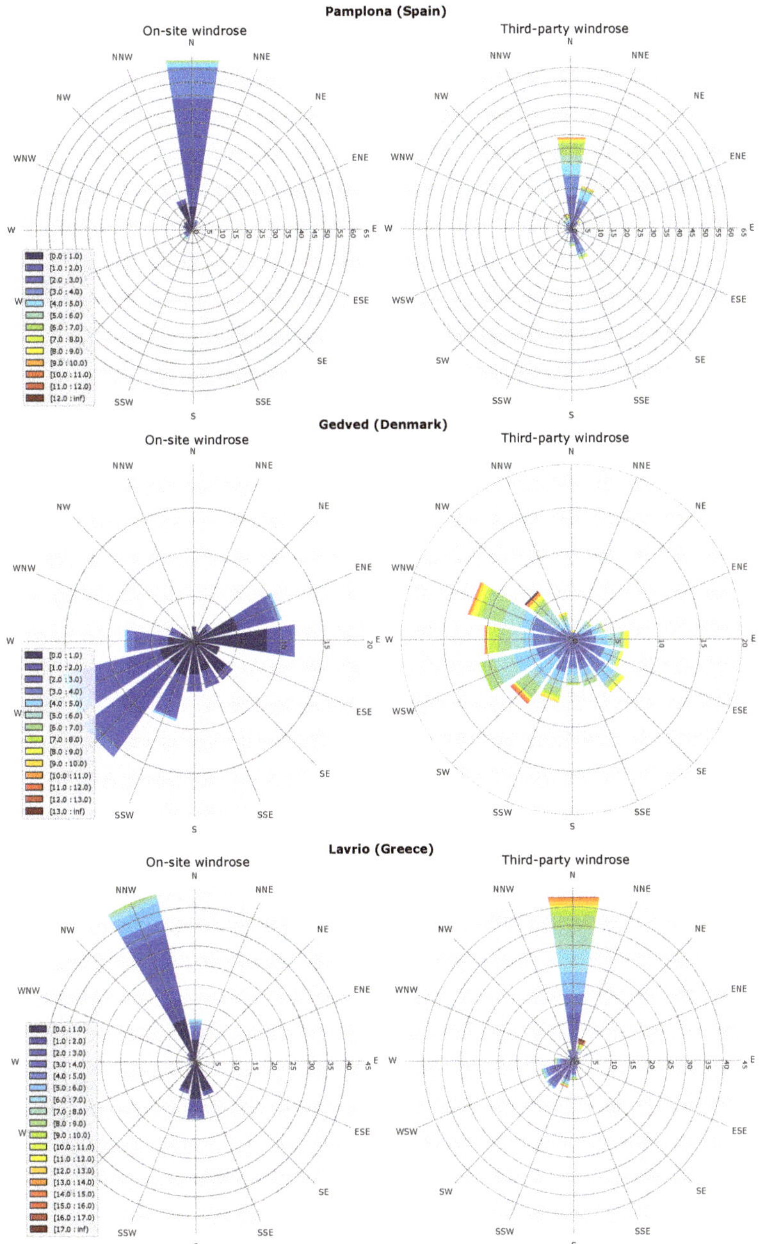

Figure 4. Wind rose comparison between the weather station and *third-party* data of the Pamplona (Spain), Gedved (Denmark), and Lavrion (Greece) locations.

3.2. Energy Analysis Results

After analyzing the variations in the different weather parameters between the *on-site* and *third-party* weather data for the three weathers, the following analysis consisted of the study of the impact produced by these variations in the test sites' building energy demands using detailed BEMs. Figure 5 shows the four test sites analyzed in this study showing a real image from the building and an image of the performed model in EnergyPlus (in colors for the different thermal zones of each building).

Figure 5. The test sites analyzed. From left to right: office building (University of Navarre, Spain); Gedved school (Denmark); H2SusBuildand administration building in Lavrio (Greece). On top, the real buildings and, on the bottom, the building energy models (SketchUp thermal zone representation, OpenStudio plugin [67]).

The first test site was the office building attached to the Architecture School at the University of Navarre in Pamplona (Spain). It hosts administration uses and classrooms for the postgraduate students. This building is a 755 square meter single-story building built in 1974. It has a concrete structure; the outdoor walls are built of red brick fabric (U value = 0.3 W/m^{2}K); the flat roof has the insulation above the deck (U value = 0.2 W/m^{2}K); and aluminum window frames were installed in situ with an air chamber. The Gedved public school (Denmark) consists of six buildings and was built in 1979 and then renewed in 2007. The library of one of the school buildings was selected as the test site. It is a one-story building with a total surface area of 1138 m^2, with a big central space—the library—and nine classrooms and serving spaces around it. The building walls consist of two brick layers with 150 mm insulation in between (U value = 0.27 W/m^{2}K). The windows are two-layer double-glazed windows with cold frames. The ceiling is insulated with 200–250 mm mineral wool for the sloping and flat ceiling, respectively (U value = 0.07 W/m^{2}K and 0.16 W/m^{2}K, respectively), and the floor is made of concrete and contains 150 mm insulation under it (U value = 0.21 W/m^{2}K).

In Lavrion (Greece), two buildings from the Lavrion Technological and Cultural Park (LTCP) were used as test sites: H2SusBuild and the administration building. H2SusBuild has a ground floor and an attic floor with a total surface area of approximately 505 m^2 . The ground floor hosts a small kitchen, toilets, the control room, and the main area. The attic also hosts two offices and a meeting room. Its envelope consists of a concrete structure with double concrete block walls and single-glazed windows with aluminum frames. It also has external masonry consisting of double brick walls with 10 cm expanded polystyrene (EPS) insulation (U value = 0.25 W/m^{2}K). The roof consists of metallic panels with a 2.5 cm polyurethane insulation layer in the middle (U value = 0.75 W/m^{2}K). The administration building hosts the LTCP managing authority and administrative services. It is a two-story renovated neoclassic building with a surface area of about 644 m^2. The building envelope is made of stone approximately 70 cm thick (U value = 1.85 W/m^{2}K) with wooden-framed double-glazed large windows. The roof consists of a wooden frame with gutter tiles placed on top (U value = 0.49 W/m^{2}K).

For each building, an individual pattern of use corresponding to the actual use of the building was implemented in the simulation model. Each building had its own calendar of use, occupancy, and internal loads of electric equipment and lighting. Regarding the HVAC systems, setpoints and usage hours were defined for each. The office building in Pamplona and H2SusBuild and administration building in Lavrion implemented both heating and cooling systems in the models, and the Gedved school had only a heating system. Table 2 shows the input data of the four models.

Table 2. Input data of the four models.

	Office Building Pamplona (Spain)	Public School Gedved (Denmark)	H2SusBuild Lavrion (Greece)	Administration Building Lavrion (Greece)
Lighting (W/m^2)	10	10	8.5	8.5
Equipment (W/m^2)	8	15	8	8
Occupation schedule	Wd 9–21 h/Sat 9–14 h	Wd 8–16 h/Sat 8–13 h	Wd 9–20 h/Sat 9–14 h	Wd 9–20 h/Sat 9–14 h
Heating setpoint (°C)	20	Day 21/Night 15.6	21	21
Cooling setpoint (°C)	26	No cooling	24	24

Wd: weekdays. Sat: Saturdays.

The results of the statistical analysis for the energy study are presented in Table 3. The table is divided into four sub-tables, one for each test site. They show the three uncertainty metrics calculated for the energy demand obtained from simulations using the *on-site* and *third-party* weather files (*TPW*), with *on-site* as the reference. The first column of each test site's table, designated as *TPW*, shows the difference in percentage (*MADP*) of the energy demand when the third-party weather file is used in the simulation with respect the the reference simulated with *on-site* weather data. With the inputs shown in Table 2, the models provide the following annual energy demand: 21.9 kWh/m^2 for the office building, 91.5 kWh/m^2 for the Gedved school, 142.2 kWh/m^2 for H2SusBuild, and 91.6 kWh/m^2 for the administration building.

The table allows performing two different analyses depending on how it is read. The vertical interpretation of the table shows the percentage results as a function of the time resolutions (from annual to hourly criteria) employed for the analysis. On the other hand, horizontally, the variations in the energy demand for the sensitivity analysis changing only one weather parameter at a time are presented (*DHI*, *DNI*, *RH*, *Temp*, *WD*, and *WS*).

Table 3. Uncertainty metrics (*MADP*, *CV*(*RMSE*), and R^2) were used in the energy analysis for the four test sites. *TPW*: the results using the weather file with all the parameters from the *third-party* weather data source. The rest changed only one parameter at a time: *DHI* (diffuse horizontal irradiation), *DNI* (direct normal irradiation), *RH* (relative humidity), *Temp* (temperature), *WD* (wind direction), and *WS* (wind speed).

Statistic Index	Time	Office Building, Pamplona (Spain)							School, Gedved (Denmark)						
		TPW	DHI	DNI	RH	Temp	WD	WS	TPW	DHI	DNI	RH	Temp	WD	WS
MADP (%)	Year	1.42	0.79	4.33	0.50	4.07	0.03	7.95	43.82	0.82	6.24	1.16	0.68	0.01	46.91
	Season	9.61	3.93	9.01	3.17	4.07	0.03	11.74	43.82	0.82	6.24	1.16	1.71	0.01	46.91
	Month	18.14	4.65	10.29	2.97	12.07	0.03	13.60	43.82	0.82	6.24	1.16	3.59	0.01	46.91
	Week	17.88	5.24	11.17	3.40	14.39	0.04	13.99	43.82	0.90	6.24	1.19	5.16	0.02	46.91
	Day	26.03	6.25	11.80	4.02	22.53	0.05	14.08	44.41	0.98	6.25	1.37	8.78	0.02	46.91
	Hour	29.96	6.64	12.27	4.71	25.58	0.08	14.31	45.17	1.03	6.27	1.50	10.10	0.02	46.91
CV(RMSE) (%)	Year	1.42	0.79	4.33	0.50	4.07	0.03	7.95	43.82	0.82	6.24	1.16	0.68	0.01	46.91
	Season	10.78	4.72	10.11	4.00	4.54	0.03	15.48	58.42	0.95	7.61	1.49	2.25	0.02	60.25
	Month	26.12	5.55	13.15	4.02	14.61	0.04	17.96	52.44	0.99	7.05	1.41	4.39	0.02	54.62
	Week	26.72	7.12	15.04	5.17	19.68	0.06	20.01	54.15	1.19	7.10	1.45	7.55	0.02	55.29
	Day	42.90	9.29	18.18	7.60	34.96	0.08	22.07	60.39	1.44	7.74	1.86	13.11	0.03	59.33
	Hour	61.42	13.14	22.37	12.56	50.71	0.20	30.55	91.79	2.75	14.04	3.54	24.99	0.05	88.81
R^2 (%)	Hour	90.86	99.48	98.64	99.53	92.54	100.00	98.43	95.18	99.99	99.82	99.98	98.55	100.00	96.84

Statistic Index	Time	H2SusBuild, Lavrion (Greece)							Administration Building, Lavrion (Greece)						
		TPW	DHI	DNI	RH	Temp	WD	WS	TPW	DHI	DNI	RH	Temp	WD	WS
MADP (%)	Year	32.86	5.95	5.22	1.97	5.21	1.05	39.40	1.29	11.40	7.27	3.45	9.62	1.93	12.19
	Season	44.58	8.14	9.49	3.21	7.25	2.16	39.40	27.75	13.83	13.40	4.20	9.62	2.73	21.63
	Month	45.83	10.32	11.20	3.95	11.82	2.94	41.66	38.70	15.56	15.80	4.41	14.73	2.99	29.97
	Week	47.80	10.50	11.15	4.05	13.02	2.93	42.14	38.70	15.69	15.80	4.52	14.78	2.99	29.97
	Day	49.46	10.75	11.40	4.46	17.17	2.94	42.49	38.91	15.72	15.80	4.91	16.60	3.00	29.97
	Hour	51.58	10.93	11.85	5.12	19.97	2.95	43.95	39.45	15.79	15.83	5.49	17.88	3.01	30.22
CV(RMSE) (%)	Year	32.86	5.95	5.22	1.97	5.21	1.05	39.40	1.29	11.40	7.27	3.45	9.62	1.93	12.19
	Season	61.65	9.69	10.71	3.91	12.40	2.58	63.61	36.21	15.57	14.26	4.78	15.61	2.93	34.39
	Month	65.24	13.41	13.37	5.04	15.66	3.44	65.89	43.53	20.89	18.63	6.20	20.23	3.83	38.06
	Week	72.40	14.11	13.97	5.46	19.34	3.52	73.29	46.18	21.35	19.09	6.62	21.59	3.85	40.61
	Day	82.65	14.77	14.74	6.37	24.90	3.68	81.45	49.81	21.93	19.63	7.76	24.17	3.97	43.01
	Hour	90.37	18.35	17.92	8.83	34.68	4.85	85.67	90.81	37.52	33.69	15.03	47.91	6.67	72.69
R^2 (%)	Hour	85.58	98.41	98.43	99.61	93.76	99.88	92.43	81.85	97.30	97.80	99.51	94.91	99.92	91.18

The first analysis obtained from Table 3 was the influence of the time resolution used in the study of the energy demand variation. In this case, the analysis was done in the vertical from the annual to hourly criteria. The percentage metrics $MADP$ and $CV(RMSE)$ allowed us to compare the results for each time grain and study its influence in the results. Both indexes were closely related; however, $CV(RMSE)$ gave a relatively high weight to large variations. It is remarkable that the differences between $CV(RMSE)$ and $MADP$ decreased as the time grain increased (from hourly to annual criteria) as, when the energy demand was accumulated, the outliers were minimized. The results for the hourly basis showed that the $CV(RMSE)$ values were around twice the $MADP$ values for the four sites and all the weather parameters. This indicates that significant outliers were present in the energy demand results when both simulations based on the *on-site* and *third-party* weather datasets were compared.

On the other hand, both the $MADP$ and $CV(RMSE)$ metrics showed how, in the four test sites, the variations in the energy demand grew as the time grain decreased, which matches with ASHRAE's statement about the energy data granularity [44]. If the results were analyzed with the accumulated energy demand for a period of time (i.e., monthly, annual, etc.), the energy variation was minimized with respect to the hourly analysis. For example, differences of $MADP$ up to $\pm38\%$ between the annual and hourly criteria are seen in the results for the administration building. In this case, the $MADP$ for the accumulated data for the year was only 1.29%; thus, the annual building energy demand simulated for the *third-party* weather file was very similar to the reference, simulated with the *on-site* weather file.

However, for the hourly basis, this variation grew up to 39.45%, which is a significant deviation. The reason is because, alternately, in some cases, the model simulation overestimated the energy demand needed by the building (the model demanded more energy than the reference), and in other cases, the model underestimated it. When the data were accumulated from the hourly basis to longer periods of time (daily, weekly, monthly, seasonal, and annual), a compensation effect occurred by canceling each other out, which resulted in the minimization of the energy demand variation. As the length of the periods increased, so did the compensation effect and, therefore, also the minimization of the variations.

It is also remarkable that for all the test sites, the $CV(RMSE)$ results showed high values for the monthly and hourly resolutions, which are the time criteria commonly used by the energy analysis standards.

The second analysis was the study of the influence of each weather parameter in the energy demand variation. In this case, the interpretation of the tables from 3 was done horizontally: the first column presents the results for the simulation with the *third-party* weather file (TPW), which had all the weather parameters changed, and the following columns show the results for the different weather parameters.

Comparing the results of the four test sites using the $MADP$ and $CV(RMSE)$ indexes, some common observations can be made. In all of them, the weather parameter that generated less impact in the simulated energy demand was WD, even though it was the weather parameter that worst fit the *on-site* weather data, as was shown in the Taylor diagram (Figure 3). The reason is because the mechanical ventilation and infiltration EnergyPlus objects used in these models did not account for WD in the simulations [68].

On the other hand, in the four test sites, when WS was analyzed, it showed an important impact in the energy demand simulations' outputs. This was mainly due to two causes: The first was the use of dynamic infiltrations introduced by using the EnergyPlus object *ZoneInfiltration:EffectiveLeakageArea*, which took into account the WS parameter in the calculations [68]. The leakage area in cm^2 was calculated by the calibration process previously developed by the authors [69–71]. The second was because the differences between the *third-party* WS data and *on-site* data (see the Taylor diagram in Figure 3 and the wind roses from Figure 4) were significant.

In both the Gedved school and H2SusBuild, the *third-party* wind speed provided faster values, which generated a significant increase in the energy demand during almost all the year, but there were a few moments with a decrease in the energy demand. Therefore, the compensation effect between the overestimated and underestimated energy demand was reduced. This explains why the variation due to WS was high for all the time grains for these test sites. This effect was especially clear in the Gedved school, which did not have a cooling system. In this case, all the time grains for WS provided the same $MADP$ because the higher WS of the *third-party* data always meant a higher heating energy demand and no energy demand compensation existed.

Regarding the *Temp* parameter, in the weather data analysis (Section 3.1), based on an hourly time grain, it was the parameter with less variation between *on-site* and *third-party* data for the three sites. However, the $MADP$ and $CV(RMSE)$ results, especially for the hourly criteria, showed that it had a significant impact on the energy demand in the four test sites. It was the second parameter of influence for the Gedved school, H2SusBuild, and administration building after wind speed and the first one in the office building with an $MADP$ of $\pm26\%$. In relation to the solar irradiation, the Taylor diagram (Figure 3) showed that DHI from the *third-party* weather data provided a better correlation than DNI for the three locations, and this was reflected in the sensitivity energy analysis. For the Gedved school, these two parameters had less impact on the energy demand than for the other three models. The reason is because the school lacked a cooling system; therefore, in summer, when more solar access was available, no energy demand was taken into account.

To conclude the explanation of Table 3, factor R^2 was analyzed. It compared the shape of the energy demand curves from the different simulations and showed that the energy demand simulated with the *third-party* weather data fit quite well with the energy demand simulated from the *on-site* data for the four test sites (with R^2 between $\pm82\%$ and $\pm95\%$). Regarding the different weather parameters, the results for each parameter matched with the analysis of the hourly percentage indexes. The parameters with lower hourly $MADP$ and $CV(RMSE)$ values had higher R^2 values.

Finally, to show in a visual way the previous analysis of the influence of the time resolution employed in the study and the sensitivity of each weather parameter, the $MADP$ index results are plotted in Figure 6. Each graph presents the $MADP$ result for each test site. In the graphs, the six time resolutions are shown on the x axis, and the dashed line presents the results for the simulation with all the *third-party* weather parameters (TPW). Each color represents the results for each weather parameter of the sensitivity analysis. The graphs show how the variations in the energy demand grew as the time grain decreased, especially *Temp*. They also show that WS was the most sensitive weather parameter for the Gedved school, H2SusBuild, and administration building. Only in the case of the office building in Pamplona was WS the most sensitive weather parameter taking into account an annual criterion; however, per hour, it changed to *Temp*.

Previous analyses showed the significant variations in the energy demand when using different *actual* weather datasets. In order to study if these differences in the energy demand were mostly due to the building architectures or to the weather, a complementary theoretical analysis was performed, and this is presented in the following section.

3.2.1. Analysis of the Buildings' Architecture Influence on the Energy Results

Since four test sites were available for this research, a complementary study was performed to analyze the influence of the building's architecture on the previous energy results. The four buildings, which were completely different in terms of the materials, construction systems, thermal mass, and window-to-wall ratio, were simulated with the same weather data (*on-site* and *third-party* weather files). For this study, we selected the most homogeneous weather when comparing the *third-party* to the *on-site* weather data: as shown in the previous analysis, energy demands are very sensitive to WS, so the Gedved weather was discarded for this analysis because its WS was the one with the worst fit to the reference (see Figure 3). *Temp* was also a sensitive parameter, and the three weathers had similar statistical metrics. Finally, for the solar radiation parameters DHI and DNI, Pamplona's weather better matched the reference compared to Lavrion. Therefore, the Pamplona weather file was chosen to develop this theoretical study, and for this reason, the four models were configured to have the same internal loads, HVAC systems, and schedules as the Pamplona office building.

Figure 7 shows the $MADP$ results for this study. The two graphs on the top present the results using the *third-party* weather file, which had all the weather parameters provided by the weather service. They show how when the test sites were simulated with their own weather files (graph on the left), the $MADP$ results and the trend of the curve were very different for the four test sites (each colored line represents one test site). However, when the test sites were simulated with Pamplona's weather file (graph on the right with dashed lines), the curves became very similar, reducing the differences in the $MADP$ values and in the trend of the curve.

Thus, the main value responsible for the variation in the energy demand was the weather dataset employed in the simulations and not the building's characteristics. This effect was also reflected in the results from the sensitivity analysis, which are also presented in Figure 7 for the *Temp, DNI, DHI,* and *WS* parameters. The curves from the graphs on the right, which are the simulations of all the test sites with Pamplona's weather file, were very similar compared to the curves from the graphs on the left, especially in the case of wind speed. This study demonstrated the great influence of the weather parameters on the variation of the building's energy demand, almost independently of the model, and this showed the importance of the selection of the weather dataset used in the BEM simulations.

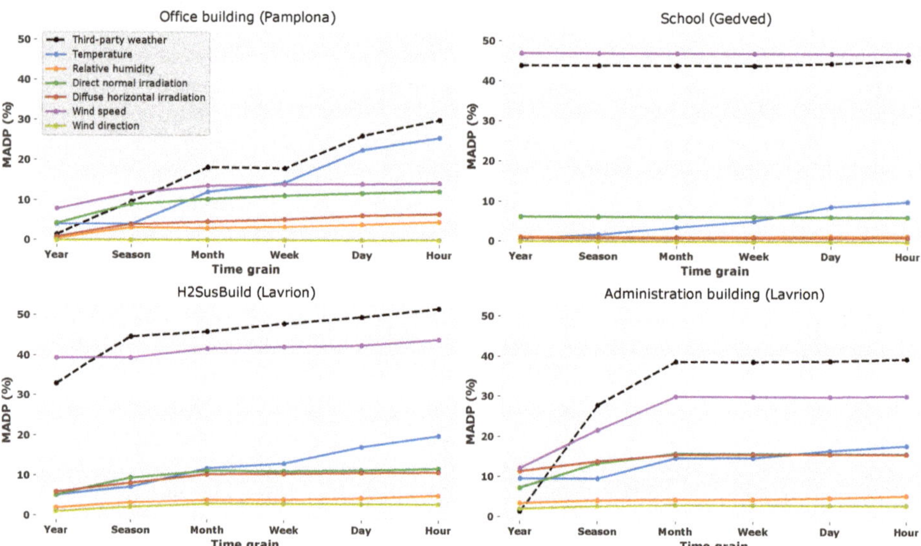

Figure 6. Representation of the $MADP$ (%) of the energy demand analysis for the four test sites and for the different time grains. The dashed black line represents the results using the weather file with all the *third-party* weather parameters. Each colored line represents each one of the weather parameter results from the sensitivity analysis.

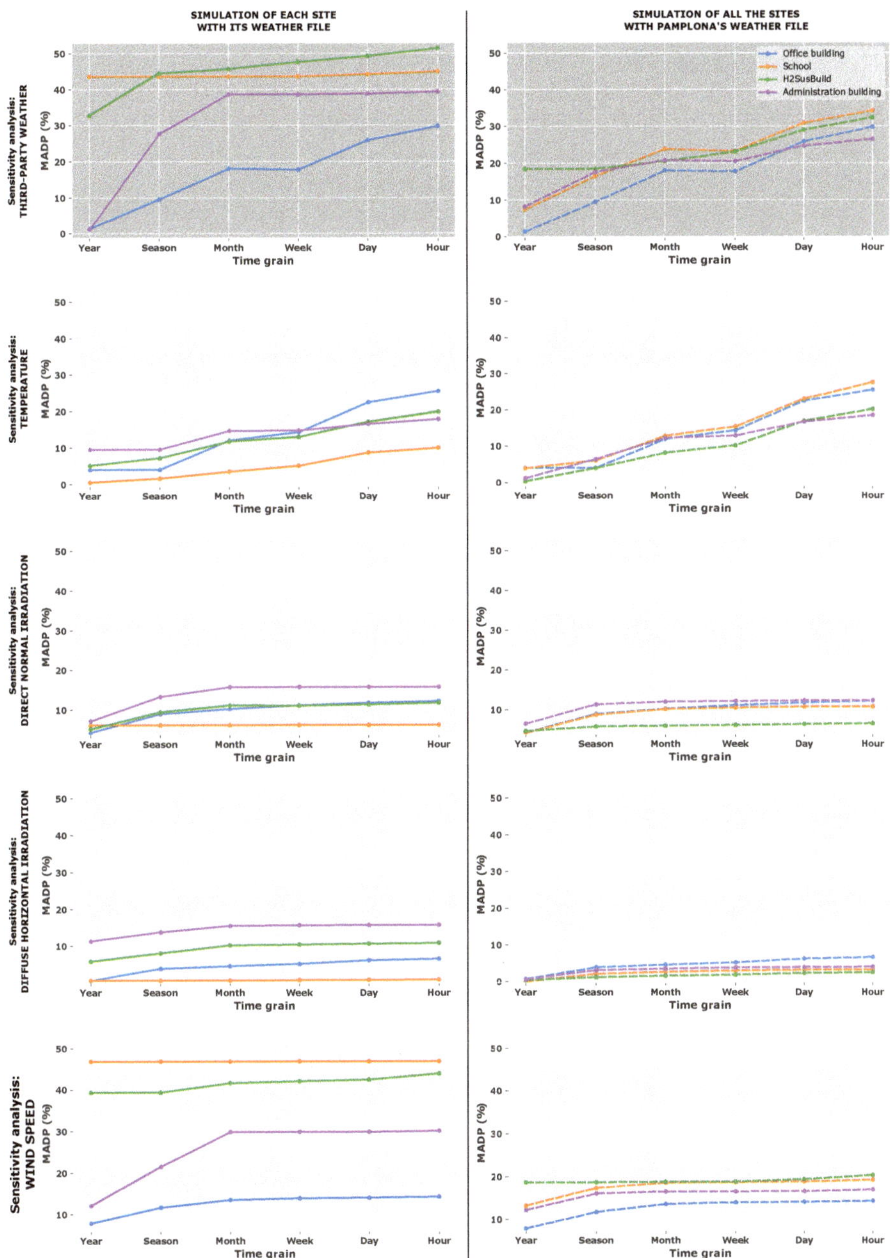

Figure 7. The comparison between the simulation results when each test site was simulated with its weather (on the left with continuous lines) and when all the test sites were simulated with Pamplona's weather (on the right with dashed lines). Each color represents a test site. The graphs show the $MADP$ for the energy demand for the different temporal resolutions. From above to below are shown: results when the *third-party* weather file was used (all the parameters were changed in the weather file) and the sensitivity analysis results for temperature, direct normal irradiation, diffuse horizontal irradiation, and wind speed.

3.3. Indoor Temperature Analysis Results

In the indoor temperature study, MAE and $RMSE$ for the hourly criteria were used in the analysis in order to measure the quantitative variation in the temperature curve, and R^2 was used to study the deviation in the temperature curve form. The results are shown in Table 4. In this case, the same energy was injected into the model for the two simulations, using the *on-site* and *third-party* weather datasets. The comparison of the thermal zones' temperature provided by the simulations provided the influence of the weather dataset employed in the indoor temperature conditions. In the table, the results are shown using three criteria: (1) for all the thermal zones (*All*), where the statistical metrics are calculated using the indoor temperature of all the thermal zones of the building; and (2, 3) for the maximum (*Max*) and minimum (*Min*) temperature, where the metrics are calculated using only the temperature of the thermal zone that provides the maximum/minimum temperature in each time step in order to compare the effect in the internal healthy conditions when using the two weather data sources. For this study, the statistical metrics were calculated only for the hourly criteria. The rest of the time grains were not considered since, for the temperature analysis, the data were not accumulated when longer periods were analyzed.

This study provided similar results to the previous energy results. Regarding the results for the temperature of all the thermal zones (*All*), the high results in the R^2 for the four test sites (from ± 87 to $\pm 95\%$) showed that the shape of the indoor temperature curve was very similar when the two weather datasets were used in the simulation. However, the quantitative statistical metrics showed a significant impact on the indoor temperature. The Gedved school was the test site with the highest MAE (1.72 °C), in line with the energy analysis where this test site reached deviations in the energy demand of $\pm 45\%$. The main reasons why the Gedved school had a higher impact on the indoor temperature were the influence of the lack of correlation of the wind speed between the *on-site* and *third-party* weather datasets and the way infiltrations were simulated based on the leakage area. The office building in Pamplona was the site with a minor impact on the indoor temperature (MAE of 0.55 °C) when the weather dataset was changed. When the maximum and minimum temperatures reached in the building were employed in the analysis (*Max* and *Min* in the table, respectively), it can be seen that the statistical metrics were similar to *All*. This means that the variation in the indoor temperature when using third-party weather data was stable and produced similar variations when the indoor conditions were minimum or maximum.

Table 4. The statistical metrics (MAE, $RMSE$, and R^2) used in the indoor temperature analysis for the four test sites. *All*: metrics calculated with the temperature of all thermal zones; *Max/Min*: metrics calculated with the temperature of the thermal zone with the maximum/minimum temperature in each time step.

Statistic Index	Office Building—Pamplona	School—Gedved	H2SusBuild—Lavrion	Administration Building—Lavrion
MAE (°C)—Min/All/Max	1.03/0.55/0.62	2.00/1.72/1.38	1.32/1.52/1.36	1.07/1.21/1.53
$RMSE$ (°C)—Min/All/Max	0.76/0.73/0.81	2.23/1.92/1.50	1.65/1.93/1.70	1.38/1.50/2.00
R^2 (%)—Min/All/Max	91.86/92.20/89.53	85.41/89.39/95.51	90.21/86.89/85.58	95.90/94.74/86.81

There was no high differences between the MAE and $RMSE$ results for the four cases, which indicated that the variations in the indoor temperature were quite homogeneous with no significant outliers. Figure 8 shows, in a visual way, the results with a scatter plot for each test site where the temperature was weighted by a thermal zone volume of air. The office building in Pamplona (above left) had fewer scattered temperature points since they were closer to the black line than the rest of the cases, and most of the points had a difference of 1 °C or less. On the other hand, the Gedved school (above right) provided the worst correlation for almost all the temperature points with a difference bigger than 1 °C due to the difference between both weather files.

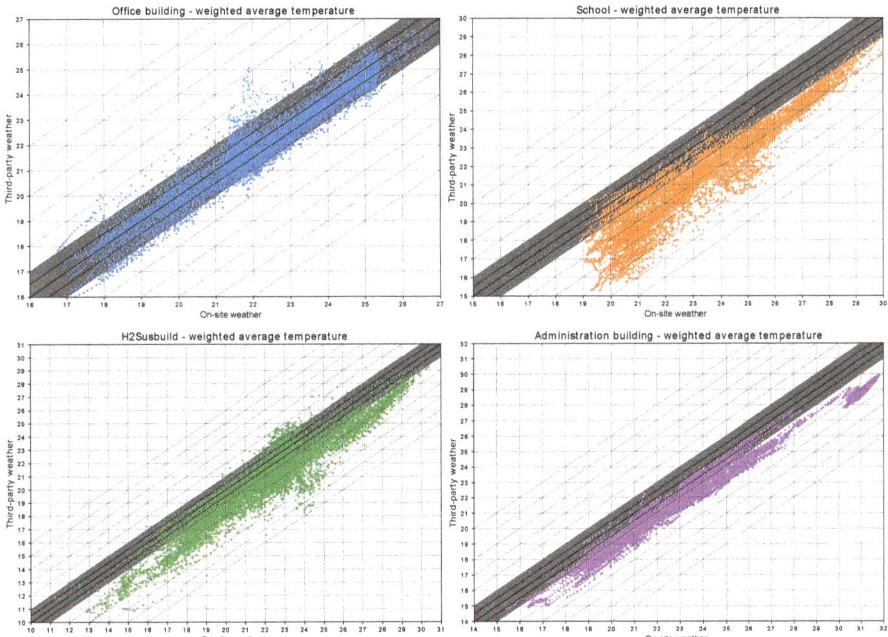

Figure 8. Scatter plots for each test site of the indoor temperature simulated using the *on-site* and *third-party* weather files.

4. Discussion

This paper shows how to study the impact of using two different *actual* weather datasets on building energy model simulations (weather data for the year 2019): one weather dataset with data measured in the building's surroundings (*on-site*), which was considered the reference weather data; and the other supplied by a weather service provider (*third-party*). Four test sites with different uses and architecture characteristics, located in three different locations, were employed in the studMor diagram, (2) an energy demand comparison, and (3) an indoor temperature comparison. In the case of the energy approach, a sensitivity analysis of the main weather parameters was also performed to study the influence of each parameter in the energy simulation. The energy results were provided with a different temporal resolution from the annual to hourly criteria in order to highlight the differences in the results.

The results for the energy analysis showed that as the time grain decreased, the impact of using different weather datasets grew, which agrees with ASHRAE [44]. The differences between the annual and hourly $MADP$ until 38% are shown in the results. This was because when the energy demand was accumulated in periods of time longer than an hour, the variation in the results was minimized due to the compensation effect of the underestimated and overestimated energy use. This must be taken into account when a weather data source is chosen according to its purpose. For instance, if the weather data will be used for model calibration purposes, it is important to take into consideration the monthly or hourly criteria, as the most used standards (*ASHRAE* [44], *FEMP* [58], and *IPMVP* [59]) employ these time grains for their recommendations, and as the study showed, the use of different weather datasets had a significant impact on the $CV(RMSE)$ results. Another application of BEM where the time grain of the analysis is relevant is model predictive control, where the hourly criteria are required.

The sensitivity analysis of the main weather parameters showed the different influence that each parameter had on the energy demand variation of each test site. In this regard, the relative humidity and wind direction had little influence on the models. In the case of the wind direction, the low influence was due to these test sites using mechanical ventilation instead of natural. On the other hand, the results also showed that the two parameters that produced a higher impact in the energy use were the wind speed and temperature. The high influence in the energy demand due to wind speed was explained by the *third-party* wind speed data having a low correspondence to the *on-site* data and because the models employed in the study used dynamic infiltrations that took into account the wind speed, instead of other models with constant values in the infiltration parameters. Therefore, the energy results for the wind speed showed that particular attention should be paid to this parameter when BEMs use dynamic infiltrations, as it has a great influence on the model's energy performance.

With the available data, the results obtained in this study suggested that for this models, some of the weather parameter data could be obtained from *third-party* weather sources, avoiding the installation of on-site sensors, as they had a low influence on the simulation results. This is the case of the relative humidity, wind direction, and even diffuse horizontal irradiation, the sensor being very expensive. On the other hand, to obtain the wind speed and outdoor temperature data, which are the weather parameters that were shown to be the most influential in the models' energy performance, we recommend the installation of an anemometer and a temperature sensor near the building. Having both *on-site* and *third-party* weather data sources would allow the verification of the data. An *on-site* sensor would also provide information regarding the micro-climate generated due to the surrounding characteristics of the building, which could be difficult to see reflected in the calculated weather data from a *third-party*.

The energy study showed that the weather dataset selected for the dynamic energy simulations had a great impact on the buildings energy performance, especially for short temporal resolutions. To emphasize the impact of the weather datasets in the building energy models, a theoretical study was performed, simulating all the test sites with the Pamplona weather file. The results showed that all the weather parameters produced similar variations in the energy demand and also a similar trend of the curve for the different time grains, independently of the model. This demonstrated the significant role played by the weather data and the importance of their correct selection when performing the building energy simulations.

In the case of the indoor temperature study, the significant impact of using different weather datasets was also shown. Although the high R^2 results for the four test sites showed that the shape of the indoor temperature curves was similar when both the *on-site* and *third-party* weather files were used, the quantitative metrics demonstrated a significant influence on the indoor temperature of the test sites with a MAE higher than 1.5 °C in some cases.

This paper showed the variation in the simulation results when two different *actual* weather datasets (*on-site* and *third-party*) were employed. In future research, it would be interesting to collect the empirical energy and temperature data from the test sites to study which of the weather datasets is closer to reality when comparing the simulation results using both weather datasets and the actual energy and temperature measurements.

5. Conclusions

The aim of this research is to show how to study the variations in energy demand and indoor temperature when using two different *actual* weather data sources with the purpose of analyzing if the data from the sensors of an *on-site* weather station (with high economic cost and maintenance) could be replaced by the data provided by a *third-party*. In this regard, it can be concluded that this research is not enough to make a general evaluation of the impact of using third-party *actual* weather data, but it has shown relevant variations in the energy demand and indoor temperature in the four test sites when both weather datasets are used in the simulations, especially for hourly criteria (used in calibration processes and in other applications such as model predictive control). The study also showed that for these types of building energy models, which employ dynamic infiltrations, wind speed's influence on the energy demand is relevant. The significant variations in the results lead us to make the recommendation for researchers to analyze in detail the impact on the building energy models using *third-party actual* weather data before employing it. For example, this analysis methodology could be performed before making an investment into a weather station by installing a rented provisional one for a period of time. This way, the most influential weather parameters for a specific building energy model and a *third-party* weather provider could be determined, and with this information, an informed choice about which sensors are worth being purchased and installed can be made.

Author Contributions: E.L.S. supervised the methodology used in the article, performed the simulations and the analysis, and wrote the manuscript. G.R.R. and C.F.B. developed the methodology and participated in the data analysis. V.G.G. and G.R.R. developed the EnergyPlus models. A.P. provided resources for the study. All the authors revised and verified the manuscript before sending it to the journal. All authors read and agreed to the published version of the manuscript.

Funding: This project received funding from the European Union's Horizon 2020 research and innovation program under Grant Agreement No. 731211, project SABINA.

Acknowledgments: We would like to thank the National Technical University of Athens for providing the data of the H2SusBuild and administration building test sites located in Lavrion (Greece) and also Insero in the case of the Gedved school test site in Denmark.

Conflicts of Interest: The authors declare no conflicts of interest.

Abbreviations

The following abbreviations are used in this manuscript:

ASHRAE	American Society of Heating, Refrigerating and Air-Conditioning Engineers
BEM	Building energy model
CV(RMSE)	Coefficient of variance RMSE
DHI	Diffuse horizontal irradiation
DNI	Direct normal irradiation
EU	European Union
FEMP	Federal Energy Management Program
HVAC	Heating, ventilation, and air-conditioning
IPMVP	International Performance Measurement and Verification Protocol
LTCP	Lavrion Technological and Cultural Park
MAE	Mean absolute error
MADP	Mean absolute deviation percentage
R^2	Coefficient of determination
RH	Relative humidity
RMS_{diff}	Centered root-mean-squared difference
RMSE	Root-mean-squared error
Temp	Temperature
TPW	*Third-party* weather data
WD	Wind direction
WS	Wind speed

Appendix A

The following figures show detailed Taylor diagrams for each one of the weathers analyzed in the study to complement Figure 3. Figure A1 shows Pamplona's weather, Figure A2 Gedved's weather, and Figure A3 Lavrion's weather. For each location, six diagrams are shown, one for each weather parameter (*Temp, RH, DNI, DHI, WS*, and *WD*). The diagrams show the statistical indexes (correlation R, centered root-mean-squared difference RMS_{diff}, and standard deviation σ_f) calculated on an hourly basis for annual, season, and monthly data.

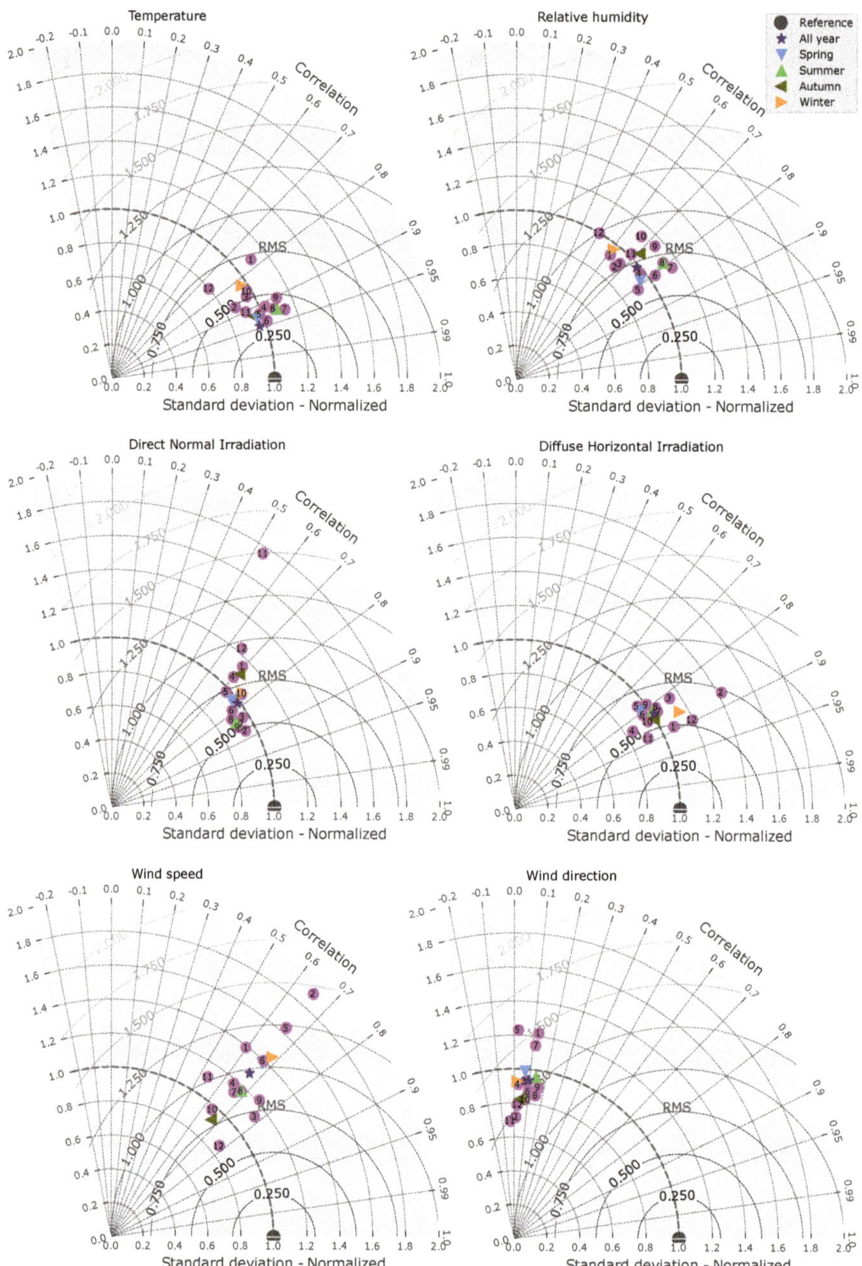

Figure A1. Normalized Taylor diagrams of Pamplona (Spain) comparing *on-site* and *third-party* weather data, showing temperature, relative humidity, direct normal irradiation, diffuse horizontal irradiation, wind speed, and wind direction.

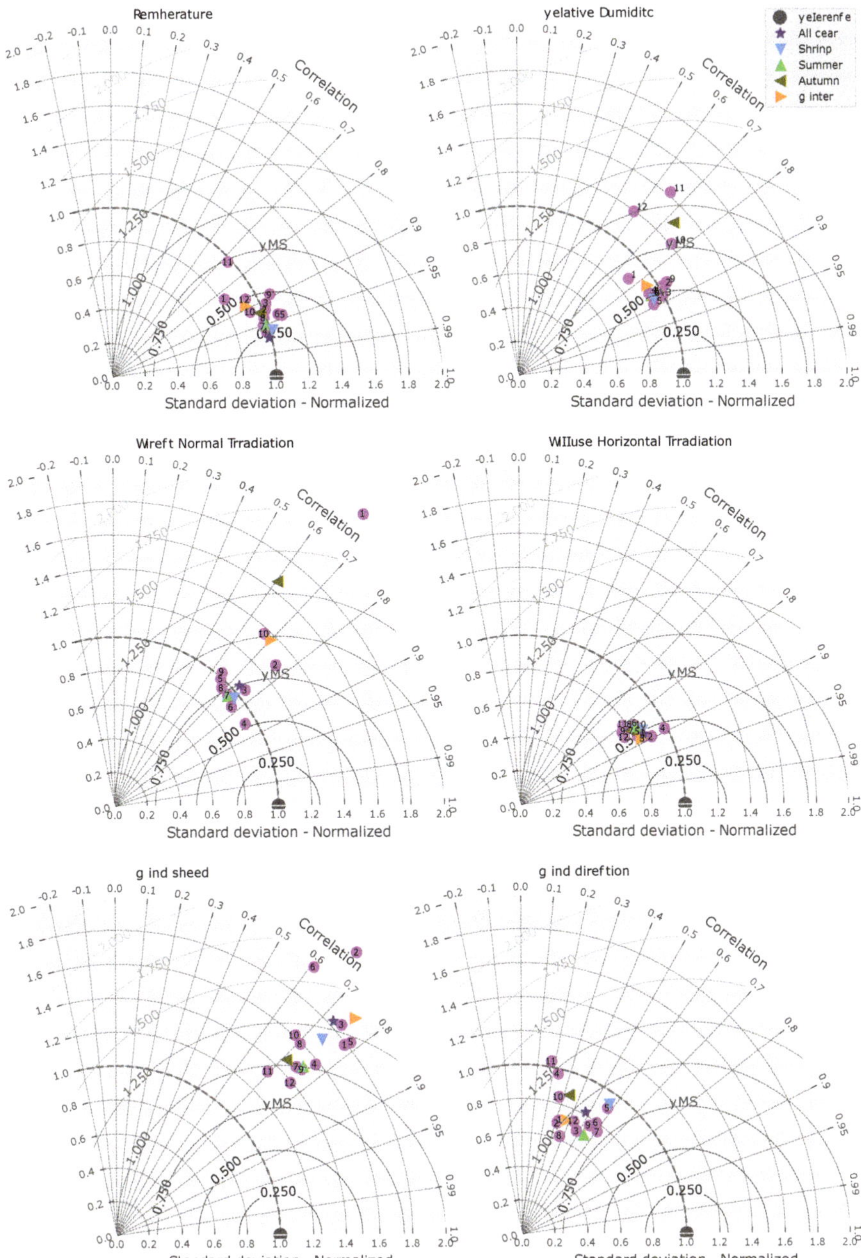

Figure A2. Normalized Taylor diagrams of Gedved (Denmark) comparing *on-site* and *third-party* weather data, showing temperature, relative humidity, direct normal irradiation, diffuse horizontal irradiation, wind speed, and wind direction.

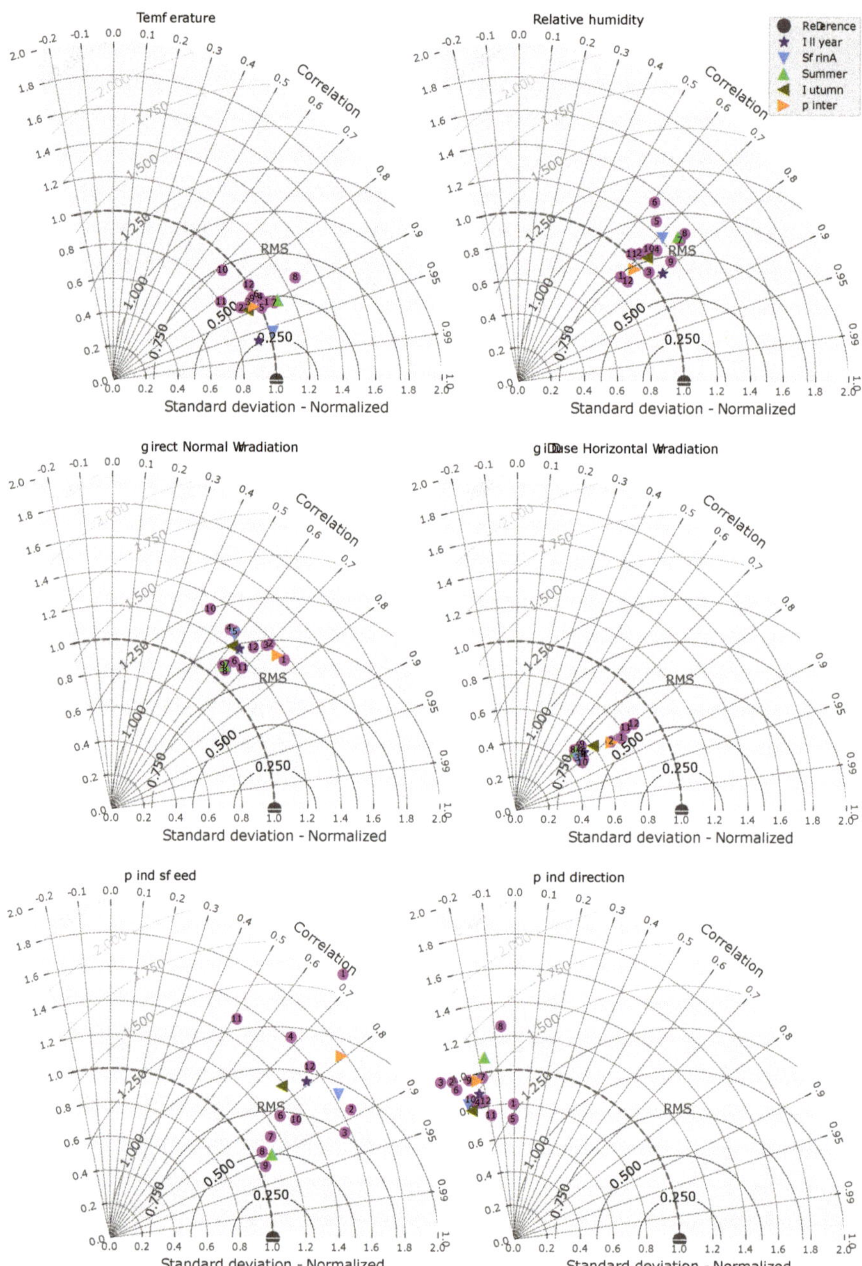

Figure A3. Normalized Taylor diagrams of Lavrion (Greece) comparing *on-site* and *third-party* weather data, showing temperature, relative humidity, direct normal irradiation, diffuse horizontal irradiation, wind speed, and wind direction.

References

1. United Nations. *The Sustainable Development Goals Report 2019*; United Nations: New York, NY, USA, 2019. Available online: https://unstats.un.org/sdgs/report/2019/The-Sustainable-Development-Goals-Report-2019.pdf (accessed on 20 April 2020).
2. UNFCCC Secretariat. *Paris Agreement*; Report of the Conference of the Parties on Its Twenty-First Session, Held in Paris from 30 November to 13 December 2015; United Nations Framework Convention on Climate Change, 2015. Available online: https://en.wikipedia.org/wiki/Paris_Agreement (accessed on 20 April 2020).
3. IEA. *Global Status Report for Buildings and Construction, towards a Zero-Emission Efficient and Resilient Buildings and Construction Sector*; IEA: Paris, France, 2019.
4. IBPSA-USA—International Building Performance Simulation Association. Building Energy Software Tools (BEST Directory). Available online: https://www.buildingenergysoftwaretools.com/software-listing (accessed on 23 April 2020).
5. Nguyen, A.T.; Reiter, S.; Rigo, P. A review on simulation-based optimization methods applied to building performance analysis. *Appl. Energy* **2014**, *113*, 1043–1058.
6. Bhandari, M.; Shrestha, S.; New, J. Evaluation of weather datasets for building energy simulation. *Energy Build.* **2012**, *49*, 109–118.
7. Marion, W.; Urban, K. *Users manual for TMY2s: Derived from the 1961–1990 National Solar Radiation Data Base*; Technical Report; National Renewable Energy Laboratory: Golden, CO, USA, 1995.
8. Wilcox, S.; Marion, W. *Users Manual for TMY3 Data Sets*; National Renewable Energy Laboratory: Golden, CO, USA, 2008.
9. Thevenard, D.J.; Brunger, A.P. The development of typical weather years for international locations: Part I, algorithms. *Ashrae Trans.* **2002**, *108*, 376–383.
10. Thevenard, D.J.; Brunger, A.P. The development of typical weather years for international locations: Part II, production/Discussion. *Ashrae Trans.* **2002**, *108*, 480.
11. Joe, Y.; Fenxian, H.; Seo, D.; Krarti, M. Development of 3012 IWEC2 Weather Files for International Locations (RP-1477). *Ashrae Trans.* **2014**, *120*, 340–355.
12. Henze, G.P.; Pfafferott, J.; Herkel, S.; Felsmann, C. Impact of adaptive comfort criteria and heat waves on optimal building thermal mass control. *Energy Build.* **2007**, *39*, 221–235.
13. Sandels, C.; Widén, J.; Nordström, L.; Andersson, E. Day-ahead predictions of electricity consumption in a Swedish office building from weather, occupancy, and temporal data. *Energy Build.* **2015**, *108*, 279–290.
14. Ramos Ruiz, G.; Lucas Segarra, E.; Fernández Bandera, C. Model predictive control optimization via genetic algorithm using a detailed building energy model. *Energies* **2019**, *12*, 34.
15. Fernández Bandera, C.; Pachano, J.; Salom, J.; Peppas, A.; Ramos Ruiz, G. Photovoltaic Plant Optimization to Leverage Electric Self Consumption by Harnessing Building Thermal Mass. *Sustainability* **2020**, *12*, 553.
16. Lazos, D.; Sproul, A.B.; Kay, M. Optimisation of energy management in commercial buildings with weather forecasting inputs: A review. *Renew. Sustain. Energy Rev.* **2014**, *39*, 587–603.
17. Crawley, D.B.; Huang, Y.J. Does it matter which weather data you use in energy simulations. *User News* **1997**, *18*, 2–12.
18. Crawley, D.B. Which weather data should you use for energy simulations of commercial buildings? *Trans. Am. Soc. Heat. Refrig. Air Cond. Eng.* **1998**, *104*, 498–515.
19. Erba, S.; Causone, F.; Armani, R. The effect of weather datasets on building energy simulation outputs. *Energy Procedia* **2017**, *134*, 545–554.
20. Voisin, J.; Darnon, M.; Jaouad, A.; Volatier, M.; Aimez, V.; Trovão, J.P. Climate impact analysis on the optimal sizing of a stand-alone hybrid building. *Energy Build.* **2020**, *210*, 109676.
21. González, V.G.; Ruiz, G.R.; Segarra, E.L.; Gordillo, G.C.; Bandera, C.F. Characterization of building foundation in building energy models. In Proceedings of the Building Simulation 2019: 16th Conference of IBPSA, Rome, Italy, 2–4 September 2019.
22. Jentsch, M.F.; James, P.A.; Bourikas, L.; Bahaj, A.S. Transforming existing weather data for worldwide locations to enable energy and building performance simulation under future climates. *Renew. Energy* **2013**, *55*, 514–524.
23. Jentsch, M.F.; Bahaj, A.S.; James, P.A. Climate change future proofing of buildings—Generation and assessment of building simulation weather files. *Energy Build.* **2008**, *40*, 2148–2168.

24. Dickinson, R.; Brannon, B. Generating future weather files for resilience. In Proceedings of the International Conference on Passive and Low Energy Architecture, Los Angeles, CA, USA, 11–13 July 2016; pp. 11–13.

25. Chow, T.T.; Chan, A.L.; Fong, K.; Lin, Z. Some perceptions on typical weather year—From the observations of Hong Kong and Macau. *Sol. Energy* **2006**, *80*, 459–467.

26. Wang, L.; Mathew, P.; Pang, X. Uncertainties in energy consumption introduced by building operations and weather for a medium-size office building. *Energy Build.* **2012**, *53*, 152–158.

27. Song, S.; Haberl, J.S. Analysis of the impact of using synthetic data correlated with measured data on the calibrated as-built simulation of a commercial building. *Energy Build.* **2013**, *67*, 97–107.

28. Silvero, F.; Lops, C.; Montelpare, S.; Rodrigues, F. Generation and assessment of local climatic data from numerical meteorological codes for calibration of building energy models. *Energy Build.* **2019**, *188*, 25–45.

29. Ciobanu, D.; Eftimie, E.; Jaliu, C. The influence of measured/simulated weather data on evaluating the energy need in buildings. *Energy Procedia* **2014**, *48*, 796–805.

30. Cuerda, E.; Guerra-Santin, O.; Sendra, J.J.; Neila, F.J. Understanding the performance gap in energy retrofitting: Measured input data for adjusting building simulation models. *Energy Build.* **2020**, *209*, 109688.

31. Du, H.; Barclay, M.; Jones, P.J. Generating high resolution near-future weather forecasts for urban scale building performance modelling. In Proceedings of the 15th IBPSA Conference, San Francisco, CA, USA, 7–9, August 2017.

32. Du, H.; Jones, P.; Segarra, E.L.; Bandera, C.F. Development of a REST API for obtaining site-specific historical and near-future weather data in EPW format. Presented at Building Simulation and Optimization, Cambridge, UK, 11–12 September 2018.

33. Du, H.; Bandera, C.F.; Chen, L. Nowcasting methods for optimising building performance. In Proceedings of the 16th IBPSA Conference, Rome, Italy, 2–4 September 2019.

34. Henze, G.P.; Kalz, D.E.; Felsmann, C.; Knabe, G. Impact of forecasting accuracy on predictive optimal control of active and passive building thermal storage inventory. *HVAC&R Res.* **2004**, *10*, 153–178.

35. Agüera-Pérez, A.; Palomares-Salas, J.C.; de la Rosa, J.J.G.; Florencias-Oliveros, O. Weather forecasts for microgrid energy management: Review, discussion and recommendations. *Appl. Energy* **2018**, *228*, 265–278.

36. Yan, X.; Abbes, D.; Francois, B. Uncertainty analysis for day ahead power reserve quantification in an urban microgrid including PV generators. *Renew. Energy* **2017**, *106*, 288–297.

37. Powell, K.M.; Sriprasad, A.; Cole, W.J.; Edgar, T.F. Heating, cooling, and electrical load forecasting for a large-scale district energy system. *Energy* **2014**, *74*, 877–885.

38. Petersen, S.; Bundgaard, K.W. The effect of weather forecast uncertainty on a predictive control concept for building systems operation. *Appl. Energy* **2014**, *116*, 311–321.

39. Zhao, J.; Duan, Y.; Liu, X. Uncertainty analysis of weather forecast data for cooling load forecasting based on the Monte Carlo method. *Energies* **2018**, *11*, 1900.

40. Haben, S.; Giasemidis, G.; Ziel, F.; Arora, S. Short term load forecasting and the effect of temperature at the low voltage level. *Int. J. Forecast.* **2019**, *35*, 1469–1484.

41. Seo, D.; Huang, Y.J.; Krarti, M. Impact of typical weather year selection approaches on energy analysis of buildings. *ASHRAE Trans.* **2010**, *116*, 416–427.

42. Radhi, H. A comparison of the accuracy of building energy analysis in Bahrain using data from different weather periods. *Renew. Energy* **2009**, *34*, 869–875.

43. SABINA SmArt BI-Directional Multi eNergy gAteway. Available online: https://sabina-project.eu/ (accessed on 20 April 2020).

44. Guideline, A. *Guideline 14-2002, Measurement of Energy and Demand Savings*; American Society of Heating, Ventilating, and Air Conditioning Engineers: Atlanta, GA, USA, 2002.

45. Taylor, K.E. Taylor Diagram Primer. 2005. Available online: http://wwwpcmdi.llnl.gov/about/staff/Taylor/CV/Taylor_diagram_primer.pdf (accessed on 20 April 2020).

46. Taylor, K.E. Summarizing multiple aspects of model performance in a single diagram. *J. Geophys. Res. Atmos.* **2001**, *106*, 7183–7192.

47. Dong, T.Y.; Dong, W.J.; Guo, Y.; Chou, J.M.; Yang, S.L.; Tian, D.; Yan, D.D. Future temperature changes over the critical Belt and Road region based on CMIP5 models. *Adv. Clim. Chang. Res.* **2018**, *9*, 57–65.

48. Chen, G.; Zhang, X.; Chen, P.; Yu, H.; Wan, R. Performance of tropical cyclone forecast in Western North Pacific in 2016. *Trop. Cyclone Res. Rev.* **2017**, *6*, 13–25.

49. Yan, G.; Wen-Jie, D.; Fu-Min, R.; Zong-Ci, Z.; Jian-Bin, H. Surface air temperature simulations over China with CMIP5 and CMIP3. *Adv. Clim. Chang. Res.* **2013**, *4*, 145–152.

50. Nabeel, A.; Athar, H. Stochastic projection of precipitation and wet and dry spells over Pakistan using IPCC AR5 based AOGCMs. *Atmos. Res.* **2020**, *234*, 104742.

51. de Assis Tavares, L.F.; Shadman, M.; de Freitas Assad, L.P.; Silva, C.; Landau, L.; Estefen, S.F. Assessment of the offshore wind technical potential for the Brazilian Southeast and South regions. *Energy* **2020**, *196*, 117097.

52. Crawley, D.B.; Lawrie, L.K.; Winkelmann, F.C.; Buhl, W.F.; Huang, Y.J.; Pedersen, C.O.; Strand, R.K.; Liesen, R.J.; Fisher, D.E.; Witte, M.J.; et al. EnergyPlus: Creating a new-generation building energy simulation program. *Energy Build.* **2001**, *33*, 319–331.

53. Crawley, D.B.; Lawrie, L.K.; Pedersen, C.O.; Winkelmann, F.C.; Witte, M.J.; Strand, R.K.; Liesen, R.J.; Buhl, W.F.; Huang, Y.J.; Henninger, R.H.; et al. EnergyPlus: An update. In Proceedings of the SimBuild, Boulder, CO, USA, 4–6 August 2004; Volume 1.

54. EnergyPlus. *Auxiliary Programs: EnergyPlus^{TM} version 8.9.0 Documentation*; U.S. Department of Energy: Washington, DC, USA, 2018.

55. Kolassa, S.; Schütz, W. Advantages of the MAD/MEAN ratio over the MAPE. *Foresight Int. J. Appl. Forecast.* **2007**, *6*, 40–43.

56. Petojević, Z.; Gospavić, R.; Todorović, G. Estimation of thermal impulse response of a multi-layer building wall through in-situ experimental measurements in a dynamic regime with applications. *Appl. Energy* **2018**, *228*, 468–486.

57. Lucas Segarra, E.; Du, H.; Ramos Ruiz, G.; Fernández Bandera, C. Methodology for the quantification of the impact of weather forecasts in predictive simulation models. *Energies* **2019**, *12*, 1309.

58. U.S. Department of Energy. *M&V Guidelines: Measurement and Verification for Federal Energy Projects Version 3.0.*; U.S. Department of Energy: Washington, DC, USA 2008.

59. IPMVP Committee. *International Performance Measurement and Verification Protocol: Concepts and Options for Determining Energy and Water Savings, Volume I*; Technical Report; National Renewable Energy Laboratory: Golden, CO, USA, 2001.

60. González, V.G.; Colmenares, L.Á.; Fidalgo, J.F.L.; Ruiz, G.R.; Bandera, C.F. Uncertainy's Indices Assessment for Calibrated Energy Models. *Energies* **2019**, *12*, 2096.

61. Zhao, J.; Liu, X. A hybrid method of dynamic cooling and heating load forecasting for office buildings based on artificial intelligence and regression analysis. *Energy Build.* **2018**, *174*, 293–308.

62. Perez, R.; Cebecauer, T.; Šúri, M. Semi-empirical satellite models. In *Solar Energy Forecasting and Resource Assessment*; Academic Press: Boston, MA, USA, 2013; pp. 21–48.

63. Kato, T. Prediction of photovoltaic power generation output and network operation. In *Integration of Distributed Energy Resources in Power Systems*; Elsevier: Amsterdam, The Netherlands, 2016; pp. 77–108.

64. Willmott, C.J.; Matsuura, K. Advantages of the mean absolute error (MAE) over the root mean square error (RMSE) in assessing average model performance. *Clim. Res.* **2005**, *30*, 79–82.

65. Meteoblue. Available online: https://meteoblue.com/ (accessed on 20 April 2020).

66. Meteoblue Weather Simulation Data. Available online: https://content.meteoblue.com/en/specifications/data-sources/weather-simulation-data/ (accessed on 6 July 2020).

67. Guglielmetti, R.; Macumber, D.; Long, N. *OpenStudio: An Open Source Integrated Analysis Platform*; Technical Report; National Renewable Energy Laboratory (NREL): Golden, CO, USA , 2011.

68. EnergyPlus. *EnergyPlus Input Output Reference*; US Department of Energy: Washington, DC, USA, 2018.

69. Ruiz, G.R.; Bandera, C.F.; Temes, T.G.A.; Gutierrez, A.S.O. Genetic algorithm for building envelope calibration. *Appl. Energy* **2016**, *168*, 691–705.

70. Ruiz, G.R.; Bandera, C.F. Analysis of uncertainty indices used for building envelope calibration. *Appl. Energy* **2017**, *185*, 82–94.

71. Fernández Bandera, C.; Ramos Ruiz, G. Towards a new generation of building envelope calibration. *Energies* **2017**, *10*, 2102.

sustainability

Article

Industrialization and Thermal Performance of a New Unitized Water Flow Glazing Facade

Belen Moreno Santamaria [1], Fernando del Ama Gonzalo [2,*], Danielle Pinette [2], Benito Lauret Aguirregabiria [1] and Juan A. Hernandez Ramos [3]

[1] Department of Construction and Architectural Technology, Technical School of Architecture of Madrid, Technical University of Madrid (UPM), Av. Juan de Herrera 4, 28040 Madrid, Spain; belen.moreno@upm.es (B.M.S.); benito.lauret@upm.es (B.L.A.)

[2] Department of Sustainable Product Design and Architecture, Keene State College, 229 Main St., Keene, NH 03435, USA; pinetteda@gmail.com

[3] Department of Applied Mathematics, School of Aeronautical and Space Engineering, Technical University of Madrid (UPM), Plaza Cardenal Cisneros 3, 28040 Madrid, Spain; juanantonio.hernandez@upm.es

* Correspondence: fernando.delama@keene.edu

Received: 9 August 2020; Accepted: 11 September 2020; Published: 14 September 2020

Abstract: New light envelopes for buildings need a holistic vision based on the integration of architectural design, building simulation, energy management, and the curtain wall industry. Water flow glazing (WFG)-unitized facades work as transparent and translucent facades with new features, such as heat absorption and renewable energy production. The main objective of this paper was to assess the performance of a new WFG-unitized facade as a high-performance envelope with dynamic thermal properties. Outdoor temperature, variable mass flow rate, and solar radiation were considered as transient boundary conditions at the simulation stage. The thermal performance of different WFGs was carried out using simulation tools and real data. The test facility included temperature sensors and pyranometers to validate simulation results. The dynamic thermal transmittance ranged from 1 W/m^{2}K when the mass flow rate is stopped to 0.06 W/m^{2}K when the mass flow rate is above 2 L/min m^2. Selecting the right glazing in each orientation had an impact on energy savings, renewable energy production, and CO_2 emissions. Energy savings ranged from 5.43 to 6.46 KWh/m^2 day in non-renewable energy consumption, whereas the renewable primary energy production ranged from 3 to 3.42 KWh/m^2 day. The CO_2 emissions were reduced at a rate of 1 Kg/m^2 day. The disadvantages of WFG are the high up-front cost and more demanding assembly process.

Keywords: building energy management; water flow glazing; unitized facade

1. Introduction

The residential and commercial building sector accounts for almost 40% of the European Union final energy consumption [1]. Thus, the goal of achieving a highly energy-efficient building stock by 2050 was set by The Energy Performance of Buildings Directive (EPBD 2018) [2]. In the United States, recent studies have shown that heating and cooling account for more than 30% of energy consumption in buildings [3]. Other non-OECD countries, including China and India, will be responsible for half of the global increase in energy consumption until 2040 [4]. The development of new materials, new heating and ventilation technologies, and energy-saving measures have improved the thermal performance of buildings in winter conditions [5]. However, in summer conditions, the increasing standards of life and the affordability of air-conditioning technologies have contributed to increasing the energy needs for cooling over the last decade [6]. Nowadays, air conditioning in office and commercial facilities accounts for 15% of the total electricity consumption in the world [7–9]. Occupants and equipment are responsible for internal heat gains, and large glass areas increase solar radiation gains, especially in warm climates, thus leading to the increased total electricity consumption for the purposes of cooling [10,11].

The design of advanced glazed facades is the most promising component in building design with the highest impact on building performance [12]. In this paper, advanced facades refer to a broad spectrum of constructive solutions that allow for the dynamic response of the building envelope. They can actively manage the heat flow and energy transfer between the building and its external environment, leading to a potentially significant reduction in heating and cooling loads [13]. Advanced glazed facades include passive solutions, such as Low-E coatings, which reflect the indoor heat when the outdoor temperature is low [14], and highly selective coatings that reflect direct and diffuse solar heat radiation in summer. Scientific literature has confirmed this potential energy reduction [15,16]. However, the most promising results can be accomplished with dynamic technologies that can adapt to different outdoor conditions. Polymer dispersed liquid crystal (PDLC), Suspended Particle Devices (SPDs), and electrochromic (EC) glass switch from transparent to colored or vary transmission and reflection parameters [17,18]. The system is limited by its high initial cost and the need for an energy management system integrated with the rest of the equipment, especially the ventilation system. Controlling the relative humidity is essential in radiant panels to prevent condensation issues, especially in summer. The integrated piping does not allow movable panels, so its use is limited to buildings with mechanical ventilation [19]. Nevertheless, the measures to improve the energy performance of buildings do not focus only on the building envelope. Building designers must consider all technical and mechanical systems in a building, such as passive elements; heating, ventilation, air conditioning (HVAC); the energy use for lighting and ventilation; and other measures to improve thermal and visual comfort [20].

Water flow glazing (WFG), as an advanced facade technology, combines passive (coatings and polyvinyl butyral (PVB) layers) and active solutions (variable water mass flow rate) to absorb or reject infrared radiation and reduce the temperature of the interior glass pane [21,22]. Flowing water captures most of the solar infrared radiation, while a significant part of the visible component goes through the glazing [23,24]. WFG radiant panels can be used as components of a heating or cooling hydronic system with little difference between the water and the indoor temperature [25]. Finally, WFG can work as an integrated solar collector to provide water heating in warm seasons, and the excess of hot water can be stored in buffer tanks [26]. The use of the facade and interior partitions as radiant heating and cooling devices have advantages compared with convective cooling systems. Using radiant ceilings or walls can reduce energy consumption between 10% to 70% compared with all-air systems [27]. This article showed some of the accomplishments of the research project: "Industrialized Development of Water Flow Glazing Systems" (InDeWag), supported by program Horizon 2020–the EU.3.3.1: Reducing energy consumption and carbon footprint by smart and sustainable use. The water flow glazing unitized facade is made of three components: glazing, circulating device, and aluminum frame [28]. The glazing comprises different layers and interfaces according to determined spectral and thermal properties. The circulating device includes a water pump moving the fluid in a closed circuit, a heat exchanger, and temperature and flow sensors to monitor and control the heat. Finally, the aluminum frame provides the unitized module with structural stability.

Despite the fast pace of product innovation, a gap has been created between the new advanced facades and the available building simulation tools to model and assess building energy performance [29]. Monitoring activities are essential to support simulation models, especially in transient state [30], when changing the boundary conditions can affect the results of dynamic simulations by more than 30% [31]. Although there are commercial building energy simulation tools that include dynamic simulations [32], very few include WFG [33,34]. The authors of this paper developed a set of equations that take into account multiple direct and diffuse reflections between the glazing surfaces, the absorptance of glass and water layers, the spectral properties of coatings, and the convective heat transfer coefficient [35,36]. Then, a simulation tool was developed to allow building designers to make decisions on the glazing type. Finally, the equations and the simulation tool were validated through the real results taken from a demonstrator placed in Sofia, Bulgaria. In addition to an accurate simulation tool, the actual challenge was to develop a monitoring system to reveal characteristic patterns of users, and to finally discover relevant design criteria that might be applicable to different orientations throughout the year. The monitoring system applied in this case study has been tested in other facilities over the last few years, and results were shown in previous articles [37,38].

This paper focused on a methodology to select and assess the performance of different WFG configurations in a test facility in Sofia, Bulgaria. Hence, to achieve this goal, it was essential to (i) simulate the steady state to select the optimum WFG in each orientation at the first stage of the design process, (ii) validate the first results by including transient boundary conditions, (iii) analyze the performance of the selected glazing in a real facility, and (iv) estimate the final energy savings, the potential renewable energy production, and CO_2 emissions in summer and winter conditions.

2. Materials and Methods

Commercial building energy simulation (BES) software does not consider WFG as a component of the heating and cooling systems. It is essential to provide building designers with a simple method to select the correct glazing and evaluate its performance at the beginning of the design stage. The first subsection defines three tested WFGs in terms of spectral properties. The second subsection shows the mathematical model that defined the dynamic thermal properties of the glazing. The third subsection describes the components and industrialization process of the unitized WFG.

2.1. Determination of Spectral Properties

The International Glazing DataBase (IGDB) is a collection of optical data for glazing products. Spectral transmittance and reflectance are measured in a spectrophotometer and contributed to the IGDB by the manufacturer of the glazing product, subject to a careful review. The IGDB currently only allows the inclusion of specular glazing materials without patterns, such as monolithic glass, plastic, laminates, applied films on glass, or thin-film coated glass. The products included in the WFG catalog are low-emissivity (Low-E) glass, high selective glass, and solar polyvinyl butyral (PVB) interlayers to increase sun energy absorbance. The solar energy spectrum can be divided into the ultraviolet (UV) light, visible light, and infrared (IR) light, depending on the wavelength. The wavelengths of the ultraviolet light range from 310 to 380 nanometers. The visible light ranges from 380 to 780 nanometers. The infrared light spectrum is transmitted as heat into a building and begins at wavelengths of 780 nanometers. Solar heat radiation has shortwave energy, and it is known as near-infrared (NIR), whereas the heat radiating off warm objects has higher wavelengths and is known as far-infrared (FIR). Three different WFGs were tested:

- Case 1 was made of the following layers: Planiclear (8 mm), 2 Saflex R solar (SG41), Planiclear (8 mm), water chamber (24 mm), Planiclear (8 mm), Planiclear (8 mm), 4 Saflex R standard clear (RB11), Planiclear (8 mm), a low-emissivity coating Planitherm XN, an argon chamber (16 mm), Planiclear (6 mm), 4 Saflex R standard clear (RB11), Planiclear (6 mm).
- Case 2 was made of the following layers: diamant glass (10 mm), an argon chamber (16 mm), a low-emissivity coating Planitherm XN, Planiclear (8 mm), 2 Saflex R solar (SG41), Planiclear (8 mm), water chamber (24 mm), Planiclear (8 mm), 4 Saflex R standard clear (RB11), Planiclear (8 mm).
- Case 3 was made of the following layers: diamant glass (10 mm), a highly reflective coating Xtreme 60.28, an argon chamber (16 mm), Planiclear (8 mm), 4 Saflex R standard clear (RB11), Planiclear (8 mm), water chamber (24 mm), Planiclear (8 mm), 4 Saflex R standard clear (RB11), Planiclear (8 mm).

Figure 1 shows the front and back reflectance (*R*), transmittance (*T*), and absorptance (*A*), as a function of the angle of incidence. It illustrates the glass panes, coatings, and the position of the air and water chambers. Case 1 showed the highest infrared absorptance (*A*), the lowest front reflectance (*R*), and the lowest infrared transmittance (*T*). It seemed the best option to heat up water. Case 2 showed a high near-infrared (NIR) absorptance, low far-infrared (FIR) absorptance, and high front far-infrared (FIR) reflectance. The absorptance was not as high as in Case 1, but its ability to reflect heat made it the right solution for large glass areas in warm climates. Case 3 showed very low infrared absorptance and very high infrared front reflectance. This case would have been the best option to reject energy and prevent heat from entering the indoor space, but it would not have been appropriate to heat up water.

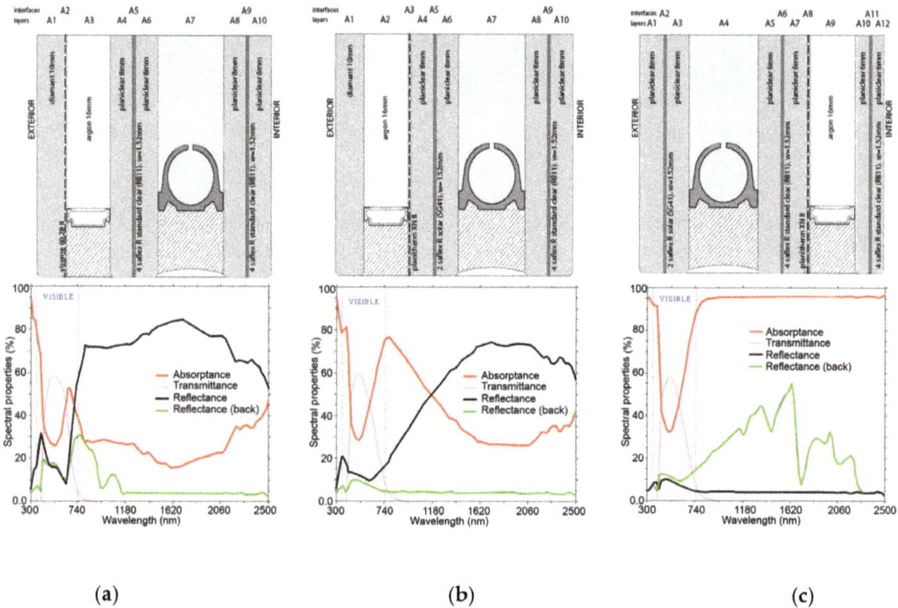

(**a**) (**b**) (**c**)

Figure 1. Spectral properties as a function of the solar wavelength. Description of the layers and coatings. (**a**) Case 1: 8 + 8/24 w /8 + 8/16 a /6 + 6. (**b**) Case 2: 10/16 a/Low-E8 + 8/24 w/8 + 8. (**c**) Case 3: 10XNII/16 a/8 + 8/24 w/8 + 8.

2.2. Determination of Heat Transfer Coefficients (h) and Dynamic Thermal Transmittance (U)

Water absorbs the solar near-infrared radiation and increases the temperature as it flows through the window. The mass flow rate and the thermal properties of glass panes must be studied to allow designers to apply energy-management strategies. Sometimes it might be interesting to increase the water temperature and store that energy for heating purposes. Other times it might be appropriate to reject as much solar radiation as possible without heating the water. The thermal transmittance (U) measures the heat transfer through the glazing and the European Standard determines its value [39,40]. Figure 2 illustrates the heat transfer coefficients (h_i, h_w, h_g, and h_e), the temperature distribution in the WFG layers (θ_i, θ_1, θ_2, θ_3, θ_w, θ_e), and the absorptance of layers (A_1, A_2, A_w, A_3).

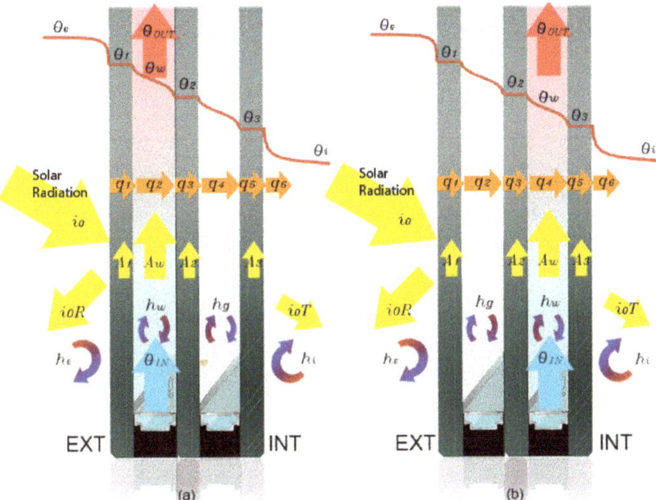

Figure 2. Temperature distribution in a triple water flow glazing (WFG) at a specific height, with heat transfer coefficients (h_i, h_w, h_g, h_e), heat fluxes through a triple WFG (q_1, q_2, q_3, q_4, q_5, q_6), solar radiation and absorptance of layers (A_1, A_2, A_w, A_3). (**a**) Triple WFG with water chamber outdoors. (**b**) Triple WFG with water chamber indoors.

The Equation (1) gives the outdoor heat flux:

$$q_e = h_e(\theta_e - \theta_1),\tag{1}$$

where h_e is the outdoor convective coefficient, θ_1 is the superficial temperature of the outermost glass pane, and θ_e is the outdoor temperature. The beam solar irradiance, diffuse irradiance, and the angle of incidence should be given. Regarding indoor boundary conditions, the indoor heat flux is given by the Equation (2):

$$q_i = h_i(\theta_3 - \theta_i),\tag{2}$$

where h_i is the indoor convective coefficient, θ_3 is the superficial temperature of the inner glass pane, and θ_i is the indoor temperature that can be a constant value or calculated solving the indoor thermal problem. Newton's law also models the heat transfer inside gas chambers. The heat flux in a gas chamber between two parallel surfaces is expressed by Equation (3):

$$q_i = h_g(\theta_i - \theta_{i+1}),\tag{3}$$

where h_g is the heat transfer coefficient of gas chambers, this coefficient considers the radiative heat transfer between the parallel glass panes and the natural convective transport. This value can be constant or calculated, knowing the emissivity of the two glass planes and an experimental correlation for the natural convective transport. In the water chamber, the situation is different. Heat flux is proportional to the temperature difference between the water temperature θ_w and the glass pane temperatures. This coefficient takes into account the heat transport mechanism forced by the water flow inside the chamber. Since the water is opaque to far-infrared, the radiative heat transfer mechanism is not present in the water chamber. Hence, the heat flux through glass panes in contact with the water chamber is expressed in Equation (4) for glass panes outside the water chamber, and Equation (5) when the glass pane is after the water chamber:

$$q_i = h_w(\theta_{i-1} - \theta_w), \tag{4}$$

$$q_{i+1} = h_w(\theta_w - \theta_i), \tag{5}$$

where h_w is the heat transfer coefficient for the water chamber. By default $h_w = 50$.

When it comes to considering the absorptance of layers (A_i), the heat flux can be expressed as in Equation (6).

$$q_{i+1} = q_i + A_i\, i_0. \tag{6}$$

Spectral and thermal properties of WFG have been explained in previous articles [35]. The authors used data from commercial software and developed equations to evaluate the influence of water flowing through glass panes. Equation (7) considers the energy balance at each layer and Newton's definition of heat flux.

$$q_i = q_{i-1} + A_w\, i_0 + \dot{m}c(\theta_{IN} - \theta_{OUT}). \tag{7}$$

Equations (8)–(17) show the heat flux of WFG with water chamber indoors.

$$q_1 = h_e\,(\theta_e - \theta_1), \tag{8}$$

$$q_2 = q_1 + A_1\, i_0, \tag{9}$$

$$q_2 = h_g\,(\theta_1 - \theta_2), \tag{10}$$

$$q_3 = q_2, \tag{11}$$

$$q_4 = h_w\,(\theta_2 - \theta_w), \tag{12}$$

$$q_4 = q_3 + A_2\, i_0, \tag{13}$$

$$q_5 = h_w\,(\theta_w - \theta_3), \tag{14}$$

$$q_5 = q_4 + A_w\, i_0 + \dot{m}c(\theta_{IN} - \theta_w), \tag{15}$$

$$q_6 = h_i\,(\theta_3 - \theta_i), \tag{16}$$

$$q_6 = q_5 + A_3\, i_0, \tag{17}$$

where h_i is the interior heat transfer coefficient; h_e, the exterior heat transfer coefficient; h_g, the air-cavity heat transfer coefficient; and h_w, the water heat transfer coefficient. The thermal transmittances (U_e, U_i) depend on the heat transfer coefficients. Equations (18) and (19) refer to WFG with the water chamber outdoors (Case 1), and Equations (20) and (21) show the expressions for WFG with a water chamber indoors (Cases 2, 3).

$$\frac{1}{U_e} = \frac{1}{h_e} + \frac{1}{h_w}, \tag{18}$$

$$\frac{1}{U_i} = \frac{1}{h_i} + \frac{1}{h_g} + \frac{1}{h_w}, \tag{19}$$

$$\frac{1}{U_e} = \frac{1}{h_e} + \frac{1}{h_g} + \frac{1}{h_w}, \tag{20}$$

$$\frac{1}{U_i} = \frac{1}{h_i} + \frac{1}{h_w}. \tag{21}$$

U represents the thermal transmittance between the room and outdoors, and U_w represents the thermal transmittance between the water chamber and indoors. The thermal transmittance (U) is almost zero when the flow rate is the design flow rate because the water chamber isolates the building from outdoor conditions. When the water flow stops, the thermal transmittance depends mainly on the air chamber. Equations (22) and (23) show the expressions for U and U_w, respectively:

$$U_w = \frac{U_i \dot{m} c}{\dot{m} c + U_e + U_i},$$ (22)

$$U = \frac{U_e U_i}{\dot{m} c + U_e + U_i},$$ (23)

where $\dot{m}$ is the mass flow rate and c is the specific heat of water.

Table 1 shows a complete description of the different layers and the selected glazing average values. Visual transmittance (T_v) is the measurement of visible light (380 to 780 nm) passing through the glazing. The thermal parameters depended on the mass flow rate. If the water was flowing, the g-factor became lower. When the water chamber was stopped, the amount of energy entering the building increased. The thermal transmittance (U) was almost zero at the design flow rate (2 L/min m^2). When $\dot{m} = 0$, U depended on the air chamber. The thermal transmittance (U_w) as defined in Equation (22) measures the heat losses or gains between the water chamber and the indoor air. $U_w = 0$ when $\dot{m} = 0$. At the working flow rate, its value was high (6.4 W/m^{2}K) when the water chamber was close to indoors, and it was low (0.9 W/m^{2}K) when the water chamber was outdoors. The first case had the water chamber outdoors and a low-emissivity coating in face 4. When $\dot{m} = 0$, the U value was 1.041 W/m^{2}K, and the g-factor was 0.25. At the operating $\dot{m}$, the g-factor became 0.22 and the U value, 0.128 W/m^{2}K. The visual transmittance ($T_v = 0.51$) was the lowest of the selected cases. The second case had a water chamber facing indoors. A low-emissivity coating was placed in face 3 and a solar PVB layer in the central pane. The U value ranged from 0.066 to 1.041 W/m^{2}K and a variable g-factor was 0.24 when the flow was ON, and 0.59 when the flow was OFF, adapting to the outdoor environment in both summer and winter conditions. The visual transmittance was 0.53. The third case had a highly selective coating in face 2. It yielded a U value of 0.995 W/m^{2}K when the flow was OFF, and 0.063 W/m^{2}K when the flow was ON. The g-value varied between 0.22 and 0.27. The visual transmittance ($T_v = 0.55$) was the highest of the selected cases. The energy transmittance (T), did not show significant variations, and it ranged from 0.20 in Case 1 to 0.21 in Cases 2 and 3.

Table 1. Spectral and thermal properties of the studied WFGs.

| Glazing | Spectral Properties [1] | | Thermal Properties [2] | | | | | |
| | | | $\dot{m} = 0$ L/min m^2 | | | $\dot{m} = 2$ L/min m^2 | | |
	T_v	T	U (W/m^{2}K)	U_w (W/m^{2}K)	g	U (W/m^{2}K)	U_w (W/m^{2}K)	g
Case 1: P8 (2SG41) P8/24water /P8 (4RB11)P8 (plaXNII) /16argon/P8 (4RB11) P8	0.51	0.20	1.041	0.0	0.25	0.128	0.977	0.22
Case 2: D10/16argon /(plaXNII) P8(2RB11) P8 /24water/P8 (4RB11) P8	0.53	0.21	1.041	0.0	0.59	0.066	6.459	0.24
Case 3: D10 Xtreme /16argon /(plaXNII) P8(2SG41) P8 /24water/P6 (4RB11) P6	0.55	0.21	0.995	0.0	0.27	0.063	6.462	0.22

[1] Visual transmittance (T_v), energy transmittance (T). [2] Thermal transmittance from water chamber to interior (U_w), Thermal transmittance of triple glazing (U), g-factor (g).

2.3. Description of the WFG Unitized Module

All the test cases were triple glazing with a 16 mm argon cavity and a 24 mm water cavity. The water cavity spacer was designed to lead the design flow rate. The circulating device was made of a plate heat exchanger and a solar water pump. The WFG units were manufactured and assembled in the glass factory and prepared for deployment on-site with an aluminum frame that enclosed the circulating device. The maximum dimension of the panel was 3000 mm high by 1300 mm width. The circulating device main components were a plate heat exchanger and a solar water pump. Its design allowed hydraulic and electrical independence of the modules with a reduced size. The operating flow rate was set to 8 l/min for a 4 m^2 module. The circulating device was made of a water pump, selected according to the desired flow rate, and a plate heat exchanger based on the glazing heat capacity. Figure 3 shows a drawing of the circulator with its materials and primary components. Inlet and outlet temperatures were measured using one-wire temperature probes, and a flow meter was placed just before the glazing inlet.

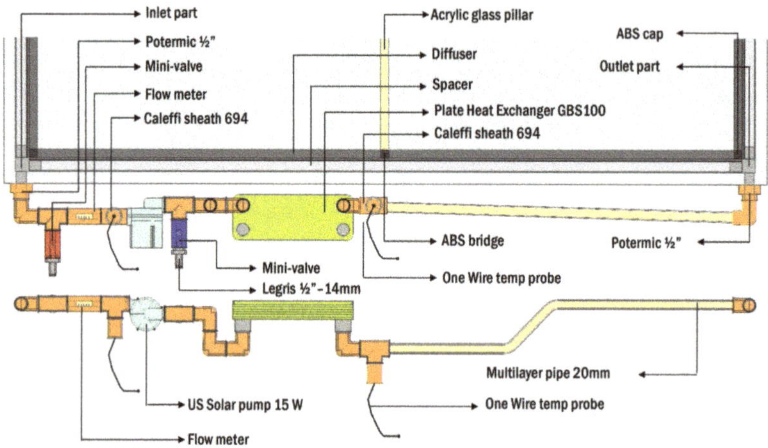

Figure 3. Description and general dimensions of the circulating device.

The modular unit hid the complexity of the hydraulic installation, so it became a plug and play product with essential advantages from the product marketing point of view. Depending on the glass selection, the WFG modules offered a broad spectrum of capabilities depending on the location, facade orientation, and use of the building.

3. Results

The objective of this section is to compare the thermal performances of different glazings using simulation tools and real data. The heat transfer coefficients (h_g, h_w) define convective heat transfer mechanisms of the air and water chambers. Optics and Window software tools were used to validate the absorptance and transmittance of commercial glass panes [41]. Optics allows analyzing the spectral properties at normal incidence, whereas Window considers spectral properties of glazing using different angles of incidence. In order to validate the simulation of the complete thermal problem, isolated glazing was considered. The assumption that indoor air temperature was constant and diffuse indoor irradiance was zero simplified the problem. Two different cases were tested: steady and transient boundary conditions.

3.1. Steady Boundary Conditions

If boundary conditions do not vary with time, the steady state does not depend on thermal mass and specific heat of components. Hence, a benchmark test case based on constant boundary conditions was the easiest way to start with validation. The study comprised three different cases of WFG. When the system was circulating, the design flow rate was 2 L/min m^2, and the inlet temperature was set at a constant value θ_{IN}. When the flow rate was stopped, the outlet temperature of the water chamber θ_{OUT} was called the stagnation temperature. The outputs of these test cases were the thermal power transported by the flow rate and the water heat gain. Figure 4 shows the dynamic U and U_w values defined in Equations (22) and (23). The red line represents the design flow rate (2 L/min m^2 = 0.029 Kg/s m^2). If the flow rate were above the design value, it would not affect the transmittances.

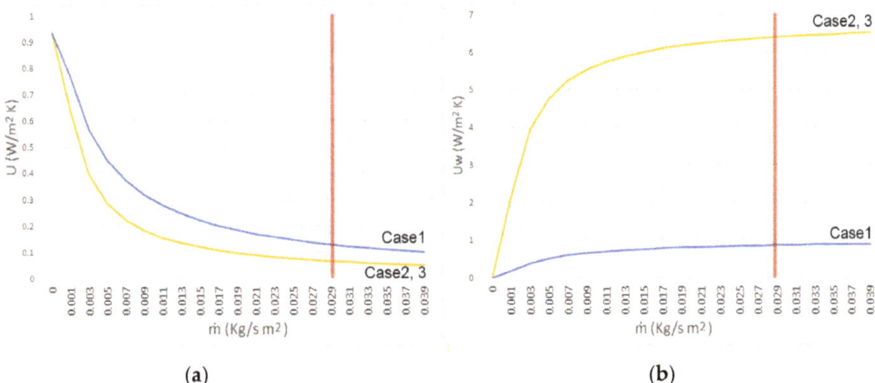

(a) (b)

Figure 4. Thermal transmittances of three WFG case studies depending on the mass flow rate, $\dot{m}$. (a) Thermal transmittance of the WFG modular units (U). (b) Thermal transmittance between the water chamber and indoors (U_w).

Equation (24) shows the absorptance, A_v, that depends on the energy absorbed by the glass panes and by the water:

$$A_v = A_1\left(\frac{U_e}{h_e}\right) + A_2\left(\frac{1}{h_g} + \frac{1}{h_e}\right)U_e + A_3\left(\frac{U_e}{h_i}\right) + A_w. \tag{24}$$

Equation (25) shows that the g-factor for WFG also depends on the mass flow rate:

$$g = \left(\frac{U_i}{\dot{m}c + U_e + U_i}\right)A_v + A_i + T, \tag{25}$$

where A_i is the secondary internal heat transfer factor. The direct solar energy transmittance (T) is related to the visible and NIR wavelengths. A_i is negligible when the water chamber is facing indoors, but if it is facing outdoors and with high values of h_w, A_i can be calculated With Equation (26).

$$A_i = A_3\left(1 - \frac{U_i}{h_i}\right). \tag{26}$$

Table 2 shows the thermal transmittance of WFG at the design flow rate. The interior or exterior convective heat transfer mechanism can be modeled by constant values or by more elaborate models given by the norm ISO 15099:2003. By default, constant values of $h_i = 8$ and $h_e = 23$ were used. However, by selecting the ISO model, h_i and h_e could be determined precisely using the European Standard [40]. A typical value for the heat transfer coefficient of the water chamber, h_w, was 50 W/m^{2}K. The heat transfer coefficient of an argon chamber was $h_c = 1.16$ W/m^{2}K. The air chamber emissivity was very low, so the heat transfer coefficient due to radiation, h_r, could be neglected. Therefore, the heat transfer coefficient of the argon chamber, $h_g = h_c + h_r$, was 1.16 W/m^{2}K. The specific heat capacity of the fluid was $c = 3600$ J/kg K. A_1 is the absorptance of the exterior glass pane, A_2, is the absorptance of the middle glass pane, and A_3 is the absorptance of the interior one. A_w is the absorptance of the water chamber. The highest A_i (0.01) is shown in Case 1, when the water chamber was placed outdoors.

Table 2. Absorptances and thermal transmittances of WFG ($\dot{m}$ = 2 L/min m^2).

Glazing	A_1	A_2	A_3	A_w	A_v	A_i	U_i (W/m^{2}K)	U_e (W/m^{2}K)	U (W/m^{2}K)	U_w (W/m^{2}K)
Case 1	0.685	0.033	0.012	0.004	0.51	0.01	0.99	15.75	0.128	0.977
Case 2	0.069	0.432	0.019	0.002	0.44	0.001	6.89	1.08	0.066	6.459
Case 3	0.291	0.028	0.019	0.001	0.06	0.001	6.89	1.08	0.063	6.462

Absorptances of glass panes (A_j), Absorptances of water layer (A_w), Total absorptance of water flow glazing (A_v), Interior thermal transmittance (U_i), Exterior thermal transmittance (U_e), Thermal transmittance of triple glazing (U), Thermal transmittance from water chamber to interior (U_w).

Equation (27) results from solving the Equations (9)–(28):

$$\theta_{OUT} = \frac{i_0 A_v + U_i \theta_i + U_e \theta_e + \dot{m}c\theta_{IN}}{\dot{m}c + U_e + U_i}. \tag{27}$$

Equation (28) shows the analytical expression of water heat gain (P).

$$P = \dot{m}c(\theta_{OUT} - \theta_{IN}), \tag{28}$$

where θ_{OUT} and θ_{IN} are the temperatures of water leaving and entering the glazing, respectively; $\dot{m}$ is the mass flow rate, and c is the specific heat of the water. By combining Equations (27) and (28), Equation (29) shows the power as a function of absorptance (A_v) and thermal transmittances (U_i, U_e).

$$P = \frac{\dot{m}c}{\dot{m}c + U_e + U_i}(i_0 A_v + U_i(\theta_i - \theta_{IN}) + U_e(\theta_e - \theta_{IN})), \tag{29}$$

where A_v comes from Equation (24). When the system reaches a steady state, the boundary conditions do not change with time. The assumption that solar radiation is perpendicular to the interfaces makes the absorptance of each layer non-dependent of the angle of incidence. When the mass flow rate is high enough ($\dot{m}c \gg U_e + U_i$), the absorbed power (P) reaches its peak value (P_{max}). Table 3 shows constant input values in winter and summer conditions. Indoor air temperature (θ_i), outdoor air temperature (θ_e), interior and exterior heat transfer coefficients (h_i, h_e), and solar irradiance (I). Equations (29)–(32) were used to calculate U_e and U_i.

Table 3. Constant input parameters in winter and summer.

Season	θ_e (°C)	θ_i (°C)	h_e (W/m^{2}K)	h_i (W/m^{2}K)	I (W/m^2)	$\dot{m}$ (L/min m^2)	θ_{IN} (°C)	h_g (W/m^{2}K)	h_w (W/m^{2}K)
Winter	0	21	23	8	600	2	21	1.16	50
Summer	35	28	23	8	800	2	17	1.16	50

Once the glazing reaches the steady state in winter and in summer, thermal performances are determined. Using Equations (27) and (29), and considering the inputs in Table 3, Table 4 shows the following outputs in winter and summer, respectively. P is the water heat gain, T is the transmittance of the glazing, θ_{OUT} is the outlet temperature when the flow rate is $\dot{m}$ = 2 L/min m^2, and θ_s is the stagnation temperature when $\dot{m}$ = 0.

Table 4. Simulation outputs. Steady state in winter and summer conditions.

Glazing	P_{winter} (W/m^2)	θ_{OUT} (°C)	θ_s (°C)	T_{winter} (W/m^2)	P_{summer} (W/m^2)	θ_{OUT} (°C)	θ_s (°C)	T_{summer} (W/m^2)
Case 1	22.9	20.78	19.41	123.7	603.3	22.77	58.8	164.9
Case 2	226.6	23.17	51.57	128.4	418.7	21.01	73.5	171.2
Case 3	11.5	19.81	4.69	129.2	131.9	18.26	34.81	172.3

If energy management in winter is based on energy harvesting, Case 2 showed the best performance. Its water heat gain in winter was ten times as high as in Case 1. On the other hand, if energy management in summer is based on energy rejection, Case 3 was the best choice. In summer, the water heat gain of Case 1 was 1.5 times as much as Case 2. When it came to cooling capacity, Case 3 showed the best performance. It had to dissipate around 131.9 W/m^2, whereas Case 2 had to dissipate 418.7 W/m^2. Case 3 showed excellent properties for energy rejection strategies in warm climates because it showed the least absorbed power in summer, whereas the transmittance (T) was not much higher than in other cases.

Considering the simulation results in a steady state, Case 2 showed the best performance for water heat absorption throughout the year. It was selected for the south elevation. Case 3 showed the best performance for energy rejection, and it was selected for east and west facades. The next subsection studies the simulation results of the selected cases in transient conditions.

3.2. Transient Boundary Conditions

Transient behavior is expected when boundary conditions such as outdoor temperature and solar irradiance vary during the day. In these following test cases, the indoor temperature was a given indoor boundary condition, and transport coefficients remained constant to avoid uncertainties in the validation process. These test cases were simulated in Sofia, and the weather file was the standard EPW file (EnergyPlus Weather). Regarding the water flow glazing, the flow rate and the inlet temperature were constant values given by Table 3. Two simulations in winter and summer were accomplished. The simulation period ran from 7 January to 11 January in winter and from 14 July to 18 July in summer. Figure 5 illustrates the thermal behavior of Case 2 and Case 3. In summer, the solar irradiance peak value was 500 W/m^2, and the maximum outdoor temperature was slightly above 26 °C on 14 July 2020. The goal of rejecting energy was met, and the water heat gain, measured by the difference between inlet and outlet temperatures, was not above 1 °C on five sample summer days. The peak solar radiation in winter on the eastern facade was 180 W/m^2 on 10 January 2020. Due to the high infrared reflectance (above 70%) of the selected glazing and the low outdoor temperature (below 5.5 °C), the water heat gains were negligible. According to the steady-state analysis, Case 2 showed the best performance to heat water, as measured by the solar irradiance on the southern facade, the outdoor temperature, and the difference between inlet and outlet temperatures in southern WFG modules. In summer, the peak solar radiation was 400 W/m^2, and the maximum temperature was above 26.5 °C on 15 July 2020. On that day, the maximum outlet water temperature was 22 °C when the inlet temperature was 20 °C, and there were water heat gains during the central hours of the day. In winter, the peak solar radiation was above 250 W/m^2, and that made the water absorb heat, although the outdoor temperature was low.

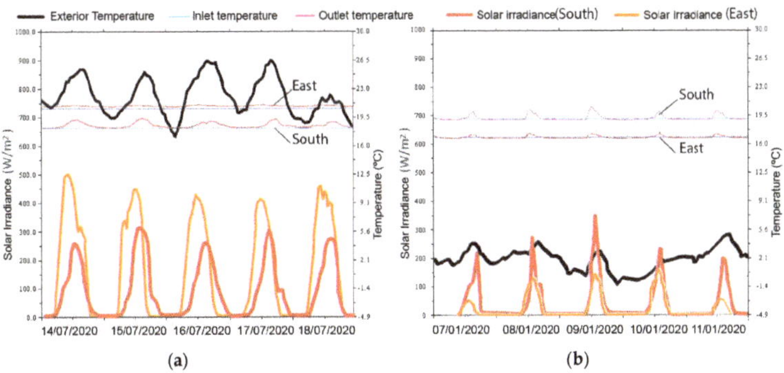

Figure 5. Simulation results of WFG Case 2 on the southern facade and Case 3 on the eastern facade with transient boundary conditions. (**a**) Summer. (**b**) Winter.

Figure 6 shows a summary of the water heat gains on two sample days in summer and winter. Case 3 was selected for eastern and western facades because it showed the least heat absorption in summer (17 KWh), whereas Case 2 showed the highest absorption in summer (34 KWh). To reject energy, the best choice for eastern and western facades was Case 3. Case 2 had the highest heat absorption on a winter day (21 KWh) and a good value in summer (30.5 KWh). Case 2 confirmed its excellent performance on southern facades.

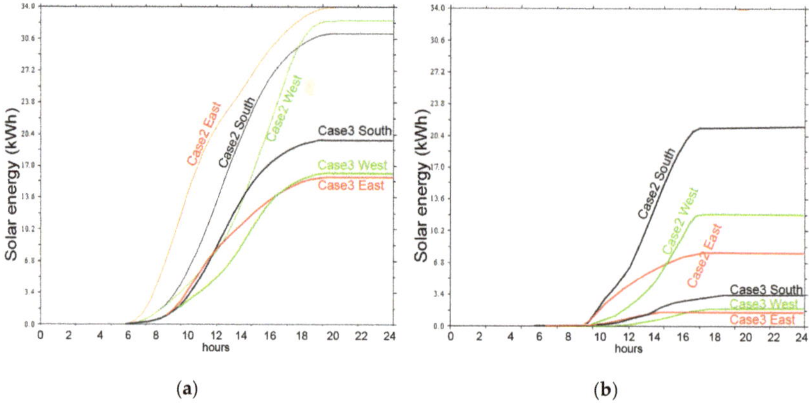

Figure 6. Accumulated energy of WFG case studies with transient boundary conditions. (**a**) Summer sample day 14 July 2020. (**b**) Winter sample day 8 January 2020.

3.3. Real Results in Sofia, Bulgaria

An experimental setup was placed in Sofia, Bulgaria (42°39′1″ North, 23°23′26″ East, Elevation: 590 m a.s.l.), to test the performance of isolated WFG modules throughout a year. Figure 7 shows the outdoor temperature. On the coldest winter days, the minimum temperature was below −10 °C, and the average daily temperature was 0 °C. During the hottest months, the maximum temperature reached 32 °C and the average temperature was 25 °C. The southern solar radiation reached a peak value of 400 W/m^2 on 21 December 2019, whereas the eastern and western were 160 W/m^2 and 270 W/m^2, respectively, on 21 June 2019. On summer days, the highest values were on the east and west facades (500 W/m^2) because the sun angle was almost perpendicular to the vertical walls. The south facade received little radiation (200 W/m^2).

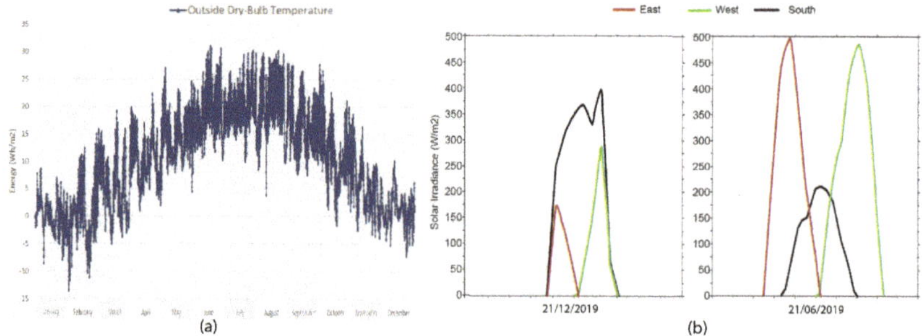

Figure 7. (**a**) Outdoor dry bulb temperature in Sofia, Bulgaria (EnergyPlus Weather file). (**b**) Eastern, western, and southern solar radiation on facades. Sample winter day 21 December 2019 and sample summer day 21 June 2019.

Based on the outdoor simulation data, and the thermal and spectral properties of the studied WFG in Table 4, the best option for the southern facade was Case 2. It showed the highest potential for heat absorption in winter (226.6 W/m^2) with the highest outlet temperature (23.17 °C). In summer, the maximum southern solar radiation was 200 W/m^2, whereas the absorption potential was 418 W/m^2, so the fluid could absorb the heat without heating the interior face of the glazing. Due to the high solar radiation in summer, the best option for eastern and western facades was Case 3. It showed the lowest absorption in summer (131.9 W/m^2) with the lowest outlet temperature (18.26 °C). Figure 8 shows the prototype plan, with five modules facing east, five modules facing west, and five more modules on the southern facade. Unitized WFG modules were placed in three different orientations (east, south, and west) with a pyranometer measuring solar radiation on each facade. Each heat plate exchanger of the circulating device was connected to inlet and outlet water distribution systems.

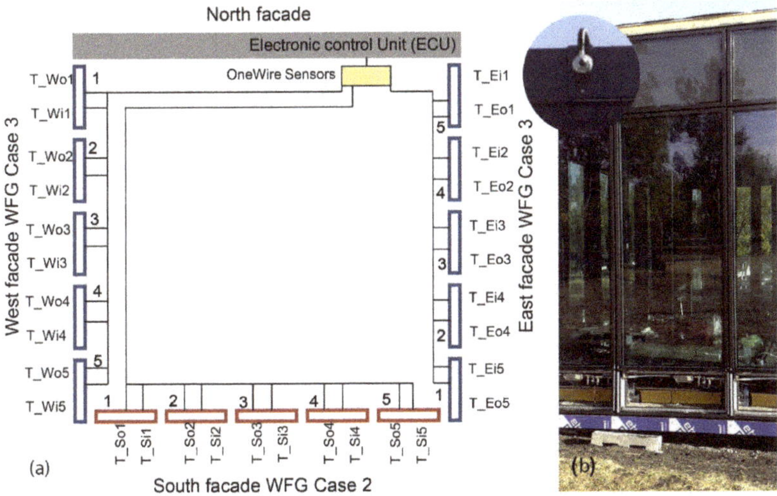

Figure 8. (**a**) Prototype plan. Position of WFG and electronic control unit. (**b**) Pictures of the unitized module in the actual facility with the pyranometer.

The output signals were collected by one-wire probes and sent to an electronic control unit (ECU), where the developed software processed the calculations and elaborated the energy outputs. The temperature sensor network was installed in both the inlet and outlet of the plate heat exchanger. Flux meters were added to the monitoring equipment to keep a steady mass flow rate through all the modules. The temperature difference in the external WFG elements could reach 10 °C, depending on the exterior conditions. Glass selection for renewable production on the southern facade (Case 2) absorbed the maximum incident solar radiation and at the same time reduced indoor solar heat gains. A heat pump was used to control the inlet temperature. Figure 9 illustrates the outdoor air temperature (*T_out*), inlet (*T_Ei5*) and outlet (*T_Eo5*) temperatures in two eastern WFG modules in summer conditions. The maximum temperature difference occurred from 7:00 a.m. to 10:00 a.m., when the solar radiation reached its peak value on the east facade. The southern modules' inlet and outlet temperatures (*T_Si5*, *T_So5*) reflected the solar radiation and outdoor temperature, and there were two peak values at 11:00 a.m. and 4:00 p.m. The maximum temperature difference between *T_So5* and *T_Si5* was 2 °C. The maximum temperature difference between the inlet (*T_Wi3*) and outlet (*T_Wo3*) temperatures in two western WFG modules occurred at 4:30 p.m., when the solar radiation reached its peak value on the west facade. The real measurements confirmed the simulation results because, despite the high solar radiation values on the eastern and western facades (700 W/m^2), the temperature difference between inlet and outlet was 1 °C. However, in the southern modules, the temperature difference was 2 °C when the maximum solar radiation was 470 W/m^2.

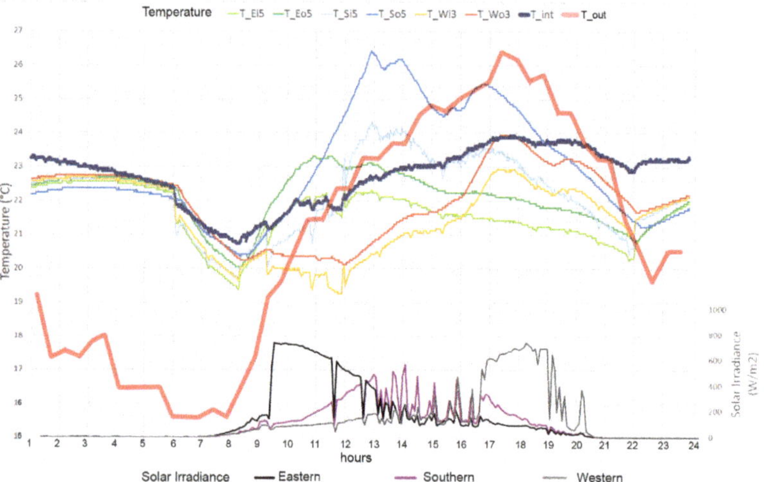

Figure 9. Inlet and outlet temperatures of eastern WFG. Sample summer day 14 July 2020. (**a**) Module E5. (**b**) Module E1.

In winter, heat absorption does not depend directly on solar radiation due to the severity of climatic conditions. The difference between indoor and outdoor temperatures affected energy performance more than the solar radiation on the eastern and western facades. Figure 10 illustrates the outdoor air temperature (*T_out*), the inlet (*T_Ei5*) and outlet (*T_Eo5*) temperatures in two eastern WFG modules. The southern WFG performance showed heat losses in the morning and in the afternoon. From 10:00 a.m. to 5:00 p.m., the outlet temperature (*T_So5*) was higher than the inlet (*T_Si5*), and the maximum difference reached 2.5 °C at 1:30 p.m. In western modules, the inlet (*T_Wi3*) and outlet (*T_Wo3*) temperatures showed that there were heat losses in the morning with no solar radiation and low outdoor temperature. The simulation results were validated with little energy absorption on eastern and western facades, and heat gains in the southern modules.

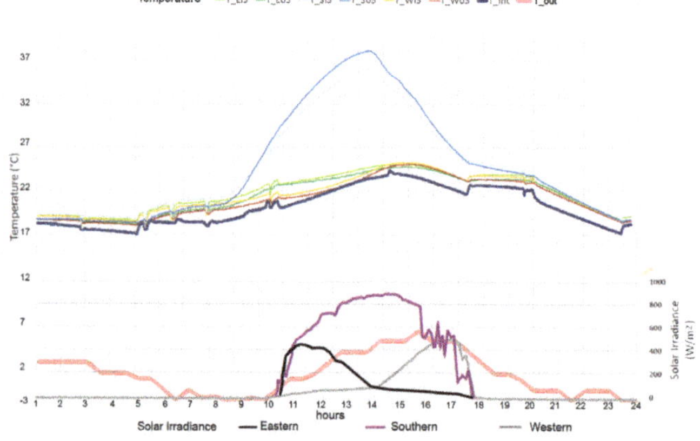

Figure 10. Inlet and outlet temperatures of eastern WFG. Sample winter day 8 January 2020. (**a**) Module E4. (**b**) Module E3.

4. Discussion

The next step was to analyze the results in terms of potential energy savings. Firstly, the results of the simulation tool were validated. Secondly, the heat absorbed by water in summer can be both subtracted from the cooling loads and considered as renewable energy production. According to the Energy Performance of Buildings Directive (EPDB 2018) recommendations [2], primary energy factors (PEFs) were used to assess the energy performance.

4.1. Validation of Energy Performance

To validate the selection of WFG, the daily absorbed energy per unit of area was calculated using Equation (29). Figure 11 shows the performance of two WFG panels in winter in three orientations. As expected, the eastern and western panels' energy absorption was not relevant. Most of the time there were heat losses due to the little radiation and the low outdoor temperature. A different performance was shown in the southern panels, where the daily absorbed energy was 21.3 kWh in 7.8 m^2, so the ratio of energy per area was 2.73 kWh/m^2.

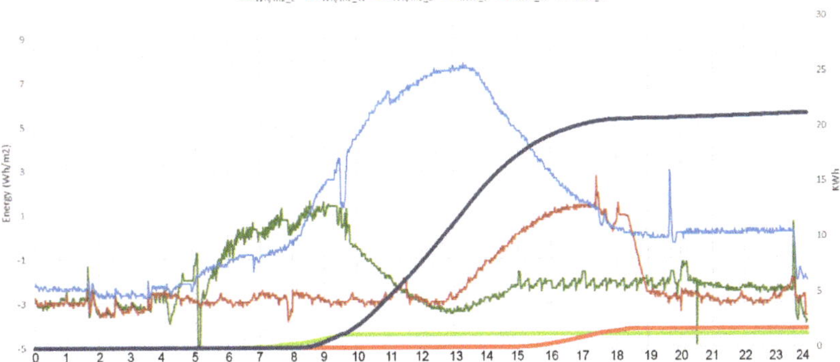

Figure 11. Energy absorption on eastern, southern, and western WFG facades. Sample winter day 8 January 2020.

Figure 12 shows the energy performance on a sample summer day (14 July 2020). The total absorbed energy was 30 KWh in two southern WFGs, 18.6 KWh in two western WFGs, and 15.9 KWh in two eastern WFGs. The energy-to-area ratio was 3.85 kWh/m^2 in the south, 2.38 kWh/m^2 in the west, and 2.04 kWh/m^2 in the east.

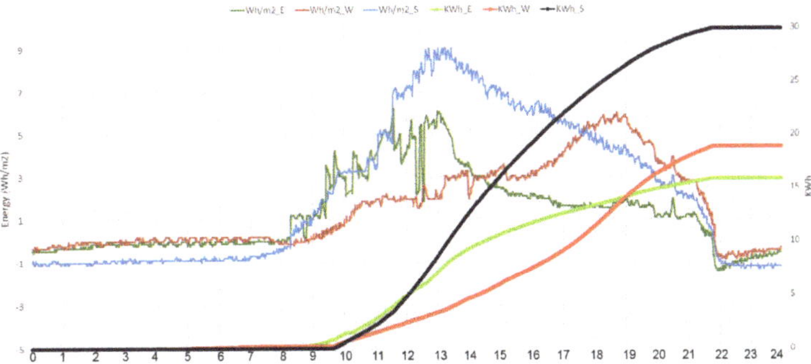

Figure 12. Energy absorption on eastern, southern, and western WFG facades. Sample summer day 14 July 2020.

The goal of rejecting energy in the east and west was met. Despite the high solar radiation, the water heated up by 2 °C, and most of the infrared energy was rejected. On the south facade, the energy absorption was similar in winter and summer, and the water heated up around 3 °C. The reliability of the simulation tool was tested by developing real prototypes. Figure 13 illustrates the comparison between the results of the real data and the simulation data. The daily energy absorption on southern facades in winter and summer were taken from Figures 11 and 12 and compared with the simulated results from Figure 6.

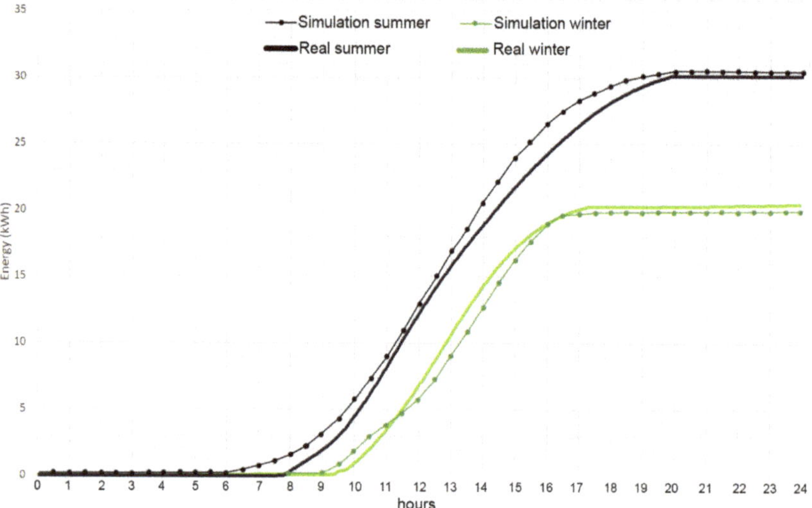

Figure 13. Accumulated energy absorption on southern WFG facades. Sample days 14 July 2020 and 8 January 2020. Comparison between simulated and real data.

The Mean Error (*ME*), shown in Equation (24), is the difference between the measured value and simulation results. The total number of measurements was $n = 2872$.

$$ME = \frac{1}{n} \sum_{i=1}^{n} |E_{Si} - E_{Ri}|, \qquad (30)$$

where E_{Si} is the simulated energy absorption, and E_{Ri} is the measured energy absorption. By computing *ME* on 14 July 2020 the value was 0.78. When it came to the energy absorption on 8 January 2020, the *ME* was 0.67. The reason for this might be the uncertainties about the inlet and indoor temperatures. Although the accumulated energy values were quite similar in the simulation and the real conditions, the intermediate values differed at some times of the day. The simulation tool could not work with variable inlet and indoor temperatures, which is the main goal for further research.

4.2. Primary Energy Consumption

Figure 14 shows the outdoor air temperature and the accumulated energy throughout five days in summer. WFG absorbed solar energy and prevented it from entering the building. The amount of energy absorbed by the water could be connected to the district heating network, geothermal boreholes, or to domestic hot water devices. In the final energy balance, the accumulated energy was subtracted from the cooling loads and added to the renewable energy production.

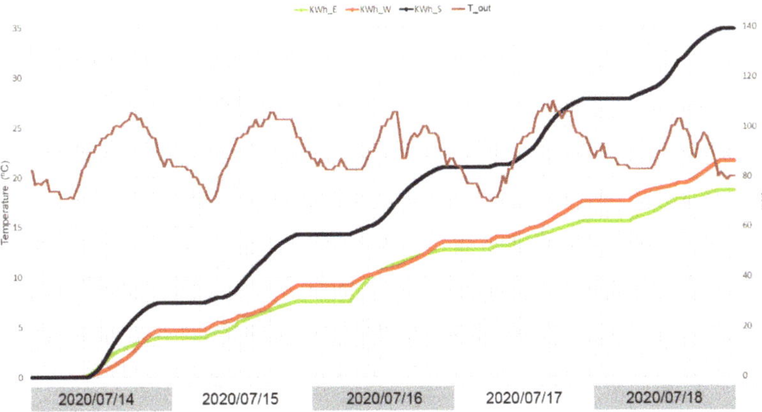

Figure 14. Energy absorption on eastern, southern, and western WFG facades. Sample summer days from 14 July 2020 to 18 July 2020.

Article 41 of the Energy Performance of Buildings Directive (EPDB 2018) recommended the use of primary energy factors to calculate the energy parameters of building envelopes [2]. Table 5 shows these energy factors, such as cooling energy demand (CED in kWh/m^2), final energy consumption (FEC in kWh/m^2), non-renewable primary energy consumption (NRPEC in kWh/m^2). The energy absorbed by the water was considered as renewable primary energy production (RPE in kWh/m^2) and CO_2 emissions (kg CO_2/m^2). The electricity consumption of the circulation water pumps was not considered because they were connected to photovoltaic panels. The circulation pump was working 24 h per day. The primary energy factor (PEF) is the inverse of electricity production efficiency from fuel source to electricity at the building, taken from official European Union documents [42,43], if the energy was produced using heat pumps, considering a Coefficient of Performance (COP) of 2.5 and a conversion factor between final energy and non-renewable primary energy (KWh NRPE/KWh FE) of 1.954. The factor of emitted CO_2 for electricity was 0.331.

Table 5. Primary energy balance of WFG in summer.

	Cooling Energy Demand (CED) kWh/m^2 day	Final Energy Consumption (FEC) kWh/m^2 day	Non-Renewable Primary Energy Consumption (NRPEC) kWh/m^2 day	Renewable Primary Energy (RPE) kWh/m^2 day	CO_2 Emissions kgCO$_2$/m^2 day
Day 1	8.27	3.31	6.46	3.42	1.09
Day 2	7.60	3.04	5.94	3.15	1.01
Day 3	8.24	3.29	6.44	3.41	1.09
Day 4	6.95	2.78	5.43	2.88	0.92
Day 5	7.25	2.9	5.67	3.00	0.96
Total	38.31	15.32	29.94	15.86	5.07

Table 6 shows the estimated winter heating loads over the working hours. Indoor (T_{int}) and outdoor (T_{ext}) temperatures were taken from Figure 10 with a surface area of 3.9 m^2. The same procedure was repeated to calculate the values on five sample days. A high-performance triple glass made of three glass panes with an argon chamber and Low-E coating was compared with the selected WFG cases. The triple-glass U value was taken from a glazing catalog [41], whereas the U value of the WFG was defined in Table 1. The total heat losses through the passive triple glazing and the WFG were 288.38 Wh/m^2 and 15.31 Wh/m^2. The energy savings per day was 273.07 Wh/m^2. In addition, the WFG was able to produce renewable energy.

Table 6. Winter heating loads on 14 January 2020.

			Triple Glass		WFG	
	T_{int} (C) [1]	T_{ext} (C) [1]	U *(triple glazing)* (W/m²K) [2]	$\sum UA(T_{int}\text{-}T_{ext})$ (Wh/m²)	U *(WFG)* (W/m²K)	$\sum UA(T_{int}\text{-}T_{ext})$ (Wh/m²)
7–8 h	18	−3	1.3	23.73	0.06	1.26
8–9 h	18	−3	1.3	23.73	0.06	1.26
9–10 h	21.5	−3	1.3	27.69	0.06	1.47
10–11 h	21.8	0	1.3	24.63	0.06	1.31
11–12 h	18.8	1.5	1.3	19.55	0.06	1.04
12–13 h	20	3.1	1.3	19.10	0.06	1.01
13–14 h	22.7	4.5	1.3	20.57	0.06	1.09
14–15 h	23	5.1	1.3	20.23	0.06	1.07
15–16 h	23.1	6	1.3	19.32	0.06	1.03
16–17 h	23.5	5.6	1.3	20.23	0.06	1.07
17–18 h	22.2	3.6	1.3	21.02	0.06	1.12
18–19 h	22	2	1.3	22.60	0.06	1.20
19–20 h	22	−1	1.3	25.99	0.06	1.38
TOTAL				288.38		15.31

[1] Values are taken from Figure 10; [2] values are taken from [41].

Table 7 illustrates the primary energy savings, the reduction of CO_2 emissions, and renewable energy production of WFG in five winter days.

Table 7. Primary energy balance of WFG in winter.

	CED kWh/m²	FEC kWh/m²	NRPEC kWh/m²	RPE kWh/m²	EM kgCO₂/m²
Day 1	0.226	0.090	0.177	2.40	0.030
Day 2	0.225	0.090	0.176	2.55	0.030
Day 3	0.222	0.089	0.174	2.46	0.029
Day 4	0.222	0.089	0.174	2.46	0.029
Day 5	0.227	0.091	0.177	2.57	0.030
Total	0.116	0.449	0.877	12.44	0.149

4.3. Cost Considerations

The ideal project for these advanced facades would be a large tower-type office building with limited site access. The facade should consist of repetitive geometry that can be divided easily into panels. The system would not be fit to have movable windows, so special modules for openings and mechanical ventilation systems are required. A literature review was carried out to assess the cost and the performance of advanced facades [44–46]. The unit costs of the components included the material, production, and assembly costs. These values are the average of the unit costs taken from two different passive curtain wall systems [47]. The cost analysis for the WFG-unitized facade considered a triple glass described in Case 2, the total estimated costs for aluminum production, module fabrication, on-site transportation, and facade assembly. The energy values were calculated with the indoor and outdoor temperatures shown in Figures 9 and 10 for sample summer and winter days, respectively. The energy parameters and construction costs calculated for all the alternatives are shown in Table 8.

Table 8. Energy and cost parameters.

System Description	Glazing	U W/m²K	FEC kWh/m²		RPE kWh/m²		EM kgCO₂/m²	Cost EUR/m²
			S	W	S	W		
Aluminum frame fixed to slab	Triple glazed	1.3 [1]	2.42	0.64	-	-	1.01	620 [1]
Unitized facade Aluminum frame	WFG (Case 2)	0.066	0.94	0.30	3.42	2.73	0.41	1375

[1] Values taken from [47].

The initial costs of the WFG made up of the triple glazing, the circulating device, and the unitized aluminum frames were high, compared to passive glazing systems. However, a holistic approach should include energy savings, energy production, and CO_2 emissions. The total final energy consumption was 3.06 kWh/m^2 for the standard triple-glazed lightweight enclosure and 1.24 kWh/m^2 for the WFG. When it came to CO_2 emissions, the standard curtain wall solution would account for 2.5 times as much CO_2 as the studied WFG.

5. Conclusions

The design of new light envelopes for zero-energy buildings must integrate different disciplines such as architectural design, building simulation, HVAC systems, and the curtain wall industry. This article developed a methodology for selecting WFG solutions for different facades and tested its performance using real data. Case 1 was a triple glazing with the water chamber facing outdoors. Case 2 was made of triple glazing with water chamber indoors and Low-E coating. Case 3 had a high reflective coating on face 2 and a water chamber indoors.

1. Dynamic properties of WFG allowed considering options for different orientations and the internal loads, which depend on the building use. The thermal transmittance (U) ranged from 1 W/m^{2}K ($\dot{m} = 0$) to 0.06 W/m^{2}K ($\dot{m} = 2$ L/min m^2). A mass flow rate ($\dot{m}$) above 2 L/min m^2 (0.029 Kg/s m^2) did not impact the water–energy absorption.
2. The position of the gas and water cavities and the spectral properties of glass panes and coatings affected the performance of the WFG. Simulation results at steady conditions showed that Case 2 had the best performance for energy absorption in winter (226.6 W/m^2). Case 1 showed the highest energy absorption in summer (603.3 W/m^2). Case 3 showed the lowest energy absorption, both in summer (131.9 W/m^2) and winter (11.5 W/m^2).
3. The primary energy factor (PEF) was used to assess building energy performance. Energy savings ranged from 5.43 to 6.46 KWh/m^2 day in non-renewable energy consumption, whereas the renewable primary energy production ranged from 3 to 3.42 KWh/m^2 day. The CO_2 emissions were reduced at a rate of 1 Kg/m^2 day.
4. In the cold winter season, the absorbing south facade heated water (2.73 kWh/m^2), whereas eastern and western facades received very little solar radiation.
5. In summer, eastern and western facades rejected most of the solar radiation, and the flowing water was heated without surpassing the comfort temperature. The maximum outlet temperature was 24 °C on 14 July 2020. The daily absorbed energy was 3.84 KWh/m^2 in the southern WFG, 2.38 KWh/ m^2 in WFG, and 2.04 KWh/ m^2 in WFG.

This article showed an industrialized water flow glazing unitized facade ready to be used in the architecture, engineering, and construction industries. The authors developed a simulation tool to be used at the first stage of the design process. The outputs were validated with an actual test facility placed in Sofia, Bulgaria. The difficulties identified were related to the limitations of the software for simulating the dynamic properties of WFG. The high initial cost and the need for an energy management system integrated with the rest of the equipment conditioned the WFG system. After the first year of monitoring, there are uncertainties and system issues that must be addressed. Firstly, the control unit must integrate the ventilation system to reduce condensation risks. Secondly, the presented simulation tool must be integrated into commercial building performance simulation software. Finally, further research on the deployment is needed to bring down payback periods.

With economies of scale a price comparable to triple-pane glazing systems equipped with automated exterior shading can be achieved.

Author Contributions: Conceptualization, B.M.S., F.d.A.G., J.A.H.R.; methodology, B.M.S., F.d.A.G.; software, J.A.H.R.; formal analysis, B.M.S., F.d.A.G.; data curation, J.A.H.R.; writing—original draft preparation, B.M.S., F.d.A.G., J.A.H.R.; writing—review and editing, F.d.A.G., D.P.; visualization, B.M.S., F.d.A.G., B.L.A.; supervision,

J.A.H.R., B.L.A.; project administration, B.M.S.; funding acquisition, F.d.A.G. All authors have read and agreed to the published version of the manuscript.

Funding: This article has been funded by a KSC Faculty Development Grant (Keene State College, New Hampshire, USA).

Acknowledgments: This work was supported by program Horizon 2020-EU.3.3.1: Reducing energy consumption and carbon footprint by smart and sustainable use, project Ref. 680441 (InDeWaG: Industrialized Development of Water Flow Glazing Systems). Special thanks to the Central Laboratory of Solar Energy and New Energy Sources of the Bulgarian Academy of Science (CL SENES–BAS) for providing measured data of the solar radiation on the different facades of the test facility in Sofia, Bulgaria.

Conflicts of Interest: The authors declare that they have no conflict of interest.

Nomenclature

Symbol	Meaning
A_j	Absorptance of glass layers
A_w	Absorptance of water
A_v	Total absorptance of water flow glazing
h_i	Interior heat transfer coefficient (W/m^{2}K)
h_w	Water heat transfer coefficient (W/m^{2}K)
h_g	Air chamber heat transfer coefficient (W/m^{2}K)
h_e	Exterior heat transfer coefficient (W/m^{2}K)
q_j	Heat fluxes through the different layers of the glazing
i_0	Normal incident solar irradiance (W/m^2)
θ_i	Interior temperature (K)
θ_e	Exterior temperature (K)
θ_j	Temperature of the glass layer (K)
θ_{IN}	Inlet temperature of the water chamber (K)
θ_{OUT}	Outlet temperature of the water chamber (K)
θ_w	Temperature of the water (K)
θ_S	Stagnation temperature of the water when $\dot{m} = 0$ (K)
U	Thermal transmittance (W/m^{2}K)
U_i	Interior thermal transmittance (W/m^{2}K)
U_e	Exterior thermal transmittance (W/m^{2}K)
U_w	Thermal transmittance (water chamber–interior) (W/m^{2}K)
T	Transmittance of the glazing
R	Reflectance of the glazing
$\dot{m}$	Mass flow rate (kg/s m^2)
c	Specific heat of the water (J/Kg K)
P	Heat gain in the water chamber (W)

References

1. Stenqvist, C.; Nielsen, S.B.; Bengtsson, P.-O. A Tool for Sourcing Sustainable Building Renovation: The Energy Efficiency Maturity Matrix. *Sustainability* **2018**, *10*, 1674. [CrossRef]
2. European Union. Directive (EU) 2018/844 of the European Parliament and of the Council of 30 May 2018. Amending Directive 2010/31/EU on the Energy Performance of Buildings and Directive 2012/27/EU on Energy Efficiency. 2018. Available online: https://eur-lex.europa.eu/legal-content/EN/TXT/PDF/?uri=CELEX:32018L0844&from=EN (accessed on 24 July 2020).
3. DOE/EIA. *International Energy Outlook*; US Energy Information Administration, US Department of Energy: Washington, DC, USA, 2016.
4. Chin, J.; Lin, S.-C. A Behavioral Model of Managerial Perspectives Regarding Technology Acceptance in Building Energy Management Systems. *Sustainability* **2016**, *8*, 641. [CrossRef]
5. Prieto, A.; Knaack, U.; Klein, T.; Auer, T. 25 Years of cooling research in office buildings: Review for the integration of cooling strategies into the building facade (1990–2014). *Renew. Sustain. Energy Rev.* **2017**, *71*, 89–102. [CrossRef]

6. Santamouris, M.; Kolokotsa, D. Passive cooling dissipation techniques for buildings and other structures: The state of the art. *Energy Build.* **2013**, *57*, 74–94. [CrossRef]

7. Sudhakar, K.; Winderl, M.; Shanmuga Priya, S. Net-zero building designs in hot and humid climates: A state-of-art. *Case Stud. Therm. Eng.* **2019**, *13*, 100400. [CrossRef]

8. Fernandez-Antolin, M.-M.; del-Río, J.-M.; Gonzalez-Lezcano, R.-A. Influence of Solar Reflectance and Renewable Energies on Residential Heating and Cooling Demand in Sustainable Architecture: A Case Study in Different Climate Zones in Spain Considering Their Urban Contexts. *Sustainability* **2019**, *11*, 6782. [CrossRef]

9. Ürge-Vorsatz, D.; Cabeza, L.F.; Serrano, S.; Barreneche, C.; Petrichenko, K. Heating and cooling energy trends and drivers in buildings. *Renew. Sustain. Energy Rev.* **2015**, *41*, 85–98. [CrossRef]

10. Bustamante, W.; Vera, S.; Prieto, A.; Vasquez, C. Solar and Lighting Transmission through Complex Fenestration Systems of Office Buildings in a Warm and Dry Climate of Chile. *Sustainability* **2014**, *6*, 2786–2801. [CrossRef]

11. Ozel, M. Influence of glazing area on optimum thickness of insulation for different wall orientations. *Appl. Therm. Eng.* **2019**, *147*, 770–780. [CrossRef]

12. López-Ochoa, L.M.; Las-Heras-Casas, J.; López-González, L.M.; García-Lozano, C. Energy Renovation of Residential Buildings in Cold Mediterranean Zones Using Optimized Thermal Envelope Insulation Thicknesses: The Case of Spain. *Sustainability* **2020**, *12*, 2287. [CrossRef]

13. Ulpiani, G.; Giuliani, D.; Romagnoli, A.; di Perna, C. Experimental monitoring of a sunspace applied to a NZEB mock-up: Assessing and comparing the energy benefits of different configurations. *Energy Build.* **2017**, *152*, 194–215. [CrossRef]

14. Hermanns, M.; del Ama, F.; Hernández, J.A. Analytical solution to the one-dimensional non-uniform absorption of solar radiation in uncoated and coated single glass panes. *Energy Build.* **2012**, *47*, 561–571. [CrossRef]

15. Ascione, F.; de Masi, R.F.; de Rossi, F.; Ruggiero, S.; Vanoli, G.P. Optimization of building envelope design for nZEBs in Mediterranean climate: Performance analysis of residential case study. *Appl. Energy* **2016**, *183*, 938–957. [CrossRef]

16. Manz, H.; Menti, U.P. Energy performance of glazings in European climates. *Renew. Energy* **2012**, *37*, 226–232. [CrossRef]

17. Allen, K.; Connelly, K.; Rutherford, P.; Wu, Y. Smart windows—Dynamic control of building energy performance. *Energy Build.* **2017**, *139*, 535–546. [CrossRef]

18. Ghosh, A.; Norton, B.; Duffy, A. Measured overall heat transfer coefficient of a suspended particle device switchable glazing. *Appl. Energy* **2015**, *159*, 362–369. [CrossRef]

19. Casini, M. Smart windows for energy efficiency of buildings. *Int. J. Civ. Struct. Eng. IJCSE* **2015**, *2*, 230–238. [CrossRef]

20. Gueymard, C.; duPont, W. Spectral effects on the transmittance, solar heat gain, and performance rating of glazing systems. *Solar Energy* **2009**, *83*, 940–953. [CrossRef]

21. Gutai, M.; Kheybari, A.G. Energy consumption of water-filled glass (WFG) hybrid building envelope. *Energy Build.* **2020**, *218*, 110050. [CrossRef]

22. Gil-Lopez, T.; Gimenez-Molina, C. Influence of double glazing with a circulating water chamber on the thermal energy savings in buildings. *Energy Build.* **2013**, *56*, 56–65. [CrossRef]

23. Chow, T.T.; Li, C.; Lin, Z. Thermal characteristics of water-flow double-pane window. *Int. J. Therm. Sci.* **2010**, *50*, 140–148. [CrossRef]

24. Li, C.; Chow, T.T. Water-filled double reflective window and its year-round performance. *Proc. Environ. Sci.* **2011**, *11*, 1039–1047. [CrossRef]

25. Chow, T.T.; Li, C. Liquid-filled solar glazing design for buoyant water-flow. *Build. Environ.* **2013**, *60*, 45–55. [CrossRef]

26. Ji, J.; Luo, C.; Chow, T.T.; Sun, W.; He, W. Thermal characteristics of a building-integrated dual-function solar collector in water heating mode with natural circulation. *Energy* **2011**, *36*, 566–574. [CrossRef]

27. Lanzisera, S.; Dawson-Haggertym, S.; Cheung, H.; Taneja, J.; Culler, D.; Brown, R. Methods for detailed energy data collection of miscellaneous and electronic loads in a commercial office building, *Build. Environ.* **2013**, *65*, 170–177. [CrossRef]

28. Nikolaeva-Dimitrova, M.; Stoyanova, M.; Ivanov, P.; Tchonkova, K.; Stoykov, R. Investigation of thermal behaviour of innovative water flow glazing modular unit. *Bulg. Chem. Commun.* **2018**, *50*, 21–27.

29. Dagdougui, Y.; Ouammi, A.; Benchrifa, R. Energy Management-Based Predictive Controller for a Smart Building Powered by Renewable Energy. *Sustainability* **2020**, *12*, 4264. [CrossRef]

30. Gan, V.J.L.; Lo, I.M.C.; Ma, J.; Tse, K.T.; Cheng, J.C.P.; Chan, C.M. Simulation Optimisation towards Energy Efficient Green Buildings. *J. Clean. Prod.* **2020**, *254*, 120012. [CrossRef]

31. Chiesa, G.; Grosso, M. The influence of different hourly typical meteorological years on dynamic simulation of buildings. *Energy Procedia* **2015**, *78*, 2560–2565. [CrossRef]

32. Loonen, R.; Favorino, F.; Hensen, J.; Overend, M. Review of current status, requirements and opportunities for building performance simulation of adaptive facades. *J. Build. Perform. Simul.* **2016**, *1493*, 1–19. [CrossRef]

33. Bambardekar, S.; Poerschke, U. The architect as performer of energy simulation in the early design stage. In Proceedings of the IBPSA 2009—International Building Performance Simulation Association 2009, Eleventh International IBPSA Conference, Glasgow, UK, 27–30 July 2009; pp. 1306–1313.

34. Fernandez-Antolin, M.; del-Rio, J.M.; del Ama Gonzalo, F.; Gonzalez-Lezcano, R. The Relationship between the Use of Building Performance Simulation Tools by Recent Graduate Architects and the Deficiencies in Architectural Education. *Energies* **2020**, *13*, 1134. [CrossRef]

35. Sierra, P.; Hernandez, J.A. Solar heat gain coefficient of water flow glazing. *Energy Build.* **2017**, *139*, 133–145. [CrossRef]

36. Chow, T.; Chunying, L.; Clarke, J.A. Numerical prediction of water-flow glazing performance with reflective coating. Proceedings of Building Simulation 2011, 12th Conference of International Building Performance Simulation Association, Sydney, Australia, 14–16 November 2011.

37. Moreno Santamaria, B.; del Ama Gonzalo, F.; Pinette, D.; Gonzalez-Lezcano, R.-A.; Lauret Aguirregabiria, B.; Hernandez Ramos, J.A. Application and Validation of a Dynamic Energy Simulation Tool: A Case Study with Water Flow Glazing Envelope. *Energies* **2020**, *13*, 3203. [CrossRef]

38. Moreno Santamaria, B.; del Ama Gonzalo, F.; Lauret Aguirregabiria, B.; Hernandez Ramos, J.A. Experimental Validation of Water Flow Glazing: Transient Response in Real Test Rooms. *Sustainability* **2020**, *12*, 5734. [CrossRef]

39. German Institute for Standardization. *Glass in Building—Determination of Thermal Transmittance (U Value)—Calculation Method*; EN 673; German Institute for Standardization: Berlin, Germany, 2011.

40. German Institute for Standardization. *Glass in Building—Determination of Luminous and Solar Characteristics of Glazing*; EN 410; German Institute for Standardization: Berlin, Germany, 2011.

41. Finlayson, E.; Arasteh, D.; Huizenga, C.; Rubin, M.; Reilly, M. *WINDOW 4.0:Documentation of Calculation Procedures*; University of California, Lawrence Berkeley Laboratory: Berkley, CA, USA, 1993.

42. Baranzelli, C.; Lavalle, C.; Sgobbi, A.; Aurambout, J.; Trombetti, M.; Jacobs, C.; Cristobal Garcia, J.; Kancs, D.; Kavalov, B. *Regional Patterns of Energy Production and Consumption Factors in Europe—Exploratory Project EREBILAND—European Regional Energy Balance and Innovation Landscape*; EUR 27697; Publications Office of the European Union: Luxembourg, 2016. [CrossRef]

43. Edwards, R.; Larivé, J.F.; Rickeard, D.; Weindorf, W. *Well-To-Tank Report Version 4.a. Well-to-Wheels Analysis of Future Automotive Fuels and Powertrains in the European Context*; Publications Office of the European Union: Luxembourg, 2014. [CrossRef]

44. Vanhoutteghem, L.; Skarning, G.C.J.; Hviid, C.A.; Svendsen, S. Impact of façade window design on energy, daylighting and thermal comfort in nearly zero-energy houses. *Energy Build.* **2015**, *102*, 149–156. [CrossRef]

45. Casini, M. 7-Advanced insulation glazing. In *Smart Buildings*; Casini, M., Ed.; Woodhead Publishing: Cambridge, MA, USA, 2016; pp. 249–277.

46. Kralj, A.; Drev, M.; Žnidaršič, M.; Černe, B.; Hafner, J.; Jelle, B.P. Investigations of 6-pane glazing: Properties and possibilities. *Energy Build.* **2019**, *190*, 61–68. [CrossRef]

47. Tam, V.W.Y.; Le, K.N.; Wang, J.Y. Cost Implication of Implementing External Facade Systems for Commercial Buildings. *Sustainability* **2018**, *10*, 1917. [CrossRef]

 *sustainability*

Article

Evaluation of Thermal Comfort and Energy Consumption of Water Flow Glazing as a Radiant Heating and Cooling System: A Case Study of an Office Space

Belen Moreno Santamaria [1], Fernando del Ama Gonzalo [2,*], Benito Lauret Aguirregabiria [1] and Juan A. Hernandez Ramos [3]

[1] Department of Construction and Architectural Technology, Technical School of Architecture of Madrid, Technical University of Madrid (UPM), 28040 Madrid, Spain; belen.moreno@upm.es (B.M.S.); benito.lauret@upm.es (B.L.A.)
[2] Department of Sustainable Product Design and Architecture, Keene State College, Keene, NH 03435, USA
[3] Department of Applied Mathematics, School of Aeronautical and Space Engineering, Technical University of Madrid (UPM), 28040 Madrid, Spain; juanantonio.hernandez@upm.es
* Correspondence: fernando.delama@keene.edu

Received: 19 July 2020; Accepted: 11 September 2020; Published: 15 September 2020

Abstract: Large glass areas, even high-performance glazing with Low-E coating, could lead to discomfort if exposed to solar radiation due to radiant asymmetry. In addition, air-to-air cooling systems affect the thermal environment indoors. Water-Flow Glazing (WFG) is a disruptive technology that enables architects and engineers to design transparent and translucent facades with new features, such as energy management. Water modifies the thermal behavior of glass envelopes, the spectral distribution of solar radiation, the non-uniform nature of radiation absorption, and the diffusion of heat by conduction across the glass pane. The main goal of this article was to assess energy consumption and comfort conditions in office spaces with a large glass area by using WFG as a radiant heating and cooling system. This article evaluates the design and operation of an energy management system coupled with WFG throughout a year in an actual office space. Temperature, relative humidity, and solar radiation sensors were connected to a control unit that actuated the different devices to keep comfortable conditions with minimum energy consumption. The results in summer conditions revealed that if the mean radiant temperature ranged from 19.3 to 23 °C, it helped reduce the operative temperature to comfortable levels when the indoor air temperature was between 25 and 27.5 °C. The Predicted Mean Vote in summer conditions was between 0 and −0.5 in working hours, within the recommended values of ASHRAE-55 standard.

Keywords: building energy management; Water Flow Glazing; mean radiant temperature; final energy consumption

---

1. Introduction

Obsolete equipment, design flaws, and inappropriate use can account for up to 20% of the energy that buildings use over the operation period [1]. Dwellings, offices, educational facilities, and commercial buildings show different consumption patterns. For example, commercial buildings exhibit high energy consumption associated with heating, ventilation, and air conditioning (HVAC) systems and lighting [2]. Office buildings have a high amount of energy use by computers and monitors, while educational buildings have significantly more energy consumption for lighting [3]. Office buildings are likely to have higher cooling demands due to the impact of internal gains from occupants and IT equipment [4]. The European air conditioning (AC) market is essential in raising awareness about primary energy utilization. Over the last two decades, all members of the European Union (EU) have been committed to increasing the production of renewable energy, decreasing greenhouse gas (GHG) emissions, and reducing the final energy consumption by 20% from 1990 levels by 2020. The goal of reducing the emissions of GHG by 40% by 2030 has been set. Furthermore, the EU members have committed to reducing GHG emissions by 80–95% by 2050, and the fulfillment of the Paris Conference of the Parties 21 agreement will require a further reduction of GHG emissions [5]. In this regard, some studies show that electricity demand for cooling is increasing, especially in colder European countries [6]. If the electricity demand exceeds the projected renewable capacity, the goal of reducing GHG emissions will not be met.

An energy management system (EMS) assures that the building's energy demand is accomplished without compromising the air quality and comfort levels of its occupants [7]. The EMS can collect measurements at a specified time interval at designated measurement points. The accurate and diverse data, deployment without affecting the building operation, communication protocol, and cost influence the selection of the EMS [8]. Engineers tend to overestimate the internal heat gains in office buildings, which results in the specification of cooling systems that exceed the needed capacity. As a result, there is an energy waste over long periods of inefficient operation [9,10]. The Energy Consumption Guide (ECG) shows patterns and benchmarks for electricity consumption in office buildings [11]. Energy consumption schedules, occupants' habits, and the diversity of electric loads have a significant impact on office building energy behavior [12,13]. An energy management system allows owners to understand building performance, improve energy efficiency, and take appropriate actions [14,15].

Power load density is used to assess expected peak power demand, taking into account internal heat gains [16,17]. The building envelope materials contribute decisively to reducing the heating and cooling loads. Windows and curtain walls play a crucial role in the energy efficiency of office buildings due to solar heat gains. Although solar radiation may help reduce heating loads in the cold season, summer heat gains have to be avoided [18,19]. In this regard, the extensive use of glass in facades in office buildings has led to an 8.7% increase in the AC market in Europe over the last decade, especially in Mediterranean countries [20,21]. Despite the growth of the market, other factors like the increasing price of electricity in the European Union (17% from 2008 to 2019) [22] and new government regulations have forced manufactures to develop energy-efficient products, such as inverter technologies and new refrigerants [23]. The energy performance of heating and AC systems is measured by the energy efficiency ratio (EER) in the cooling mode and the coefficient of performance (COP) in the heating mode. The seasonal energy efficiency ratio and the seasonal coefficient of performance (SEER, SCOP) designate the total heat supplied or removed from areas ($Q_{heat/cold,\ season}$) divided by the total work input over the same period ($W_{electricity,\ season}$) [24]. By product type, split systems, coupled with air-to-air heat pumps, account for the majority of AC units per type [25]. Air-to-water and water-to-water heat pumps can be coupled with fan coil units (FCUs) and radiant panels in walls, floors, and ceilings. The EER and COP of heat pumps depend on the source and load side temperatures, so assessing the energy performance of each type requires analyzing the outdoor and indoor operating temperatures [26].

When it comes to defining thermal comfort conditions, six main factors must be taken into account: metabolic rate, clothing insulation, air temperature, radiant temperature, airspeed, and humidity [27]. Fanger's Predicted Mean Vote (PMV) method was developed to consider the different variables that influence the comfort assessment in a working environment [28,29]. The international organization for standardization document ISO 7726 defined local thermal discomfort as the thermal dissatisfaction caused by unwanted cooling or heating of one particular part of the body. It mainly affects people developing light sedentary activities [30]. The mean radiant temperature (MRT) has a strong influence on human thermal comfort because occupant bodies transfers heat to hot or cold surfaces [31]. In office buildings with convective heating and cooling systems, such as split units, and facades with extensive glazing, users experience a lack of comfort caused by the inhomogeneity in indoor surfaces and air temperatures [32]. For example, windows with high thermal transmittance and without Low-E coatings can lead to high radiant temperature asymmetry and the local dissatisfaction of some body parts [33]. An effective way to improve the comfort conditions in these buildings would be to use temperature-controlled surfaces or radiant panels as the principal source of sensible cooling and heating in the conditioned space. Radiant panels provide a comfortable indoor environment without lowering the room air's moisture content. Occupants in an area heated or cooled by radiant panels are comfortable at lower air temperatures in winter and higher air temperatures in summer. Indoor partitions and facades with Water-Flow Glazing (WFG) are considered active radiant panels that control their temperature by circulating water, and can be used to control the surface temperatures and provide an acceptable thermal environment [34]. In facades exposed to solar radiation, the water flows between two glass panes and captures most of the solar infrared radiation, and the visible component enters the building [35]. Since the WFG is a dynamic envelope, the solar heat transmitted through the material depends on the flow rate. When the water flows, the transmitted solar heat is low, and when the water flow stops, the solar radiation enters the building [36]. In interior partitions, WFG panels can supply the needed power at a rate of 120 W/m^2 if the difference between the circulating water and the indoor air temperature is 10 °C [37].

This paper focused on assessing the performance of WFG envelopes in commercial buildings by analyzing power demand patterns through measured data obtained from a testing facility. Hence, to achieve this goal, it was essential to: (i) validate the energy management system to enhance the thermal performance of the building, (ii) estimate the final energy consumption of the office space in summer and winter conditions, and (iii) evaluate the comfort conditions and the influence of the mean radiant temperature in the Predicted Mean Vote over the office space working hours.

2. Materials and Methods

Commercial building energy simulation (BES) tools do not include Water-Flow Glazing as an option, so it is necessary to validate its behavior in real facilities. This section described the office space layout, the description of the envelope and the Water-Flow Glazing, the energy management system components, and the electronic control unit logic operation.

2.1. Description of the Facility

The testing facility was an office space of the Department of Applied Mathematics in the School of Aeronautics and Space Engineering in Madrid, Spain (40,44389° N, −3,7261972° E). Two faculty members occupy the room from 8:00 a.m. to 8:00 p.m., and there are meetings with students during office hours. The occupancy is limited to six people at a time. The facility validated the WFG behavior as a component of the heating and cooling system. Figure 1 illustrates the floor plan. Four transparent WFG panels (WFG1, WFG2, WFG3, and WFG4) separated the corridor from the office. The thermal and spectral properties of these transparent panels were carefully selected to absorb the maximum heat from the beam solar radiation, which entered through the main glazed facade, impinging into the WFG in the afternoon for four to five hours, depending on the season. The northeast facade was an insulated opaque wall, and the rest of the interior partitions were translucent WFG (WFG_TP01 to WFG_TP09). In all, there were thirteen WFG panels of 1500 mm height by 1300 mm width. The energy management system is placed outdoors, in the north-east facade. The electronic control unit (ECU) monitored the temperatures of the WFG and the indoor, corridor, and outdoor temperatures.

Table 1 shows the thermal transmittance and areas of the office envelope. The opaque internal partitions were modular walls with melamine panel finish (0.5 cm) and rock–wool acoustic insulation (3 cm). The northeast facade was an insulated opaque wall made up of a zinc plate external finish (1 mm), ventilated air chamber (3 cm), a brick wall (11 cm), rock–wool thermal insulation (6 cm), air chamber (5 cm), and a plaster board (12 mm). The roof was composed of a zinc plate external finish (1 mm), ventilated air chamber (3 cm), metal deck with concrete (10 cm), air chamber (10 cm), rock–wool thermal insulation (6 cm), and a plaster board (12 mm). The thermal transmittances met the requirements of the Spanish Building Code [38].

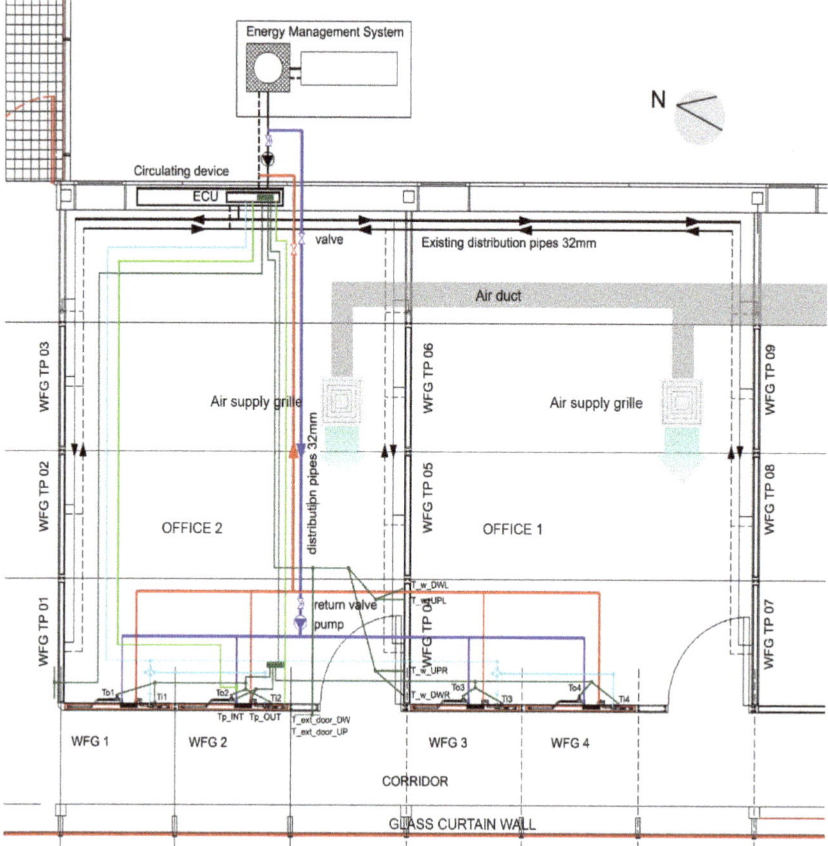

Figure 1. Plan view of the office spaces. The transparent Water-Flow Glazing (WFG) was connected directly to the primary circuit. The translucent interior partitions were connected in parallel to the circulating device.

Table 1. Parameters of the office envelope.

Thermal and Geometric Parameters	Roof	N-E Wall	N-E Window	Int Wall	Int Glass	Floor
U (W/m^{2}K) [1]	0.3	0.3	2.9	0.7	5.2	0.6
A (m^2)	40	22.5	2.2	19.6	7.8	40
$\sum UA$ (W/K)	12	6.75	6.38	13.72	40.56	24

[1] Values meet the Spanish Building Code (CTE DB HE1) requirements [38].

Figure 2 shows the space with transparent WFG (a) facing south-west and translucent interior partitions (b). The former was double glazed; each glass pane was composed of 8 mm planiclear, 1.54 mm saflex Rsolar SG41, 8 mm planiclear, and a 20 mm water chamber. The latter was double glazed; each glass pane was formed of 10 mm planiclear, 1 mm translucent Polyvinyl butyral (PVB) 000A CoolWhite, 3 mm planiclear, and a 16 mm water chamber. The mass flow rate through the transparent WFG was set to be 2 L/min, and through the translucent glazing, it was 1 L/min. The transparent panes were exposed to western solar radiation and had to absorb a large amount of heat. In contrast, the translucent panes were designed to deliver heat or cold in winter or summer.

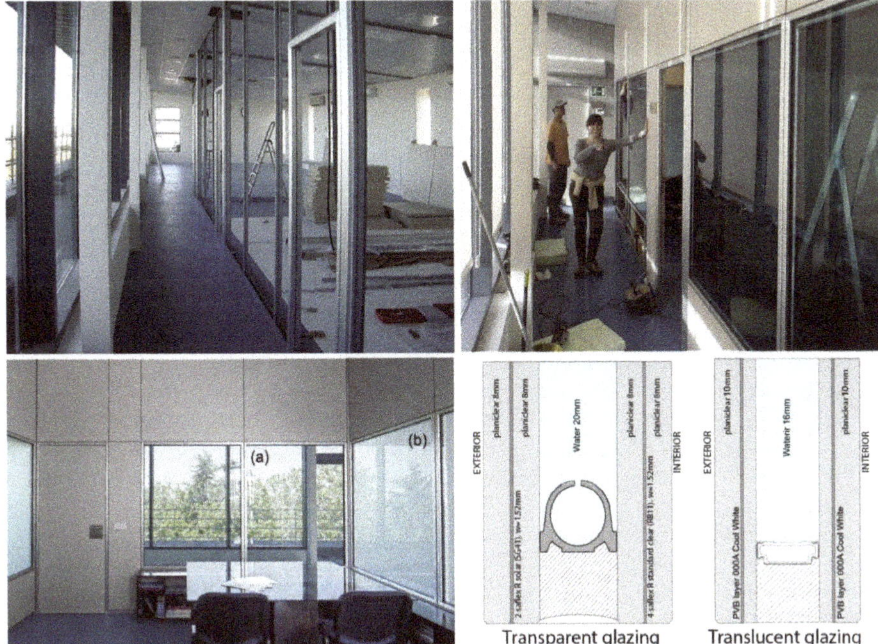

Figure 2. Top: View from the access corridor during the construction process. Bottom: Glass configuration of the office space. (**a**) Transparent Water-Flow Glazing (corridor), (**b**) translucent Water-Flow Glazing (interior partitions).

Table 2 shows the estimated heating and cooling loads in the office space. Ventilation loads (*Vent*) and internal loads (*IL*) were calculated for an occupancy of six people and average office equipment [38,39]. The total glazed surface was 7.8 m^2 of transparent WFG and 17.55 m^2 of translucent interior partitions. The wall-to-window ratio of the wall exposed to solar radiation was 40%. The total area of WFG radiant panels was 25.35 m^2, with a floor area of 40 m^2. The expected power delivered by WFG was 130 W/m^2 when the difference between the circulating water and the indoor air temperature was 13 °C. The dew point temperature for the indoor air temperature was at 27 °C and relative humidity at 40% was 12 °C. Therefore, keeping the WFG inlet temperature above 12 °C, the indoor air temperature at 27 °C, and the average water temperature at 14 °C, the delivered cooling power would be 130 W/m^2. The total WFG surface area was 25.35 m^2 and the total cooling power was 3295 W, which was above the predicted cooling loads shown in Table 2. The cost of the system depends on a few different factors, including the dimensions of the glass, thickness, and distance between the energy management system and the panels. A typical 2 m^2 double glass panel costs 900 USD (around 450 USD/m^2), including the piping and individual circulating devices. Installation of WFG requires a professional team, which could run 50 to 70 USD per hour.

Table 2. Estimation of heating and cooling loads in the offices.

Operating Condition	T_{int} (°C)	T_{ext} (°C) [1]	T_{ext_C} (°C)	n	$\sum UA(T_{int}\text{-}T_{ext})$ (W)	$\sum UA(T_{int}\text{-}T_{ext\_C})$ (W)	Vent (W)	IL (W)	SR (W)	Total (W)
Heating	22	4	22	2	452.34	-	1620	-	-	2018.16
Cooling	23	35	30	6	301.56	379.26	1080	1400	82.5	3243.42

[1] Values are taken from CTE DB HE1 [38].

Figure 3 shows the schematics of both circuits. The energy management circuit consisted of a 370 L buffer tank, an expansion tank, an air-to-water heat pump, and an air heat exchanger. The heat pump's (Saunier Duval Genia Air 8/1 Power A7/W35 = Power A35/W18) nominal power was 7.60 kW in winter (at an outdoor air temperature of 7 °C and inlet water temperature of 35 °C) and summer (at an outdoor air temperature of 35 °C and inlet water temperature of 18 °C). The heat pump was selected for commercial reasons, regarding availability and budget constraints. Some malfunctions and operating issues related to the oversized cooling and heating power are addressed in the following sections. The air heat exchanger works when the outdoor air temperature is low enough to cool down water. This cooling mode can only be used when outdoor ambient air temperatures are below 12 °C. When the air heat exchanger is used for free cooling, the control system uses valves to isolate the heat pump from the rest of the loop, and the heat exchanger is used like a chiller. Once the buffer tank is heated or cooled down, the water flows to transfer heat or cold to the circulating device. Then, the secondary circuit transports the heated or cooled water to thirteen radiant WFG units. A control system with a thermostat based on the indoor temperature turned the heat pump and the flow rate ON and OFF. The secondary circuit was made up of two branches—one that transferred heat or cold to translucent partitions and another one for the transparent WFG modules. Each transparent WFG module had a circulating device (CDi). The mass flow rate through the transparent modules was set to $\dot{m} = 2$ L/min m^2 when the system was ON. All the translucent WFG panels were connected to the same circulating device ($CD\ TP$), and the flow rate was $\dot{m} = 1$ L/min m^2. The influence of the mass flow rate on the ability to deliver or absorb heat and the recommended values have been studied in previous articles [37]. Transparent WFG panels are exposed to solar radiation, so the mass flow rate had to be higher to absorb heat in summer and keep the water temperature within acceptable values. The electronic control unit actuated the WFG circulating devices, the heat pump, and the air heat exchanger using the basic commands of ON and OFF, with the control logic explained in Table 3. There was a mechanical ventilation system that met the requirements of the Spanish Regulation of Thermal Installations in Buildings (RITE) for ventilation of office spaces (12.5 L per second per person) [40]. The mechanical ventilation provided conditioned air and operated over the working hours (8:00 a.m. to 8:00 p.m.) at a constant air volume. However, it was not a component of the controlled energy management system. The lack of control of the ventilation device was one of the system's uncertainties because high relative humidity can cause condensation in radiant panels when operating in cooling mode, and can affect the latent loads.

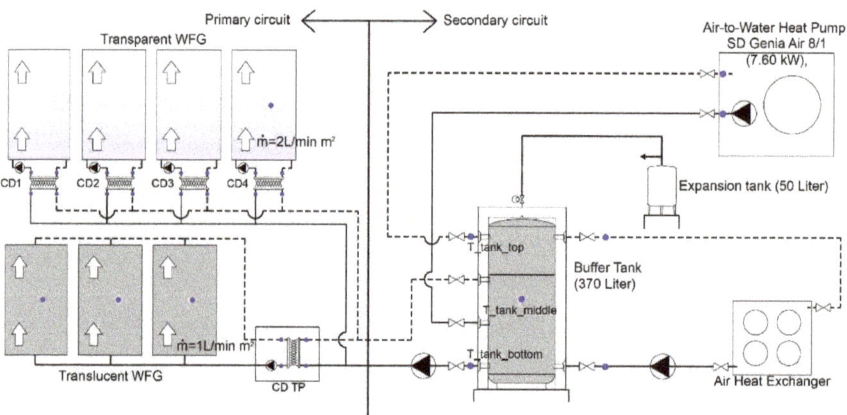

Figure 3. Schematics of the testing facility. The primary circuit connects the energy management devices (heat pump, air heat exchanger, and buffer tank). The secondary circuit goes from the buffer tank to the WFG.

Table 3. Energy management system in cooling mode.

Device	HP	AHX	WFG	WFG_HP
Condition 1	$T_{\_int} > 25\,°C$	-	$T_{\_int} > 25\,°C$	
Condition 2	$(T_{\_tank\_top} - T_{\_ext}) < 10\,°C$	$(T_{\_tank\_top} - T_{\_ext}) > 10\,°C$	$(T_{\_int} - T_{\_tank\_bottom}) > 10\,°C$	
8:00 p.m.–7:00 a.m.	-	ON	ON	ON
7:00 a.m.–8:00 p.m.	ON	-	ON	ON

Tables 3 and 4 show the proposed energy management strategy in the heating and cooling modes. The heat pump (*HP*) was set to operate during working hours, whereas the air heat exchanger (*AHX*) operated only in cooling mode during non-working hours. The first condition was related to the indoor air temperature ($T_{\_int}$) and the second condition depended on the difference between the outdoor air temperature ($T_{\_ext}$) and the tank temperatures ($T_{\_tank\_top}$, $T_{\_tank\_bottom}$).

Table 4. Energy management system in heating mode.

Device	HP	AHX	WFG	WFG_HP
Condition 1	$T_{\_int} < 20\,°C$		$T_{\_int} < 20\,°C$	
Condition 2	-		$(T_{\_tank\_bottom} - T_{\_int}) > 10\,°C$	
8:00 p.m.–7:00 a.m.	-	-	ON	ON
7:00 a.m.–8:00 p.m.	ON	-	ON	ON

2.2. Description of the Sensors

To measure the water heat gain of the WFG panels, flow meters and inlet and outlet digital thermometers were installed in the primary and secondary circuits. The DS18B20-PAR digital thermometer communicated over a one-wire bus with the energy control unit (ECU). They had an operating temperature range of −55 to +100 °C and an accuracy of ±0.5 °C. A pyranometer Delta Ohm LP PYRA 03, placed on the vertical south-western facade, allowed measurement of the solar irradiance. It is a second-class pyranometer according to ISO 9060 standards and the World Meteorological Organization (WMO); it had to be placed outdoors because obstacles and reflections can affect the measurements. The same monitoring equipment has been described in other articles [37]. Figure 4 shows the position of the temperature sensors in the WFG and the circulating device. The flow meter (s) measures the flow rate at the inlet of the WFG panels. The flow meter (p) measures the flow rate of the primary circuit. The temperature sensors, T_{i2} and T_{o2}, measure the inlet and outlet temperatures in the WFG 2, respectively, and T_{p2} and T_{p2} measure the temperatures at the primary circuit.

Every WFG module had a circulator that comprised a water pump, a plate heat exchanger, and two one-wire sensors inserted into two pocket wells to measure the inlet and outlet temperatures of the glazing. In addition, one module was monitored with a digital flow meter for the primary circuit and another digital flow meter for the secondary circuit. Together with the inlet and outlet temperatures, these flow meters allowed validation of the design flow rate of the glazing as well as having precise actual values for the water heat gain of each WFG panel. The one-wire digital thermometers were inserted into the pocket wells. Each sensor had a unique 64-bit serial number etched into it, and allowed the housing of a considerable number of sensors to be used on one data bus. There were four transparent WFG modules and two thermometers per module, plus the inlet and outlet temperatures for the primary circuit, measured with the same data bus. Thermostats and timers controlled the heating and cooling system. All indoor temperatures were measured 150 cm above the floor level. The main objective of this strategy was to maintain a comfortable indoor temperature and to minimize energy consumption using solar energy harvesting and free cooling. Table 5 presents a description of the sensors and parameters that have been measured. The WFG transparent panels were located in a corridor with south-west orientation. When the solar radiation impinged on the glazing, the water absorbed the energy. After analyzing the indoor temperature, the EMS decided whether to store the heat or to distribute it through the rest of the translucent interior partitions. The energy surplus could be stored in the buffer tank. If there was no solar energy to harvest or there was not enough energy harvested in the buffer tank, the heat pump would work to satisfy the demand. Generally, an office building demands cold throughout the year due to its high internal heat load. In winter, the outdoor temperature is low enough to dissipate the internal heat load utilizing an air heat exchanger. The heat pump electricity consumption was not measured. The electricity consumption was estimated with the heat pump thermal power, the coefficient of performance, and the energy efficiency ratio provided by the manufacturer.

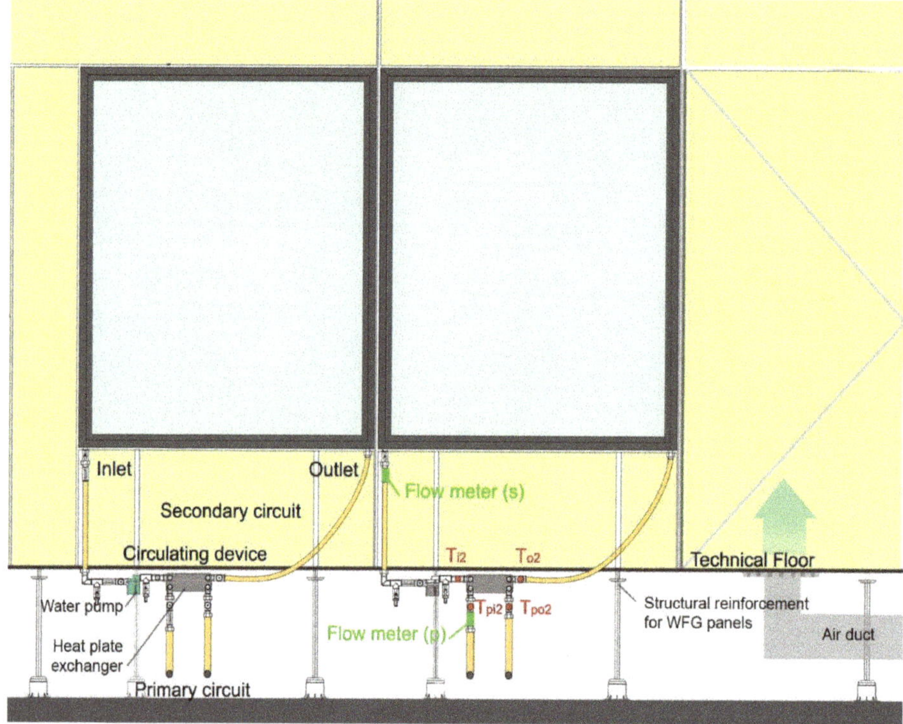

Figure 4. Front view of the transparent WFG in the corridor facing south-west. Location of the circulating device underneath the technical floor. Location of inlet and outlet temperature probes and flow meters for the primary and secondary circuits of the WFG.

Table 5. Nomenclature and description of sensors.

Sensor	Description
T_{ik}	Inlet temperature of the transparent WFG [1] (°C)
T_{ok}	Outlet temperature of the transparent WFG [1] (°C)
T_{s_IN}	Inlet temperature of the translucent WFG partitions (°C)
T_{s_OUT}	Outlet temperature of the translucent WFG partitions (°C)
$T_{\_ext}$	Outside temperature (°C)
$T_{\_int}$	Indoor air temperature (°C)
$T_{\_tank\_bottom}$	Temperature at the bottom of the buffer tank (°C)
$T_{\_tank\_middle}$	Temperature in the middle of the buffer tank (°C)
$T_{\_tank\_top}$	Temperature at the top of the buffer tank (°C)
$T_{\_ext\_C}$	Temperature of the corridor right outside the door (°C)
T_{floor}	Surface temperature of the floor (°C)
T_{wall}	Surface temperature of opaque walls (°C)
$T_{ceiling}$	Surface temperature of the ceiling (°C)
T_{WFG}	Surface temperature of transparent WFG (°C)
T_{WFG_TP}	Surface temperature of WFG translucent partitions (°C)
Sun_rad	Solar irradiance on vertical surface (W/m^2)
kWh_HP	Thermal heating/cooling energy by the heat pump (kWh)
kWh_AXH	Thermal cooling energy by the air heat exchanger (kWh)
kWh_WFG	Heating/cooling energy delivered by transparent WFG (kWh)
kWh_WFG_TP	Heating/cooling energy delivered by WFG translucent partitions (kWh)
RH	Indoor Relative Humidity (%)

[1] k is the module number from 1 to 4.

3. Results

This section presents monitoring temperatures and the power efficiency of WFG modules. The implementation of different energy strategies was validated. By analyzing the system's performance, the energy strategy is improved, achieving significant energy savings. Finally, the power performance of the WFG module is obtained by measuring the inlet and outlet temperatures and flow rate of each WFG panel.

3.1. Analysis in Summer Conditions

Figure 5 shows the system temperatures and the irradiance curve of a sample summer week from 10 July 2019 to 16 July 2019. T_{i2} and T_{o2} illustrate the inlet and outlet temperatures of the WFG. $T_{\_int}$ is the indoor temperature, and $T_{\_ext}$ is the exterior temperature. $T_{\_ext\_C}$ corresponds to the temperature in the corridor between the office and the exterior. The first day, 10 July 2019, was clear, with some evolution clouds between 16:30 and 18:00. On clear days, direct beam radiation prevailed over diffuse radiation. The typical irradiance curve (*Sun_rad*) reached maximum levels above 700 W/m^2. From 9:00 a.m. to 1:00 p.m., the south-west facade was shaded due to geometrical obstructions, and the irradiance was mainly diffuse, reaching values around 200 W/m^2. However, in the afternoon, the facade was exposed to direct solar radiation, and the corridor temperature rose to 35 °C. On 11 July 2019, the indoor and outdoor temperatures showed a similar performance, although the oscillations of the inlet and outlet temperatures were different from those of the previous day. On 12 July 2019, the solar irradiance showed irregular values because of clouds, and it affected the temperature of the corridor, which was slightly above 30 °C. Over the weekend, on 13 July 2019 and 14 July 2019, the mass flow rate was 0 and the heat pump did not operate. Inlet and outlet temperatures of the WFG (T_{i2} and T_{o2}) did not show any difference and reached peak values of 32 °C. The indoor air temperature reached a maximum of 34 °C, whereas the temperature in the corridor ($T_{\_ext\_C}$) was 39 °C. A WFG circuit is a closed loop and there are two cases, mass flow rate $\dot{m} = 0$ or $\dot{m}$ = design flow rate. Over the weekend, the mass flow rate was 0 and the heat pump was not in operation. After two weekend days, the indoor temperature rose to 32 °C, making it necessary to cool down the office temperature. Figure 5 shows that the inlet and outlet temperatures dropped on Sunday 14/07/2019 before 7:00 a.m., although the heating pump did not operate that day. The same behavior was shown on Monday, 15/07/2019, before 7:00 a.m. The reason was that the air heat exchanger operated both days for two hours when the difference between the top tank water temperature ($T_{\_tank\_top}$) and the outdoor air temperature ($T_{\_ext}$) was above 10 °C.

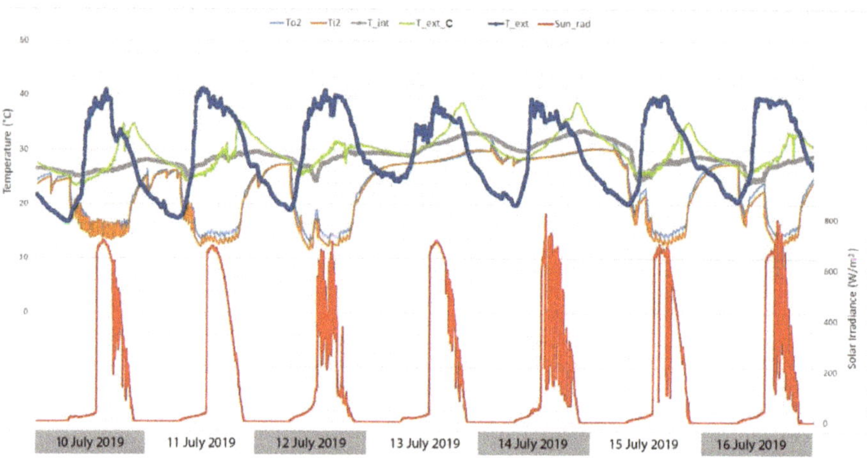

Figure 5. Solar irradiance and indoor and outdoor temperatures—sample summer week of 10 July 2019 to 16 July 2019.

Figure 6 shows the detailed evolution of temperatures on two consecutive days. Figure 6a shows that the irradiance curve on 10 July 2019 had some oscillations in the afternoon, and the outdoor temperature declined, which indicated the existence of clouds. The inlet and outlet temperatures ($T_{\_i2}$, $T_{\_o2}$) showed that the heat pump worked at three cycles per hour. The heat pump parameters were fixed to meet the manufacturer's requirement for minimum cycle times. Figure 6b showed that the minimum time between starts was, at least, forty minutes. On 10 July 2019 at 7:00 p.m., there was a peak in the corridor temperature ($T_{\_ext\_C}$), and this peak did not occur on 11 July 2019. The corridor had a cooling system that was not monitored or controlled by the studied energy management system, and its temperature was a boundary condition of the studied space. The indoor air temperature rose to 27 °C on 10 July 2019 and to 29.5 °C on 11 July 2019. Although the temperatures might seem too high, due to the effect of radiating panels and a low mean radiant temperature, there is thermal comfort in the space, as shown in the discussion section.

3.2. Analysis in Winter Conditions

Figure 7 shows the temperatures and the irradiance curve of a sample winter week from 08 January 2020 to 14 January 2020. On sunny working days (from 08 January 2020 to 10 January 2020), the outdoor temperature showed typical values of winter in Madrid, with a minimum temperature slightly below 0 °C and a maximum temperature between 10 and 15 °C. The solar radiation impinged on the south-west facade as of 11:00 a.m. with a peak value of 300 W/m². The indoor air temperature ($T_{\_int}$) was below comfort until 7:00 p.m. because the heating system was off. In the morning, the heat pump started working, and the radiant WFG panels were delivering heat. In the afternoon, the temperature in the corridor ($T_{\_ext\_C}$) rose to 30 °C, which helped to heat the office space air temperature ($T_{\_int}$) to 22 °C. The solar radiation in the afternoon made the heat delivered by the WFG unnecessary. Over the weekend (11 January 2020 and 12 January 2020), the heat pump was not operating, and the indoor air temperature declined and reached its lowest value (14 °C) on Monday 13 July 2020 at 7:00 a.m. Due to the solar radiation, the temperature in the corridor rose to 28 °C. On weekend days, the heat pump did not operate in the morning, so the indoor temperature continued to drop until the afternoon, when the solar radiation and the corridor overheating contributed to raising the indoor air temperature from 17 to 19 °C on 11 January 2020 and from 15 to 17 °C on 12 January 2020. Nevertheless, the indoor temperature on 11 January 2020 at 7:00 a.m. was 18 °C, and on 13 January 2020, it was 14 °C after two days without operating the heat pump. On working days, the indoor air temperature was above 18 °C at the beginning of the working hours. On 13 July 2020 and 14 July 2020, the solar irradiance was low, and the outdoor temperature variation over the day was only 5 °C. The heat pump operated most of the working hours, unlike on sunny days, when it operated only in the morning.

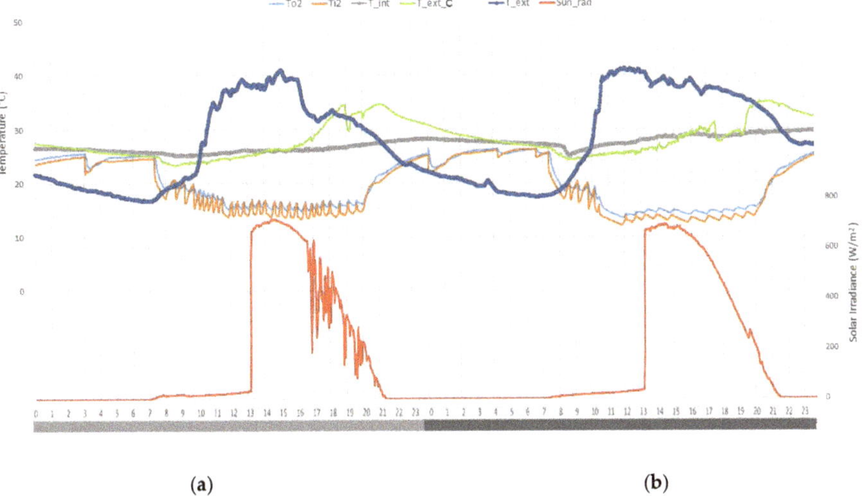

(a) (b)

Figure 6. Solar irradiance and indoor and outdoor temperatures—summer sample days (**a**) 10 July 2019 and (**b**) 11 July 2019.

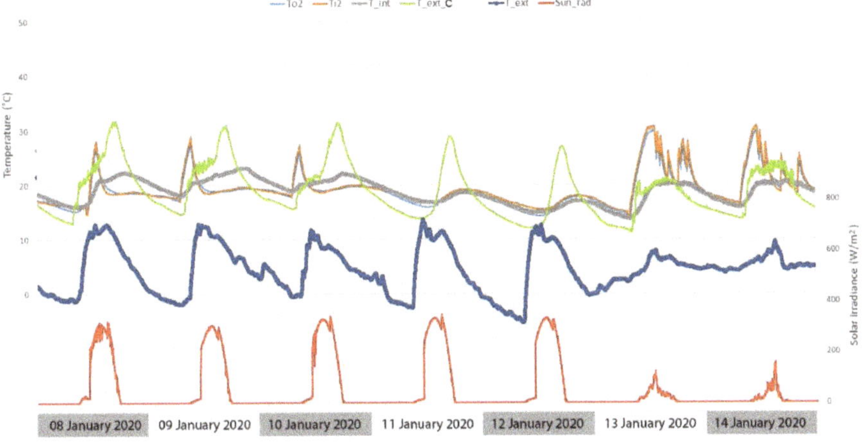

Figure 7. Solar irradiance and indoor and outdoor temperatures—sample winter week of 08 January 2020 to 14 January 2020.

Figure 8 details the parameters on two winter days with different outdoor conditions. Figure 8a illustrates a sunny winter day when the solar irradiance reached a peak value of 300 W/m^2, and the outdoor temperature ranged from −1 to 11 °C. The WFG started working in heating mode from 7:00 a.m., when the indoor air temperature was 18 °C, to 9:30 a.m., when the indoor air temperature reached 20 °C. The indoor air temperature continued to rise to 22 °C because the corridor air temperature reached a peak of 30 °C. Figure 8b shows a winter day with little solar radiation and an outdoor temperature that ranged from 4 to 10 °C. The WFG started working in heating mode at 7:00 a.m., when the indoor air temperature was 16 °C. It took the system four hours to increase the indoor air temperature to 20 °C. The heat pump was connected to the buffer tank, so the heating time seemed too long due to the thermal inertia. Starting the heat pump four hours before the working hours would be an excellent strategy to improve comfort conditions on winter days after the holidays.

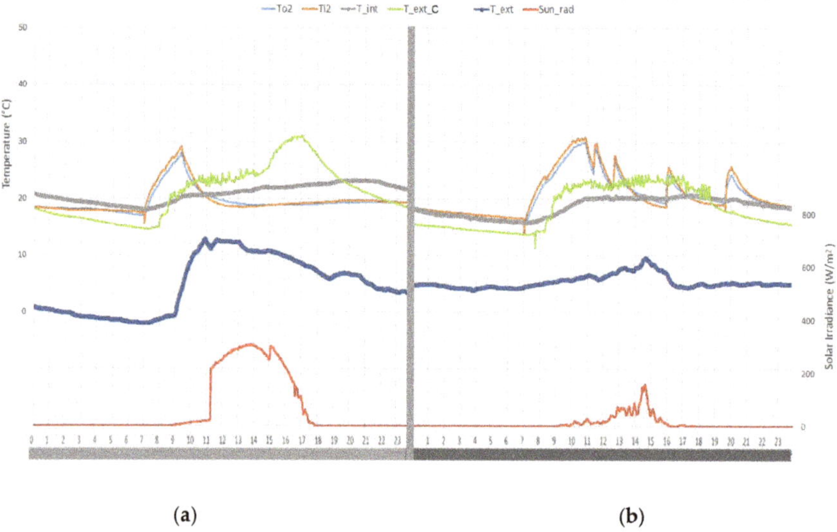

Figure 8. Solar irradiance and indoor and outdoor temperatures—sample winter days (**a**) 09 January 2020 and (**b**) 14 January 2020.

Figure 9 presents a sample week of February, from 19 February 2020 to 25 February 2020. The minimum outdoor air temperature was 0 °C on 20 February 2020, and the maximum temperature was 21 °C on 24 February 2020. The indoor air temperature ($T_{\_int}$) in the office space maintained comfortable conditions operating in a free-floating temperature regime with zero energy consumption. The WFG circuit was never empty. During the free-floating regime, the mass flow rate was 0 and the heat pump was not in operation. Temperature in the corridor ($T_{\_ext\_C}$) showed peak values above 32 °C in the afternoon. The solar irradiance on the west facade (Sun_rad) reached a peak of 480 W/m².

Figure 10 illustrates the performance on two consecutive February days. Although the minimum outdoor air temperature was 0 °C on 19 February 2020 and 20 February 2020, the peak solar radiation (440 W/m²) increased the temperature inside the studied office in the afternoon. When the indoor air temperature reached 25 °C, the water inlet temperature dropped, and the outlet temperature was above the inlet. As stated in Table 3, the heat pump was set to operate in cooling mode when indoor temperature was above 25 °C. The heat pump cooled down water three times between 5:00 p.m. and 7:00 p.m. on 19 February 2020, and only once at 6:30 p.m. on 20 February 2020.

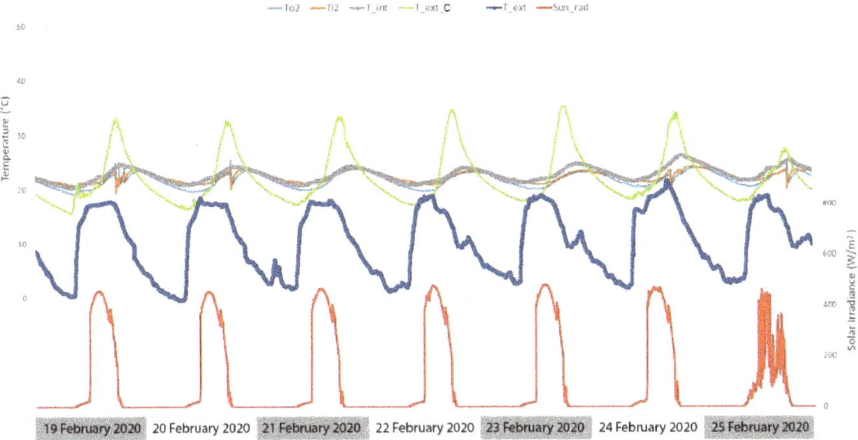

Figure 9. Solar irradiance and indoor and outdoor temperatures—sample week in February from 19 February 2020 to 25 February 2020.

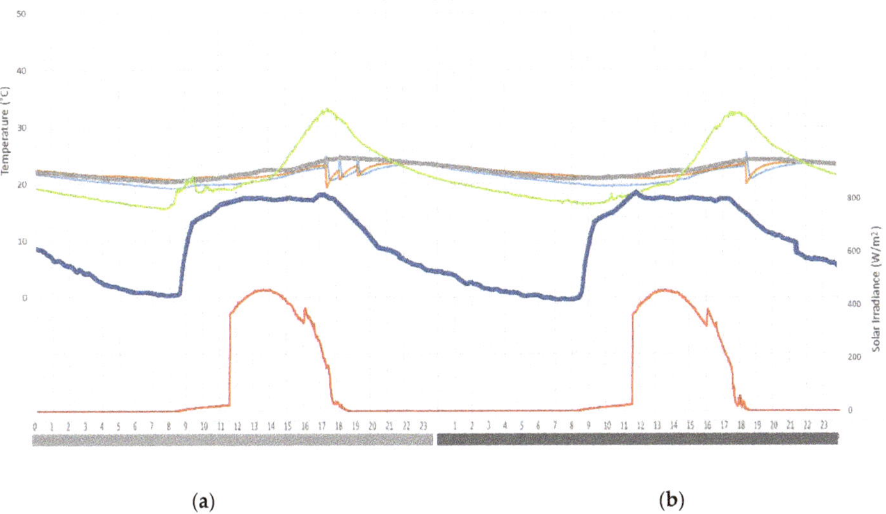

(a)   (b)

Figure 10. Solar irradiance and indoor and outdoor temperatures—sample February days (**a**) 19 February 2020 and (**b**) 20 February 2020.

Table 6 shows a summary of the energy performance on four days in different seasons. On 10 July 2019, the system was working in cooling mode. The heat removed from the office space by the transparent WFG (*kWh_WFG*) and by the translucent partitions (*kWh_WFG_TP*) was 4.9 KWh. The transparent WFG absorbed the most significant amount of heat during the working hours because of the high mass flow rate ($\dot{m}$ = 2 L/min m^2), whereas the translucent interior partitions performed better during the night. The contribution of the air heat exchanger (*kWh_AXH*) during the night was negligible compared with the heat pump, which operated from 7:00 a.m. to 8:00 p.m.

Table 6. Thermal energy summary on four sample days.

Date	10 July 2019			09 January 2020			14 January 2020			20 February 2020		
hour	0–7	7–20	20–24	0–7	7–20	20–24	0–7	7–20	20–24	0–7	7–20	20–24
kWh_WFG [1]	−0.3	−4.3	−0.3	3.4	-	-	-	8.6	-	-	−2.3	-
kWhWFGTP [1]	−1.6	−2.0	−1.3	3.2	11.3	4.1	1.7	17.4	5.6	−2.5	−1.1	−1.2
kWh_HP [1]	-	−21	-	-	7.12	-	-	15.0	-	-	−0.6	-
kWh_AXH [1]	−0.7	-	-	-	-	-	-	-	-	−0.1	-	−0.3

[1] Energy values in kWh.

On 09 January 2020, the heat delivered by the translucent WFG (*KWh_WFG_TP*) was 18.7 kWh, whereas the total amount of energy delivered by the transparent WFG (*KWh_WFG*) was 3.4 kWh. In the afternoon, the transparent WFG circuit was stopped to allow solar radiation to enter the office space. The translucent WFG supplied most of the heat during the working hours. The thermal energy delivered by the heat pump (*kWh_HP*) was 7.12 kWh from 7:00 a.m. to 11:00 a.m. In the afternoon, the thermal inertia of the tank and the solar radiation made it unnecessary to operate the heat pump again. On 14 January 2020, the contribution of the transparent WFG was higher because there was little solar radiation in the afternoon. The heat pump operated over the working hours and released twice as much thermal energy as on 09 January 2020.

On 20 February 2020, the system was working in cooling mode. The air heat exchanger (*kW_AXH*) was cooling down the buffer tank during the night, and the heat pump operated during the working hours. The heat removed by the translucent WFG (*kWh_WFG_TP*) was 4.8 kWh. The energy delivered by the heat pump was 0.6 kWh, and the thermal inertia of the buffer tank was enough to keep indoor temperature between 20 and 26 °C. In Section 4.4, these conditions are assessed to evaluate the occupants' comfort.

4. Discussion

Radiant WFG panels were part of the heating and cooling system. They impact the indoor air temperature and help reduce the mean radiant temperature and, therefore, the operative temperature. The thermal problem of the glazing is coupled with the thermal problem of the room, and the indoor temperatures should be measured.

4.1. Validation of Energy Management System

The power released or absorbed by the water (*P*) is measured in watts per square meter (W/m^2), and is shown in Equation (1).

$$P = \dot{m}c(T_O - T_i), \tag{1}$$

where $\dot{m}$ is the mass flow rate (Kg/s m^2), c (J/Kg °C)is the specific heat of the water, and T_o and T_i are the temperatures of water leaving and entering the glazing, respectively (°C). The mass flow rate is the mass of a fluid passing by a point over time. In summer conditions, the transparent WFG was set to operate during working hours. It had to absorb most of the solar radiation impinging on the glazing. Figure 11 illustrates the buffer tank temperatures and the thermal energy provided by the heat pump in a sample summer week. The top tank temperature (T_{tank_top}) showed that the heat pump was set to work when T_{tank_top} was between 15 and 18 °C. On 10 July 2019, it worked at three cycles per hour. The following days, it was fixed to operate at a minimum time between starts of forty minutes. Over the weekend, the heat pump did not operate, and the buffer tank temperature reached 35 °C. The maximum power delivered by the heat pump (31.13 kWh) took place on 11 July 2019, when the solar irradiance reached its maximum value without any obstructions, according to Figure 5.

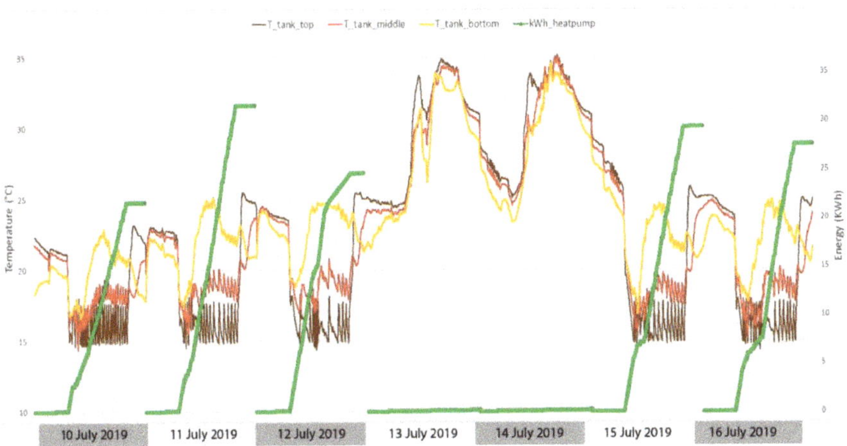

Figure 11. Tank temperatures and heat pump thermal power—sample summer week from 1 July 2019 to 16 July 2019.

When the heat pump was working in the heating mode in winter conditions, the transparent WFG was set to operate in the morning. It did not operate in the afternoon because the solar radiation on the south-west partition helped reduce the heating load. Figure 12 shows the tank temperatures ($T_{\_tank\_top}$, $T_{\_tank\_middle}$, $T_{\_tank\_bottom}$) and the thermal consumption of the heat pump ($kWh\_heatpump$) measured with the water flow rate and the difference of water temperature between the inlet and outlet in the heat pump. On sunny days, the heat pump operated mainly in the morning because the solar radiation heated up the office space in the afternoon. On 09 January 2020, when the outdoor air temperature ranged from −1 to 11 °C and a peak solar radiation of 300 W/m^2, the heat pump heated the buffer tank from 7:00 a.m. to 9:00 a.m. The thermal inertia of the tank and the solar radiation in the afternoon made it unnecessary to operate the heat pump again. The total energy consumption per day was 7.12 kWh. The average heat pump thermal energy was 7 kWh on 08 January 2020, 09 January 2020, and 10 January 2020, whereas on Monday 13 January 2020, a cloudy winter day after non-working days, the total energy consumption was 20.05 kWh. The warm-up response was too low, and it took four hours to raise the temperature to comfort conditions. Over the weekend, the tank temperature dropped, and this made it necessary to increase the energy supplied by the heat pump. The lack of solar radiation in the afternoon was the reason to operate the heat pump until the end of the working hours.

Figure 13 shows the tank temperatures ($T_{\_tank\_top}$, $T_{\_tank\_middle}$, $T_{\_tank\_bottom}$) and the thermal consumption of the heat pump ($kWh\_heatpump$) on six February days. The heat pump operated in cooling mode and cooled down the top tank temperature in the afternoon. On 21 February 2020 and 22 February 2020, the heat pump did not operate, and the buffer tank was in a free-floating regime. The average energy consumption per day was 1 kWh on working days. The difference between the heat pump consumption on 14 January 2020 (15 kWh) and on 20 February 2020 (1.1 kWh) can be explained because the peak solar radiation on 09 January 2020 was below 300 W/m^2, and the outdoor temperature was above 10 °C for four hours. On 20 February 2020, the peak solar radiation was 450 W/m^2, and the outdoor temperature was close to 18 °C for 7 h.

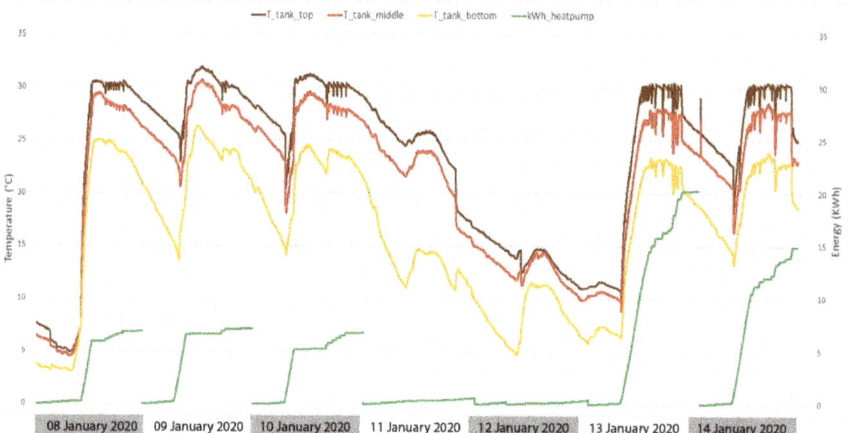

Figure 12. Tank temperatures and heat pump thermal power—sample winter week from 08 January 2020 to 14 January 2020.

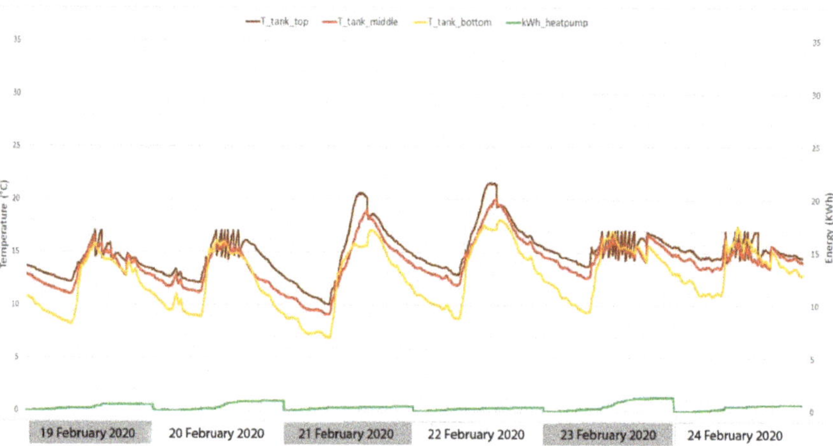

Figure 13. Tank temperatures and heat pump thermal power—ample February week from 19 February 2020 to 25 February 2020.

4.2. Estimation of Final Energy Consumption

Tables 7 and 8 show the estimated heating (positive) and cooling (negative) loads. Ventilation loads (*Vent*) were calculated with the number of occupants (*n*) at each hour. Internal loads (*IH*) are calculated with the number of occupants, the metabolic rate of typical office activity, and 20 W/m^2 for lighting and equipment. Solar radiation (*SR*) was taken from Figure 6a with a surface area of 7.8 m^2 and a solar heat gain coefficient of 0.5. The same procedure was repeated to calculate the values on five sample days.

Tables 9 and 10 compare the thermal energy consumption of the air-to-water heat pump with the calculated cooling and heating loads. The values are taken from Figures 11 and 12 (*kWh_heatpump*) and Tables 7 and 8 by adding the heating and cooling loads over the working hours.

Table 7. Summer cooling loads on 10 July 2019.

Hour	$T_{\_int}$ (C)	$T_{\_ext}$ (C) [1]	$T_{\_ext\_C}$ (C) [1]	n	$\sum UA(T_{\_int}\text{-}T_{\_ext})$ (Wh)	$\sum UA(T_{\_int}\text{-}T_{\_ext\_C})$ (Wh)	Vent (Wh)	IL (Wh)	SR (Wh)	Total (Wh)
7–8	23	18	24	0	−125.65	54.18	0	800	0	728.53
8–9	23	20	23.8	2	−75.39	43.34	−90	1000	0	877.95
9–10	23	21.5	23.7	2	−37.69	37.92	−45	1000	0	955.23
10–11	23	28	23.9	2	125.65	48.76	150	1000	0	1324.41
11–12	23	36	24	6	326.69	54.18	1170	1400	0	2950.87
12–13	23	38	24.2	6	376.95	65.01	1350	1400	0	3191.96
13–14	23	38.5	26	2	389.51	162.54	465	1000	510	2527.05
14–15	23	39.7	28	2	419.67	270.9	501	1000	521.2	2712.82
15–16	23	39.8	30	2	422.18	379.26	504	1000	510	2815.44
16–17	23	35	34	6	301.56	595.98	1080	1400	390	3767.54
17–18	23	32	33	6	226.17	541.8	810	1400	281.2	3259.22
18–19	23	31	34	2	201.04	595.98	240	1000	202.5	2239.52
19–20	23	29.5	32	2	163.34	487.62	195	1000	82.5	1928.45

[1] Values are taken from Figure 6.

Table 8. Winter heating loads on 14 January 2020.

Hour	$T_{\_int}$ (C)	$T_{\_ext}$ (C) [1]	$T_{\_ext\_C}$ (C) [1]	n	$\sum UA(T_{\_int}\text{-}T_{\_ext})$ (Wh)	$\sum UA(T_{\_int}\text{-}T_{\_ext\_C})$ (Wh)	Vent (Wh)	Total (Wh)
7–8 h	22	4	14	0	452.34	433.44	0	885.78
8–9 h	22	5	20	2	427.21	108.36	510	1045.57
9–10 h	22	6	22	2	402.08	0	480	882.08
10–11 h	22	7	23	2	376.95	−54.18	450	772.77
11–12 h	22	7	23	6	376.95	−54.18	1350	1672.77
12–13 h	22	8	24	6	351.82	−54.18	1260	1557.64
13–14 h	22	9	24	2	326.69	−54.18	390	662.51
14–15 h	22	10	24	2	301.56	−54.18	360	607.38
15–16 h	22	8	24	2	351.82	−54.18	420	717.64
16–17 h	22	4	24	6	452.34	−54.18	1620	2018.16
17–18 h	22	4	23	6	452.34	−54.18	1620	2018.16
18–19 h	22	5	21	2	427.21	54.18	510	991.39
19–20 h	22	5	18	2	427.21	216.72	510	1153.93

[1] Values are taken from Figure 8.

Table 9. Sample summer week energy consumption (kWh).

Date	10 July (kWh)	11 July (kWh)	12 July (kWh)	15 July (kWh)	16 July (kWh)	Total (kWh)
Cooling loads (WFG) [1]	21.60	31.13	23.90	29.40	27.6	133.63
Cooling loads (Aid-to-Air)	27.20	35.75	28.38	32.72	31.56	155.61

[1] Values are taken from Figure 11.

Table 10. Sample winter week energy consumption (kWh).

Date	08 January (kWh)	09 January (kWh)	10 January (kWh)	13 January (kWh)	14 January (kWh)	Total (kWh)
Heating loads (WFG) [1]	6.93	7.15	6.79	20.05	15.01	55.93
Heating loads (Ait-to-Air)	8.72	9.11	8.39	18.27	16.53	61.02

[1] Values are taken from Figure 12.

Final energy (FE) consumption, non-renewable final energy (NRFE) consumption, and the CO_2 emissions in kg are primary energy factors in calculating the energy performance of buildings, according to the Energy Performance of Buildings Directive (EPDB 2018) [39]. The Spanish regulation of building thermal systems (RITE) recommends a conversion factor between final energy (FE) and non-renewable final energy (NRFE) of 1.954 [40]. The factor of emitted CO_2 for electricity is 0.331. The final energy consumption and CO_2 emissions were calculated with two different heat pumps. Table 11 illustrates the performance of the air-to-water heat pump in cooling and heating mode. The performance depends on the outlet temperature of the WFG (T_o = 15 °C in summer, T_o = 30 °C in winter) and the source inlet temperature in the heat pump ($T_{s,i}$ = 20–35 °C in summer, $T_{s,i}$ = 15–20 °C in winter). The outdoor temperature, T_{ext}, is shown in Figures 5 and 7, respectively. $T_{s,i}$ values were taken from the top tank temperatures (T_{tank_top}) shown in Figures 11 and 12. The air-to-water heat pump shows a better coefficient of performance (COP) when the water temperature is close to 35 °C and a better energy efficiency ratio (EER) when the water temperature is close to 18 °C. The top tank temperatures (T_{tank_top}) in Figures 11 and 12 confirmed the range of optimal operating temperatures. Although the actual heat pump electrical energy consumption has not been measured, the estimated COP and EER have been taken from [41].

Table 11. Final energy analysis. Air-to-water heat pump.

	Air-to-Water Heat Pump			
T_{ext_db} (°C)	Cooling 35 °C		Heating 7 °C	
$T_{s,i}$ (°C)	7 °C	18 °C	35 °C	45 °C
Energy consumption (kWh)	133.63	133.63	55.93	55.93
EER [1]/COP [2]	2.90 [1]	3.62 [1]	4.50 [2]	3.50 [2]
FE consumption (kWh)	46.08	36.91	12.43	15.98
NRFE consumption (kWh)	90.04	72.13	24.29	31.22
CO_2 emissions (KgCO$_2$)	15.25	12.22	4.11	5.29

[1] Energy efficiency ratio (EER)/[2] coefficient of performance (COP) values are taken from [41].

Air-to-air heat pumps were also analyzed using the cooling and heating loads from Tables 9 and 10. The parameters that influence air-to-air heat pump performance are the dry bulb exterior air temperature (T_{ext_db}) and the dry bulb interior return air temperature (T_{ri_db}). Table 12 shows the final energy (FE), non-renewable final energy (NRFE), and the emitted CO_2 for electricity of the air-to-air heat pump.

Table 12. Final energy analysis—air-to-air heat pump.

	Air-to-Air Heat Pump	
T_{ri_db} (°C)	Cooling 23 °C	Heating 22 °C
T_{ext_db} (°C)	35 °C	7 °C
Energy consumption (kWh)	155.61	61.02
EER [1]/COP [2]	3.25 [1]	3.72 [2]
FE consumption (kWh)	47.88	16.40
NRFE consumption (kWh)	93.56	32.05
CO_2 emissions (KgCO$_2$)	15.85	5.43

[1,2] EER/COP values are taken from [41].

The radiant WFG panel system coupled with a buffer tank and air-to-water heat pump showed non-renewable final energy (NRFE) consumption of 72.13 kWh in cooling mode and 24.29 kWh in heating mode, whereas the expected values of an air-to-air system were 93.56 kWh and 32.05 kWh in the studied summer and winter weeks. This resulted in a final energy savings of 23% in summer and 24% in winter. The reductions of CO_2 emissions were 3.63 kg/week in summer and 1.32 kg/week in winter. As stated in Section 2.1, the ventilation device was not a component of the energy management system, and its performance was not controlled. The ventilation load was estimated by multiplying the air flow by the specific enthalpy (kJ/kg) difference between indoor and outdoor conditions. In summer, the specific enthalpy of outdoor air at 31.3 °C with 35% relative humidity was 58.8 kJ/kg. At 26 °C and 36% relative humidity, the indoor air specific enthalpy was 46.7 kJ/kg. At a ventilation air flow rate of 75 L per second, the total ventilation cooling load over 12 h was 10.8 kWh. In winter, the indoor and outdoor specific enthalpy were 37.11 kJ/kg and 16.36 kJ/kg, respectively, and the ventilation load over the working hours was 13.24 kWh. The electrical consumption of the ventilation device, including the engine and the fan, was 3.24 kWh [42].

The non-renewable energy consumption was 72 kWh in a summer week and 24 kWh in a winter week. The expected energy consumption projection throughout the year was 1700 kWh with a floor area of 40 m^2. Therefore, the yearly heating and cooling energy consumption per m^2 was 42.5 kWh/m^2 per year. If the average energy savings compared to an air-to-air heat pump with multi-split were 23%, the total non renewable energy consumption (NREC) savings accounted for 391 kWh/year. The average price of electricity in Spain is 0.12 EUR/kWh [22], and the system overcosts compared to traditional indoor wall partitions plus the split system can be 50 EUR/m^2. For 24 m^2 of radiant WFG panels, the expected payback period would be 20 years. WFG technology is not competitive nowadays, so future research is needed in industrialization and standardization to bring down the initial costs.

4.3. Mean Radiant and Operative Temperatures

Mean radiant temperature (MRT) expresses the influence of surface temperatures in the room on occupant comfort. The area-weighted method shown in Equation (2) is a simple way to calculate MRT, but it does not reflect the geometric position, posture, and orientation of the occupant, ceiling height, or radiant asymmetry [29]. In Equation (3), the calculation of mean radiant temperature from surrounding surfaces considers the surface temperatures of the surrounding elements and the angle factor. The angle factor is a function of shape, size, and the position concerning the occupant standing or being seated. The surfaces of the room are assumed as black, with high emissivity and no reflection. In this case, the angle factors weight the enclosing surface temperatures to the fourth power [28].

$$T_{mr} = \frac{T_1 A_1 + T_2 A_2 + \ldots + T_N A_N}{A_1 + A_2 + \ldots + A_N}, \tag{2}$$

$$\overline{T}_{mr}^4 = T_1^4 F_{p-1} + T_2^4 F_{p-2} + \ldots + T_N^4 F_{p-N}, \tag{3}$$

where

T_{mr} = mean radiant temperature, °C,
T_N = surface temperature of surface N, °C (calculated or measured),
A_N = area of surface,
F_{p-N} = is the angle factor between the person and surface N.

The angle factors quantify the amount of radiation energy that leaves the human body and reaches each surface. They were calculated according to Figures B.2 to B.5 in [28]. If the difference between the indoor surface temperatures is relatively small (<10 °C), Equation (4) can be used.

$$\overline{T}_{mr} = T_1 F_{p-1} + T_2 F_{p-2} + \ldots + T_N F_{p-N} \tag{4}$$

The MRT is calculated as the average value of the surrounding temperatures weighted according to the angle factors. If the temperature difference between indoor surfaces is below 10 °C, then the MRT error calculated with Equation (4) will be less than 0.2 °C [28]. Equation (5) shows the formula to calculate the angle factor [43].

$$F_{p-N} = F_{max}\left(1 - e^{-(a/c)\tau}\right)\left(1 - e^{-(b/c)\gamma}\right), \tag{5}$$

where

$\gamma = A + B(a/c)$,
$\tau = C + D(b/c) + E(a/c)$.

Parameters *a*, *b*, and *c*, defined in Figure 14, are related to dimensions and distances between the occupant and the envelope. Table 13 shows the parameters *A*, *B*, *C*, and *D* to calculate angle factors for seated persons and walls, floors, and ceilings.

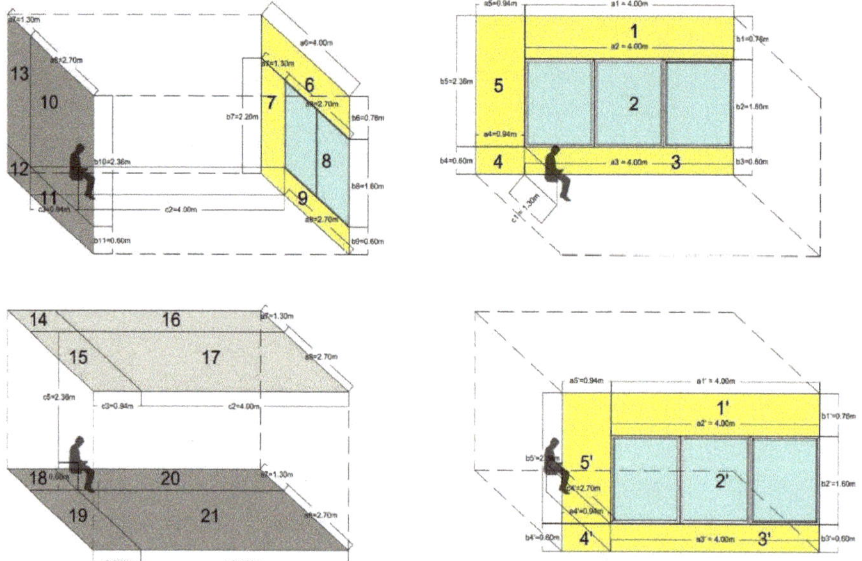

Figure 14. Geometry of the office space and occupant's position and dimensions to calculate the angle factor.

Table 13. Parameters for calculating angle factors [1].

	F_{max}	*A*	*B*	*C*	*D*	*E*
Seated person (Wall/window facing person)	0.118	1.216	0.169	0.717	0.087	0.052
Seated person (Floor/ceiling facing person)	0.116	1.396	0.130	0.951	0.080	0.055

[1] Values are taken from [28].

Figure 14 illustrates the dimensions and geometry of the office space and the different surfaces considered to calculate the MRT for a seated person. The facing direction was ignored for simplification. The temperature of each rectangle (1 to 21) was measured to calculate the MRT. Due to small differences, only five temperatures have been taken into account. $T_1 = T_6 = T_{1'} = T_{14} = T_{15} = T_{16} = T_{17} = T_{ceiling}$; $T_3 = T_{3'} = T_4 = T_{4'} = T_9 = T_{11} = T_{12} = T_{18} = T_{19} = T_{20} = T_{21} = T_{floor}$; $T_7 = T_5 = T_{5'} = T_{wall}$; $T_8 = T_{WFG}$; $T_2 = T_{2'} = T_{WFG\_TP}$.

A rough approximation to obtain the operative temperature may be to use the arithmetic average of the mean radiant temperature (MRT) of the heated space and dry-bulb air temperature if air velocity is less than 0.2 m/s and MRT is less than 50 °C. In cases where the air velocity is between 0.2 and 1 m/s, or where the difference between mean radiant and air temperature is above 4 °C, the ASHRAE 55 provides a formula, shown in Equation (6), to calculate operative temperature [27].

$$T_{op} = A\, T_a + (1 - A)\, T_{mr},$$ (6)

where

T_{op} = operative temperature (°C),

T_a = indoor air temperature (°C), and

T_{mr} = mean radiant temperature (°C).

The value of A can be found in Table 14 as a function of the relative air speed, v_r.

Table 14. Parameters of WFG.

vr	<0.2 m/s	0.2 to 0.6 m/s	>0.6 m/s
A	0.5	0.6	0.7

Figure 15 illustrates the indoor relative humidity (*RH*), the surface temperature of indoor surfaces, and the MRT calculated according to Equation (4). The WFG panel temperatures (T_{WFG}, $T_{WFG\text{-}TP}$) contribute to cooling the mean radiant temperature down to 20 °C when the energy management system is in operation. T_{WFG} was lower than $T_{WFG\text{-}TP}$ because the mass flow rate through the transparent panels was set to $\dot{m}$ = 2 L/min m^2, and through the translucent interior partitions, it was $\dot{m}$ = 1 L/min m^2. Another reason to explain the temperature difference was that each transparent WFG has its circulating device, whereas the translucent panels share the same circulating device. The former has proven to be more effective in delivering the cold from the heat pump than the latter. Floor, opaque walls, and ceiling temperatures (T_{floor}, T_{wall}, $T_{ceiling}$) are taken into account with their angle factors, which are calculated according to Equation (5).

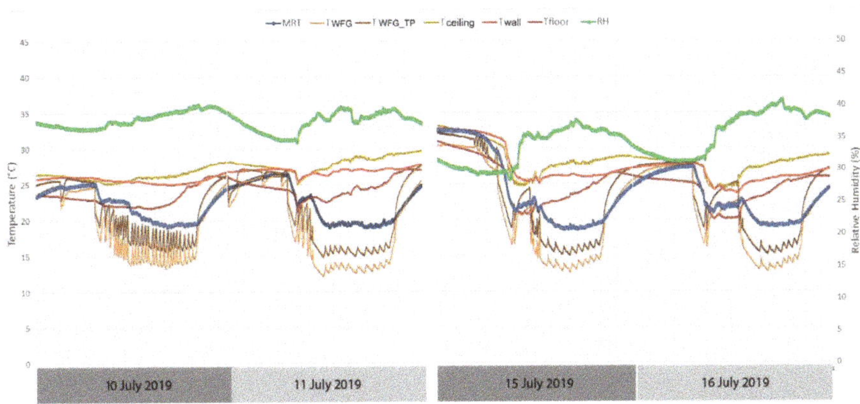

Figure 15. Surface temperatures of indoor surfaces, mean radiant temperature (*MRT*), and indoor relative humidity (*RH*)—sample days 10 July 2019 to 11 July 2019 and 15 July 2019 to 16 July 2019.

4.4. Predicted Mean Vote (PMV)

The Predicted Mean Vote (PMV) model uses six key factors to address thermal comfort: metabolic rate, clothing insulation, air temperature, radiant temperature, airspeed, and humidity. These factors may vary with time; however, in this article, the airspeed, metabolic rate, and clothing insulation are considered steady. Compliance with the ASHRAE-55 standard is tested using the CBE Thermal Comfort Tool. This tool, developed at the University of California at Berkeley, allows designers to calculate thermal comfort according to ASHRAE Standard 55-2017. The indoor air temperature and the MRT were taken from Figure 15 during operating hours. Clothing was set as 0.8 Clo (typical office indoor clothing); the metabolic rate was set as 1 Met (sedentary activity), the relative humidity was taken from Figure 14, and air velocity was set as 0.10 m/s (mean air velocity of the day). The ASHRAE-55 Comfort Zone, shaded in gray in Figure 16, represents the recommended predicted mean vote, between −0.5 and +0.5, for buildings where the occupants have metabolic rates of between 1.0 met and 1.3 met, and clothing provides between 0.5 clo and 1.0 clo of thermal insulation. Figure 16 illustrates the variations of the predicted mean vote (PMV), mean radiant temperature (MRT), indoor air temperature ($T_{\_int}$), and operative temperature (T_{op}) on four summer days. The PMV over the working hours ranged from −0.04 to −0.42 on 10/07/2019, while the MRT ranged from 23.0 to 19.3 °C, and the indoor air temperature ranged from 25.2 to 27.4 °C. During the working hours, the highest indoor temperature was on 11 July 2019 at 8:00 p.m., when the MRT was 20.1 °C and the predicted mean vote was 0.1, very close to the optimum value. Similar values are shown over the four days.

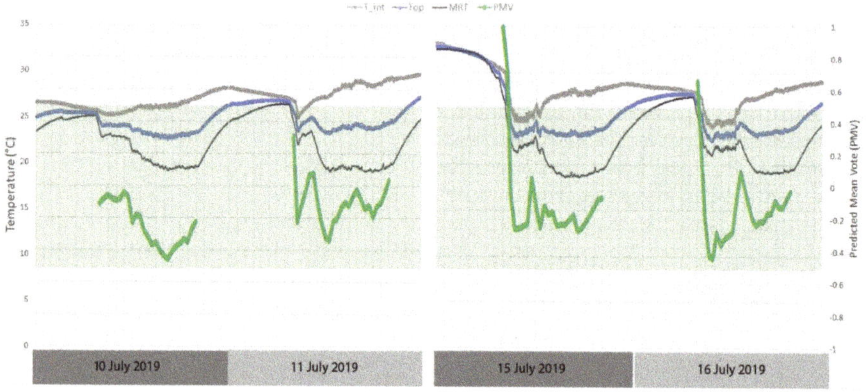

Figure 16. Operative temperature (T_{op}) and Predicted Mean Vote (*PMV*) in summer—sample days 10 July 2019 to 11 July 2019 and 15 July 2019 to 16 July 2019.

The comfort zone is defined by the combinations of the six key factors for thermal comfort. The PMV model is calculated with the air temperature and mean radiant temperature in question along with the applicable metabolic rate, clothing insulation, airspeed, and humidity. If the resulting PMV value generated by the model is within the recommended range, the conditions are within the comfort zone. Table 15 defines the PMV range for the thermal sensation scale. For 1.1% of the working hours, the PMV was above +0.4; for 6.7%, the PMV was from 0 to +0.2; for 32.3%, the PMV was from 0 to −0.2; for 54.3%, the PMV was from −0.2 to −0.4; and for 5.6% of the time, the PMV was below −0.4. Despite the high indoor air temperature, the PMV showed that occupants would describe their comfort conditions as "Slightly Cool" and always within the recommended limits specified by ASHRAE-55 (−0.5 < PMV < +0.5). The transparent WFG provided the partition exposed to solar radiation with a temperature that prevented thermal asymmetry and a lack of comfort. Hence, the results in Figure 15 indicated that the system gave consistent performance and provided comfortable conditions.

Table 15. ASHRAE thermal comfort scale [1].

Cold	Cool	Slightly Cool	Neutral	Slightly Warm	Warm	Hot
−3	−2	−1	0	+1	+2	+3

[1] Values are taken from [27].

The same comfort analysis was carried out in February. Figure 17 illustrates the variations of the predicted mean vote (PMV), mean radiant temperature (MRT), indoor air temperature ($T_{\_int}$) and operative temperature (T_{op}), and relative humidity (RH) on four February days. As shown in Figure 9, the conditions on sunny winter days are required to operate the heat pump in cooling mode in the afternoon. The indoor air temperature dropped to 20.5 °C on 19 February 2020 at 8:00 a.m., and reached 27 °C on 24 February 2020 at 7:00 p.m. The relative humidity ranged from 35% to 40%. The PMV over the working hours ranged from −1 on 19 February 2020 to 0.8 on 24 February 2020. Both values are out of the comfort range. In the morning, the PMV on the four days was below −0.5, so the occupants would describe their comfort conditions as "Slightly Cool" or "Cool". The heat pump was set to operate in heating mode when the indoor temperature was below 20 °C, and that condition was not met. On 24 February 2020, the PMV was above 0.5 from 5:00 p.m. to 8:00 p.m. Even though the heat pump was operating in cooling mode, the occupants would describe their comfort conditions as "Slightly Warm" or "Warm". For 45% of the working hours, the predicted mean vote was below −0.5, out of the shaded area representing the recommended comfort range. For 8% of the working hours, the predicted mean vote was above the comfort range when the indoor temperature surpassed 25.5 °C, and the WFG temperature was not low enough to bring down the mean radiant temperature.

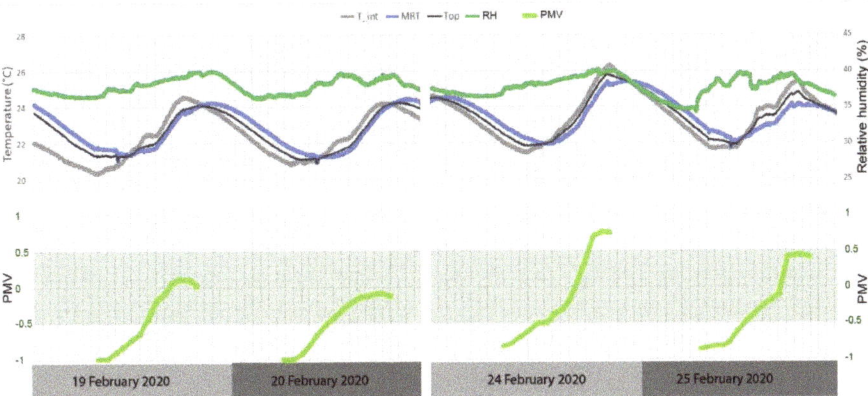

Figure 17. Operative temperature (T_{op}), Predicted Mean Vote (*PMV*), and indoor relative humidity (*RH*) in winter—sample days 19 February 2020 to 20 February 2020 and 24 February 2020 to 25 February 2020.

5. Conclusions

This paper has studied the energy performance of innovative building envelopes (facade and internal partitions), such as water flow glazing (WFG), coupled with an energy management system, as well as the relationships with steady and transient parameters. The energy strategies varied from a free-floating temperature regime on sunny winter days to the air-to-water heat pump, air heat exchanger, and buffer tank in summer conditions. A simple logic energy management system received inputs from temperature and relative humidity sensors. It controlled the heat pump and the air heat exchanger to deliver heat or cold to the buffer tank. The results included actual indoor air and glazing temperatures, heating and cooling energy consumption, and the influence of WFG in the mean radiant temperature and comfort.

Water-Flow Glazing was evaluated as a component of a hydronic radiant heating and cooling system. It showed final energy-saving potential, provided thermal comfort, and may be considered a valid option for office retrofitting. On the hottest day of the year, when the temperature ranged from 18 to 40 °C and the peak solar radiation was above 700 W/m², the energy system consumed 32 kWh (0.8 kWh/m²) and the WFG managed to keep the indoor air temperature between 25 and 27 °C. The contribution of the air heat exchanger was negligible over the year because it was set to work for cooling only when the difference between the tank top temperature and outdoor temperature ($T_{\_tank\_top} - T_{\_ext}$) was above 10 °C. It complicated the piping and the control logic and did not improve the energy performance.

Radiant panels improve the performance of air-to-water heat pumps. The energy efficiency ratio (EER) reached 3.62 when the water temperature was 18 °C, and the coefficient of performance (COP) was 4.5 when the water temperature was 35 °C in heating mode. Using WFG as a radiant cooling facade and indoor partitions effectively reduced the operative temperature to comfortable levels when the indoor air temperature was between 25 and 27.5 °C.

The Predicted Mean Vote (PMV) in summer conditions was between 0 and −0.5 in working hours, within the recommended values of ASHRAE-55 standard. The MRT ranged from 19.3 to 23 °C, and the indoor air temperature ranged from 25.2 to 29.1 °C. In winter conditions, the electronic control unit was set to operate in heating mode if the indoor air temperature was below 20 °C. Then, for 45% of the working hours, the predicted mean vote was below −0.5, out of the comfort range, so the occupants would describe their comfort conditions as "Slightly Cool" or "Cool". The control unit logic should be fixed to start operating the heating mode if the indoor temperature drops below 21 °C. On mild sunny winter days, when the outdoor temperature reached 17 °C in the afternoon, the heat pump cooled down the buffer tank, but the WFG failed to deliver enough cooling power. The predicted mean vote was above 0.5, and the conditions could be described as "Warm" and out of the comfort range for more than three hours. There were two conditions to activate WFG in the cooling mode; first, indoor air temperature should be above 25 °C, and second, the difference between indoor air temperature and the bottom tank temperature should be more than 10 °C.

Water-Flow Glazing was evaluated as a component of a hydronic radiant heating and cooling system. It showed final energy-saving potential, provided thermal comfort, and may be considered a valid option for office retrofitting. The system is limited by its high initial cost and the need for an energy management system integrated with the rest of the equipment, especially the ventilation system and the heat pump. The ventilation system is an essential aspect of comfort. Controlling the relative humidity is indispensable in radiant systems to avoid condensation issues. Therefore, a more advanced ventilation device could help optimize the whole system's performance. Including a heat recovery and variable airflow would reduce the sensible and latent thermal loads and control the dew point temperature. There were uncertainties with the air-to-water heat pump operation. Although the radiant WFG panels could improve the heat pump COP and EER, there were issues with the operating cycles that could affect its performance. The selected heat pump was oversized, and frequently started and stopped because it prematurely detected that it had reached the target temperature.

After the first year of monitoring, there are uncertainties, misfunctions, and system issues that must be addressed. Firstly, due to the complexity of the elements involved in human comfort, the control unit must integrate the ventilation device. The operation logic should be able to modify the water mass flow rate and ventilation air heat flow. Secondly, the devices must be adequately dimensioned to avoid misfunctions, especially the air-to-water heat pump. Further research must include heat pump electricity monitoring to compare the actual thermal and electricity consumption and assess energy performance more accurately. Finally, further research on the standardization of its manufacturing process and deployment is needed to bring down initial costs and payback periods. Another research line would be to integrate WFG into commercial building performance simulations.

Author Contributions: Conceptualization, B.M.S., F.d.A.G., and J.A.H.R.; methodology, B.M.S. and F.d.A.G.; software, J.A.H.R.; formal analysis, B.M.S. and F.d.A.G.; data curation, J.A.H.R.; writing—original draft preparation, J.A.H.R.; writing—review and editing, F.d.A.G. and B.L.A.; visualization, B.M.S., F.d.A.G., and B.L.A.; supervision, J.A.H.R. and B.L.A.; project administration, B.M.S.; funding acquisition, F.d.A.G. All authors have read and agreed to the published version of the manuscript.

Funding: This article has been funded by the KSC Faculty Development Grant (Keene State College, New Hampshire, USA).

Acknowledgments: This work was supported by program Horizon 2020-EU.3.3.1: Reducing energy consumption and carbon footprint by smart and sustainable use, project Ref. 680441 InDeWaG: Industrialized Development of Water Flow Glazing Systems.

Conflicts of Interest: The authors declare that they have no conflict of interest.

References

1. Kamilaris, A.; Kalluri, B.; Kondepudi, S.; Kwok Wai, T. A literature survey on measuring energy usage for miscellaneous electric loads in offices and commercial buildings. *Renew. Sustain. Energy Rev.* **2014**, *34*, 536–550. [CrossRef]
2. Balaras, C.A.; Droutsa, K.; Argiriou, A.A.; Asimakopoulos, D.N. Potential for energy conservation in apartment building. *Energy Build.* **2000**, *31*, 143–154. [CrossRef]
3. Del Ama Gonzalo, F.; Ferrandiz, J.; Fonseca, D.; Hernandez, J.A. Non-intrusive electric power monitoring system in multi-purpose educational buildings. *Int. J. Power Electron. Drive Syst.* **2019**, *9*, 51–62. [CrossRef]
4. Menezes, A.C.; Cripps, A.; Buswellb, R.A.; Wright, J.; Bouchlaghem, D. Estimating the energy consumption and power demand of small power equipment in office buildings. *Energy Build.* **2014**, *75*, 199–209. [CrossRef]
5. Europe 2020 Indicators—Climate Change and Energy. Available online: http://ec.europa.eu/ (accessed on 15 June 2020).
6. Andreou, A.; Barrett, J.; Taylor, P.G.; Brockway, P.E.; Wadud, Z. Decomposing the drivers of residential space cooling energy consumption in EU-28 countries using a panel data approach. *Energy Built Environ.* **2020**, *1*, 432–442. [CrossRef]
7. Bae, W.-B.; Mun, S.-H.; Huh, J.-H. Real-Time Occupant Based Plug-in Device Control Using ICT in Office Buildings. *Energies* **2016**, *9*, 143. [CrossRef]
8. Masoso, O.T.; Grobler, L.J. The dark side of occupants' behaviour on building energy use. *Energy Build.* **2010**, *42*, 173–177. [CrossRef]
9. Muhammad, W.; Monjur, M.; David, M.; Mario, S.; Yacine, R. Building energy metering and environmental monitoring. A state-of-the-art review and directions for future research. *Energy Build.* **2016**, *120*, 85–102. [CrossRef]
10. Komor, P. Space cooling demands from office plug loads. *Ashrae J.* **1997**, *39*, 41–44.
11. BRECSU. *Energy Consumption Guide 19: Energy Use in Offices*; Building Research Energy Conservation Support Unit: Watford, UK, 2000.
12. Dunn, G.; Knight, I. Small power equipment loads in UK office environments. *Energy Build.* **2005**, *37*, 87–91. [CrossRef]
13. Izquierdo, M.; Moreno-Rodríguez, A.; González-Gil, A.; García-Hernando, N. Air conditioning in the region of Madrid, Spain: An approach to electricity consumption, economics and CO_2 emissions. *Energy* **2011**, *36*, 1630–1639. [CrossRef]
14. Menezes, A.C.; Cripps ABuswell, R.A.; Bouchlaghem, D. Benchmarking small power energy consumption in office buildings in the United Kingdom: A review of data published in CIBSE Guide F. *Build. Serv. Eng. Res. Technol.* **2013**, *34*, 73–86. [CrossRef]
15. Pezzutto, S.; Fazeli, R.; De Felice, M.; Sparber, W. Future development of the air-conditioning market in Europe: An outlook until 2020. *Wiley Interdiscip. Rev. Energy Environ.* **2016**, *5*, 649–669. [CrossRef]
16. Lanzisera, S.; Dawson-Haggertym, S.; Cheung, H.; Taneja, J.; Culler, D.; Brown, R. Methods for detailed energy data collection of miscellaneous and electronic loads in a commercial office building. *Build. Environ.* **2013**, *65*, 170–177. [CrossRef]
17. Pezzutto, S.; De Felice, M.; Fazeli, R.; Kranzl, L.; Zambotti, S. Status quo of the air-conditioning market in Europe: Assessment of the building stock. *Energies* **2017**, *10*, 1253. [CrossRef]
18. Li, D.H.W.; Wong, S.L.; Tsang CLCheung, G.H.W. A study of the daylighting performance and energy use in heavily obstructed residential buildings via computer simulation techniques. *Energy Build.* **2006**, *38*, 1343–1348. [CrossRef]
19. Hermanns, M.; del Ama, F.; Hernández, J.A. Analytical solution to the one-dimensional non-uniform absorption of solar radiation in uncoated and coated single glass panes. *Energy Build.* **2012**, *47*, 561–571. [CrossRef]
20. Taleghani, M.; Kleerekoper, L.; Tenpierik, M.; van den Dobbelsteen, A. Outdoor thermal comfort within five different urban forms in the Netherlans. *Build. Environ.* **2015**, *83*, 65–78. [CrossRef]
21. Tullie Circle, N.E. (Ed.) *Ashrae Handbook-Heating, Ventilating, and Air-Conditioning Systems and Equipment*; ASHRAE: Atlanta, GA, USA, 2012; Available online: https://app.knovel.com/web/toc.v/cid:kpASHRAEA2/viewerType:toc/ (accessed on 15 June 2020).

22. Europe 2020 Indicators—Electricity Price Statistics. Available online: https://ec.europa.eu/eurostat/statistics-explained/index.php?title=Electricity_price_statistics (accessed on 15 August 2020).

23. Mikulik, J. Energy Demand Patterns in an Office Building: A Case Study in Kraków (Southern Poland). *Sustainability* **2018**, *10*, 2901. [CrossRef]

24. Lee, S.H.; Jeon, Y.; Chung, H.J.; Cho, W.; Kim, Y. Simulation-based optimization of heating and cooling seasonal performances of an air-to-air heat pump considering operating and design parameters using genetic algorithm. *Appl. Therm. Eng.* **2018**, *144*, 362–370. [CrossRef]

25. Priarone, A.; Silenzi, F.; Fossa, M. Modelling Heat Pumps with Variable EER and COP in EnergyPlus: A Case Study Applied to Ground Source and Heat Recovery Heat Pump Systems. *Energies* **2020**, *13*, 794. [CrossRef]

26. Fanger, P.O. *Thermal Comfort, Analysis and Application in Environmental Engineering*; Danish Technical Press: Copenhagen, Denmark, 1970.

27. ASHRAE-55. *ANSI/ASHRAE Standard 55-2017: Thermal Environmental Conditions for Human Occupancy*; ASHRAE: Atlanta, GA, USA, 2017.

28. ISO 7730:2005. *Ergonomics of the Thermal Environment-Analytical Determination and Interpretation of Thermal Comfort Using Calculation of the PMV and PPD Indices and Local Thermal Comfort Criteria*; International Organization for Standardization (ISO): Geneva, Switzerland, 2005.

29. Marino, C.; Nucara, A.; Pietrafesa, M. Mapping of the indoor comfort conditions considering the effect of solar radiation. *Sol. Energy* **2015**, *113*, 63–77. [CrossRef]

30. Godbole, S. *Investigating the relationship between Mean Radiant Temperature (MRT) and Predicted Mean Vote (PMV)*; A case study in a university building; School of Architecture and the Built Environment: Stockholm, Sweden, 2018.

31. Chow, T.; Chunying, L.; Zhang, L. Thermal characteristics of water-flow double-pane window. *Int. J. Therm. Sci.* **2011**, *50*, 140–148. [CrossRef]

32. Gueymard, C.; duPont, W. Spectral effects on the transmittance, solar heat gain, and performance rating of glazing systems. *Sol. Energy* **2009**, *83*, 940–953. [CrossRef]

33. Chow, T.T.; Li, C. Liquid-filled solar glazing design for buoyant water-flow. *Build. Environ.* **2013**, *60*, 45–55. [CrossRef]

34. Gil-Lopez, T.; Gimenez-Molina, C. Environmental, economic and energy analysis of double glazing with a circulating water chamber in residential buildings. *Appl. Energy* **2013**, *101*, 572–581. [CrossRef]

35. Gutai, M.; Kheybari, A.G. Energy consumption of water-filled glass (WFG) hybrid building envelope. *Energy Build.* **2020**, *218*, 110050. [CrossRef]

36. Moreno Santamaria, B.; del Ama Gonzalo, F.; Pinette, D.; Gonzalez-Lezcano, R.-A.; Lauret Aguirregabiria, B.; Hernandez Ramos, J.A. Application and Validation of a Dynamic Energy Simulation Tool: A Case Study with Water Flow Glazing Envelope. *Energies* **2020**, *13*, 3203. [CrossRef]

37. Moreno Santamaria, B.; del Ama Gonzalo, F.; Lauret Aguirregabiria, B.; Hernandez Ramos, J.A. Experimental Validation of Water Flow Glazing: Transient Response in Real Test Rooms. *Sustainability* **2020**, *12*, 5734. [CrossRef]

38. Spanish Ministry of Development. BasicDocument on Energy Saving of the Technical Building Code (Documento Básico de Ahorro de Energía del Código Técnico de la Edificación, CTE-DB-HE). 2019. Available online: https://www.codigotecnico.org/images/stories/pdf/ahorroEnergia/DBHE.pdf (accessed on 16 August 2020).

39. European Union. Directive (EU) 2018/844 of the European Parliament and of the Council of 30 May 2018. Amending Directive 2010/31/EU on the Energy Performance of Buildings and Directive 2012/27/EU on Energy Efficiency. 2018. Available online: https://eur-lex.europa.eu/legal-content/EN/TXT/PDF/?uri=CELEX:32018L0844&from=EN (accessed on 16 August 2020).

40. Spanish Regulation of Thermal Installations in Buildings (RITE). Factores de Emisión de CO_2 y Coeficientes de Paso a Energía Primaria de Diferentes Fuentes de Energía Final Consumidas en el Sector de Edificios en España. Agencia Estatal Boletín Oficial del Estado: Madrid, Spain. 2016. Available online: https://energia.gob.es/desarrollo/EficienciaEnergetica/RITE/Reconocidos/Reconocidos/Otros%20documentos/Factores_emision_CO2.pdf (accessed on 16 August 2020).

41. Saunier Duval. Manuel d'installation. Available online: https://www.saunierduval.fr/france/download/genia-air-1/genia-air-2/saunier-duval-genia-air-6-8-12-15-unite-exterieure-notice-installation-0020117808-03-12-2012-291831.pdf (accessed on 16 August 2020).

42. Ventilación Técnica Para la Edificación. Available online: https://www.casals.com/es/productos-casals/ventilacion-tecnica-para-la-edificacion/ventiladores-in-line-y-en-caja-insonorizada/box-bd-plus// (accessed on 4 September 2020).
43. Cannistmro, G.; Franzitta, G.; Giaconia, C.; Rizzo, G. Algorithms for the calculation of the view factors between human body and rectangular surfaces in parallelepiped environments. *Energy Build.* **1992**, *19*, 51–60. [CrossRef]

 *sustainability*

Article

Photovoltaic Power Prediction Using Artificial Neural Networks and Numerical Weather Data

Javier López Gómez [1,*], Ana Ogando Martínez [1], Francisco Troncoso Pastoriza [1],
Lara Febrero Garrido [2], Enrique Granada Álvarez [1] and José Antonio Orosa García [3]

[1] GTE Research Group, School of Industrial Engineering, University of Vigo, Campus Lagoas-Marcosende, 36310 Vigo, Spain; aogando@uvigo.es (A.O.M.); ftroncoso@uvigo.es (F.T.P.); egranada@uvigo.es (E.G.Á.)
[2] Defence University Centre, Spanish Naval Academy, 36920 Marín, Spain; lfebrero@cud.uvigo.es
[3] Department of Nautical Science and Marine Engineering, Universidade da Coruña, 15011 A Coruña, Spain; jaorosa@udc.es
*   Correspondence: javilopez@uvigo.es

Received: 5 November 2020; Accepted: 7 December 2020; Published: 9 December 2020

Abstract: The monitoring of power generation installations is key for modelling and predicting their future behaviour. Many renewable energy generation systems, such as photovoltaic panels and wind turbines, strongly depend on weather conditions. However, in situ measurements of relevant weather variables are not always taken into account when designing monitoring systems, and only power output is available. This paper aims to combine data from a Numerical Weather Prediction model with machine learning tools in order to accurately predict the power generation from a photovoltaic system. An Artificial Neural Network (ANN) model is used to predict power outputs from a real installation located in Puglia (southern Italy) using temperature and solar irradiation data taken from the Global Data Assimilation System (GDAS) sflux model outputs. Power outputs and weather monitoring data from the PV installation are used as a reference dataset. Three training and testing scenarios are designed. In the first one, weather data monitoring is used to both train the ANN model and predict power outputs. In the second one, training is done with monitoring data, but GDAS data is used to predict the results. In the last set, both training and result prediction are done by feeding GDAS weather data into the ANN model. The results show that the tested numerical weather model can be combined with machine learning tools to model the output of PV systems with less than 10% error, even when in situ weather measurements are not available.

Keywords: Artificial Neural Network (ANN); Global Data Assimilation System (GDAS); Numerical Weather Prediction (NWP); photovoltaic power; weather data

---

1. Introduction

Anthropogenic climate change and increasing levels of pollution are critical issues that continue to motivate the transition from the use of fossil fuels to renewable energy sources. The Paris Agreement, signed by 196 countries during the 2015 United Nations Climate Change Conference, still drives most United Nations' members to reduce their dependency on non-renewable energy sources. After stalling in 2018, installed renewable power is expected to increase to 1200 GW between 2019 and 2024. Solar PV energy sources are expected to account for more than half of this increase, followed by onshore wind systems [1]. Despite the disruptions caused in global energy markets by the Covid-19 pandemic, the total generation of renewable energy is expected to increase by up to 5% in 2020 [2].

Most renewable energy technologies rely on atmospheric conditions to generate electric power. Wind speed and direction determine the performance of wind turbines. Solar irradiation is the key factor that controls the power output of photovoltaic (PV) and thermal solar systems, with other variables like air temperature and humidity also affecting their performance. Reliable weather data is required to either evaluate the best placement for a new renewable installation or predict the future output of an existing system [3,4]. However, meteorological data is not always monitored at the location of renewable energy production systems, so historical local datasets may not be available. When in situ weather measurements are not available, physical or statistical models for computing the required variables [5], spatial interpolations of data from regional networks of weather stations [6], or Numerical Weather Prediction (NWP) models [7] can be used.

Several studies have attempted to forecast weather related variables using Artificial Neural Network (ANN) models. In [8], nonlinear autoregressive (NAR) neural networks were used to predict the fluid flow rate in shallow aquifers. In [9], changes in precipitable water vapour were estimated by a nonlinear autoregressive approach with exogenous input (NARX). In [10], temperature and wind speed were predicted using different upgraded versions of convolutional neural networks (CNN). In [11], the performance of multilayer perceptron (MLP) and multigene genetic programming (MGGP) neural networks for estimating the solar irradiance in PV systems are compared.

In many publications, different regression and machine learning algorithms have also been used in combination with weather data from NWP to forecast future PV power. In [12], the GPV-MSM mesoscale model (5 km of horizontal resolution) of the Japanese Meteorological Agency was combined with a support vector regression algorithm. In [13], the Global Forecast System (GFS) 0.5 product (0.5° horizontal resolution) of the National Oceanic and Atmospheric Administration of the United States was used to feed a multivariate adaptive regression splines model. In [14], an ANN model was trained with a numerical model from the European Centre for Medium-Range Weather Forecasts, although the name of the particular numerical model was not specified. ANNs are a family of machine learning techniques which are widely used to predict PV system output. An extensive review on forecasting techniques applied to PV power is presented in [15]; in that publication, ANN models are described as the most widely used option for PV forecasting among statistical, physical, and hybrid techniques. Another review on this issue is presented in [16], where ANN forecasting models are again the most widely represented algorithms.

The Global Forecast System is one of the most widely used global-scale NWP models [17–19], providing public, freely-licenced, hourly weather forecasts over different grids of horizontal and vertical points covering the entire planet. Four GFS products are currently operated: the 1.00°, 0.50°, and 0.25° horizontal grid resolution models, and the surface flux (sflux) model, with a horizontal resolution of roughly 13 km. The Global Data Assimilation System (GDAS) is another NWP model. It is used to process observational measurements (aircraft, surface, satellite, and radar) which are scattered and irregular in nature, and place them into a gridded, regular space. Those gridded measurements are then used by other models like GFS as a starting point to develop weather forecasts. Four GDAS products are currently active, with each one feeding initialisation data to one of the four aforementioned GFS products. Both the GFS and GDAS models are developed and maintained by the National Oceanic and Atmospheric Administration (NOAA), a governmental agency belonging to the United States Department of Commerce. More details about GFS and GDAS physics and subsystems can be found in [20].

One issue with GFS products is the lack of open, long-term data repositories offering access to their data outputs. Although all GFS and GDAS products are publicly available at the NOAA Operational Model Archive and Distribution System (NOMADS) near-real-time repository of NWP input and output data, each individual file is only stored there for 10 days [21]. For long-term storage, the NOAA maintains the Archive Information Request System (AIRS) [22], but only the coarser resolution GFS products (1.00° and 0.50°) are stored. However, this is not the case for GDAS products; outputs from the 0.25° product have been stored at the AIRS repository since June 2015, while sflux files have been stored since February 2012.

In this study, a real photovoltaic installation located in the South of Italy is modelled using ANN, and its PV power production is predicted. Two data sources are used to feed the neural network models. The first one includes experimental weather and power production data gathered in [23] from the real PV system during 2012 and 2013. The second source uses weather data from the GDAS sflux model, one of the NWP models from the NOAA with the highest spatial resolution, and the one with the highest resolution outputs that are publicly available for the temporal span of the experimental data campaign. Three scenarios are designed with different combinations of monitoring and GDAS weather data to train and test the neural network model. PV power output data from the monitoring campaign are used in all scenarios.

The fundamental idea behind the present study (shared with previous works by the same authors) is to evaluate the performance of global-scale data sources as either complements or replacements of more local- or regional-scale data sources (either forecasts or measurements) for different uses. This approach implies a lower quality of the available particular values. Performance differences between global-scale models and local sources may be larger or smaller depending on the quality of said models, but the former are not expected to be better except for outlier cases (e.g. using faulty or non-representative measurements, or ill-tuned local models). However, local data sources have relevant issues related to the lack of standardisation (requiring ad hoc retrieving and processing solutions) and the lack of historic data (e.g., the unavailability or excessive cost of the required amounts of data). If the differences in data accuracy and precision for global-scale models can be kept within an acceptable threshold level, they may be outweighed by the benefits of said models regarding the aforementioned issues.

Multiple studies have already combined different NWP models with statistical algorithms in order to forecast PV generation outputs, including ANN algorithms. To the best of the authors' knowledge, however, very few studies have combined weather data from the GDAS products with ANN models, and none has used GDAS products for photovoltaic power forecasting. Hence, the novelty and main objective of this study is to evaluate the performance of the GDAS sflux model as a replacement for in situ weather measurements with the aim of predicting photovoltaic power generation.

2. Materials and Methods

2.1. Modelled System and Study Conditions

The photovoltaic system analysed in this article is located in the campus of the University of Salento in Monteroni di Lecce, Puglia, located in south Italy. The installation, fully described in [24], is grid-connected and covers a net area of 4710 m^2 on the roofs of the parking lots of the aforementioned university. It is comprised of 3000 monocrystalline silicon modules connected in series, with a total nominal power of 960 kW$_p$. The azimuth of all modules is $-10°$, since they are oriented to the south east. However, two groups of modules can be differentiated according to their slope within the PV system: 3° (PV1); and 15° (PV2). Table 1 includes the main specifications of both groups.

Table 1. Main features of the modelled PV system.

PV1 Module Group		PV2 Module Group	
Nominal power [kW$_p$]	353.3	Nominal power [kW$_p$]	606.7
Number of modules	1104	Number of modules	1896
Net Surface of the modules [m^2]	1733.3	Net Surface of the modules [m^2]	2976.7
Azimuth [°]	-10	Azimuth [°]	-10
Tilt [°]	3	Tilt [°]	15

The location of the present study matches the coordinates of the campus where the PV system is placed, i.e., 40°19′32″16 N, 18°5′52″44 E. This study area is characterised by a Mediterranean climate, with warm winters and dry summers. It is located 18 km from the nearest shore line, with an average elevation of 35 m above sea level. The starting and ending dates of this study are conditioned by the duration of the weather and PV power monitoring campaign done in [23] over said installation, as explained on the next subsection. Hence, the chosen temporal span of the study starts on 6th March 2012 at 00:00:00 UTC and ends on December 30th, 2013 at 23:59:59 UTC.

2.2. Weather and Photovoltaic Data Variables

2.2.1. Monitored Data

The first data source used in this research is the measurement acquisition work carried out in [23] regarding the same PV system that is modelled on the present study. Said installation is monitored by means of two types of sensors for weather variables: an irradiance sensor LP-PYRA02 to measure the hourly mean solar irradiance on the two inclined surfaces (3° and 15°), and a PT100 temperature sensor to gauge the hourly mean air temperature and the temperature of the PV modules. Hourly mean photovoltaic power data for the entire installation is extracted from three inverters. The air temperature, the two solar irradiations, and the output power compose the monitoring dataset, the first of two used in the present study.

2.2.2. Numerical Weather Model Data

The second data source used in this study comes from one of the NWP models belonging to the NOAA. The GDAS sflux model, which assimilates scattered observational measurements into a regular grid of data used to initialise the GFS sflux forecast model, was chosen. The GDAS sflux, like its GFS counterpart, has been regularly evolving since its creation, improving its resolution and its physics and numerical components. The version used in this study applies a hybrid Eulerian grid (technical identifier T574) with 23 km of horizontal resolution and 64 vertical levels. Like all GFS and GDAS products, it is solved four times each day, with cycles starting at 00, 06, 12 and 18 h UTC. Each cycle, it generates data outputs with three-hourly resolution up to 9 hours. Only the most recent forecast hours (00 and 03) of each cycle are used, as weather conditions from the farther hours (06 and 09) can be more accurately described by the first hours of the next cycle. The output of this product, starting from 13th February 2012, is available from the AIRS online repository. Thus, the same time span of the monitoring dataset can be covered with GDAS data.

The nearest neighbour point of the GDAS sflux horizontal grid is located at coordinates 40°22′29″21 N, 18°0′1″56 E. The horizontal distance between said point and the location of the modelled PV system is roughly 9.9 km. This distance is considered small enough to directly use the data from the GDAS point, as in a nearest neighbour strategy, instead of applying more complex spatial interpolation algorithms.

The output files generated by the different GDAS models contain results for multiple weather variables for all the points of their horizontal grids and vertical levels. For temperature, different variables are available for instantaneous, average, maximum, and minimum values at soil-, distance-, and pressure-based vertical levels. For solar irradiation, two fluxes (short- and long-wave) are available with two propagation directions (downwards and upwards) at the local surface level, along with other levels. The instantaneous air temperature predicted at 2 m over the local surface and the instantaneous downward short-wave radiation flux at local surface level are the chosen variables, as they are most analogous to air temperature and global solar irradiation as measured by a weather station [25]. Those two variables compose the GDAS dataset, the second one used in this study.

2.2.3. Auxiliary Data

Temperature and solar irradiation values at each time step are the key components that decide the power output of photovoltaic panels. However, there are other variables that may modulate the effect of those weather variables. In particular, the position of the Sun affects the amount of solar radiation flux that can be used by the solar panels. Said position follows a daily cycle between dawn and dusk, but also a yearly cycle due to the relative positions of the Earth and the Sun, leading to differences in solar flux incidence angles between winter and summer, and between early morning, noon, and late afternoon.

To include the aforementioned differences in the prediction models, two additional input variables are used: the hour of the day, ranging from 0 to 23, and the hour of the year, ranging from 0 to 8759.

These two auxiliary variables are inspired by a previous study by one of the authors [26] where they were both used (in addition to the hour of the week) to characterise the evolution of thermal demands of a large building during different time periods. However, the nature of the system modelled in the present study does not have a direct dependency on the day of the week, so the hour of the week is not included as an input variable. The two selected variables are added to both monitoring and GDAS datasets.

2.3. Data Preprocessing

Real data measured with automatic sensors are often incomplete or contain errors and inconsistencies, so data preprocessing is crucial. Cleaning and organisation techniques prepare the data and make them suitable for use with machine learning models. In particular, modifications applied to the monitoring dataset are focused on fixing minor inconsistences and removing erroneous or missing data. The dataset contains many empty records. In some of the night-time slots (between 10 p.m. and 3 a.m., both inclusive), no data is available since no measurements are collected. The solar irradiation is known to be zero (by definition) during night, but that is not the case for air temperature. However, the fact that no temperature data are available at night is not an issue; since there is no photovoltaic power production at night, it is therefore an irrelevant period. Adding night-time data would provide redundant information that would increase the complexity of the model and the calculation time needed without producing relevant results.

To avoid empty records that could negatively affect learning models, rows containing null data are removed. The same procedure is used when duplicated values or incomplete records are detected, where at least one of the variables is missing. Some discrepancies are also corrected, i.e., some records have a time shift of a few hours, probably due to an error in the data dumping. To correct this, the times of sunrise and sunset are taken into account as a reference. The times of sunrise and sunset are determined as indicated in [27], considering the dates, latitude, and longitude of the location of the facility and calculating the declination by Spencer's method and the sunset hour angle. Reference dawn, noon, and dusk hours for 2013 at the study location are shown in Figure 1.

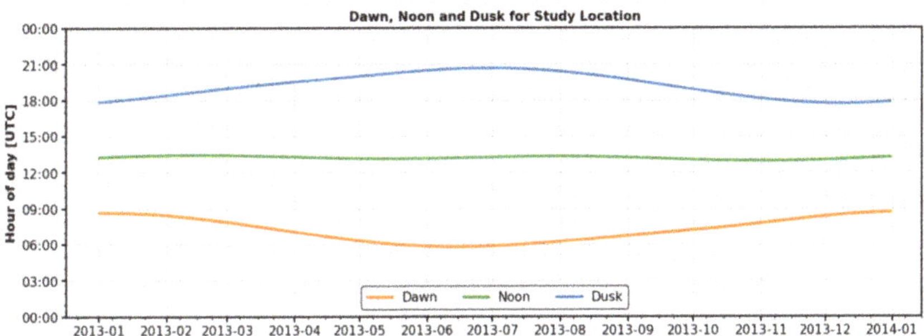

Figure 1. Dawn, noon and dusk for the study location.

The GDAS dataset is initially composed of discrete values for one in every three hours only (i.e., hours in the [00:00, 03:00,..., 21:00] group, which will hereafter be called 'GDAS-hours'), while the monitoring dataset contains hourly measured values. A fast correction for ensuring compatible time coordinates between datasets would be to filter the monitoring dataset to remove 'non-GDAS-hours' (i.e., any hour not belonging to the GDAS-hours group). However, this coarse time discretisation would fail to capture relevant points for the solar irradiation and PV power variables. As shown in Figure 1, neither noon (maximum irradiation in clear sky conditions) nor most dawn (irradiation start) or dusk (irradiation end) times are close to any of the GDAS-hour instants.

To avoid losing information from the monitoring dataset, values from the GDAS dataset must be interpolated over their temporal coordinates to an hourly resolution. Second order B-splines are used to generate continuous piecewise polynomial functions for both the global horizontal irradiation and air temperature variables of the GDAS dataset. These piecewise curves are fitted to the available GDAS-hours values of temperature and irradiation. For the latter, additional fitting points are included for the dawn and dusk instants of each day (with a 0 W/m^2 irradiation value). After generating the fitted curves, values for the non-GDAS-hours instants are extracted and added into the dataset. As all the original GDAS data correspond to either forecasts of 0 or 3 h, it can be stated that the interpolated GDAS dataset only contains forecasts inside the 0–5 hourly range.

As mentioned, the monitoring dataset contains two solar irradiation variables, each measured at a different tilt angle (3° and 15°). The GDAS dataset, however, only contains values of irradiation on a horizontal plane (same convention as most general-purpose automatic weather stations). As increasing differences on tilt angle cause increasing divergences on irradiation values, care must be taken when comparing irradiation variables from the two datasets. After some preliminary analyses, it was found that the behaviour of irradiation variables with tilts of 3° (the first irradiation variable from the monitoring dataset) and 0° (the one from the GDAS dataset) were sufficiently similar. Thus, the irradiation measured at 15° is removed from the monitoring dataset, while that measured with a tilt of 3° is used for comparisons against the global horizontal irradiation from GDAS directly.

Once both datasets are preprocessed, they are compared to find faulty time instants. If at least one of the variables has a missing value at a given instant, that instant is entirely removed for all variables of both datasets. This reduces the final available number of hours of data, but ensures that all hours fed into the prediction models are complete. After this and previous filters have been applied, 11,132 valid hours of data remain. This value corresponds to 69.6% of the total hours belonging to the temporal span of the study, including all night hours that are not present in the monitoring data included in [23].

2.4. Estimation of Photovoltaic Power: Artificial Neural Network Model

The ANN models used in this study are multi-layer perceptrons, a class of neural networks composed of multiple layers of interconnected artificial neurons. The first is the input layer, which ingests the different input variables fed into the neural network model. The last is known as the output layer, which yields the predicted values generated by the model. Between these two layers, there is a variable number of hidden layers (zero or more) which connect the input and output layers. The number of artificial neurons in the output layer is determined by the number of output variables. However, the number of neurons for the input and hidden layers is not predetermined, but rather a variable parameter to be optimised.

All the different MLP models built for this study share a common set of configuration parameters. The neurons in both the input and hidden layers use the rectifier linear unit (ReLU) function as the activation function [28]. Batch learning is used as the training methodology [29]. The chosen optimisation algorithm is Adaptive Moment Estimation (Adam), a stochastic gradient descent method [30]. The training stop criterion uses a separated fraction of the training data (a randomly chosen 10% of the training hours) to evaluate the current stage of the trained model, using the mean absolute error (MAE) metric shown below. An additional patience-based stop criterion is included, where training stops if the performance of the model does not improve for 100 consecutive epochs.

$$MAE = \sum_{i=1}^{N} \frac{|X_i - Y_i|}{N} \tag{1}$$

In the next subsections, the monitoring and GDAS datasets are combined into three different scenarios to train and test the different MLP models. At the same time, the available days of data are split into training and testing samples. These samples, applied over the different scenarios, are used to select the best MLP model and to evaluate its performance when fed with monitoring and/or GDAS meteorological data.

2.4.1. Training and Testing Scenarios

Three different scenarios, with combinations of the two data sources, are used in this study. For the first one, the ANN model is both trained and tested using weather input data from the monitoring dataset. For the second scenario, the model uses the same training with measured data as the first one, but the testing is done using weather inputs from the GDAS dataset. For the third scenario, both training and testing are done using GDAS weather input data. PV power outputs from the monitoring dataset are used in the training stages for all three scenarios. A schematic of the different scenarios is presented in Figure 2.

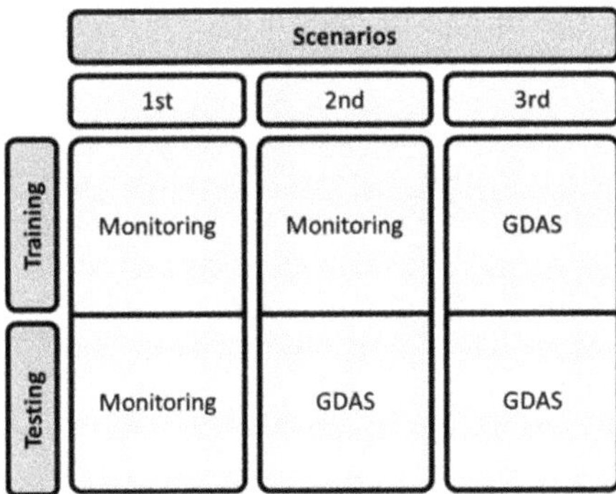

Figure 2. Training and testing scenarios used on the study.

The differences between the first and second scenarios lie exclusively in the testing part of the method. Their ANN models are trained using the same input data, following the same chain of operations, and have virtually the same values on their weight matrices. The results from the first scenario provide a way of evaluating the ability of the ANN to model the studied photovoltaic system. The predictions done by the model of the second scenario are affected by both the nature of the GDAS data and the modelling of the photovoltaic system done by the neural network. Hence, a comparison between the predictions of the first and second scenarios enables an independent analysis of the effects of the GDAS data. The third scenario assumes that no local measured weather data are available, and is able to determine whether the GDAS dataset is adequate to entirely replace these local measurements in the weather data.

2.4.2. Training and Testing Dates

The days comprising the temporal interval of the study are divided into a training sample and three testing samples. The training sample is fed into the ANN algorithms to build the photovoltaic power estimation models, while the testing samples are used to evaluate the performance of the estimation models.

First, all complete available days are filtered. Here, a complete day is defined as any day containing at least all its daytime hours (hours between dawn and dusk instants, computed as indicated in the Data Preprocessing section) and with no missing data for all its available hours (in the monitoring or the GDAS datasets). From this pool of complete days, two days are selected at random from each month and added to the first and second test samples, respectively.

The random selection algorithm ensures that all complete days from each month have the same probability of being included on either test sample. It also ensures that no day belongs to both random test samples. The first and second testing samples each comprise 22 individual days. These samples can be considered representative of the entire temporal span of the study, as all months are guaranteed to be equally represented. This means that all seasonal weather conditions are taken into account when testing the performance of the estimation models.

Two complete weeks of data are handpicked to form the last testing sample from the remaining complete days not belonging to any of the two random test samples. They span from 28th April to 5th May 2012, and 12th April to 19th April 2013. This sample is to be used to graphically represent the performance of the weather and power output variables. The two complete weeks have enough duration to capture the multidaily-resolution features of the studied variables, while still being able to spot hourly-resolution patterns. However, days belonging to this testing sample are not to be considered representative of the entire temporal span of the study, as they were manually chosen.

Finally, the training sample includes all days not belonging to any of the three testing samples, and comprises the majority of the available data. This training sample is fed into the different ANN architectures to build the estimating models.

2.4.3. Selection of Best Model

Once the training and testing temporal samples have been selected, the different ANN models are built. The number of hidden layers (varying between zero and two) and the number of neurons inside each of the input and hidden layers are the only configuration parameters that differ between models. The number of neurons of the output layer is fixed by the number of output variables (only one, i.e., PV power production), as already stated. All other configuration parameters are kept constant.

With the required parameters defined for all neural network models, the training dataset of the second scenario is fed into each ANN model, using hours of data from the training sample. Then, the testing dataset from the same scenario is fed into each trained model to generate PV power estimations, using the first testing sample of random days. These estimations are compared against the corresponding monitored values of PV outputs, and mean nRMSE values are obtained for each tested ANN model. The same process is done again, training and testing each model from zero using data from the third scenario.

The best neural network model is chosen considering the combination of the nRMSE mean errors from the second and third scenarios. This best ANN model is comprised of an input layer of five neurons and a single hidden layer of 30 neurons, in addition to the single-neuron output layer. This selected architecture is shown in Figure 3. The second and third test samples are finally used to evaluate the performance of the chosen model for all three scenarios, as explained on the Results and Discussion section.

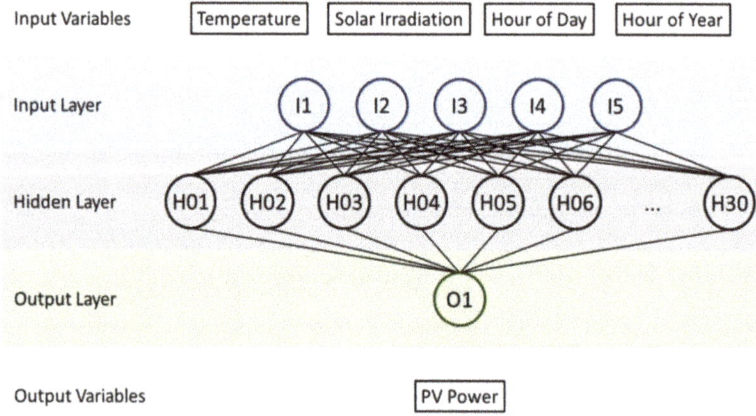

Figure 3. Layer architecture of the best ANN model.

2.5. Error Measurements

Three error metrics are used to evaluate the performance of the ANN model when predicting photovoltaic power, and also to compare the two datasets used on each of its input variables: mean bias error (MBE), root mean square error (RMSE), and normalised root mean square error (nRMSE). They are defined using the three following equations.

$$MBE = \sum_{i=1}^{N} \frac{X_i - Y_i}{N} \tag{2}$$

$$RMSE = \sqrt{\sum_{i=1}^{N} \frac{(X_i - Y_i)^2}{N}} \tag{3}$$

$$nRMSE = \frac{\sqrt{\sum_{i=1}^{N} \frac{(X_i - Y_i)^2}{N}}}{Y_{max}} \tag{4}$$

Where X_i is an individual value from one of the variables of the GDAS dataset (estimations), Y_i is the corresponding value from the monitoring dataset (observations), Y_{max} is the largest value of the entire monitoring dataset (43.53 °C for temperature, 1045.48 W/m^2 for solar irradiation and 848.66 kW for PV production), and N is the number of data hours (sample size).

The MBE provides a measurement of the general bias of a given variable, while the RMSE provides more information about individual discrepancies with reality for a large set of estimations. Both of them compute error values in physical units. The nRMSE is a nondimensional version of the RMSE. It is useful for evaluating the relative performance between unrelated magnitudes. In the case of power generation systems, it can be used to directly compare the output errors from modelled systems with different nominal capacities [16].

3. Results and Discussion

In order to properly evaluate the performance of GDAS numerical data when coupled with an ANN model for predicting the power generation of a PV system, both the input weather and the output power variables are analysed. First, input values from the GDAS dataset are compared with those of the monitoring dataset, for both temperature and solar irradiation inputs. Then, PV power predictions obtained from the ANN model under the three training and testing scenarios are compared with the real PV measurements taken from the monitoring dataset.

3.1. Analysis of Weather Inputs

Figure 4 compares solar irradiation values from the monitoring and GDAS datasets, for each of the two manually picked weeks included in the test samples. For each time period, both the real (monitoring) and estimated (GDAS) signals, and the difference between those signals, are represented. Temperature graphs for manual weeks, and both temperature and irradiation graphs for random days, are omitted for the sake of brevity.

Table 2 shows the MBE, RMSE, and nRMSE metrics results for temperature and solar irradiation in the GDAS dataset. Errors are computed using the corresponding values from the monitoring dataset as a reference. The results are computed for three different periods: the entire temporal span of the study, the 22 random chosen days from the second test sample, and the two manually picked weeks from the third test sample. Individual error values are calculated for each day of each period, and then the mean and standard deviation values of those errors are computed.

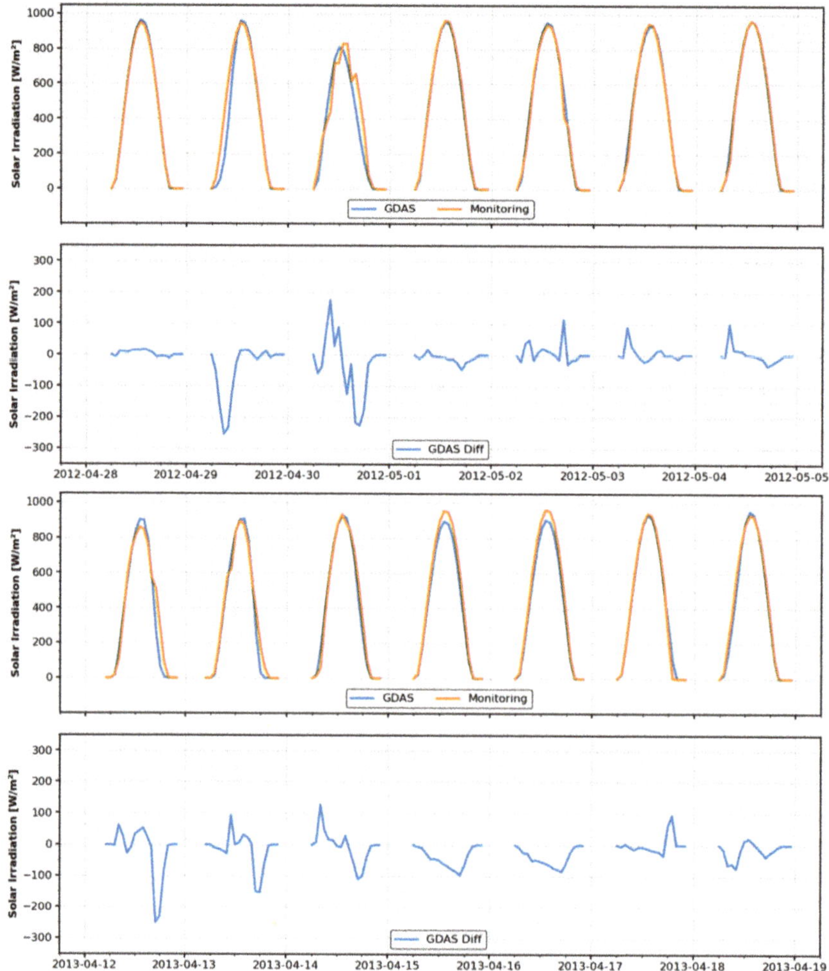

Figure 4. Comparison of solar irradiation inputs for manually chosen weeks.

Table 2. Error metrics of weather inputs.

Input Variable	Sample	MBE		RMSE		nRMSE [%]	
		Mean	**St Dev**	**Mean**	**St Dev**	**Mean**	**St Dev**
Temperature [°C]	All study span	−3.30	6.04	7.04	3.65	16.17	8.39
	Random days	−3.03	5.49	6.63	3.16	15.23	7.26
	Manual weeks	−5.76	2.51	7.34	2.38	16.86	5.47
Solar Irradiation [W/m²]	All study span	−12.98	89.89	121.84	85.92	11.65	8.22
	Random days	−15.81	54.84	100.00	48.01	9.56	4.59
	Manual weeks	−14.46	17.44	47.72	28.53	4.56	2.73

It is clear that GDAS has a moderate tendency to underestimate both mean air temperature and solar irradiation variables for the studied location and temporal interval. However, MBE values are much closer to zero than their RMSE counterparts for the entire study span and the 22 randomly chosen days, especially for solar irradiation. For the two manually chosen weeks, which cover similar dates of 2012 and 2013, distances towards zero are not so divergent between both metrics. This behaviour shows that underestimations and overestimations of weather variables for individual days tend to almost compensate one another for periods of time that comprise a larger variety of weather conditions. Mean and standard deviation values of nondimensional errors for solar irradiation are lower than their temperature equivalents on all temporal samples.

Key instants on the daily behaviour of irradiation curves, namely, dawn and dusk hours and the hour of maximum value, match closely between the monitoring and GDAS variables, as can be seen in Figure 4. The same behaviour is found for the randomly picked days; for almost all days, the GDAS estimations and the monitoring data have their maximum value at the same hour or with a one-hour difference, while dawn and dusk hours both match for most days with maximum differences of one hour.

When comparing sample periods, errors for the randomly chosen days are more similar to those of the entire study span when placed against errors for the manually chosen weeks. This indicates that the former sample is more representative of the global behaviour of both weather variables for the study span. Although the sample of two weeks provides an efficient way of visualising specific features of the solar irradiation inputs, numerical error metric values for said sample should be treated as examples, and should not be generalised.

The main objective of the present study is related to the prediction of PV power outputs, and not to air temperature or solar irradiation. However, both of these weather variables are key inputs of the ANN model used for predicting photovoltaic power, particularly solar irradiation. As the GDAS dataset contains interpolated weather values for every two out of three hours, it is sensible to compare the error values in this dataset with those of similar studies. For example, [31] uses an ANN model trained with the LMA algorithm to forecast surface irradiation from extraterrestrial solar irradiation in the South of China, among other inputs, using both historical data series and statistical feature parameters to feed the models. Their RMSE results are ~43 W/m^2 for sunny days, and ~85–255 W/m^2 for cloudy days (using statistical parameters or historic series as inputs). The RMSE irradiation errors from Table 2 are larger than the statistical-based-predictions from [31], but lower than historic-based-predictions, which are more comparable.

It must be highlighted that GDAS is a NWP model, meaning that the derived meteorological data are estimated (forecasted) values. For the particular case of GDAS, this estimation includes the interpolation of scattered real observations into a regular grid of points, and also a very-short-term forecasting of future weather conditions. These spatial interpolations and temporal extrapolations imply a certain level of uncertainty. In addition, GDAS forecasts have a three-hour resolution. Instantaneous air temperature and solar irradiation values at each three hours are taken from GDAS outputs and interpolated to a one-hour temporal resolution using second order B-splines. This means that short-term phenomena cannot be captured by the GDAS model (e.g. transient shadows due to passing clouds). Even though these issues with GDAS data remain, the resulting weather values (particularly solar irradiation) are shown to have a reasonably good match with the monitored data. The nRMSE errors for temperature and irradiation are lower than a 17% and 12%. RMSE errors for solar irradiation are in accordance to those found in related bibliography, and estimated key instants of the daily behaviour of these variables (dawn, noon, and dusk) closely match those of the monitored data.

3.2. Analysis of PV Power Predictions

Figure 5 shows photovoltaic power outputs for the same two weeks used in Figure 4, again including real (Monitoring) and estimated (Scenarios 1–3) signals, and the differences between them. Output values for the three analysed scenarios are included, as well as values from the monitoring dataset for reference. Again, temperature graphs are omitted.

A comparison between solar irradiation inputs and PV outputs for four of the randomly chosen days is shown in Figure 6. They cover situations of good and bad matching between monitoring and GDAS solar irradiation for both 2012 (first and second subplots, corresponding to days of March and April 2012) and 2013 (third and fourth subplots, for days of August and October 2013).

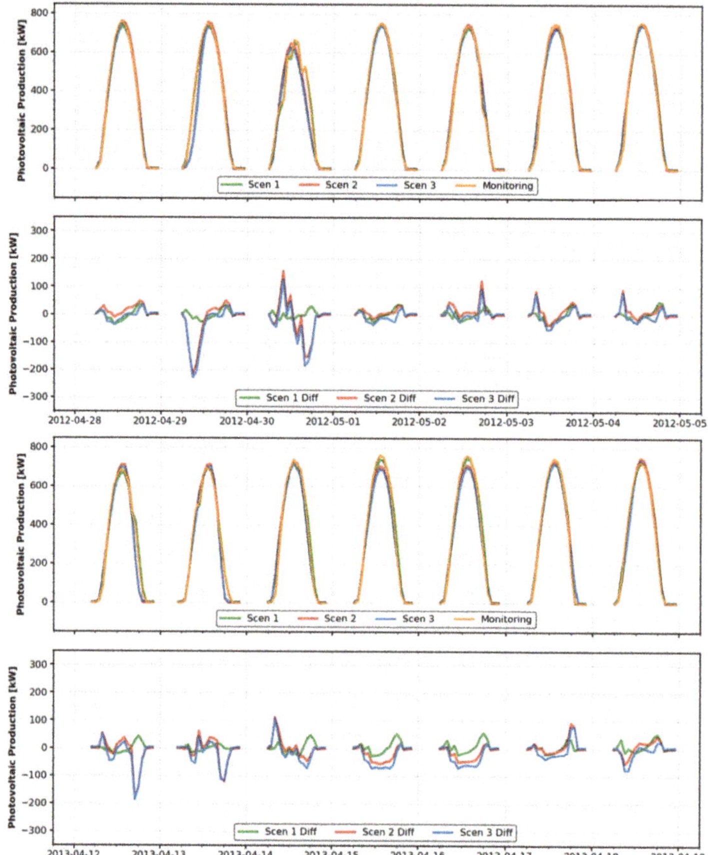

Figure 5. Comparison of PV power outputs for manually chosen weeks.

Table 3 shows the error results for photovoltaic power predictions for the three scenarios. The same considerations from Table 2 apply here. In this new table, only the random days and manual weeks are used. The entire temporal span of the study is not considered in order to avoid including data values already used during the training of the ANN model.

The performance of the predicted PV output variable is reasonably good on all scenarios and for both samples of random days and manual weeks. As with the input variables, MBE results for power predictions are much closer to zero than their RMSE counterparts, meaning that power biases for individual days and hours tend to be nullified for more complete periods of time. RMSE errors on power production are lower than 25 kW (first scenario) and 85 kW (second and third scenarios). The real magnitude of these errors is easier to evaluate when compared with the 960 kW_p of peak production installed for the monitored PV system, and the actual maximum value of PV production from the monitoring dataset, i.e., 849 kW. In fact, nRMSE values are lower than 3% for the first scenario, and lower than 10% for the other two, with maximum standard deviation values of 5%.

Graphs of power outputs and irradiation inputs for the 22 randomly chosen days tested (as well as those of the two manually chosen weeks) show almost identical patterns. This can be seen in the examples in Figure 6, in which power prediction curves from the second and third scenarios perfectly match the solar irradiation curve predicted by GDAS, while the power prediction curve from the first scenario replicates the solar irradiation curve provided by the monitoring dataset. This behaviour is to be expected due to the high correlation between the solar irradiation and the power output of a PV system [16,32]. Key instants on the daily behaviour of PV production curves (starting, peak, and ending production hours) for the monitoring data match closely with the predictions of the ANN model in all three scenarios, in the same fashion as with solar irradiance.

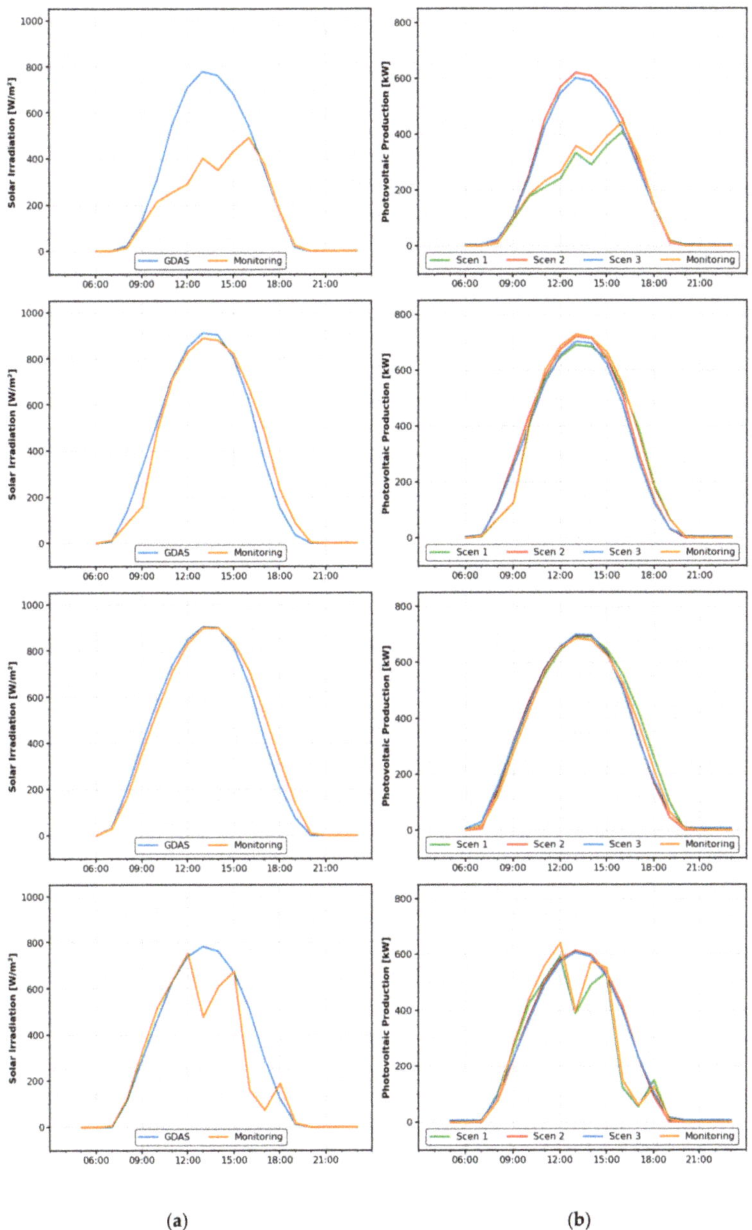

Figure 6. Comparison of irradiation and power prediction. (**a**) Monitoring and GDAS solar irradiation for four randomly chosen test days. (**b**) Monitoring and scenarios PV power prediction for the same four days.

Table 3. Error metrics of PV power outputs.

Output Variable	Sample	MBE		RMSE		nRMSE [%]	
		Mean	St Dev	Mean	St Dev	Mean	St Dev
PV power 1st	Random days	3.12	18.84	24.91	29.77	2.94	3.51
scenario [kW]	Manual weeks	2.03	3.17	17.59	3.74	2.07	0.44
PV power 2nd	Random days	−8.61	45.01	83.66	42.71	9.86	5.03
scenario [kW]	Manual weeks	−2.13	13.84	38.09	18.82	4.49	2.22
PV power 3rd	Random days	−10.52	44.58	84.71	40.13	9.98	4.73
scenario [kW]	Manual weeks	−13.21	13.64	40.97	20.43	4.83	2.41

Power outputs from the first scenario are most similar to measurements from the monitoring dataset, with very low errors (mean RMSE lower than 25 kW) and almost identical curves for all random days and manual weeks. These similarities are to be expected, as the neural network model uses monitoring data for both training and testing on this first scenario. The performance of PV predictions from the second and third scenarios are also quite similar, in terms of both error metrics (except mean MBE values for manual weeks) and graphics. This is a remarkable result, as the ANN models from these scenarios are trained using weather inputs from different datasets, even when they both use the same weather data to test their performances. Although the performance of the second and third scenarios is not as good as that of the first one, it is still quite significant, with mean bias errors of less than 14 kW below the real outputs, and mean nRMSE errors lower than 10% of the maximum power output measured for the entire study span.

As with solar irradiation, it is sensible to compare results for PV production with those available in the literature. In [33], next-day forecasts of power outputs from a 264 kW$_\mathrm{p}$ PV plant in the North of Italy were generated using an ANN model, trained with the error back-propagation method, and coupled with a clear sky solar radiation model. The error analysis took into account three different significant days with sunny, partially cloudy, and cloudy weather conditions, obtaining nRMSE values of 12.5%, 24%, and 36.9% respectively. In [32], nRMSE values in the 10.91–23.99% range were achieved when predicting power output forecasts in a horizon of 1–24 h for the same PV system as that modelled in the present study (a 960 kW$_\mathrm{p}$ installation located in the South of Italy), using an Elman ANN model. In [34], a 1 MW$_\mathrm{p}$ PV plant in California was modelled using a feed-forward ANNs model and a genetic algorithms/ANN (GA/ANN) hybrid model, among others. The reported RMSE errors for forecasts 1 and 2 h forward were 88.23–142.74 kW for the ANN model, and 72.86–104.28 kW for the GA/ANN model. The reported nRMSE values were computed using a different definition, thus making comparisons with the present study unreliable.

The nRMSE values from Table 3 for random days in the second and third scenarios are equivalent to those of the best-performing cases in [33] (sunny days) and [32] (one-hour-horizon forecasts). The corresponding RMSE values are slightly lower than those obtained in [34] using their ANN model for a forecast horizon of one hour, and slightly worse than those from their GA/ANN model. It is worth noting that no distinctions are made between prediction performance for sunny or cloudy days in the present study. However, the selection of test days ensures that both high-irradiation days (typical of the summer season and clear sky conditions) and low-irradiation days (typical of the winter season and clouded sky conditions) are represented in the random test sample. Also, no distinctions between forecast hourly-horizons are made here, but the nature of the GDAS dataset implies that all data belong to a forecast horizon from 0 and up to 5 h, as explained in the Data Preprocessing section. The modelled PV system presented in [34] is not identical to the one in the present study. However, the peak output productions from both installations are similar enough, so at least a soft comparison of dimensional metrics like the RMSE can be made.

3.3. Comments on Training and Testing Scenarios

After presenting and commenting on the relevant results for photovoltaic power predictions in the three different training and testing scenarios, a brief discussion about the implications of each scenario is in order.

The first proposed scenario both trains and tests the performance of the chosen ANN model using data from the monitoring dataset. No inaccuracies due to the forecast nature of GDAS data are fed into the neural network model here. Hence, all differences between predictions and actual values are due to the mathematical modelling of the PV system done by the ANN model. As all the tested error metrics have great performance in this scenario, it can be concluded that the proposed neural network model is indeed adequate to model the studied installation.

The second proposed scenario uses the same training as that of the first scenario, and thus, the resulting trained model is virtually identical, and the same conclusion about its modelling performance applies. The testing part here is done using GDAS data. As already stated, differences in the prediction performance between the first and second scenarios can be attributed to uncertainties in the GDAS data. The error values are indeed higher for this second dataset, but relative differences are not drastic. They can still match (and even outmatch) the performance of other forecast strategies reported in the literature, as stated in the previous subsection. For a generic monitored PV installation, this scenario would be translated as having an ANN model trained with a historic dataset of in situ measurements of weather variables. In this case, GDAS data could still be used to predict power production for the next hours based on GDAS forecasts, or to fill missing days in historic datasets.

The third and last scenario completely replaces the weather data from the monitoring dataset with those from the GDAS dataset, keeping only the measured PV power outputs from the former. Here, the trained ANN model differs from that of the other two scenarios (although the same basic configuration is used), so the prior conclusion about the performance of the ANN model does not apply. However, error results for this scenario are most similar to those of the second one, with only slightly worse bias errors and almost identical RMSE and nRMSE errors. This implies that a historical dataset of in situ weather measurements may not be a crucial item when aiming to predict future power outputs with GDAS. Further studies targeting different locations and facilities would be required to confirm this. Nevertheless, it seems that even if the sensors for monitoring relevant weather variables were never available at the location of the PV system, GDAS forecasts could still be used to train an ANN model which is able to predict power production.

4. Conclusions

The main objective of this study was to evaluate the applicability of the GDAS sflux numerical weather model as a replacement for in situ weather measurements to model the power outputs of a photovoltaic system. Three training and testing scenarios, with different combinations of monitoring and GDAS weather data, were used to feed and evaluate the performance of one prediction model using a multilayer perceptron ANN algorithm. Solar irradiation and air temperature were the main input variables, while PV power production was the only predicted output.

Bias errors on individual days tended to be compensated when considering more complete temporal samples. This happened both on the weather inputs and the PV power outputs.

Mean nRMSE values of 2.9% and 9.9% on PV outputs were achieved for the most representative testing sample in the first and second scenarios, respectively. A comparison between those values led to the conclusion that most of the power prediction errors were due to the approximate nature of the GDAS solar irradiation data. However, the 100.00 W/m^2 mean RMSE error achieved for this weather variable was in accordance with other solar irradiation forecast methodologies included in the bibliography. The neural network model used was shown to model the real power system with solid accuracy.

An analysis of the second scenario indicated that the GDAS sflux product is a reliable source of weather data for forecasting future PV power outputs, when an ANN model built with past in situ weather measurements is already available. The analysis of the third scenario, on the other hand, showed that even when said historic dataset of local weather measurements is not available, GDAS data can be effectively used to train the ANN model, with a minimal loss in the accuracy of PV power predictions.

Less than 10% mean nRMSE errors in PV power outputs were achieved for both the second and third scenarios. A comparison with other relevant studies showed that the errors in the photovoltaic power predictions for all scenarios in this study were analogous to those presented in the solar forecasting literature. The use of GDAS weather data in combination with ANN algorithms makes it possible to predict PV power with a performance that matches or even outmatches other PV forecast methods.

Future research expanding this work could focus on tackling some aspects which were not fully analysed in the present study. The influence of other weather variables could be studied (like wind effects on the cooling of the solar modules, or rainfall removing possible depositions of fine dust and dirt), if reference measured data were available for said variables. Second order B-splines were used here to interpolate weather values for the hours when GDAS data were not generated, but other temporal interpolation methods could be tested. Finally, the performance of the GDAS model could be tested against other NWP models, like a high resolution version of the Global Forecast System.

In conclusion, the present study shows that the GDAS sflux numerical weather model is a reliable source of weather data for photovoltaic power prediction when combined with Artificial Neural Network algorithms. Estimative data from this numerical model can be fed into an existing ANN model, already trained with local weather measurements, or can entirely replace said measurements and be used to train the model when historical local weather data are not available.

Author Contributions: Conceptualization, J.L.G., A.O.M. and L.F.G.; Data curation, J.L.G. and A.O.M.; Formal analysis, J.L.G. and A.O.M.; Funding acquisition, E.G.Á. and J.A.O.G.; Investigation, J.L.G. and A.O.M.; Methodology, J.L.G., A.O.M. and F.T.P.; Project administration, E.G.Á.; Resources, L.F.G. and J.A.O.G.; Software, J.L.G., A.O.M. and F.T.P.; Supervision, E.G.Á. and J.A.O.G.; Validation, J.L.G., A.O.M. and F.T.P.; Visualization, J.L.G., A.O.M. and L.F.G.; Writing—Original draft, J.L.G. and A.O.M.; Writing—Review & editing, F.T.P. and L.F.G. All authors have read and agreed to the published version of the manuscript.

Funding: This investigation article was partially supported by the University of Vigo through the grant Convocatoria de Axudas á Investigación 2018: Axudas Predoutorais UVigo 2018 (grant number 00VI 131H 641.02). This investigation article was also partially supported by the Ministry of Universities of the Spanish Government through the grant Ayudas para la Formación de Profesorado Universitario: Convocatoria 2017 (grant number FPU17/01834). This investigation article was also partially supported by the Ministry of Universities of the Spanish Government by means of the SMARTHERM (Project: RTI2018-096296-B-C2).

Acknowledgments: The authors would like to than to the National Centres for Environmental Information (NCEI) and the National Oceanic and Atmospheric Administration (NOAA) for providing free public access to their historic repositories of GDAS outputs through the Archive Information Request System (AIRS).

Conflicts of Interest: The authors declare no conflict of interest.

Abbreviations

AIRS	Archive Information Request System
ANN	Artificial Neural Network
GDAS	Global Data Assimilation System
GFS	Global Forecast System
MBE	Mean Bias Error
MLP	Multi-Layer Perceptron
NOAA	National Oceanic and Atmospheric Administration
nRMSE	Normalised Root Mean Square Error
NWP	Numerical Weather Prediction
PV	Photovoltaic
RMSE	Root Mean Square Error
sflux	Surface flux
UTC	Coordinated Universal Time

References

1. IEA. *Renewables 2019*; IEA: Paris, France, 2019. Available online: https://www.iea.org/reports/renewables-2019 (accessed on 22 October 2020).
2. IEA. *Global Energy Review 2020*; IEA: Paris, France, 2020. Available online: https://www.iea.org/reports/global-energy-review-2020 (accessed on 22 October 2020).
3. Alharthi, Y.Z.; Siddiki, M.K.; Chaudhry, G.M. Resource assessment and techno-economic analysis of a grid-connected solar PV-wind hybrid system for different locations in Saudi Arabia. *Sustainability* **2018**, *10*, 3690. [CrossRef]
4. Richardson, W.; Krishnaswami, H.; Vega, R.; Cervantes, M. A low cost, Edge computing, all-sky imager for cloud tracking and intra-hour irradiance forecasting. *Sustainability* **2017**, *9*, 482. [CrossRef]
5. Kim, K.H.; Oh, J.K.-W.; Jeong, W. Study on solar radiation models in South Korea for improving office building energy performance analysis. *Sustainability* **2016**, *8*, 589. [CrossRef]
6. Dirksen, M.; Knap, W.H.; Steeneveld, G.-J.; Holtslag, A.A.M.; Tank, A.M.G.K. Downscaling daily air-temperature measurements in the Netherlands. *Theor. Appl. Climatol.* **2020**, *142*, 751–767. [CrossRef]
7. López Gómez, J.; Troncoso Pastoriza, F.; Granada Álvarez, E.; Eguía Oller, P. Comparison between geostatistical interpolation and numerical weather model predictions for meteorological conditions mapping. *Infrastructures* **2020**, *5*, 15. [CrossRef]
8. Taherdangkoo, R.; Tatomir, A.; Taherdangkoo, M.; Qiu, P.; Sauter, M. Nonlinear autoregressive neural networks to predict hydraulic fracturing fluid leakage into shallow groundwater. *Water* **2020**, *12*, 841. [CrossRef]
9. Rahimi, Z.; Mohd Shafri, H.Z.; Norman, M. A GNSS-based weather forecasting approach using Nonlinear Auto Regressive Approach with Exogenous Input (NARX). *J. Atmos. Solar Terr. Phys.* **2018**, *178*, 74–84. [CrossRef]
10. Mehrkanoon, S. Deep shared representation learning for weather elements forecasting. *Knowl. Based Syst.* **2019**, *179*, 120–128. [CrossRef]
11. De Paiva, G.M.; Pimentel, S.P.; Alvarenga, B.P.; Marra, E.G.; Mussetta, M.; Leva, S. Multiple site intraday solar irradiance forecasting by machine learning algorithms: MGGP and MLP neural networks. *Energies* **2020**, *13*, 5. [CrossRef]
12. Junior, J.G.d.S.F.; Oozeki, T.; Ohtake, H.; Shimose, K.-i.; Takashima, T.; Ogimoto, K. Forecasting regional photovoltaic power generation—A comparison of strategies to obtain one-day-ahead data. *Energy Proc.* **2014**, *57*, 1337–1345. [CrossRef]
13. Massidda, L.; Marrocu, M. Use of multilinear adaptive regression splines and Numerical Weather Prediction to forecast the power output of a PV plant in Borkum, Germany. *Solar Energy* **2017**, *146*, 141–149. [CrossRef]
14. Mellit, A.; Massi Pavan, A.; Lughi, V. Short-term forecasting of power production in a large-scale photovoltaic plant. *Solar Energy* **2014**, *105*, 401–413. [CrossRef]
15. Antonanzas, J.; Osorio, N.; Escobar, R.; Urraca, R.; Martinez-de-Pison, F.J.; Antonanzas-Torres, F. Review of photovoltaic power forecasting. *Solar Energy* **2016**, *136*, 78–111. [CrossRef]
16. Das, U.K.; Tey, K.S.; Seyedmahmoudian, M.; Mekhilef, S.; Idris, M.Y.I.; Van Deventer, W.; Horan, B.; Stojcevski, A. Forecasting of photovoltaic power generation and model optimization: A review. *Renew. Sustain. Energy Rev.* **2018**, *81*, 912–928. [CrossRef]
17. Geer, A.J.; Lonitz, K.; Weston, P.; Kazumori, M.; Okamoto, K.; Zhu, Y.; Liu, E.H.; Collard, A.; Bell, W.; Migliorini, S.; et al. All-sky satellite data assimilation at operational weather forecasting centres. *Q. J. R. Meteorol. Soc.* **2018**, *144*, 1191–1217. [CrossRef]
18. Kato, T. Chapter 4—Prediction of photovoltaic power generation output and network operation. In *Integration of Distributed Energy Resources in Power Systems*; Funabashi, T., Ed.; Academic Press: Cambridge, MA, USA, 2016. [CrossRef]
19. Mengaldo, G.; Wyszogrodzki, A.; Diamantakis, M.; Lock, S.-J.; Giraldo, F.X.; Wedi, N.P. Current and Emerging Time-Integration Strategies in Global Numerical Weather and Climate Prediction. *Arch. Comput. Methods Eng.* **2019**, *26*, 663–684. [CrossRef]
20. The Global Forecast System (GFS)—Global Spectral Model (GSM). Available online: https://www.emc.ncep.noaa.gov/emc/pages/numerical_forecast_systems/gfs/documentation.php (accessed on 22 September 2020).

21. NOMADS. NOAA Operational Model Archive and Distribution System. Available online: https://nomads.ncep.noaa.gov/ (accessed on 14 October 2020).

22. Archive Information Request System (AIRS). Available online: https://www.ncdc.noaa.gov/has/HAS.DsSelect (accessed on 19 October 2020).

23. Malvoni, M.; De Giorgi, M.G.; Congedo, P.M. Data on photovoltaic power forecasting models for Mediterranean climate. *Data Brief* **2016**, *7*, 1639–1642. [CrossRef]

24. Congedo, P.M.; Malvoni, M.; Mele, M.; De Giorgi, M.G. Performance measurements of monocrystalline silicon PV modules in South-eastern Italy. *Energy Conver. Manag.* **2013**, *68*, 1–10. [CrossRef]

25. López Gómez, J.; Troncoso Pastoriza, F.; Fariña, E.A.; Eguía Oller, P.; Granada Álvarez, E. Use of a Numerical Weather Prediction model as a meteorological source for the estimation of heating demand in building thermal simulations. *Sustain. Cities Soc.* **2020**, *62*, 102403. [CrossRef]

26. Martínez Comesaña, M.; Febrero-Garrido, L.; Troncoso-Pastoriza, F.; Martínez-Torres, J. Prediction of building's thermal performance using LSTM and MLP neural networks. *Appl. Sci.* **2020**, *10*, 7439. [CrossRef]

27. Duffie, J.A.; Beckman, W.A.; Blair, N. *Solar Engineering of Thermal Processes, Photovoltaics and Wind*; John Wiley & Sons: Hoboken, NJ, USA, 2020.

28. Eckle, K.; Schmidt-Hieber, J. A comparison of deep networks with ReLU activation function and linear spline-type methods. *Neural Netw.* **2019**, *110*, 232–242. [CrossRef] [PubMed]

29. Guresen, E.; Kayakutlu, G.; Daim, T.U. Using Artificial Neural Network models in stock market index prediction. *Expert Syst. Appl.* **2011**, *38*, 10389–10397. [CrossRef]

30. Kingma, D.P.; Ba, J.L. Adam: A method for stochastic optimization. In Proceedings of the 3rd International Conference on Learning Representations, ICLR 2015—Conference Track Proceedings, San Diego, CA, USA, 7–9 May 2015.

31. Wang, F.; Mi, Z.; Su, S.; Zhao, H. Short-term solar irradiance forecasting model based on Artificial Neural Network using statistical feature parameters. *Energies* **2012**, *5*, 1355–1370. [CrossRef]

32. Giorgi, M.G.D.; Congedo, P.M.; Malvoni, M. Photovoltaic power forecasting using statistical methods: Impact of weather data. *IET Sci. Meas. Technol.* **2014**, *8*, 90–97. [CrossRef]

33. Leva, S.; Dolara, A.; Grimaccia, F.; Mussetta, M.; Ogliari, E. Analysis and validation of 24 hours ahead neural network forecasting of photovoltaic output power. *Math. Comput. Simul.* **2017**, *131*, 88–100. [CrossRef]

34. Pedro, H.T.C.; Coimbra, C.F.M. Assessment of forecasting techniques for solar power production with no exogenous inputs. *Solar Energy* **2012**, *86*, 2017–2028. [CrossRef]

Publisher's Note: MDPI stays neutral with regard to jurisdictional claims in published maps and institutional affiliations.

 *sustainability*

Article

Challenges of the Facilities Management and Effects on Indoor Air Quality. Case Study "Smelly Buildings" in Belgrade, Serbia

Milena Vukmirovic [1,*], Alenka Temeljotov Salaj [2,*] and Andrej Sostaric [3]

1   Faculty of Forestry, University of Belgrade, 11000 Belgrade, Serbia
2   Department of Civil and Environmental Engineering, Faculty of Engineering,
    Norwegian University of Science and Technology (NTNU), 7491 Trondheim, Norway
3   Institute of Public Health of Belgrade, 11000 Belgrade, Serbia; andrej.sostaric@zdravlje.org.rs
*   Correspondence: milena.vukmirovic@sfb.bg.ac.rs (M.V.); alenka.temeljotov-salaj@ntnu.no (A.T.S.);
    Tel.: +381-11-3053913 (M.V.); +47-46445072 (A.T.S.)

Abstract: One of the key objectives and challenges nowadays is to live in safe and healthy cities. Accordingly, maintaining good air quality is one of the preconditions for achieving this goal, which is not a simple task given the various negative impacts. This paper deals with a phase of the construction process that is a cause of extreme indoor air pollution in the newly built facilities of the Dr Ivan Ribar settlement in Belgrade, popularly known as "smelly buildings." Indoor air pollution is observed from the aspect of indoor air quality (IAQ) prevention and facilities management (FM) in order to define recommendations for future prevention of these and similar situations. The research indicates the existence of specific sources of indoor pollutants, as well as the need to pay special attention to indoor air as an aspect that affects the health, comfort and well-being of individuals who permanently or temporarily use a particular space, and to point out additional costs. The paper will also consider the potential of the FM approach in preventing negative issues related to IAQ, especially in the field of public construction and social and affordable housing.

Keywords: indoor air quality; facility management; construction materials; "smelly buildings"; Belgrade; Serbia

 check for updates

Citation: Vukmirovic, M.; Salaj, A.T.; Sostaric, A. Challenges of the Facilities Management and Effects on Indoor Air Quality. Case Study "Smelly Buildings" in Belgrade, Serbia. *Sustainability* **2021**, *13*, 240. https://doi.org/10.3390/su13010240

Received: 26 November 2020
Accepted: 23 December 2020
Published: 29 December 2020

Publisher's Note: MDPI stays neutral with regard to jurisdictional claims in published maps and institutional affiliations.

1. Introduction

A safe, clean, healthy and sustainable environment is necessary for the full enjoyment of a vast range of human rights, including the rights to life, health, food, water and development. At the same time, the exercise of human rights, including the rights to information, participation and remedy, is vital to the protection of the environment [1]. Based on these values, the United Nations (UN) Human Rights Council has developed a framework of principles on human rights that addresses the right to a healthy environment and looks forward to the next steps in the evolving relationship between human rights and the environment.

Based on recent World Health Organization (WHO) estimates, 22% of the global burden of disease is due to the environment [2]. Noncommunicable diseases (NCDs), including heart disease, stroke, cancer, diabetes and chronic lung disease, are collectively responsible for almost 70% of all deaths worldwide [3]. Within NCDs, air quality is a risk factor for several of the world's leading causes of death, including heart disease, pneumonia, stroke, diabetes and lung cancer [4], as well as impaired mental health [5]. Indoor air risk includes different kinds of gas pollutants, including volatile organic compound (VOC) and the radioactive gas radon [6]. Several studies show that long-term exposure to road traffic noise and ambient air pollution is associated with increased cardiovascular risk factors [7]. Air pollutant concentration is increased in urban built environments, making air pollution from human activities the largest environmental health risk in Europe [8]. The exposure to environmental factors plays an important role in the prevalence of NCDs and is even more problematic in the current pandemic period [9]. The review of the main epidemiological studies that evaluate the respiratory effects of indoor air pollutants show the consistent short-term and long-term effects on asthma, chronic bronchitis and chronic obstructive pulmonary disease in indoor settings with poor air quality [10].

On the other side, Sustainable Development Goals (SDGs) 3 and 11 also emphasize the importance of living in a safe and healthy environment [10]. SDG 3—*Good health and wellbeing*—endeavors to ensure healthy lives and promote well-being at all ages as essential to sustainable development, while SDG 11— *Sustainable cities and communities*—advocates the future in which cities provide opportunities for all, with access to basic services, energy, housing, transportation and more.

Safe Community movement (SC) and Healthy City program (HC) are two major initiatives encouraged by the WHO to promote safety and health in communities [11]. After merging similar safety and health promotion topics, Tabrizia et al. [11] reviewed their relevancy and determined perfect matches which are reflected in risk-groups' health and safety, child safety, disaster preparedness and response, home and buildings' safety and health, and healthy and safe urban planning and design. In addition, relative matches were seen within traffic safety, violence prevention, work safety, safe public places, water safety, school safety, tobacco-free environment, mental well-being, addiction and substance abuse prevention, physical environmental quality and social support. Air quality is recognized only within the Healthy City program.

Based on this study, Tabrizia et al. [11] developed the Safe and Health Promoting Community (SHPC) model comprising seven values and 14 main dimensions to improve safety and health in the community. Two of these deal with the outdoor and indoor environment and air quality: healthy and safe environment (healthy waste, clean air, healthy food, safe waste management and access to sewage system) [5] and healthy and safe urban planning (safe urban furniture, safe public places, safe leisure places and safe home) [10]. It is important to mention that public empowerment and participation are concepts that are highly focused on new theories of safety and health promotion, where health is seen as a political choice. It is about the kind of society we want to live in [12].

Even though the basic function of a building is to shelter occupants from outdoor elements and provide a healthy, comfortable environment for productive activity, according to Lozano Patino and Siegel [13], conditions in social housing units are usually substandard. This generally associates with higher exposure to indoor pollutants and, ultimately, negative health effects. The authors [13] acknowledge the significance of this problem considering that social housing populations are generally more vulnerable due to age and/or socioeconomic status. Therefore, it is important to research the indoor air quality (IAQ) in these environments, especially because several studies found correlations between poor housing conditions and negative health outcomes [14–17].

Considering that people in urban areas spend up to 90% of their time indoors [18] and there are specific sources of pollutants indoors in relation to the environment [19,20], there is a clear need to pay special attention to indoor air. Considering this, problems of indoor air pollution are increasingly recognized as important risk factors for human health, requiring different management approaches from those used for outdoor air pollution [21].

Accordingly, by presenting an extreme example of bad IAQ in "smelly buildings" in Belgrade, Serbia, the purpose of this paper is to look at this specific and extreme case of indoor air pollution from an aspect of IAQ and facilities management (FM) in order to define recommendations for future prevention of these and similar situations. In accordance with this objective, the paper is structured as follows. The first part of the paper deals with the comparative analysis of the activities and services of FM and phases of critical building design control in achieving IAQ, as observed from the aspect of the linear model of the construction process. A brief overview of the context of public construction and the social housing complex in Serbia is presented in the second part of the paper to clarify the circumstances under which they were built. The third part of the paper includes a case study of "smelly buildings" in the Dr Ivan Ribar settlement in Belgrade, using the chronological course of events to indicate critical phases of the process. The final discussion part of the paper seeks to point out the need for the presence of FM during the entire construction process, in order to prevent certain situations that result in deterioration of IAQ.

2. Material and Method

The research presented in this paper includes the following methods:

- Critical analysis of sources and views of various authors dealing with the topics of the FM approach and process of ensuring IAQ. This part of the research covers 23 sources dealing with the FM approach and particular phases within the construction process and eight sources dealing with the process of achieving IAQ.
- Comparative analysis of five frameworks that identify different activities and competencies of FM and the process of achieving the desired IAQ, observed from the aspect of different phases of the construction course of the facility. Critical analysis of primary and secondary sources related to the character of public construction and construction of social housing in Serbia covers seven primary (laws, plans and decrees) and three secondary sources concerning public housing construction.
- Case study of the "smelly buildings" in Dr Ivan Ribar social and affordable housing settlement in Belgrade, Serbia. Using the chronological course of events, the aim of this part of the research is to indicate critical phases of the process, as well as to point out the various consequences that have arisen, which can be overcome by improving certain stages of the process and the competencies of individual actors. For the purposes of the case study, the reports and results of the research of the Vinca Institute of Nuclear Sciences, Laboratory of Physical Chemistry and the Institute of Public Health of Belgrade from August 2014 were used. In addition, the research relied on articles from the daily press, as well as on the contents of the Beobuild blog on the topic, "State Housing—Settlement in Block 72 (Dr Ivana Ribara)," in order to reconstruct the course of events.

The research location, Dr Ivan Ribar's settlement, is located in New Belgrade's Block 72, about 7.5 km away from the city center at the location planned for social and affordable housing. The land on which the complex was built is the public property of the Republic of Serbia. The facilities were built from the state budget and the settlement was built in two phases. In the period from 2010 to 2012, a part of the settlement was built according to the project of the awarded competition work from 2006. The second part of the settlement, where the disputed facilities are located, was built during 2013, according to the awarded competition project from 2011. This settlement includes six buildings with 770 housing units of different types. After moving in, disputed buildings No. 1 and 3 (see Figure 1) were found to have been built by the same company, hired by the Construction Directorate of Serbia.

Figure 1. Dr Ivan Ribar settlement in Belgrade, Serbia. Source: Authors and Geo Serbia Map portal.

In addition, the situation with the social and affordable housing in Serbia is briefly described with focus on the legal framework and regulations relating to this type of construction, bearing in mind that social housing is still considered cheap. This type of construction was done with minimal costs and at the cheapest price, i.e., following the logic of "least cost" mentality [22].

3. Facility Management (FM), Sustainability Issues and Indoor Air Quality (IAQ)

Considering the general definition that facility management (FM) is a profession that encompasses multiple disciplines to ensure functionality, comfort, safety and efficiency of the built environment by integrating people, place, process and technology [23], its possibilities in creating an environment that will meet the parameters of IAQ should be reconsidered. More specific, ISO/FIDIS 41001:2018 standard gives an updated emphasis on health: "FM integrates multiple disciplines to have an influence on the efficiency, productivity and economies of societies, communities and organizations as well as the manner in which individuals interact with the built environment. FM affects the health, well-being and quality of life of the world's societies and population through services, management and deliveries" [24]. This is especially important when comparing the phases and subcategories of critical building design control [25] in delivering IAQ and services [26–29] covered by FM. In addition, according to Hodges [30], "the facility manager is in a unique position to view the entire process and is often the leader of the only group that has influence over the entire life cycle of a facility." This also includes the responsibility to manage individual and community health [31–33].

At present, the central issues of FM practice consist of place or facility, people or user of the building, and process or activities in the facility [26,34,35]. On the other hand, FM can involve several strategic issues such as property asset portfolio management, strategic property decision, and facility planning and development, which are related to policy and strategic planning of the organization [26,36–38]. Rondeau et al. [39] state that facility managers' "role is to ensure that the customer and the corporation have an on-time and on-budget project with the best possible site, space, facilities, furnishings and support systems to serve their needs today and tomorrow." Accordingly, they are responsible for identifying, securing and working with qualified and high-quality service and product providers. The benefits of sustainability and green building practices in FM are well established [30], however, some researchers [40–43] state that although awareness of sustainability is high, the participants' efforts in implementation are low, due to different barriers [44,45]. Looking from the perspective of the sustainability 'triple bottom line', social and environmental factors often take a back seat to the overall strategy of the organization, due to the need to build at the lowest cost.

Keeping in mind the entire life cycle of the facility, facility managers are in the position to view the entire process [30]. In this regard, different authors [30,46–48] indicate the strategic role of FM in achieving sustainability. This implies an integrative role [48] of the facility manager in understanding the needs of the end-users of the facility and acting as an advisor to owners, designers, consultants and contractors about sustainability requirements in a project. In line with the sustainability aspects of the facilities, one group of the FM engagements includes activities related to the prevention and maintenance of IAQ.

In recent years, the more proactive approach of Urban FM has contributed to an increase in health and well-being [49], which is important in helping the physical causes and symptoms of poor health, as well as the social, economic and environmental components of individual, community, and overall well-being. With community-based participation and collaborative governance processes, the co-creation processes capitalize on a local community's assets, capital, inspiration and potential, resulting in both the enhancement of housing quality and the creation of quality public spaces [50,51]. This in turn contributes to people's health, happiness and well-being, and the community's resilience and sustainability [50]. In addition, health-directed design interventions in cities and facilities are related and create scope for crossovers among different professions on the urban level [33].

The American Society of Heating, Refrigerating and Air-Conditioning Engineers (ASHRAE) defines acceptable IAQ in its Standard 62-1989 as "air in which there are no known contaminants at harmful concentrations as determined by cognizant authorities and with which a substantial majority (90 percent or more) of the people exposed do not express dissatisfaction." In order to act effectively in this area, buildings need to be viewed as a "habitat" in which the IAQ "ecosystem" model consists of the occupants, their activities, the air pathway and the heating, ventilation and air conditioning (HVAC) system, the building envelope, and its environmental setting [22]. All the aforementioned elements comprise an interlinking model which cannot be dealt with individually.

Problems with the IAQ usually start with the building occupants' complaints on discomfort, headaches, nausea, dizziness, sore throats, dry or itchy skin, sinus congestion, nose irritation or excessive fatigue [22]. Depending on the character of the complaints, they can be divided into two groups, namely: sick building syndrome (SBS) and building related illness (BRI). SBS is a building that has a condition that can make its occupants uncomfortable, irritated or even ill, while BRI is deemed as a building associated with a clinically verifiable and diagnosable disease. According to Bas [52], "buildings may have as many as 900 contaminants indoors with thousands of sources—including new furniture, cleaning agents, smoking, new building materials, pesticides and even perfume and other cosmetics." Many contaminants are microbiological or otherwise organic, triggering asthma and allergies.

To prevent the indoor air pollution that could be related to the building construction process, Levin [25] identified mayor phases and subcategories of critical building design control that cover site planning and design, overall architectural design, ventilation and climate control, materials selections and specifications and construction process and initial occupancy.

Comparing selected frameworks in prevention of indoor air pollution [25] and FM services and responsibilities developed by Chotipanich [26], Rondeau et al. [39], Barret and Finch [37] and Jensen et al. [53], individual FM activities were identified which have a decisive role in preventing the occurrence of indoor air pollution. These include:

- Identification of user needs and definition of air quality performance measures within the briefing phase of the construction process;
- Architectural, engineering and construction management design, planning and performance as well as users' needs and preferences within the design phase;
- Purchasing, contact control and negotiations, delivering performance targets and creation of trust in the contract phase; and
- New construction/reconstruction and construction management, applying the building codes/standards and FM's role as "facilitator" within the construction phase (see Figure 2).

In addition to pollutants from the external environment, construction materials and materials used for interior design and furnishing, the so-called subsidiary means used for construction, also have an impact. Dealing with the issues of the construction and subsidiary materials, Burroughs and Hansen [22] identified building materials of particular concern divided in accordance with the construction phase (see Figure 3).

Linear model of construction process	**AIR QUALITY** Major phases and subcategories of critical building design control **(Levin, 1990)**	**FACILITY MANAGEMENT** Cluster of support services **(Chotipanich, 2004)**	Nine Facility Management related job responsibilities **(Rondeau, et al., 2006)**	Typical facility management activities **(Barrett & Finch, 2014)**	10 Questions concerning sustainable buidling renovation **(Jensen et al., 2018)**
BRIEFING	Pollutant-generating activities Mix of users in building Standards development	Real estate/Property portfolio strategy Location searches and selection Acquisition and disposal of sites and buildings Space allocation, utilisation and relocation Long-term resource planning Mid-term resource planning Work programming Development planning	Long-range facility planning Facility financial forecasting and management	Strategic space planning Set (corporate) planning standards and guidelines **Identify user needs** **Define performance measures**	Understanding the **users' needs and drivers** for renovation. Renovation planning with focus on durability, economy, environment, comfort and aesthethics.
DESIGN	Vehicle access Building openings Operable windows Envelope and structural materials Penetration of volumes Basement dehumidification Smoking lounges	Facility planning/master planning Space planning Space configuration and reconfiguration Landscaping **Health and safety**	Real estate acquisition and/or disposal Interior space planning, work specification and installation and space management **Architectural and engineering planning and design**	Furniture layouts **New building design and management**	Pre-evaluation of **building's design, condition and performance** in the planning and renovation. **People's needs and preferences** Understand the existing building design and architectural expression.
CONTRACT	Low-emitting materials Preventive installation procedures In-place curing	Administration and management Budget and cost control **Purchasing and Contract control and negotiation** Office furniture and stationary provision		**Service level agreement (SLA)** **Negotiations and management of leases**	**Putting the performance targets** Evaluating performance improvements. Measuring users' satisfaction Focus on Relational contracting* **Creation of trust**, using strategic, tactical and operational tools.
CONSTRUCTION	Design documentation and commissioning	**New building** Extending and Alteration Demolition	**New construction and/or renovation work**	**New building and construction management** Advice on property investment Control of capital budget	Performance based renovation targets and requirements. **Applying the building codes/standards to existing buildings** High upgrading requirements **FM's role as a 'renovation facilitator'** to navigate different policies and stakeholder groups
OCCUPANCY	Ambient air quality Local source control Special ventilation Initial occupancy period Dilution by outdoor air Air intakes Exhaust locations Air cleaning Space air distribution Heat recovery Microbial control	Space use audit and monitoring Landscape maintenance Cleaning and housekeeping M&E/operating/run plant Energy distribution and management Waste disposal and environment management Pest control Disaster prevention and recovery **Health and safety** Security	Annual facility planning (tactical planning) Maintenance and operations of the physical plant Telecommunications integration, security, and general administrative services	Monitor space use Select and control use of furniture Computer-aided facility management Run and maintain plant Maintain building fabric Manage and undertake adaptation Energy management Security Voice and data communication Control operating budget Monitor performance Supervise cleaning and decoration Waste management and recycling	Evaluating technical condition Preparing renovation plans Focus on sustainable building technologies Post-occupancy analysis

* Integrated projec delivery, Partnering and Strategic partnerships/alliences

Figure 2. Services and categories of facilities management (FM) approach and delivering indoor air quality (IAQ) compared and redistributed according to the phases of the linear model of the construction process.

Site preparation and foundations covers	**Envelope**	**Mechanical systems**	**Interiors and finishes**
Soil treatment insecticides Foundation waterproofing Especially oil derivates High levels of dirt and dust	Wood preservatives Concrete sealers Curing agents Caulking Sealants Glazing compounds Joint filters.	Duct sealants and Mastics	Subfloor or underlayment Floor or carpet adhesives Carpet backing or pad Carpet or resilient flooring Wall coverings Both vinyl and fabric Adhesives Paints Stains Panelling Partitions Furnishings Ceiling tiles Gypsum dust Window coverings

Figure 3. Building materials of particular concern [22].

It is important to mention that the effects of many of chemicals emitted from the products are still not fully understood, but many are known or suspected human irritants, and some are suspected human carcinogens. Considering all those mentioned, the activities of FM must cover the process of commissioning, which is emerging as a critical component of successful completion of the construction process and involves the aggressive overview of each stage of the construction project in order to assure the conformance of the project to the design, resulting in a building that performs according to intent [22]. On the other side, every case is specific and may indicate distinct situations and circumstances that lead to the occurrence of indoor air pollution, the knowledge of which could improve the construction process and prevent future cases and similar situations.

4. Social and Affordable Housing Legislative in Serbia

In Serbia, social housing is defined in the Law on Social Housing [54] as "housing of an appropriate standard that is provided with state support to households that for social, economic and other reasons cannot provide housing under market conditions." It has a "housing of an appropriate standard that is provided with state support to households that for social, economic and other reasons cannot provide housing under market conditions." It has a "residual" role [55] which means it is primarily intended for the most socially vulnerable categories, unlike in some developed European countries. Master Plan of Belgrade (MPB) 2021 from 2003 [56] played a key role in introducing and defining the concept of social housing in Serbia, following the example of developed European countries. This Plan identifies socially vulnerable categories of society that need assistance in finding housing and provides certain guidelines for design and construction standards, as well as criteria for determining locations. The Plan has defined a list of 58 locations with a total area of 228.6 ha planned for social housing in Belgrade.

The adoption of the Law on Social Housing in 2009 was followed by the National Social Housing Strategy [57] and The Decree on Standards and Norms for Planning, Design, Construction and Conditions for the Use and Maintenance of Apartments for Social Housing [58]. This is important to mention because in the period from 2001 to 2014, several social housing programs were implemented in Serbia, despite the lack of a defined comprehensive housing system, as well as appropriate technical regulations for planning and designing this type of housing [55,59].

The program of construction of solidarity apartments 2001–2005 was implemented by the Assembly of the City of Belgrade with the public fund for financing the construction of apartments. The project of constructing 2000 socially non-profit apartments in Belgrade started in 2005 by adoption of the Decision on the conditions of sale of 2000 socially non-profit apartments in Belgrade [60]. Since social housing is seen as a good of public interest, a number of public architectural and urban competitions were announced in cooperation with the Society of Architects of Belgrade. On the other side, public competitions are considered directly related to the public interest, including the widest professional public, and therefore require the highest level of professional and social responsibility. However, in the competitions conducted by the City of Belgrade, the topic of social housing was mostly problematized from the aspect of economy, without considering other specific requirements in that housing category [55].

This is not surprising because, according to Djokic et al. [59], "social housing has been often seen as a measure of the social care system, a tool of poverty reduction, in achieving social justice and ensuring the fundamental human right to housing, but seldom as an instrument of economic development." However, this should not limit the possibility of implementing a strategic framework for improving environmental sustainability in housing with increased resilience and adaptability of housing, the provision of healthy living conditions and a healthy environment, and the reduction of waste from the use of heating and cooling energy, coupled with carbon dioxide emissions, reduction of water and soil pollution, adequate use of materials and waste recycling.

The existing technical regulations for the planning, design and construction of social housing in Serbia do not sufficiently consider specific guidelines in the field of energy optimization and IAQ that would be usable in the development of planning solutions, which is a significant limitation [61]. This is seen in the gap between the recommendations of sustainability in the planned development of the social housing sector in Serbia and their practical application. In addition, "there is a large gap between the inherited experiences of "quantitative" satisfaction of housing needs through mass state housing and "qualitative", long-term goals of sustainable development" [61]. Furthermore, there are no regulations aimed at improving and maintaining indoor air quality. Unlike outdoor air quality, the monitoring and assessing of IAQ are not regulated by legal acts, only guidelines and recommendations related to certain pollutants.

It is also necessary to mention that public buildings constructed within the period between 2010 and 2012 were built in accordance with the special conditions, applying the Law on Encouraging the Construction Industry in Serbia in the Conditions of Economic Crisis [62]. This Law had a fixed term with emphasis on encouraging the development and employment of domestic construction companies and providing liquidity to this sector to strengthen the development and employment of domestic companies engaged in the production of construction materials. The buildings covered by this Law are fully or partially financed from the budget of the local self-government unit, the autonomous province (i.e., the Republic of Serbia) and include schools and kindergartens, hospitals and other health facilities, flats, sport objects, facilities for the purpose of performing cultural activities, highways and other state roads and other objects of public importance. In accordance with the above, the construction of facilities of public interest were completed exclusively by domestic construction companies and with domestic construction materials (70% of the total amount of construction materials and equipment).

5. Dr Ivan Ribar Settlement and "Smelly Buildings"

At the initiative of the City of Belgrade in 2010, the realization of the Program for the construction of social and affordable apartments at the location Dr Ivan Ribar in New Belgrade continued. As was the case with other locations planned for this purpose, an architectural-urban competition was announced by the Association of Belgrade Architects for this location in 2011. In total, 53 works were submitted to the competition, and the first prize was awarded to the RAUM architects from Belgrade [63]. According to the jury, RAUM's proposal (see Figure 4) met all the criteria set by the tender conditions, especially in terms of the expected structure, square footage and organization of housing units in the buildings.

Figure 4. Awarded competition proposal by RAUM architects for the Dr Ivan Ribar social and affordable housing settlement in New Belgrade. Source: RAUM architects, 2011.

Housing units are organized into six identical buildings of two-tract type, considered a rational solution from the aspect of project implementation. By forming two types of inner courtyards, a high quality of housing was achieved in the settlement, namely green areas with rest areas and a modified hilly relief that regulates the views and provides privacy and peace [63]. The free area on the western part of the plot, which is unfavorable for construction, is arranged as a green park area where sports and recreational facilities and sports fields are located.

The project was realized with minor changes (see Figure 5) by the Construction Directorate of Serbia according to the first-prize winning competition solution, and the first apartments were occupied in October 2013. As planned, the apartments in this complex belong to the categories of social and affordable housing but are built so that there is no difference in terms of quality of construction.

Figure 5. Urban housing block Dr Ivana Ribara, Block 72, 707 apartments. Completed: autumn 2013. Source: RAUM Architects Facebook page. Author of the photography: Dragan Babovic.

The main investor of this project was the Construction Directorate of Serbia (CDS), which was also responsible for the sale of the apartments, as well as for guaranteeing their quality. According to the Law [62], the CDS engaged several domestic construction companies to construct particular buildings within the complex. The engaged companies independently carried out the stages of the process on individual facilities that were the subject of their work, which included the procurement and purchase of construction materials and equipment produced in Serbia (70% of the total amount of construction materials and equipment). The domestic construction company, Koto d.o.o., was engaged and responsible for the building of Buildings 1 and 3, which would later be known as "smelly buildings."

5.1. "Smelly Buildings"

Following the archives of digital editions of the Serbian daily press and the Beobuild forum [64], the first complaints about the unpleasant smell in the apartments appeared in July 2014. These were preceded by announcements of satisfied apartment owners who moved in during October and November 2013. At that time, most of the discussions at the forum with the topic, "State housing—Settlement in Block 72 (Dr Ivana Ribara)" [64], were about the construction process, the characteristics of apartments, the aesthetics of the accepted solution and the procedure for redistribution of apartments to different categories of tenants.

The problem occurred in 220 apartments within Buildings 1 and 3 (No. 80, 82, 100 and 102), in which the tenants complained of an extremely strong, unpleasant smell of unknown origin [65]. The mentioned buildings were named "smelly buildings." After addressing the Construction Directorate of Serbia, which was responsible for the construction and sale of these apartments, there was a need for a quick response to prevent potential negative effects on the health of tenants. To determine the causes and character of the harmful vapors, the Vinca Institute of Nuclear Sciences, Laboratory of Physical Chemistry and the Institute of Public Health of Belgrade were engaged.

The analyses were performed during July and August 2014. Sampling and testing of air quality in 24-h samples were made at the following sampling sites: indoors in twenty apartments in the subject buildings, one grocery store, control measurements in two apartments where the presence of unpleasant odors was not detected and outdoors at two sampling sites near the subject buildings [18]. Qualitative and quantitative analysis of IAQ in apartments and buildings in which the presence of unpleasant odors was found, as well as control measurements, included the following parameters: formaldehyde, acrolein, polycyclic aromatic hydrocarbons (PAH) of solid and gaseous phases, volatile organic compounds, benzene, toluene, ethylbenzene and isomers of xylene (BTEX) and phenolic compounds. Key results arose from the analysis of the polyurethane foam used for gaseous phase PAH sampling by the GC-MSD technique. This analysis showed the presence of phenols and phenolic compounds, namely phenol, 3-ethyl phenol and 3-methyl phenol, only in the samples taken from the apartments where characteristic odors where reported.

Tests carried out at the Vinca Institute for Nuclear Research showed that the pollution comes from the oil used to coat the formwork boards used for pouring concrete elements [66]. A part of the report of the Vinca Institute, which illustrates the origin of pollution, is shown in Figure 6. Mass spectrum of tested samples confirms the presence of phenol, 3-ethyl phenol and 3-methyl phenol in oil and concrete samples, and as fingerprint, confirms the source of contamination.

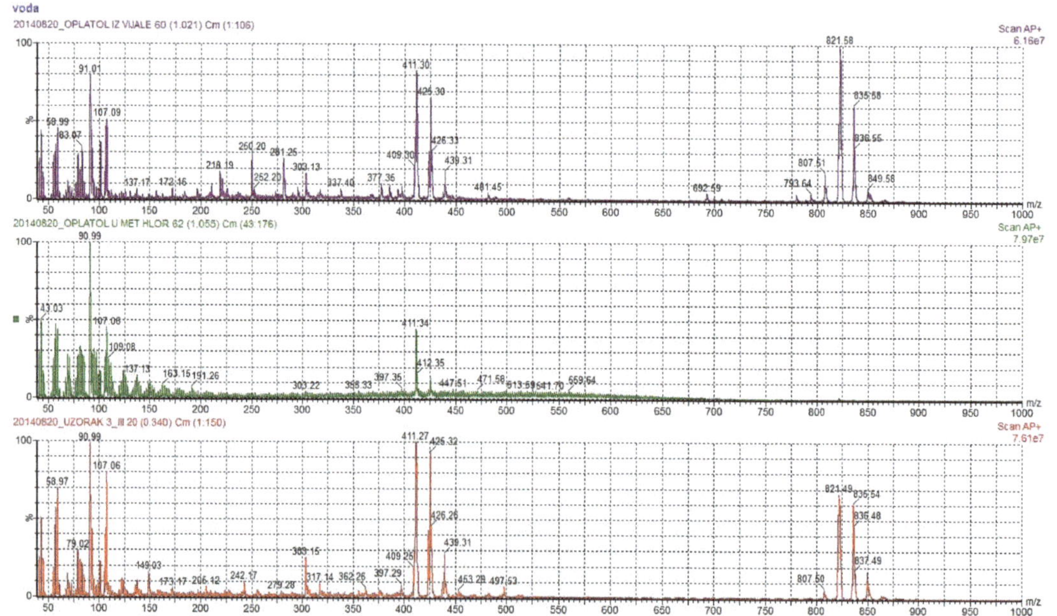

Figure 6. Mass spectrum of tested oil oplatol in water-up, in methanol-middle and mass spectrum of concrete-down sample. Source: [66].

Comparison with the control measurements showed that the presence of phenolic compounds in the indoor air, as well as the significant presence of phenanthrene, stand out from all the examined parameters. The presence of formaldehyde, acrolein, toluene and PAH of gaseous phases, except phenanthrene, was observed in slightly higher concentrations than in control measurements in ambient air, but, although somewhat higher, the values did not deviate significantly from those observed in control apartments where characteristic odors were not observed. Higher concentrations of these pollutants in indoor air are characteristic of newly built facilities equipped with new interiors [18].

The analysis of the content of the disputed oplatol oil and concrete samples, as stated in the report of the Vinca Institute, unequivocally established that the source of the unpleasant odors was concrete poured into the disputed facilities, while the origin of the pollution was oplatol oil used to coat the formwork boards used for pouring concrete elements. Unlike phenanthrene, which almost certainly originates from contaminated concrete but may have other sources, the presence of 3-ethyl phenol and 3-methyl phenol is characteristic of the indoor air of buildings in which the disputed oplatol was used and can be used as a kind of marker (fingerprint) for oplatol-induced pollution.

Based on the results of the conducted examinations, the expert team at the Institute of Public Health of Belgrade assessed that the apartments in which unpleasant odors were felt did not meet the sanitary-hygienic conditions for permanent residence of people [18].

The Toxicology Department of the Military Medical Academy (MMA) in Belgrade was also engaged to conduct an analysis of the health condition of the occupants of these facilities. The testing covered one-third of the tenants and the results showed that the values of phenol and hippuric acid in urine samples were significantly increased in samples from several tenants, especially children, as well as in samples taken from workers who were exposed to these substances at their workplaces [67].

A similar problem was reported in several other apartment buildings in Belgrade, Novi Sad, Niš and other cities in Serbia, as well as the Bora Stanković theatre in Vranje and a stem cell bank in Belgrade. All these buildings were built around the same time when a disputable batch of "oplatol" was on the market. This 'epidemic' spread all over Serbia during the few months when the notorious batch of oplatol was used. Even though only one batch of oplatol was contaminated, it caused enormous damage which illustrates how a little negligence can produce massive financial and emotional (tenants) damage.

5.2. Problem Solving Process

Immediately after receiving the results of the mentioned analyses, the Construction Directorate of Serbia (CDS) initiated a series of procedures intended to solve the problem. Everything was done in cooperation with the tenants, in order to meet their demands. For the needs of internal communication, the citizens of this settlement launched a website [65] as well as a blog related to the topic of smelly buildings, called "poisonous buildings" [68].

Given the situation, two problems had to be resolved—resettlement and temporary or permanent accommodation of the occupants of the problematic facilities, and the establishment of a procedure for further treatment of these facilities. Negotiations between the tenants and the Directorate covered a number of topics that needed to be agreed upon, in order to further treat the tenants as clients of the Directorate. Accordingly, the Directorate offered to finance the rental of the apartment until the buildings were rehabilitated (five euros per square meter of the purchased apartment, with the possibility of increasing the amount), to pay relocation costs, to pay a deposit for renting an apartment and an agency commission and all overhead costs in the purchased apartment in two smelly buildings. On the other side, the tenants asked that the problem in the buildings not be treated as an unpleasant odor, but as a danger to human health, to determine whether "oplatol" or some other substances such as concrete additives really evaporate in the apartments, and to ensure that the Directorate offered clear criteria and methods of rehabilitation to convince the tenants that their apartments will be safe. In order to find the solution for treating the subjected buildings, the CDS has consulted several scientific institutions such as University of Belgrade—Faculty of Technology, Vinca Institute of Nuclear Sciences, the Royal Institute in Belfast and the Heidelberg Institute. However, it was determined that there is no ready solution to eliminate phenol. A special problem in the search for a solution was the fact that the chemical transformation of detected pollutants over time is very possible and that each treatment, as well as the passage of time, carries with it the possibility of a new generation of pollutants originating from the initial pollution [18].

After a two-year process of negotiating with apartment owners and seeking solutions for the rehabilitation of buildings, the CDS decided to build new facilities for property owners, and to renovate the disputed buildings by 'extracting' pollution from them based on the methodology developed by Vinca Institute of Nuclear Sciences. The rehabilitation plan includes chemical treatment with hydrogen peroxide, treatment with UV lamps and ozone of all concrete walls where the presence of pollutants was detected, which originated from the coating oil [67].

Research has also shown that the state institutions, the Ministry of Construction, Transport and Infrastructure of the Republic of Serbia and the CDS, have accepted responsibility because the harmful insulating material oplatol was used in the construction of that part of the settlement. The core of the problem is the retention of harmful materials on the market and the possibility of their trade, but also the issue of the work of inspection bodies and bodies that control the trade and issue certificates for certain types of materials.

Six years after the occurrence of the problem, the works on the rehabilitation of the "smelly buildings" are nearing completion. The management of the CDS has not yet decided on the purpose of these buildings, but it is assumed [69] that one building will function as a hostel, while apartments in another building will be rented.

6. Discussion

Dr Ivan Ribar settlement in New Belgrade and its "smelly buildings" is one of many examples of the so called "least cost" mentality, which has led to construction decisions that prioritized cost savings, but whereby multiple prices of the entire investment were ultimately paid due to omissions. In this case, it was a legal obligation to purchase certain material from a domestic manufacturer whose production was not adequately controlled. This led to the appearance of a certain amount of material that did not correspond to the quality for which there was a certificate but possessed substances that later turned out to be toxic and could endanger the health of the inhabitants.

Considering the linear model of the construction process, the problem occurred within the contract phase, during the purchase of the specific type of product. The case of smelly buildings indicates that stricter control of products used in the construction of facilities and interior equipment, and especially the so-called byproducts, is needed. The significance of the case of the smelly building is in the fact that it drew the attention of the scientific and professional public to the importance of IAQ control, but also to the importance of outdoor air quality control because these two are inextricably linked.

These experiences indicate the need to further work on improving the existing legislation governing the field of air quality control, especially in light of the emergence of a new generation of pollutants, as well as the adoption of legislation and development of guidelines for IAQ. Special attention should be paid to the critical group of materials used in and during construction, in order to investigate the various effects, they may have on IAQ, and thus on endangering the health of residents.

Having in mind the complexity of the construction process and the domain of IAQ, FM with its scopes and activities have the potential to accompany all phases, including the preparation of short-term and long-term FM maintenance plans. Facility managers and those with facility-related responsibilities must make the risk analysis, which also addresses indoor environmental risks. This includes audit reports that trigger new maintenance needs, but, more importantly, to change the protocols and start immediate actions to prevent IAQ related incidents from occurring in the first place. Auditing health and safety issues should create measures to fulfil the regulatory requirements and control regular measures.

The research indicates the importance of the engagement of facility managers during the entire construction process and the building life cycle as an independent actor. This is especially important in contract and construction phases. In the contract phase, FM could control purchasing, contracting and negotiation processes to achieve adopted building and quality performance targets. On the other side, independent construction management could include control in the application of building codes and standards, as well as checking the quality of materials and equipment.

This is especially important for public investments such as social and affordable housing complexes like in Belgrade's case of the Dr Ivan Ribar settlement, which is not an isolated case of public construction facing this problem. In the period from 2012 to 2015, cases of reconstruction of the theatre and maternity hospital in Vranje, a residential complex in Novi Sad and a stem cell bank in Belgrade were recorded. On the one hand, the situation can be seen from the aspect of irresponsible disposal and spending of public funds, and on the other hand, it is necessary to allocate additional funds to solve the problem. The example of the Dr Ivan Ribar residential complex pointed out the following groups of additional costs and material losses:

- Researching the problem and determining the methodology that will be applied for its remediation, including engagement of domestic and foreign scientific and professional institutions
- Assistance to aggrieved tenants until the final solution is found, including
    - costs of the rentals
    - relocation costs
    - deposit for renting an apartment
    - and agency commission costs
    - all overhead costs in the purchased apartment
- New building design and construction, into which a certain number of damaged tenants will move
- Rehabilitation and the remediation of the problem in existing facilities
- Object conversion and new usage

In addition to the above costs, one should keep in mind the creation of a bad reputation for the entire settlement, which caused a reduction in the market price of real estate, but also stress and suffering for tenants and their families over a period of several years (endangered health, relocation, postponement of the final solution, negotiations, etc.).

On the other side, FM practice is likely to be case-specific by nature, dealing with the diversities of facility, organization, business sector, surrounding environment and context, and circumstance [26]. From that perspective, Urban FM is very important to improve citizens' health and well-being, especially from the perspectives of healthy buildings, accessibility and services to the vulnerable population. Using the community-based approach of FM is important to recognize the risks, enable the enhancement and contribute to a healthier society. Similarly, understanding the needs of neighborhoods requires engaging with citizens, leading to their empowerment to understand, recognize and report the sickness parameters, and consequently, to better support a healthy environment.

Author Contributions: Conceptualization, M.V., A.T.S. and A.S.; methodology M.V., A.T.S. and A.S.; validation, M.V., A.T.S. and A.S.; investigation M.V. and A.T.S.; resources, M.V., A.T.S. and A.S.; writing—original draft preparation, M.V.; writing—review and editing, M.V., A.T.S. and A.S.; visualization, M.V.; supervision, A.T.S. and A.S. All authors have read and agreed to the published version of the manuscript.

Funding: This research received no external funding.

Institutional Review Board Statement: Not applicable.

Informed Consent Statement: Not applicable.

Data Availability Statement: Used data belong to the domain of publicly available information, and in this paper their systematization and critical interpretation is performed.

Acknowledgments: We take this opportunity to express our gratitude to Roberto Alonso González Lezcano, Guest Editor for inviting us to publish our research within Sustainability Special Issue entitled: "Sustainable Building and Indoor Air Quality".

Conflicts of Interest: The authors declare no conflict of interest.

References

1. UN General Assembly. *Report of the Special Rapporteur on the Issue of Human Rights Obligations Relating to the Enjoyment of a Safe, Clean, Healthy and Sustainable Environment*; UN: New York, NY, USA, 2018.
2. Prüss-Ustün, A.; Wolf, J.; Corvalan, C.; Bos, R.; Neira, M. *Preventing Disease through Healthy Environments: A Global Assessment of the Burden of Disease from Environmental Risk*; World Health Organisation: Geneva, Switzerland, 2016.
3. WHO-World Health Organisation. *The Impact of the COVID-19 Pandemic on Noncommunicable Desease Resources and Services: Results of Rapid Assesment*; World Health Organisation: Geneve, Switzerland, 2020.
4. WHO-World Health Organisation. *Fact Sheet No. 292-Household Air Pollution and Health*; World Health Organisation: Geneva, Switzerland, 2014.
5. Zijlema, W.L.; Wolf, K.; Emeny, R.; Ladwig, K.-H.; Peters, A.; Kongsgård, H.; Hveem, K.; Kvaløy, K.; Yli-Tuomi, T.; Partonen, T.; et al. The association of air pollution and depressed mood in 70,928 individuals from four European cohorts. *Int. J. Hyg. Environ. Health* **2016**, *219*, 212–219. [CrossRef] [PubMed]
6. Tran, V.V.; Park, D.; Lee, Y.-C. Indoor Air Pollution, Related Human Diseases, and Recent Trends in the Control and Improvement of Indoor Air Quality. *Int. J. Environ. Res. Public Health* **2020**, *17*, 2927. [CrossRef] [PubMed]
7. Cai, Y.; Hansell, A.L.; Blangiardo, M.; Burton, P.R.; de Hoogh, K.; Doiron, D.; Fortier, I.; Gulliver, J.; Hveem, K.; Mbatchou, S.; et al. Long-term exposure to road traffic noise, ambient air pollution, and cardiovascular risk factors in the HUNT and lifelines cohorts. *Eur. Heart J.* **2017**, *38*, 2290–2296. [CrossRef] [PubMed]
8. European Environmental Agency EEA. *The European Environment—State and Outlook 2020: Knowledge for Transition to a Sustainable Europe*; Publication Office of European Union: Luxembourg, 2020.
9. Hulin, M.; Simoni, M.; Viegi, G.; Annesi-Maesano, I. Respiratory health and indoor air pollutants based on quantitative exposure assessments. *Eur. Respir. J.* **2012**, *40*, 1033–1045. [CrossRef] [PubMed]
10. UN Department of Economic and Social Affairs. *Sustainable Development. The 17 Goals*; United Nations: New York, NY, USA, 2020; Available online: https://sdgs.un.org/goals (accessed on 28 November 2020).
11. Tabrizia, J.S.; Bazargani, H.S.; Ardakani, M.A.; Saadati, M. Developing safe community and healthy city joint model. *J. Inj. Violence Res.* **2020**, in press.
12. De Leeuw, E.; Simos, J. (Eds.) *Healthy Cities The Theory, Policy, and Practice of Value-Based Urban Planning*; Springer: New York, NY, USA, 2017.
13. Patino, E.D.L.; Siegel, J.A. Indoor Environmental Quality in Social Housing: A Literature Review. *Build. Environ.* **2018**, *131*, 231–241. [CrossRef]
14. Thompson, H.; Petticrew, M.; Morrison, D. Health effects of housing improvement: Systematic review of intervention studies. *BMJ* **2001**, *323*, 187–190. [CrossRef]
15. Saegert, S.C.; Klitzman, S.; Freudenberg, N.; Cooperman-Mroczek, J.; Nassar, S. Healthy housing: A structured review of published evaluations of US interventions to improve health by modifying housing in the United States. *Am. J. Piublic Health (AJPH)* **2003**, *93*, 1471–1477. [CrossRef]
16. Rauh, V.; Landrigan, P.J.; Claudio, L. Housing and health—Intersection of poverty and environmental exposures. *Ann. N. Y. Acad. Sci.* **2008**, *1136*, 276–288. [CrossRef]
17. Bashir, S.A. Home Is Where the Harm Is: Inadequate Housing as a Public Health Crisis. *Am. J Public Health* **2002**, *92*, 733–738. [CrossRef]
18. Milutinovic, M.; Sostaric, A.; Mladenovic, S.; Slepcevic, V. Ispitivanje Kvaliteta Vazduha Zatvorenog Prostora. Slucaj "Smrdljive Zgrade". In *Dani Zavoda 2016. 26. Strucna Konferencija Kvalitet Vazduha. Monitoring, Modelovanje, Unapredjenje*; Beograd, Serbia, 2016.
19. WHO. *WHO Guidelines for Indoor air Quality: Dampness and Mould*; WHO Regional Office for Europe: Copenhagen, Denmark, 2009.
20. WHO. *WHO Guidelines for Indoor air Quality: Selected Pollutants*; WHO Regional Office for Europe: Copenhagen, Denmark, 2010.
21. WHO Regional Office for Europe. *Evolution of WHO Air Quality Guidelines: Past, Present and Future*; WHO Regional Office for Europe: Copenhagen, Denmark, 2017.
22. Burroughs, H.; Hansen, S.J. *Managing Indoor Air Quality*, 5th ed.; Taylor & Francis Ltd.: London, UK, 2011.
23. IFMA. What is Facility Management? 2018. Available online: https://www.ifma.org/about/what-is-facility-management (accessed on 10 April 2020).
24. International Organisation for Standardisation. *Facility Management—Management Systems—Requirements with Guidance for Use (ISO standard No. 41001:2018)*; ISO: Geneva, Switzerland, 2018.
25. Levin, H. Critical Building Design Factors for Indoor Air Quality and Climate: Current Status and Predicted Eends. *Indoor Air* **1990**, *1*, 79–92. [CrossRef]
26. Chotipanich, S. Positioning facility management. *Facilities* **2004**, *22*, 364–372. [CrossRef]
27. Atkin, B.; Brooks, A. *Total Facility Management*; John Wiley and Sons: Hoboken, NJ, USA, 2015.
28. Baricic, A.; Salaj, A.T. The impact of office workspace on the satisfaction of employees and their overall health-research presentation. *Zdr. Vestn.* **2014**, *83*, 217–231.
29. Tolman, A.; Parkkila, T. FM tools to ensure healthy performance based buildings. *Facilities* **2009**, *27*, 469–479. [CrossRef]

30. Hodges, C.P. A facility manager's approach to sustainability. *J. Facil. Manag.* **2005**, *3*, 312–324. [CrossRef]
31. Alexander, K.; Brown, M. Community-based facilities management. *Facilities* **2006**, *24*, 250–268. [CrossRef]
32. Lindkvist, C.; Salaj, A.T.; Collins, D.; Bjorberg, S.; Haugen, T.B. Exploring urban facilities management approaches to increase connectivity in smart cities. *Facilities* **2020**. Vols. ahead-of-print, no. ahead-of-print. [CrossRef]
33. Nijkamp, J.E.; Mobach, M.P. Developing healthy cities with urban facility management. *Facilities* **2020**, *38*, 819–833. [CrossRef]
34. Abdullah, S.; Sulaiman, N.; Latiffi, A.A.; Baldry, D. Integration of Facilities Management (FM) Practices with Building Information Modeling (BIM). December 2013. Available online: https://www.researchgate.net/publication/260036097_Integration_of_Facilities_Management_FM_Practices_with_Building_Information_Modeling_BIM (accessed on 10 July 2020).
35. Mudrak, T.; van Wagenberg, A.; Wubben, E. Assessing the innovative ability of FM teams: A review. *Facilities* **2004**, *22*, 290–295. [CrossRef]
36. Van der Voordt, T. Facilities management and corporate real estate management:FM/CREM or FREM? *J. Facil. Manag.* **2017**, *15*, 244–261. [CrossRef]
37. Barrett, P.; Finch, E. *Facilities Management The Dynamics of Excellence*, 3rd ed.; Willey Blackwell: Oxford, UK, 2014.
38. Jensen, P.A.; van der Voordt, T.J. Healthy workplaces: What we know and what else we need to know. *J. Corp. Real Estate* **2019**, *22*, 95–112. [CrossRef]
39. Rondeau, E.P.; Brown, R.K.; Lapides, P.D. *Facility Management*, 2nd ed.; Wiley: Hoboken, NJ, USA, 2006.
40. Son, H.; Kim, C.; Chong, W.K.; Chou, J.-S. Implementing sustainable development in the construction industry: Constructors' perspectives in the US and Korea. *Sustain. Dev.* **2011**, *19*, 337–347. [CrossRef]
41. Abidin, N.Z. Investigating the awareness and application of sustainable construction concept by Malaysian developers. *Habitat Int.* **2010**, *34*, 421–426. [CrossRef]
42. Chong, W.K. Understanding and interpreting baseline perceptions of sustainability in construction among civil engineers in the United States. *J. Manag. Eng.* **2009**, *25*, 143–154. [CrossRef]
43. Annunziata, E.; Testa, F.; Iraldo, F.; Frey, M. Environmental responsibility in building design: An Italian regional study. *J. Clean. Prod.* **2016**, *112*, 639–648. [CrossRef]
44. Ástmarsson, B.; Jensen, P.A.; Maslesa, E. Sustainable renovation of residential buildings and the landlord/tenant dilemma. *Energy Policy* **2013**, *63*, 355–362. [CrossRef]
45. Lindkvist, C.; Karlsson, A.; Sørnes, K.; Wyckmans, A. Barriers and Challenges in nZEB Projects in Sweden and Norway. *Energy Procedia* **2014**, *58*, 199–206. [CrossRef]
46. Aaltonen, A.; Maattanen, E.; Kyro, R.; Sarasoja, A.-L. Facilities management driving green building certification: A case from Finland. *Facilities* **2013**, *31*, 328–342. [CrossRef]
47. Nousianien, M.; Junnila, S. End-user requirements for green facility managemen. *J. Facil. Manag.* **2008**, *6*, 266–278. [CrossRef]
48. De Paula, N.; Arditi, D.; Melhado, S. Managing sustainability efforts in building design, construction, consulting, and facility management firms. *Eng. Constr. Archit. Manag.* **2017**, *24*, 1040–1050. [CrossRef]
49. Salaj, A.T.; Lindkvist, C. Guest editorial. *Facilities* **2020**, *38*, 761–763. [CrossRef]
50. Vukmirovic, M.; Gavrilovic, S. Placemaking as an approach of sustainable urban facilities management. *Facilities* **2020**, *38*, 801–818. [CrossRef]
51. Salaj, A.T.; Gohari, S.; Senior, C.; Xue, Y.; Lindkvist, C. An interactive tool for citizens' involvement in the sustainable regeneration. *Facilities* **2020**, *38*, 859–870. [CrossRef]
52. Bas, E. *Indoor Air Quality. A Guide for Facility Managers*, 2nd ed.; The Fairmont Press: Lilburn, GA, USA, 2004.
53. Jensen, P.A.; Meslesa, E.; Berg, J.B.; Thuesen, C. 10 questions concerning sustainable builidng renovation. *Build. Environ.* **2018**, *143*, 130–137. [CrossRef]
54. Law on Social Housing. In *Official Gazette of Republic of Serbia No. 72/2009*; Official Gazette: Belgrade, Serbia, 2009.
55. Bajic, T.; Manic, B.; Kovacevic, B. Social Housing in Belgrade: Practice in Architecture and Urban Planning Competititons (2003-2014). *Arhiteltura i Urban.* **2014**, *39*, 29–43. [CrossRef]
56. Master Plan of Belgrade 2021. In *Official Gazette of the City of Belgrade. No. 27-2003*; Official Gazette: Belgrade, Serbia, 2003.
57. Low on Social Housing. In *Official Gazette of RS. 13/2012*; Official Gazette: Belgrade, Serbia, 2012.
58. The Decree on Standards and Norms for Planning, Design, Construction and Conditions for the Use and Maintenance of Apartments for Social Housing. In *Official Gazette of the RS. 26/2013*; Official Gazette: Belgrade, Serbia, 2013.
59. Djokic, V.; Gligorijevic, Z.; Damnjanovic, V.M.C. Towards Sustainable Development of Social Housing Model in Serbia–Case Study of Belgrade. *Spatium* **2015**, 18–26. [CrossRef]
60. Decision on the Conditions of Sale of 2000 Socially Non-Profit Apartments in Belgrade. In *Official Gazette of the City of Belgrade No. 7/2005*; Official Gazette: Belgrade, Serbia, 2005.
61. Bajic, T. Enargy Efficiency Criteria in Urban Planning of Social Housing. In *Spatial, Ecological, Enargetic and Social Aspects od Settlement Developing and Climate Changes*; Pucar, M., Riznic, M.N., Eds.; IAUS: Belgrade, Serbia, 2016.
62. Law on Encouraging the Construction Industry in Serbia in the Conditions of Economic Crisis. In *Official Gazette of RS 45/2010, 99/2011 and 121/2012*; Official Gazette: Belgrade, Serbia, 2012.
63. Association of Belgrade Architects. Naselje Ivana Ribara-Novi Beograd-izložba konkursnih Radova. 21 November 2011. Available online: http://www.dab.rs/projekti/item/274-naselje-ivana-ribara (accessed on 10 July 2020).

64. Beobuild. Državna Stanogradnja-Naselje u Bloku 72 (Dr Ivana Ribara). 25 October 2013. Available online: http://beobuild.rs/forum/viewtopic.php?f=22&t=777&start=50 (accessed on 12 September 2020).
65. Predstavnicko telo Naselja Mileve Maric Anstajn. Naselje Mileve Maric Anstajn 26-78, Blok 72 Novi Beograd. 2013. Available online: https://trade.in.rs/ko-smo-mi-o-nama/ (accessed on 10 July 2020).
66. Nikolic, V.; Tasic, G.; Kaninski, M.M.; Nastasijevic, B.; Nikolic, Z. *Uzrik i Sanacija Neprijatnih Mirisa Detektovanih u Stambenim Jedinicama Naselja "Dr Ivan Ribar"*; INN Vinca: Beograd, Serbia, 2014.
67. Conic, I. Kako Sanirati Smrdljive Zgrade. 29 September 2014. Available online: https://www.gradnja.rs/kako-sanirati-smrdljive-zgrade/ (accessed on 10 July 2020).
68. Otrovne Zgrade. 2014. Available online: http://otrovnezgrade.blogspot.com (accessed on 10 July 2020).
69. Business Telegraf. *Gotova Sanacija Čuvene "Smrdljive Zgrade": Da Li Ovo Znači da je Već Useljena?* Internet Group: Belgrade, Serbia, 2020.

 *sustainability*

Article

Building Energy Model for Mexican Energy Standard Verification Using Physics-Based Open Studio SGSAVE Software Simulation

Andrés Jonathan Guízar Dena [1], Miguel Ángel Pascual [2] and Carlos Fernández Bandera [1,*]

[1] School of Architecture, University of Navarra, 31009 Pamplona, Spain; aguizardena@alumni.unav.es
[2] EFINOVATIC Certificación Energetica SL, 31192 Pamplona, Spain; mapascual@efinovatic.es
*   Correspondence: cfbandera@unav.es; Tel.: +34-948-425-600 (ext. 803189)

Abstract: The aim of the project detailed in this article was the development of an energy model for verifying Mexican energy standard compliance using the energy simulation engine EnergyPlus through Open Studio SGSAVE software. We aimed to improve the tool's ability to increase the comfort of social housing through the implementation of the standard in a practical digital tool. The project followed a four-stage methodology. The first stage was the development of climatic zoning for the country. The second stage involved the research and classification of the main traditional construction systems. The third stage was extensive research on the actual state of Mexican energy verification and its legal framework. The standard studied was NOM-020-ENER-2011. The final stage was testing the verification method by introducing the energy Mexican rule into the proposed software with the zoning and construction systems catalogue. A base model of a social housing type was developed in the software. Then, this model was improved to respond to each representative climate zone. Both models were simulated and we verified if they met the requirements. The results were contrasted for determining if there were energy savings. As a conclusion, we found that the actual energy standard of Mexico needs to be changed and we suggest the implementation of the energy simulation engine Energy Plus for creating more complete reports. This will help with the practical improvements in social housing conditions.

Keywords: Mexico; energy simulation; building energy model; Open Studio; SGSAVE; NOM-020-ENER-2011; climate zoning; traditional construction systems; social housing; verification method

 check for **updates**

Citation: Guízar Dena1, A.J.; Pascual, M.Á.; Fernández Bandera, C. Building Energy Model for Mexican Energy Standard Verification Using Physics-Based Open Studio SGSAVE Software Simulation. *Sustainability* 2021, *13*, 1521. https://doi.org/10.3390/su13031521

Received: 23 November 2020
Accepted: 26 January 2021
Published: 1 February 2021

Publisher's Note: MDPI stays neutral with regard to jurisdictional claims in published maps and institutional affiliations.

1. Introduction

The low thermal comfort in social housing is becoming a severe cause of house abandonment in Mexico. The general director of the National Fund for Social Housing of Mexico (INFONAVIT), David Penchyna, declared for the digital news portal El Financiero: "There are 100 abandoned social houses. The causes are diverse, like non-payment, not enough public services, and houses that did not meet the minimum conditions of comfort" [1]. This social housing problem is the result of the lack of a comfort model, with studies only focusing on construction quality and cost. Taking, for example, the situation in another context, like the European Union, we observe that the public regulations for building construction and restoration consider the energy performance. To address the energy demands and the comfort needs for interior spaces, the European standards provide reference values for ensuring the optimal habitat conditions of buildings. Because not all the social housing owners have enough economical sources for implementing a specialized design and optimization for their particular project, it is crucial that the authorities establish a guide for building based on proper studies and customized reference values.

Some studies focused on building residential projects considering energy efficiency optimization. Griego and Krarti highlighted the importance of implementing energy optimization strategies for residential projects in the Mexican context: "The need to reduce

231

domestic energy consumption is highly urgent, particularly as the number of homes in Mexico continues to increase. The awareness of this need has initiated the development of sustainability and energy efficiency guide-lines in the national residential building code, CEV. Findings from this study indicate that greater emphasis should be placed on implementing the minimum thermal insulation levels" [2].

For the last twelve years, Mexico has made a remarkable efforts to create and implement the principles of sustainability and energy savings in their laws and codes for construction and the commercialization of equipment and household appliances. In 2011, the government of Mexico published in "Diario Oficial de la Federación" (Official Journal of Mexican Federation), the first energy standard of energy efficiency for buildings: NOM-020-ENER-2011 [3]. The objectives of the norm were defined in the document: "In Mexico, the thermal conditioning of buildings has a great impact on the peak demand of the electrical system. It is greater in the northern and coastal areas of the country, where the use of cooling equipment is more common than heating. In this sense, this standard optimizes the design of the thermal behavior of the envelope, obtaining benefits like energy savings due to the reduction of the capacity of the cooling equipment" [3]. The role of building performance simulation (BPS) in electrical grid stability has been widely studied by different authors [4–6], showing the importance of making these models available from building design to operation.

The Mexican energy standards focus the scope of testing on the heat gains of the thermal envelope of the building. However, given the instability due to the constant change in political parties in charge of government leadership, the Mexican energy standard stopped its development and implementation in the regional building codes of the country. The last actualization of the standard was approved on 2013. The only tools available for the verification of the standard are a digital guide for its application and a digital tool (made in Excel) for automatic calculations by entering the numeric data in the corresponding spaces [7]. Although this tool fulfills its purpose of checking if a building satisfies the Mexican energy requirements, it only considers the total surface area and some climatology variables defined inside the documentation of the rules.

The energy standard should be analyzed by understanding the characteristics of the Mexican housing scheme. In the essay Cuantificando la clase media en México en la primera mitad del siglo XXI: un ejercicio exploratorio (Quantifying the middle class in Mexico at the first half of twenty-first century: an exploratory exercise), the authors conducted a study supported by statistics from the National Institute of Statistics and Geography (INEGI) to analyze the composition of economic classes in depth. They concluded that the highest% of the Mexican population is identified as lower class, at 55.1% of the total. The second group is composed of the middle class at 42.4%, with the upper class being 2.5% of the total. Notably, according to the essay, 80% of the Mexican population lives in cities and metropolitan areas. Therefore, 78% of the population is the target market for social housing developers [8]. The construction industry is one of the strongest and most profitable economic activities in Mexico. Residential and mixed-used constructions are some of the most wanted business models for private and government investment (through the national housing fund, INFONAVIT, and other public institutes).

Although NOM-020-ENER-2011 is the only official standard designed for mandatory application, other public tools are available for enhancing energy performance and environmental strategies for sustainable architecture. The NMX-AA-164-SCFI-2013 is a code developed by the Secretariat of Agriculture and Environment (SEMARNAP) and the Mexican Chamber of Construction Industry (CMIC) for promoting environmentally friendly techniques, strategies, and technologies in construction. It is a document with several categories of different ecological aspects (water, soil, energy, materials, landscape, interior comfort, and social responsibility) and a scheme of credit fulfillment [9]. Mexico also offers economical alternatives to implementing sustainable architecture and technologies for energy savings INFONAVIT runs a public program for social housing constructors called Hipoteca Verde (Green Mortgage). With this program, the constructor can obtain a public loan for real estate projects if they support the application with eco-technologies: design strategies and technology for energy savings, like photovoltaic panels, solar thermal collectors, thermal insulation, and water recycling [10]. As such, there is an interest in and potential market for the implementation of energy simulation and building efficiency models (BEMs). According to the statistics published by INEGI in 2018, the biggest energy expense of the country's households is electricity. In the northern states, 40% of families use air conditioning in their houses, providing a solution to their thermal comfort needs [11]. Due to the need for mechanical air conditioning, the evaluation of the energy performance of buildings and indoor living space quality provides an opportunity reducing electrical consumption and costs while improving the well-being of users.

Given the lack of practical implementation methods for the standard (providing an opportunity to create a different and attractive proposal for the Mexican market), the collaborating enterprise decided to investigate and develop their own model for energy verification using their tools and software. The aim of this investigation was to develop a model for verifying Mexican energy standard compliance for social housing models, and to research the benefits of using the Energy Plus building energy program [12]. Demonstrating the potential for building energy savings by designing an efficient thermal envelope and applying bio-climatic strategies to the design will be a useful tool for national construction.

The first three sections explain the procedures followed and the primary results. Each section describes the principal sources studied. Then, the procedure used for developing each of the method's aims is explained in detail, and the section ends with primary conclusions. The fourth section begins with an explanation of the testing method used. Then, a quick explanation of the developed models is reported. Finally, the results section reports the simulation results with a brief analysis of each one. In the conclusions section, a list of the pending aims and topics is provided, with a final reflection on the research results.

2. Methodology

2.1. Design and Phases

Creating a new model of energy verification for Mexico seems to be a complex project. Our principal research aims were the development of a model for verifying the model's compliance with the Mexican standard and conducting a comparison with the Energy Plus simulation model for testing the benefits of including thermal comfort analysis in the verification model. For a new scheme, profound studies are required to support the coefficients and reference values of the standard. For practical purposes, the use of an actual verification method was suggested, but with the customization of several issues. Due to the lack of an international regulation code at the time (2010), the actual energy standard NOM-020-ENER-2011 was developed using the practices and guidelines quoted in publications, like the ASHRAE (American Society of Heating, Refrigerating and Air Conditioning Engineers) Fundamentals Manual, 1998 and 2001 editions. The standard presents its own method of verification, pursuing energy savings by controlling the thermal envelope heat transfer with a mathematical calculation method [13]. So, as a first approach for a new verification model, new reference values were researched and created. To obtain the planned results of the initial hypothesis, the working plan was divided into four stages,as displayed on Figure 1: (1) The development of climatologic zoning of Mexico according to its different climatic severities; (2) the study of Mexican energy standard NOM-020-ENER-2011 and its application; (3) constructing a traditional building system catalogue for the country; (4) the development of an energy testing exercise with the previous materials and tools, applied for the selected social housing model. The fourth stage, the testing exercise, was used to test if the existing model complied with the standard. If not, improved models for each climatic zone were developed until standard compliance was achieved. Finally, for verifying the applicability of energy efficiency strategies on different traditional construction systems, a comparison test was performed.

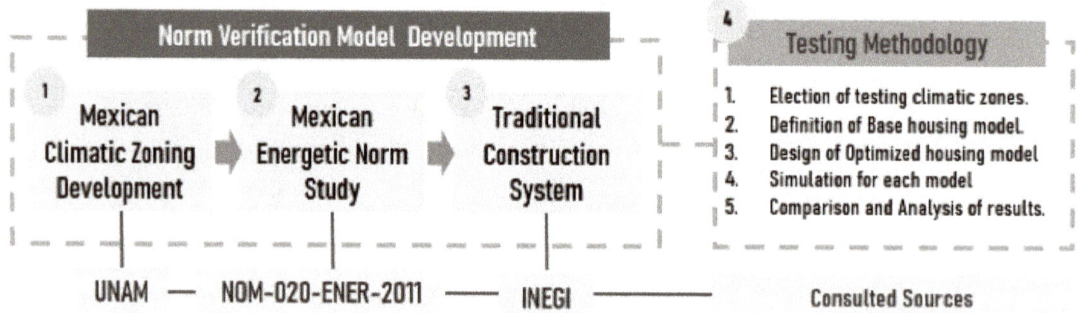

Figure 1. A general scheme of the four research stages for the energy verification model.

2.2. Climatic Zoning Development

Weather files are an important element for energy simulation because of their influence on the building energy performance results. The article entitled Impact Assessment of Building Energy Models Using Observed vs. Third-Party Weather Data Sets clarified that: "The sensitivity analysis of the main weather parameters showed the different influence that each parameter had on the energy demand variation of each test site". In this regard, the relative humidity and wind direction had little influence on the models; the two parameters with the strongest influence on the models were wind speed and temperature [14]. The variations in elemental climatic characteristics result in different thermal envelope behaviors. Then, the importance of having a classification of different climatic zones, with their corresponding weather reference values, becomes a reason for considering climatic zoning. There are some examples of how weather conditions and climate affect the thermal envelope and HVAC system performance. Hang et al. explained the importance of considering weather statistics for improving the energy performance of HVAC systems. They tested an HVAC system control with the new set-point temperatures calculated from the derived equation, improving thermal comfort by 38.5%. This study confirmed that a cooling set-point temperature that considers both the thermal characteristics of a building and the weather conditions effectively enhances the indoor thermal comfort during summer. They also noted the need for a continuous update of the existing hourly weather data files that are used in dynamic simulations or for the prediction of the peak thermal loads of buildings, as they affect the capacity of HVAC equipment to respond to current climate change [15,16].

Why does climatic zoning need to be developed for Mexico? The only available zoning was developed by the Institute of Geography of the Mexican Autonomous University (UNAM) in 2005. In the introduction, we explained how they delimited the zones. To delimit the domains with potentially similar climates, a regionalization was proposed based on the country's orography, hydrology, and elevation. Figure 2, for example, explains the relationship between cities elevation and their climatic characteristics (higher elevations are associated with cold weather, and lower elevations with warm weather). The geographical characteristics of rainfall, humidity, and temperature change even between short distances. As a consequence, a wide diversity of climates is created [17]. This zoning classifies the national territory into eleven zones. Each of these zones was determined by the study of several factors: dominant winds, rainfall patterns, annual temperature records, thermal annual oscillations, relative humidity, and average temperature rates. There are similarities between different climatic zones. For example, the states located on the coast line of the Pacific Ocean (Guerrero, Jalisco, Michoacán, and Oaxaca) present iso-thermal climates, meaning they have a thermal oscillation of 5 °C or less throughout the year.

The impact of climate change and warming on the climatic characteristics of Mexico must be highlighted. Liberman and O'Brien explained how climate change produces variations in moisture, rainfall, and mean temperatures. By analyzing the regional impacts of global warming in the country, the results always tend to be warmer and drier. Whichever model is used, it seems that potential evaporation will increase, and, in most cases, moisture availability will decrease, even where the models project an increase in precipitation. Owing to the effects of global warming, the temperature, moisture and humidity data have variances over time [18]. From this statement, weather data and climatic files are necessary for obtaining more accurate energy simulations.

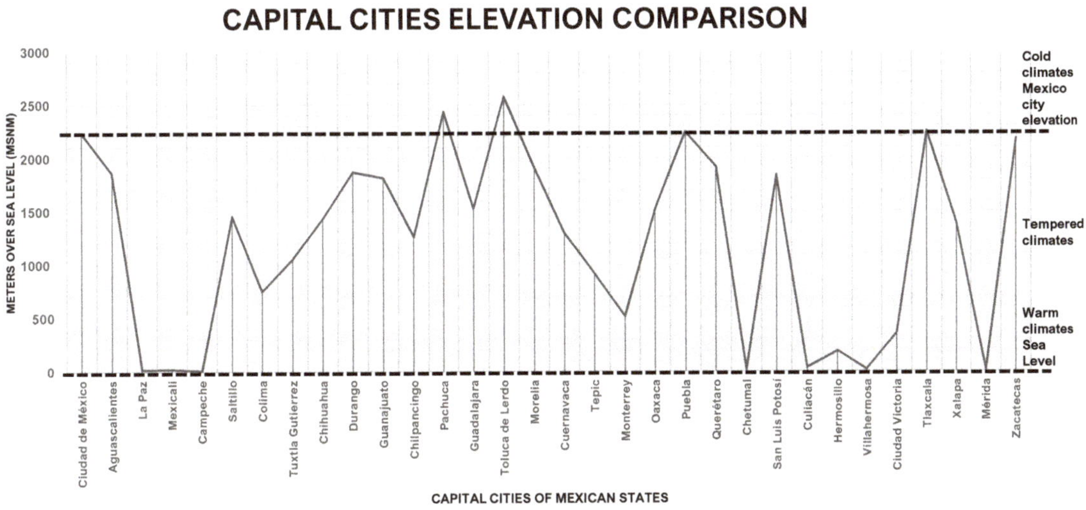

Figure 2. Comparison of Mexican capital cities elevations, and the relationship of elevation with cold and warm climates. The coldest climates are located on mountains and at high elevations, whereas the warmest climates are located at low elevations close to sea level.

For constructing a building energy model (BEM) and a verification/certification model, a simplified and more practical climatic zoning is needed. A similar climatic zoning design procedure was followed by Bai and Wang for defining climatic zoning in China. They used building simulation to analyze the potential impact of re-assignment of cities to new thermal climate zones based on public building energy consumption and recent meteorological data..The defects of current climate zones were identified and new updated zones were defined [19]. For each climatic zone, several reference and nominal values were considered for their entry in the software calculations (average temperatures, average humidity, winter and summer severity, etc.) Therefore, we decided to design a new zoning based on thermal demands. For the proposed classification, several housing models that are common in the majority of Mexican cities were studied. Three models were selected: economic house, medium-level house, and apartment block. The three models were selected from the web-page catalogue of a real estate broker affiliated with INFONAVIT [20]. These models were chosen because of their constant replications in all the Mexican states. For the energy simulation, Open Studio software with the extension SGSAVE was used. SGSAVE is a complement developed by the Spanish energy testing software developer EFINOVATIC [21]. Since 2018, it has served as an official verification tool of the energy requirements of the Spanish building normative. Using the Energy Plus simulation engine (with local parameters and building geometry), the software conducts an energy simulation, providing an analytic report of different environmental results [22].

Using this software, the project continued with simulations for each house model. The simulation output the energy demands for heating and cooling for each housing type (expressed in kWh/m^2). The simulation period considered for the results was one year.

The Energy Plus Weather File (EPW) file or weather data file of capital cities [23] was used of each of the 32 states of the country, including the nation's capital city (Mexico City). Different simulations were performed by changing the position of the principal façade oriented to the four cardinal points (north, south, east, and west) for each house type. For a new simulation, we also considered the use of three different window-to-wall ratios: 10%, 30%, and 60% of openings (Figure 3).

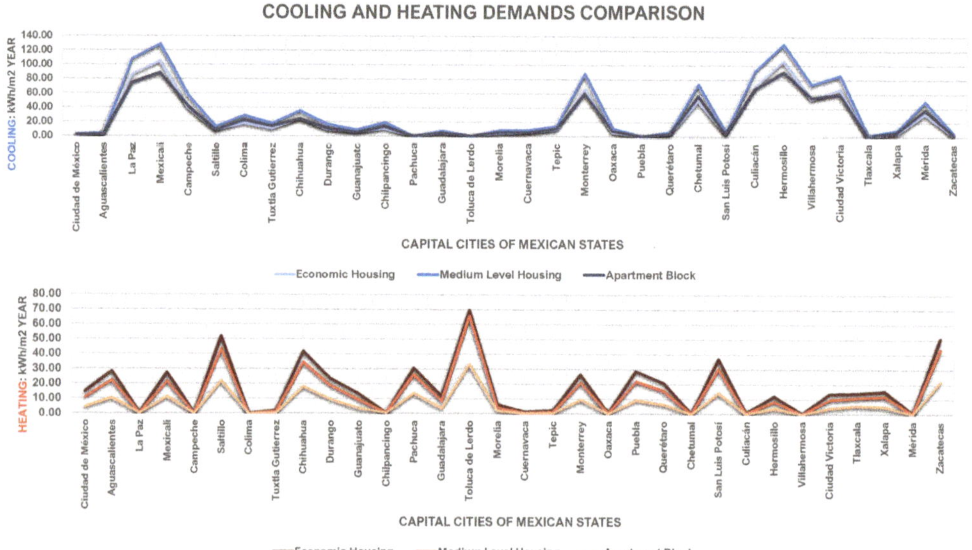

Figure 3. Cooling and heating comparison for each of the housing model types, showing that medium-level housing registered the highest cooling demands in all climates. Economic houses registered the highest heating demands; however, scaling the results, the heating demands are lower (70 kWh/m^2 for the highest value) than the cooling demands (125 kWh/m^2 for the highest value).

Adding all the procedures, 384 simulations were conducted for both the heating and cooling demands for each house type. Once all the energy demand results were compiled, the four possible orientations were averaged for each capital city and each house type. The next step was averaging all the energy demands for all the capital cities, obtaining a mean value for heating demand and other for cooling demand. For the last step, an average value of all the energy demand for the three housing types was generated, obtaining final energy demand values for each capital city. The next step was attaining normalized values by scaling all cities' values for the collection of a unique scale of comparable values. We normalized the values by calculating an average value from all cities' values. This global mean value was assigned a numeric value of 1. So all cities' values were scaled in comparison with this numeric value. This procedure resulted in a scale ranging from 1 to 3 for three winter severities and a from V to Z for five summer severities. The final scale included a total of ten climatic zones, displayed on Figure 4 map.

The ten climatic zones were divided in three groups. The first group corresponds to cold and temperate climates, containing V1, V2, and V3 zones. The V zones contain the majority of the national territory, except for Estado de Mexico, which has the coldest climatic conditions in the country. This group has a temperature range of 0 to 21 °C 73% of the time [24]. The next group corresponds to warm and tropical climates, containing W1, W2, and X1 zones. The W and X zones are located in states next to the Pacific Ocean (Michoacán and Colima), and Yucatan Peninsula. This group presents a temperature range from 21 to 27 °C 40% of the time, and from 27 to 38 °C the other 40% of the time [24].

The last group corresponds to arid and desert climates, containing Y1, Y2, Z1, and Z2 zones. The Y zones are located in the northern coastal states (Baja California Sur, Baja California Norte, Sinaloa and Tamaulipas, and Sonora) and Nuevo León. This group presents a temperature range from 27 to 38 °C 50% of the time, and temperatures higher than 39 °C the 10–20% of the time [24].

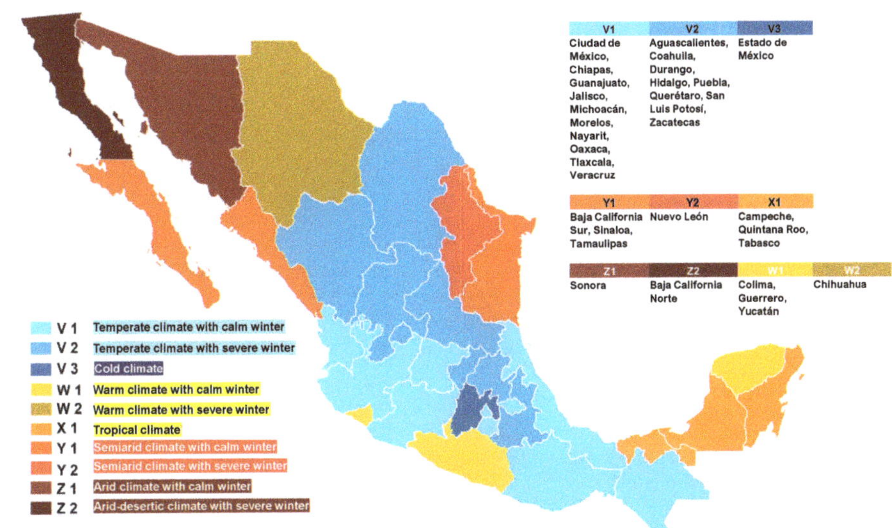

Figure 4. Climatic zoning designed for the proposed verification method using winter and summer severities for analysis and comparison.

2.3. Study of Mexican Energy Standard

For the second aim of the methodology, we conducted an extensive search and review of the existing Mexican energy standard. Notably, there is no mandatory energy standard in Mexico. This situation is common for North America. In the United States, for example, some states have regional regulations that include an energy performance test. Others states only incorporate energy strategies in their own construction codes. The American energy codes rely on ASHRAE guidelines and reference values. In Mexico, there are some energy guidelines and a group of non-mandatory standards [25]. For this project, we used NOM-020-ENER-2011, published by the National Committee for the Efficient Use of Energy (CONUEE) [3]. With voluntary application, this norm uses the comparison of energy savings for a normal case and a saving-based case as its principal verification method (designed by the optimization of the thermal envelope of buildings and the heat gains analysis). The verification calculation begins with the definition of a reference model for the building using reference values (provided by attached tables and an application manual contained in the standard's documentation). The next step is the calculation of the same variables, but from the projected model for the building. This information is obtained by analyzing the building geometry and consulting the available data on the plans and material catalogue for the project [7].

With this previous information, the calculations provided by the standard provide the heat gains by radiation and conduction for the reference and projected models. If the heat gains obtained by the projected model are equal to or less than the heat gains of the reference model, then the building complies with the standards. Therefore, the standard focuses on the verification results only in terms of the performance of the thermal envelope, neglecting other results obtainable from a simulation. A verification standard is not an energy performance study. The use of elements, like internal gains, occupation calendars, and energy expenses of electricity and sanitary hot water, is difficult to implement in a standard because they are dependent ob the user. The Spanish Building Technical Code (CTE) does not consider them, either.

In this proposed model, the HVAC performance is important. The inclusion of HVAC design and performance in building energy performance for an energy certification is complex. HVAC design requires a specific study of demands and an input of actual user variables, like occupancy, space use, and schedules. The HVAC equipment design is also dependent the unique characteristics of the project, for which it is difficult to determine an exact result for a final energy certification for standard verification. Due to the functionality of the Mexican verification (applicable for preliminary architectural projects), the project can be simulated with ideal loads for HVAC equipment, because the standard work with heat gains, not energy demands. The simulation tools available on the market, like Energy Plus, can simulate and describe energy demands by defining certain HVAC equipment characteristics, but the customization options result in a large number of possible results. The simulator requires the collaboration with an expert in HVAC installations and sizing.The simulation parameters increase in complexity as the HVAC design evolves and requires more precise information. This is complicated because it implies a level of updating, which will increase the cost. It is recommended to first verify if a project complies with the standard and then perform a personalized energy simulation of the particular building project.

In summary, the principal focus of the NOM-020-ENER-2011 is the heat transfer control of the thermal envelope for improving energy demands. Instead of optimizing or pursuing the goal of a Nearly Zero Energy Building (like recent energy codes and certification models), the Mexican standard focuses on improving energy demands for cooling systems due to the prevalence of arid, warm, and humid climatic zones in the country. The principal energy consumption rates for thermal conditioning is due to cooling (40%) [11], The standard requires low improvement levels for maintaining low construction costs and the competitive housing affordability for most of the Mexican population. Nine of ten Mexican citizens are planning to acquire a house, but 45% of them cannot afford it [26]. The standard establishes several reference coefficients for calculating the energy efficiency: U-values, temperatures, glazing coefficient, etc. (Figure 5). However, as the scope of the standard is thermal envelope performance for heat transfer, it ignores thermal comfort considerations. This means that the the standard does not provide a comfort model for simulation, following only its own calculation method.

Consignas para Temperaturas (exteriores – interiores – superficiales – Factor Ganancia Solar)																							
				CONDUCCIÓN														RADIACIÓN					
				Opaca									Transparente					Transparente					Barrera para vapor
Estado	Ciudad			Temperatura equivalente promedio Te (°C)														Factor Ganancia Solar Promedio FG (W/m2)					
		Temp. Interior	Superficie Interior	Techo	Muro Masivo				Muro Ligero				Tragaluz y domo	Ventanas				Tragaluz y domo					
					Norte	Este	Sur	Oeste	Norte	Este	Sur	Oeste		Norte	Este	Sur	Oeste		Norte	Este	Sur	Oeste	
Aguascalientes	Aguascalientes	24	26	37	24	27	26	25	30	33	32	32	22	23	24	24	24	274	91	137	118	146	
Baja California Sur	La Paz	25	31	45	31	34	32	33	36	40	38	39	26	27	28	29	29	322	70	159	131	164	Si
Baja California Norte	Mexicali	25	34	50	36	40	37	38	41	45	43	45	29	30	32	32	32	322	70	159	131	164	
	Ensenada	24	25	35	22	25	24	23	28	31	30	30	20	22	22	22	23	322	70	159	131	164	Si
Campeche	Campeche	25	31	45	31	35	33	33	36	40	38	40	26	27	29	29	29	284	95	152	119	133	Si
Coahuila	Saltillo	25	27	38	25	28	26	26	30	34	33	33	22	24	24	24	25	322	70	159	131	164	
	Torreón	25	30	43	30	33	31	31	35	39	37	38	25	27	28	28	28	322	70	159	131	164	
Colima	Colima	25	29	42	28	32	30	30	34	38	36	37	24	26	27	27	27	274	91	137	118	146	Si
Chiapas	Tuxtla Gutiérrez	25	29	42	29	32	30	30	34	38	36	37	24	26	27	27	27	272	102	140	114	134	Si
	San Cristóbal de las Casas	23	22	31	19	20	20	20	25	27	27	26	18	20	20	20	20	272	102	140	114	134	
	Arriaga	25	31	46	32	35	33	33	37	41	39	40	26	28	29	29	29	272	102	140	114	134	Si
Chihuahua	Chihuahua	25	28	41	27	30	29	29	33	36	35	36	24	25	26	26	26	322	70	159	131	164	
Ciudad de México		23	23	33	20	22	22	21	26	29	28	28	19	21	21	21	21	272	102	140	114	134	

Figure 5. Example table of the reference values quoted in NOM-020-ENER-2011. It contains interior and exterior temperatures, U-values, and other coefficients.

2.4. Traditional Construction Systems

For the third aim of the methodology, we researched the main traditional construction systems in Mexico. This included official information sources, like the Home Survey of the National Geography and Statistics Institute (INEGI) [27], and two scholarly publications about construction systems used in Latin American countries [8,28]. Each year, INEGI publishes the results obtained from an annual population census. Among the different result categories listed, one is focused on housing. The systems listed on the report were studied, and we compiled a list of those most used in the country. The listed systems were divided by the building elements to which they are applied (roof, walls, or ground), and the principal material present in their component elements.

As shown in Figure 6, the principal materials used in Mexican traditional construction systems were determined. For the construction of ceilings, the principal material used is concrete slab or joist and beams slab, with a 70% incidence. The second most popular material for ceilings is metallic sheet with a 16% incidence, followed by wood at 4% [27]. For the construction of floors, the principal material used is concrete slab, at 55% incidence, followed by covering (wood and mosaic) at 53% and pure soil at 3% [27]. For the construction of walls, the principal material used is fabrics (brick, block, stone, and partition) at an 85% incidence, followed by adobe (vernacular technique with mud and straw) at 9% and wood at 5% [27]. The survey results showed that the main construction system used in Mexican buildings is reinforced concrete with an incidence between 55% and 85% for all components). Mexican constructors used local vernacular techniques inherited by their ancestors, like adobe and bahareque (construction technique that uses adobe, straw, bamboo, and metal or fiber cement sheets for walls).

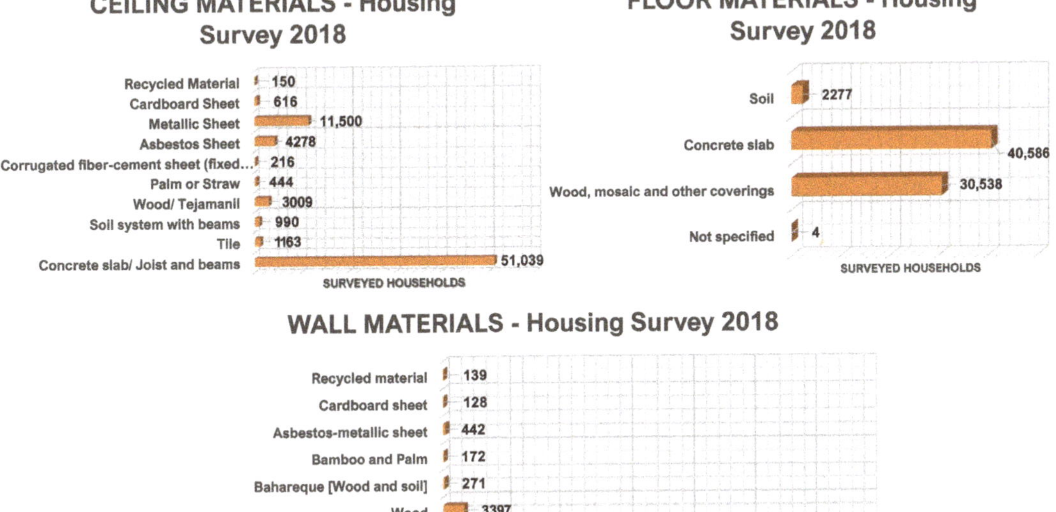

Figure 6. The principal materials used in Mexican housing grouped by building components (walls, ceiling, and floor). A total of 73,405 houses were considered by the National Survey of Housing and Population, National Institute of Statistics and Geography (INEGI) 2018.

By analyzing the results, two undesirable practices were denoted. First is the use of fiber–cement in construction. This material contains asbestos. Constant asbestos exposure poses serious human health risks for diseases, like lung cancer, asbestosis, and mesothelioma. It is not dangerous when packed or sealed in elements, like tiles, panels, or cabinet tops, but they are hazardous when those elements are in bad condition and the interior material is not sealed (resulting in the release of asbestos fiber in the air). Asbestos is forbidden in the majority of world construction standards due to its dangerous effects [29]. Knowing its hazardous characteristics, the common use of fiber–cement materials in Mexican housing is unsafe. Notably, many houses do not have floors, using the local soil (3%). In Mexico, almost 14 million houses do not meet the minimum acceptable conditions for a comfortable house due to the dirt floor and lack of access to quality water supply and sanitation services. In Mexico, 2 of 10 citizens lack access to one public service in their homes [30]. This is an opportunity for improving social housing conditions; improving energy performance and internal comfort with affordable materials could contribute to this improvement.

From each of the construction systems, we broke down the component elements, listing the different material layers' compositions. Finally, the four main characteristics of every material listed in the catalogue (layer width, conductivity coefficient, density, and measure unit) were researched and reported. With all the data compiled, the final construction system catalogue was completed.

2.5. Testing Methodology

For the fourth aim of the methodology, the testing of the proposed verification model of the Mexican standard, we followed two fundamental steps. First, we needed to learn and understand the digital software SGSAVE provided by EFINOVATIC [21]. Next, we designed a personalized testing methodology for the verification of the model to achieve the desired results. The digital tool used for the testing methodology was the energy simulation software Open Studio, using the complementary software called SGSAVE.

For understating how SGSAVE works, we reviewed the energy performance simulation process in Open Studio [31]. Open Studio is software contained inside the Google Sketch-up interface. The energy simulation process is composed of four principal phases: modeling, definition, simulation, and reporting. For the first phase, the 3D modeling of the evaluated building project, the user works in the Google Sketch-up interface. In this phase, the user establishes the physical values of the building (height, length, width, opening geometry, etc.). For the second phase, the user establishes the parameters and reference values for the modeled building project (materials, construction systems, schedules and calendars, HVAC definitions, space types, etc.). For the third phase, the user defines the simulation parameters before running the simulation engine (run period, desired report results, etc.). The final phase is the report, where the user analyzes the generated simulation report with the corresponding results (verification of the standard, energy demands, etc.). Open Studio functions are delimited from the first to the last phase.

SGSAVE is software that offers a direct verification of the Spanish energy standard and generates a complete report of all energy simulation results, including the European Union energy label designed by the EU Directive 92/75/EC [32]. SGSAVE introduces the reference values, parameters, and directives of Spanish energy standards (CTE) into Open Studio interface by: (1) entering Spanish/European reference values and parameters and (2) generating the Spanish/European verification certification report of compliance. The parameters and reference values are introduced by the user using a SGSAVE tool panel in the Google Sketch-up interface. With this panel, the user can introduce values, like thermal bridges, U-value limit verification, thermal space configuration for reference occupation schedules, window and door configurations, etc. It also includes tools to clean and refine the three-dimensional model, HVAC installation design, and location customization tools. For the report final phase, SGSAVE generates a certification report that declares whether the building project complies with the standard. The report also includes a graphic of detail heat gains, projected energy demands for ideal loads, and physical characteristics of the model (square meters per thermal zone, for example). Figure 7 describes the workflow.

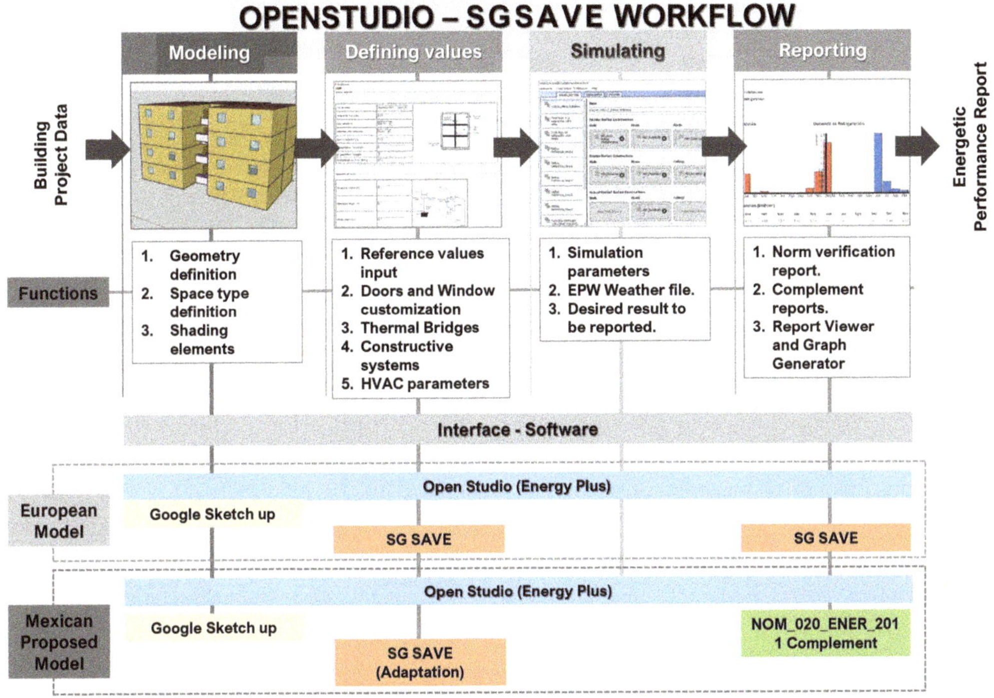

Figure 7. Open Studio and SGSAVE workflow diagram explaining the work process for entering the building project in the software interface and using the digital model for energy simulation.

By understanding how SGSAVE works in refining and adapting the European/Spanish standard requirements for Open Studio parameters, our aim was to adapt the software for the Mexican standard. As the standard's development in Mexican is in its infancy, the technical data were insufficient for fulfilling all SGSAVE data requirements. For the testing exercise, the data provided by the Mexican standard was used, and the Spanish/European values were used for the remaining fields to enable testing the Mexican data's adaptability to the software.

For each component (roof, openings, walls, doors, and shading elements), the software allows the user to assign values, entering the data corresponding to each requested variable, and then the user assigns a construction system for each element. Each construction system was entered into the software, determining the width of each of its composition layers and their respective conductivity coefficients and characteristics, like density and measurement unit. For the last arrangements of the thermal envelope's geometrical model, the user enters the remaining data needed, like occupation calendar, zone uses (kitchen or bedroom, for example), external shading components, and immediate context elements (trees, roads, location, or other adjacent buildings) [31]. A base model was introduced and configured with Openstudio tools, following the geometry of a typical social housing model, replicated on different Mexican cities (Figure 8).

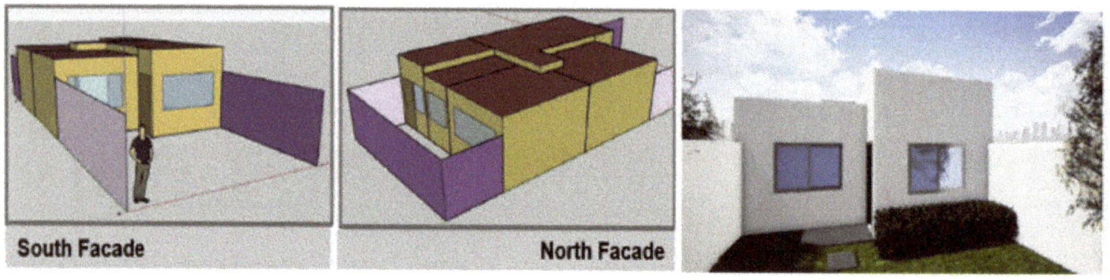

Figure 8. The basic model of the economic housing type. Geometry modeled in the Open Studio interface.

Before launching the simulation, the user must enter the weather file EPW of the city's project location and choose the desired simulation period. For the next step in the fourth aim, we determined the application of the following testing methodology. First, the most representative climatic cities of Mexico were chosen in accordance with summer severities (V, W, X, Y, and Z, returning to Figure 4) and those with the most extreme weather: Toluca de Lerdo for the coolest city and Mexicali for the warmest city. In the next step, the housing models were used and different simulations were performed by changing the weather file. Before running simulation processes for the testing method, a Building Component Library (BCL) report-generation add-on for Mexican standard verification was installed in the software. This add-on was developed by the Mexican agency ITÖM. It is a public-access complement provided at no cost. The add-on takes the geometry and the simulation parameters of the tested model and generates a report with the verdict (if the project complies with the standards) [33].

Using the verification complement of NOM-020-ENER-2011 for Open Studio, we tested if the building complied with the standards. If not, changes and bio-climatic design strategies for the optimization of the economic housing thermal envelope were applied to increase the energy savings (improving the score) of each housing model. After verifying the improved models for economic housing with personalized content for each tested climatic zone, new simulations were run for the improved models to verify if the model complied with the norm standards. With the analysis of the results of radiative and conductive heat gains, we compared the simulated energy savings between the normal case and the improved case. Finally, new simulations were run, but with the Open Studio report manager, for evaluating the differences between both reports.

Although these test simulations provided results for improving building energy performance, remember that the simulation needed other factors that are strictly related to the potential user. David Bienvenido Huertas stated: "Energy consumption simulations are directly related to six factors: three technical and physical factors (climate, building envelope, and building equipment) and three social factors (operation and maintenance, occupant behavior, and indoor environment conditions). So, the energy performance of a building depends not just on technical characteristics (e.g., the thermal performance of the facade) but also on users' behavior" [34]. For the simulation, the Mexican user operation, occupant behavior, and indoor environment conditions characteristics were needed. Some information was provided by NOM-020-ENER-2011 guide, but, for the remaining variables, the CTE (Spanish Building Technical Code) calendars and reference values [35] provided by SGSAVE were used. The reference values used from CTE were: occupation calendars, thermal bridges, average consumption rates, etc. The simulations were run with an HVAC configuration for ideal loads.

2.6. Testing Application

After reviewing the steps followed for the testing of the verification model, the climates and model characteristics were explained. For the test, six different climatic zones were chosen. For the cold-climate group, V3 and V1 zones were chosen. V3 has the coldest climatic conditions, with Toluca de Lerdo as the studied city. We selected V1 because it represents the temperate climatic conditions, with Mexico City as the studied city. For the warm-tropical group, W1 and X1 were chosen: W1 because it represents warm climatic conditions, with Merida as the studied city, and X1 because it represents the tropical climatic conditions of Yucatan Peninsula, with Campeche as the studied city. Finally, for the arid desert group, zones Y1 and Z2 were chosen: Y1 because it represents the semiarid climatic conditions of he northern states, with Culiacan as the studied city, and Z2 because it has the warmest climatic conditions, with Mexicali as the studied city.

2.6.1. Basic Model: Definition and Test

Recalling the information provided in the Introduction, INFONAVIT is the organization that manages social housing for workers. As portrayed in its official housing catalogue, they used to build the same housing model in different cities and states due to having the most economic design and construction [20]. The principal aim of our project was applying the Mexican standard to social housing, so the first step in the testing stage is the definition of a universal basic model, using the analogous example from the INFONAVIT housing catalogue. This model is the most replicated and affordable house in the catalogue.

The construction system used for the basic model was a reinforced concrete structure, with brick walls with plaster (17 cm wide). The floor was a 30 cm foundation bed/slab, and the ceiling was a joist and block concrete slab. After designating the tested climatic cities for the testing exercise, the base model was simulated for each of the EPW weather files (corresponding to each city). The add-on for Open Studio was run, and the model compliance with NOM-020-ENER-2011 was tested. None of the climatic zone models complied with the standard because all resulted in a negative energy savings of –151%. In other words, if the reference model had an allowed heat gain limit of 1519 Watts for thermal envelope, the base model had a calculated heat gain of 3816 Watts for thermal envelope. It exceeds the limit almost 150%. These results showed the unsuitability of using the same housing model for all cities without considering the climatic characteristics.

2.6.2. Improved Model: Definition and Test

Gumbarevic explained that the main goal of model improvement, is to minimize the heat transfer through the building envelope. For that reason, it is important to pay attention to the design and construction details in all delivery phases—from schematic and design phases up to the construction phase [36]. Because of the poor results obtained with the basic model simulations, the next step was the implementation of bio-climatic strategies for each climatic zone. For choosing the correct bio-climatic strategies, we used weather analysis software Climate Consultant V6 [24]. Using the ASHRAE 55 model for defining thermal comfort parameters, Climate Consultant software generated Givoni's diagrams for describing the most adjustable bio-climatic strategies for each climate (Figure 9); with the base construction systems, several changes were implemented to construct an improved model that should comply with the Mexican energy standard.(A graphic display of implemented strategies is showed on Figure 10)

For the V3 improved model, a continuous thermal insulation layer was proposed, using expanded polystyrene (EPS) with a thermal conductivity of 0.029 W/m^{2}K, and a width of 6 cm. By using insulation, the thermal masses of the brick and concrete elements were enhanced. Several shading elements were added to the windows. For the improved V1 model, the implemented changes were: a continuous thermal insulation layer using expanded polystyrene (EPS) with a thermal conductivity of 0.029 W/m^{2}K and a width of 3 cm; by using the insulation, the thermal mass of the brick and concrete elements retained heat gains for nocturne diffusion; horizontal shading elements on the south façade for sun heat control; shading elements for windows; and openings for sun heat control.

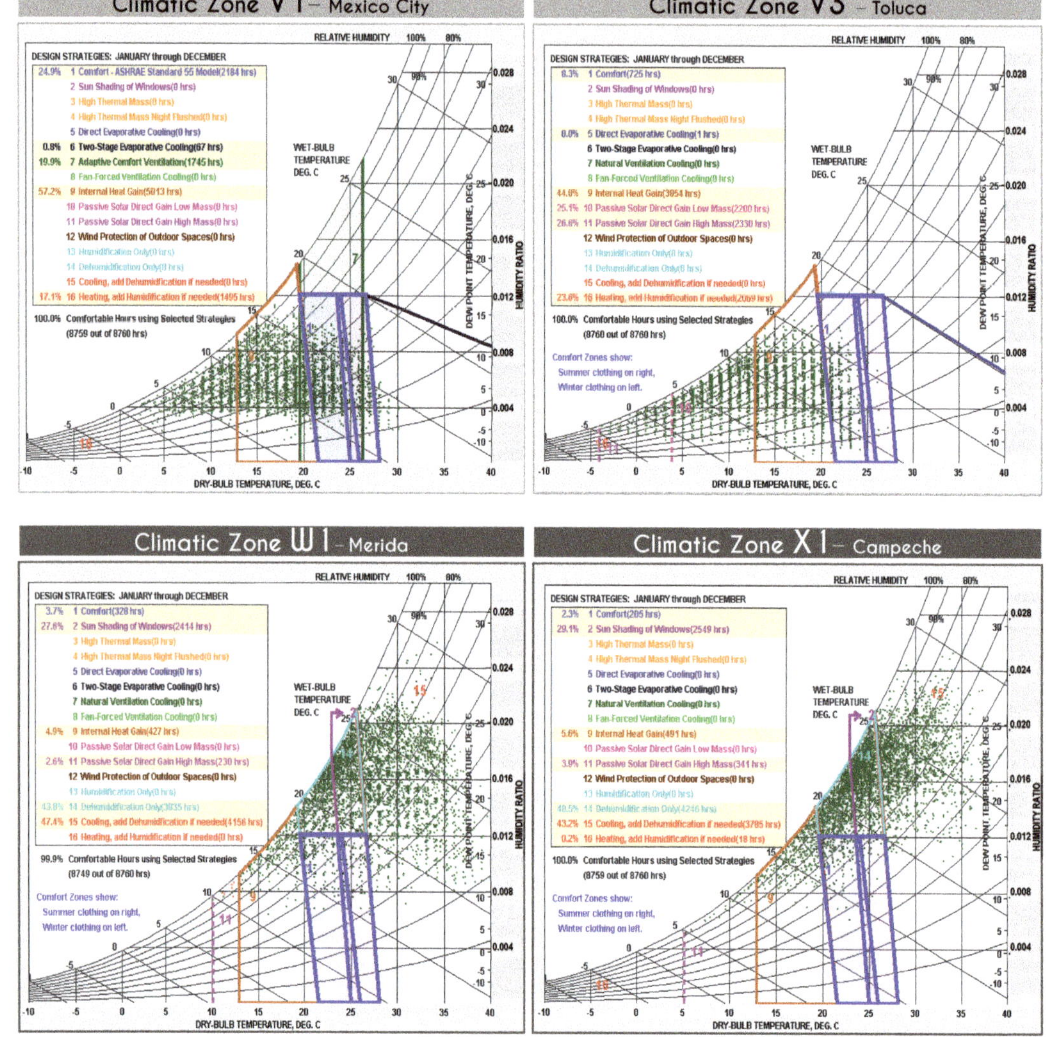

Figure 9. *Cont.*

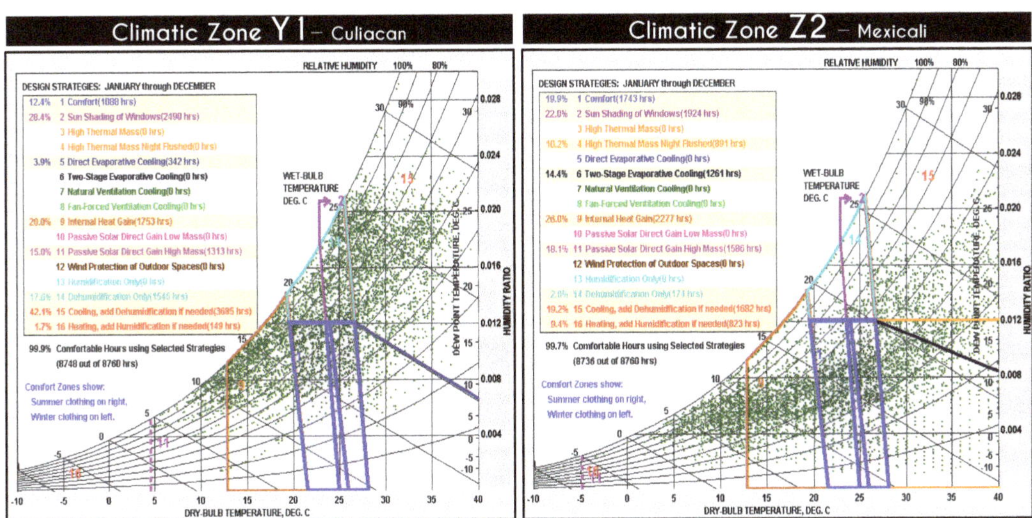

Figure 9. Givoni's diagrams produced by Climate Consultant V6 software [24]. Each diagram describes the suggested bio-climatic strategies for reaching thermal comfort inside the improved models for each climatic zone.

For the improved W1 model, the construction system of the ceiling was changed from the previous joist and block concrete slab to a lightweight EPS-blocks concrete slab for thermal insulation. In addition, a continuous thermal insulation layer was proposed using EPS expanded polystyrene with a thermal conductivity of 0.029 W/m^{2}K and a width of 3 cm. Using the insulation, the thermal masses of the brick and concrete elements were enhanced. The walls were modified, adding an extra layer of brick with a 3 cm air gap between them. For better ventilation flow inside the building, 30 cm was added to the interior height. The façade windows were modified, adding 30 cm tall operable ventilation openings for cross-ventilation. Shading elements were added to windows and openings. For the improved X1 model, the construction system and strategies were quite similar to the W1 model, including the EPS-blocks concrete slab, a thermal insulation layer, thermal mass, windows configuration, shading, and blinds. The difference between both models is the air gap between the two-brick layer. For X1, the air gap is 5 cm wide. The reason for the similarities between W1 and X1 is that they share climatic characteristics in terms of temperature and other variables. They have many isotherm areas (areas that have an annual temperature variation from 0 to 5 °C). The difference between them is the high relative humidity in the tropical X1 area.

For the improved Y1 model, the construction system of the ceiling was changed from joist and block concrete slab to a reticular lightweight EPS-blocks concrete slab for thermal insulation. A continuous thermal insulation layer was also proposed using EPS expanded polystyrene with a thermal conductivity of 0.029 W/m^{2}K and a width of 6 cm. Using the insulation, the thermal masses of the brick and concrete elements were enhanced. The walls were modified, adding an extra layer of brick, with a 5 cm air gap between them. For better ventilation flow inside the building, 30 cm was added to the interior height. The façade windows were modified, adding 30 cm tall operable ventilation openings for cross-ventilation. Because of the high radiation in this location, the building receives unnecessary heat gains from the northern façade. To solve this problem, the improved model included shading elements for northern windows. Shading elements were added in windows and openings. Finally, for the improved Z2 model, the construction system of the ceiling was similar to the Y1 model. However, due to the extremely warm climate, a second ceiling was added, with a ventilated air gap of 15 cm. The second concrete ceiling receives the direct radiation, and the cross-ventilation in the air gap dissipates the overheat. The ventilation method of this model relies on bulk airflow measures, driven by wind, to promote natural ventilation. Ventilation was enhanced also by promoting the stack effect by positioning ventilation openings near the roof (which is 20 cm higher than the other improved models.

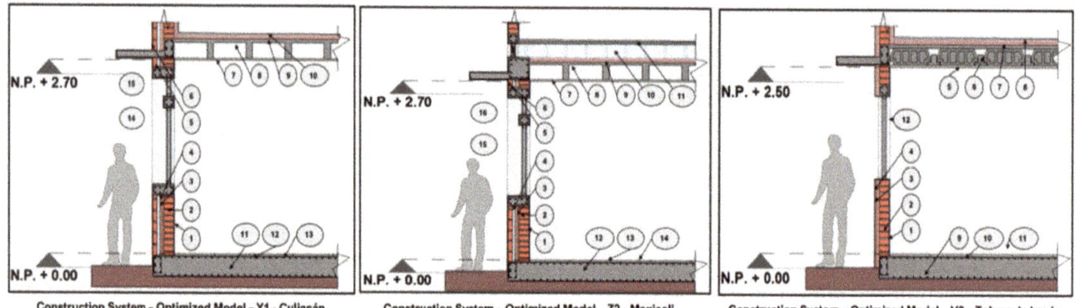

Figure 10. Architectural details of the construction systems used in some of the improved models for testing. For more detailed information and the key for the numbered circles, see Appendix A.

This improved model included all the mentioned strategies, like 6 cm thermal EPS insulation, double brick layer with a 5 cm air gap, shading elements in windows, etc. The south façade windows were covered with a stationary during summer, and its leaves fall in winter (allowing direct solar heat gains). The interior height of the model was increased 60 cm, creating a 3.30 m height to allow the warm air to concentrate in the upper part of the interior space.

Using the SGSAVE complementary tools, certain special characteristics were introduced to all improved models:double-glass windows with a 4-cm air gap, and windows blinds for the south façade, with an operation calendar of 30% aperture in the summer. These improved models were constructed for testing the energy performance of the most-used construction system in Mexico (reinforced concrete elements with red brick walls and a foundation bed) in the different climatic zones in the country. To obtain a wider view of the desired results, we created improved models with different construction systems. For comparing the results with the traditional system, the two most-used traditional systems after reinforced concrete and fabric were chosen: the wood construction system and the adobe–metal sheets construction system. All the bio-climatic strategies that were used for the improved models were taken from a practical guide of 101 basic rules for low energy consumption [37].

The wood construction system (Figure 11) was composed of 1.8-cm wide Triplay OSB (Oriented Strand Bond) panels and timber structure in the walls and ceiling. The floor was a foundation bed of concrete, with 60 × 60 cm ceramic tile. The walls had a double Triplay OSB panel layer with a wood-cork thermal insulation 5-cm wide with a thermal conductivity of 0.04 W/m^{2}K and a vinyl waterproof sheet for stopping condensations [38]. The adobe–metal sheet construction system (Figure 11) was composed of a reinforced concrete structure with walls with adobe blocks (mud and straw regional block). Because of the low bearing capacity of these walls, the model needs a lightweight structure for the ceiling. So, a wood-timber ceiling was proposed, with a floor composed of a foundation bed of concrete slab. The adobe–metal sheet construction is an experimental construction technique, inspired by bahareque vernacular system [39]. The bahareque system is composed of a bamboo frame structure, with a medium-height wall of adobe and a second wall of bamboo panels. The ceiling should be a lightweight structure, like straw or a light wood or bamboo latticework. For this option, the bamboo structure was replaced by a reinforced concrete structure (for seismic resistance), and the walls were covered by a metallic sheet (corrugated galvanized steel), which has high heat reflectance.

The technical information of the traditional materials used in the proposed model construction elements was obtained from Appendix D of NOM-020-ENER-2011 [3]. For the remaining materials, the data were obtained from the digital catalogue of the Mexican construction store Home Depot [40].

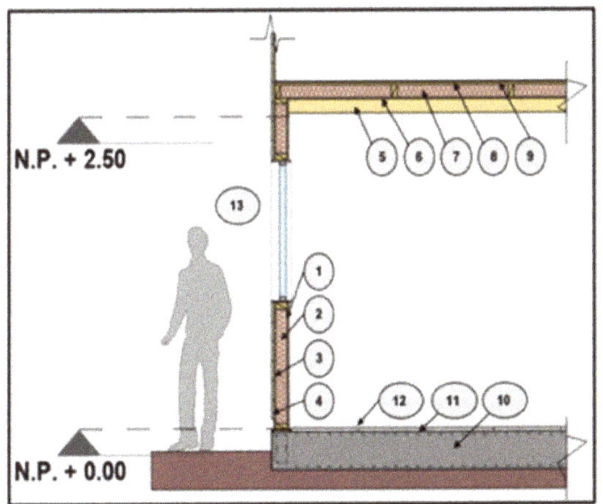

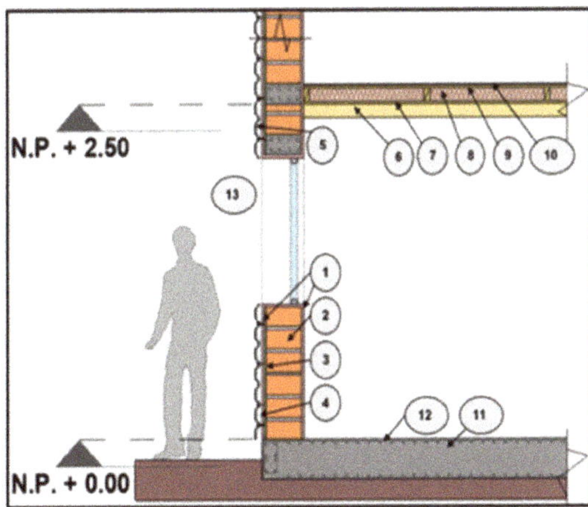

Figure 11. Architectural details of the two construction systems used to compare the results with the traditional reinforced concrete and fabric walls system. For more detailed information and the key for the numbered circles, see Appendix A.

3. Results

The final results of the testing produced several conclusions. All models were simulated in the software for verifying the standards. The following improved models met the Mexican standard: V3 and V1 with an energy savings of 20%; V1, X1, and Y1, with an energy savings of 19%; and the Z2 model, with an energy savings of 20%. Compared with the −151% of the base model, the 19–20% improvement with the the updated model is significant. Howeer, these results only show the heat gains and loses for the thermal envelope as requested by the Mexican energy standard. Open Studio SGSAVE software, using Energy Plus, simulated and calculated a wide variety of results. For example, it provides a broken-down report of the different heat gains and losses of the building. The report includes heat gains for two different season groups (winter and summer) and a classification according to transference channel.

As reported Figure 12, there are different heat gains and losses that should be analyzed. In summer, the largest heat gains of the thermal envelope occur in the external walls, windows through radiating heat gains, air infiltration, ventilation, and internal gains. For winter, the situation for heat losses also occurs for the same channels. These gains and losses can result in larger cooling demands. These results are important because they indicate the critical aspects to consider when optimizing the thermal envelope design. The Mexican standard only reports heat gains through the thermal envelope. However, the verification methods used by other entities, like the European Union, or private certifications, like LEED (Leadership in Energy and Environmental Design), consider energy demands for heating and cooling. Fortunately, Energy Plus also reports the heating and cooling demands for the evaluated project for a simulation period of one year. These are the results obtained, grouped by basic model demands, improved model demands, and energy savings percentages.

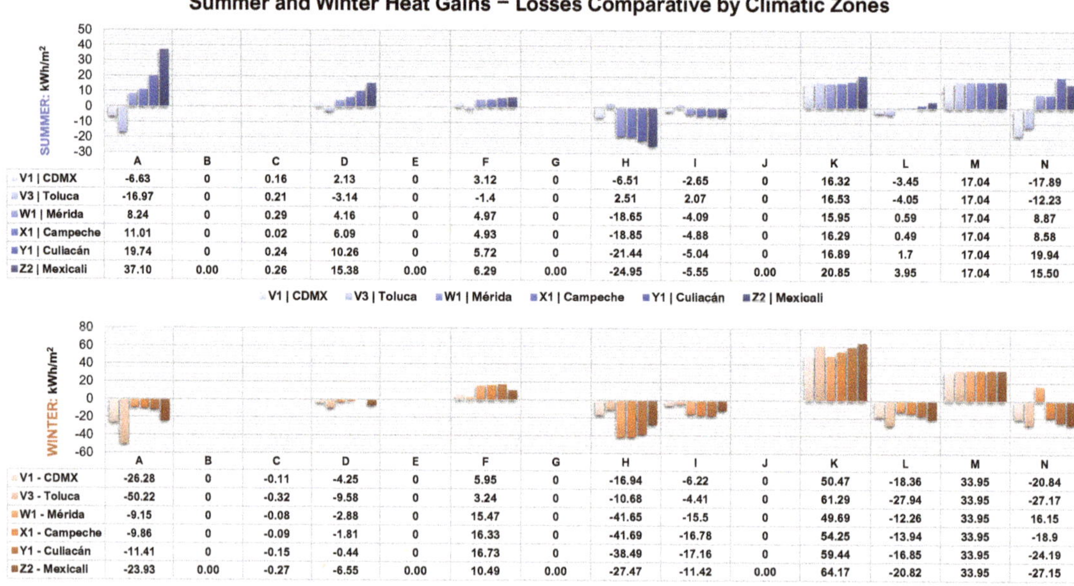

	A	B	C	D	E	F	G	H	I	J	K	L	M	N
V1 | CDMX	-6.63	0	0.16	2.13	0	3.12	0	-6.51	-2.65	0	16.32	-3.45	17.04	-17.89
V3 | Toluca	-16.97	0	0.21	-3.14	0	-1.4	0	2.51	2.07	0	16.53	-4.05	17.04	-12.23
W1 | Mérida	8.24	0	0.29	4.16	0	4.97	0	-18.65	-4.09	0	15.95	0.59	17.04	8.87
X1 | Campeche	11.01	0	0.02	6.09	0	4.93	0	-18.85	-4.88	0	16.29	0.49	17.04	8.58
Y1 | Culiacán	19.74	0	0.24	10.26	0	5.72	0	-21.44	-5.04	0	16.89	1.7	17.04	19.94
Z2 | Mexicali	37.10	0.00	0.26	15.38	0.00	6.29	0.00	-24.95	-5.55	0.00	20.85	3.95	17.04	15.50

	A	B	C	D	E	F	G	H	I	J	K	L	M	N
V1 - CDMX	-26.28	0	-0.11	-4.25	0	5.95	0	-16.94	-6.22	0	50.47	-18.36	33.95	-20.84
V3 - Toluca	-50.22	0	-0.32	-9.58	0	3.24	0	-10.68	-4.41	0	61.29	-27.94	33.95	-27.17
W1 - Mérida	-9.15	0	-0.08	-2.88	0	15.47	0	-41.65	-15.5	0	49.69	-12.26	33.95	16.15
X1 - Campeche	-9.86	0	-0.09	-1.81	0	16.33	0	-41.69	-16.78	0	54.25	-13.94	33.95	-18.9
Y1 - Culiacán	-11.41	0	-0.15	-0.44	0	16.73	0	-38.49	-17.16	0	59.44	-16.85	33.95	-24.19
Z2 - Mexicali	-23.93	0.00	-0.27	-6.55	0.00	10.49	0.00	-27.47	-11.42	0.00	64.17	-20.82	33.95	-27.15

Figure 12. Heat gains and losses of the thermal envelope for summer and winter. The heat gains are reported by heat transference channel: (A) exterior walls, (B) ground contact walls, (C) interior partitions, (D) exterior ceilings, (E) ground ceilings, (F) interior ceilings, (G) open-air floors, (H) ground-contact floors, (I) interior floors, (J) thermal bridges, (K) windows radiation (solar) heat gains, (L) windows' conductive heat gains, (M) internal gains, and (N) air infiltration and ventilation.

3.1. Testing Results 1: Improved Models for Climatic Zones

For the V1 (Table 1) cooling demand, the base model needed 0.40 kWh/m², the improved model needed 0 kWh/m², and the savings was 100%. For V1 heating demand, the base model needed 14.87 kWh/m², the improved model needed 17.10 kWh/m², and the savings was –16%. In this case, the V1 model meets the Mexican standards, but the improved model resulted in larger heating demands due to thermal insulation. For V3 (Table 1) cooling demand, the base model needed 0 kWh/m², the improved model needed 0 kWh/m², and the savings was 0%, because there was no cooling demand due to the effectiveness of the thermal envelope. For V3 heating demand, the base model needed 76.26 kWh/m², the improved model needed 58.69 kWh/m², and the savings was 23%. As shown, the energy savings are similar in both the Mexican standard and in the heating/cooling demands.

For W1 (Table 1) cooling demand, the base model needed 33.30 kWh/m², the improved model needed 2.80 kWh/m², and the savings was 91%. For W1 heating demand, the base model needed 0.46 kWh/m², the improved model needed 0 kWh/m², and the savings was 100%. Although the Mexican standards report energy savings of only 19%, the heating/cooling demand reports savings from 91% to 100%.

Table 1. Energy simulation results of basic and improved models for cooling and heating demands, with their correspondent savings percentage.

Testing Results for Cooling Demands—Improved Models		
Climatic Zone	**Basic Cooling Demand (kWh/m^2)**	**Improved Cooling Demand (kWh/m^2)**	**Energy Saving Percentage (%)**
V1	0.40	0.00	100
V3	0.00	0.00	0
W1	33.30	2.80	91
X1	40.34	2.78	93
Y1	70.83	12.06	83
Z2	104.77	24.16	77

Testing Results Comparison for Heating Demands—Improved Models		
Climatic Zone	**Basic Heating Demand (kWh/m^2)**	**Improved Heating Demand (kWh/m^2)**	**Energy Saving Percentage (%)**
V1	14.87	17.10	−16
V3	76.26	58.69	23
W1	0.46	0.00	100
X1	0.00	0.00	0
Y1	0.59	0.20	66
Z2	26.73	20.68	23

For X1 (Table 1) cooling demand, the base model needed 40.34 kWh/m^2, the improved model needed 2.78 kWh/m^2, and the savings was 93%. For X1 heating demand, the base model needed 0 kWh/m^2, the improved model needed 0 kWh/m^2, and the energy savings was 0% because this climatic zone is tropical and does not require energy for heating.

For Y1 (Table 1) cooling demand, the base model needed 70.83 kWh/m^2, the improved model needed 12.06 kWh/m^2, and the savings was 83%. For the Y1 heating demand, the base model needed 0.59 kWh/m^2, the improved model needed 0.20 kWh/m^2, and the savings was 66%. Analyzing this semi-arid zone with high temperature range (27 to 35 °C for 50% of the year), we found that the cooling demand became more relevant and expensive than heating demand. So, 66% energy savings translates into lower electricity cost for the model.

For the Z2 (Table 1 cooling demand, the base model needed 104.77 kWh/m^2, the improved model needed 24.16 kWh/m^2, and the savings was 77%. For the Z2 heating demand, the base model needed 26.73 kWh/m^2, the improved model needed 20.68 kWh/m^2, and the savings was 23%. This climatic zone with the most extreme conditions had high energy demands for cooling. The heating is needed at night due to being located in a desert area. Although the energy savings are the lowest compared with the other climatic zones, they represent significantly lower energy consumption.

The energy label, graphic proposal and energy demand results, for each climatic zone, are displayed on Figure 13.

	Cooling Demand kWh/m²	Heating Demand kWh/m²
Base Model	0.40	14.87
Improved Model	0.00	17.10
Energy Savings	0.40	-2.27

	Cooling Demand kWh/m²	Heating Demand kWh/m²
Base Model	0.00	76.26
Improved Model	0.00	58.69
Energy Savings	0.00	17.59

	Cooling Demand kWh/m²	Heating Demand kWh/m²
Base Model	33.30	0.46
Improved Model	2.80	0.00
Energy Savings	30.50	0.46

	Cooling Demand kWh/m²	Heating Demand kWh/m²
Base Model	40.34	0.00
Improved Model	2.78	0.00
Energy Savings	37.56	0.00

	Cooling Demand kWh/m²	Heating Demand kWh/m²
Base Model	70.83	0.59
Improved Model	12.06	0.20
Energy Savings	58.77	0.39

	Cooling Demand kWh/m²	Heating Demand kWh/m²
Base Model	104.77	26.73
Improved Model	24.16	20.68
Energy Savings	80.61	6.05

Figure 13. Testing results for the proposed geometric model, Mexican standard certification label, and comparison of heating and cooling demand simulations between the base model and the improved model for Mexican standard certification.

3.2. Testing Results 2: Comparison between Different Construction Systems

The previous results only apply to the traditional reinforced concrete structure fabric walls system. New simulations were run to test the other two proposed construction systems: wood and adobe–metal sheet. Because the goal of this second testing was to quickly compare the systems, the most extreme climatic zones were chosen: V3, Toluca de Lerdo, for the coolest climate zone; Z2, Mexicali, for the warmest climate zone; and V1, Mexico City, for the temperate climate zone. Before running the simulations, there was a problem of units that needed to be addressed. The Mexican energy standard NOM-020-ENER-2011 reports the heat gains of the thermal envelope in Watts. The Energy Plus engine reports the heat gains of the thermal envelope in kilowatts hour per square meter (kWh/m^2). For precise comparison, the results of both methods, the complementary simulation report was modified to obtain a wider range of results. For this comparison, two indicators were taken: window total heat loss rate (for radiation heat gains) and surface inside face conduction heat transfer rate (for conductive heat gains). Both indicators are expressed in Watts. The results obtained from the simulations are depicted in Figure 14.

Energy Plus considers other heat sources and heat gains for the projected building. The internal gains, for example, are an important source of heat for building interiors. In accordance with the conclusions of Turley et al., occupancy heating plays an important role in internal heating and energy performance: "Occupancy-aware heating, ventilation, and air conditioning (HVAC) control offers the opportunity to reduce energy use without sacrificing thermal comfort. Residential HVAC systems often use manually-adjusted or constant set-point temperatures, which heat and cool the house regardless of whether it is needed. By incorporating occupancy-awareness into HVAC control, heating and cooling can be used for only those time periods it is needed" [41]. Turley et al. found that occupancy-aware control of HVAC equipment produces important energy savings due to the contribution of internal gains and no-waste-energy intervals.

This sensitive control theory of HVAC equipment for the proposed new building is supported by Jonghoon's findings about achieving an equilibrium between thermal comfort and the energy use of a building: "Despite the improvement of mechanical thermal models associated with advanced statistical tools have been performed, there is a necessity of the investigation of sensitive control models for supply heating and cooling energy into a single space scale which can be closely related to users' workability and productivity" [42]. Constructing a simulation of HVAC equipment with sensitive control of set-point temperature could be achieved using Energy Plus with a detailed model and advance parameters definition.

The behaviors of the three systems were similar for the three representative climatic zones. The construction system with the lowest heat gains was the traditional system of reinforced concrete and fabric walls. The construction system with the worst performance and the highest heat gains in the three climatic zones was the wood system.This phenomenon occurs because of wood high insulation properties. Wood restricts heat transfer from inside to outside, promoting heat accumulation for interior spaces. For cold climatic zones, this phenomenon helps for lowering heating demand. But for warm climatic zones, the effects are different. In summer, wood does not dissipate heat properly, and the interior spaces have unwanted heat accumulation. The same effect occurs for arid warm and desertic climatic zones. The heat accumulation increases building cooling demands. And for Mexican normative, by absorbing and not dissipating heat gains, wood reported high heat gains balance. In contrast with concrete and adobe systems, because of the breath-ability characteristics of their principal materials.

Notably, the aim of this second testing exercise was to perform a quick comparison between the different systems. For detailed results and exact behaviors, we strongly recommend additional, separate research. Some important findings to highlight from the previous graphs include: The adobe-metal sheet construction system showed remarkable performance in the three climatic zones, although it did not have the best performance.

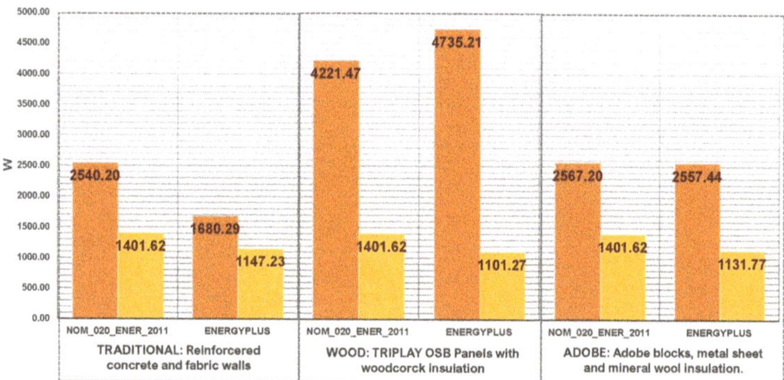

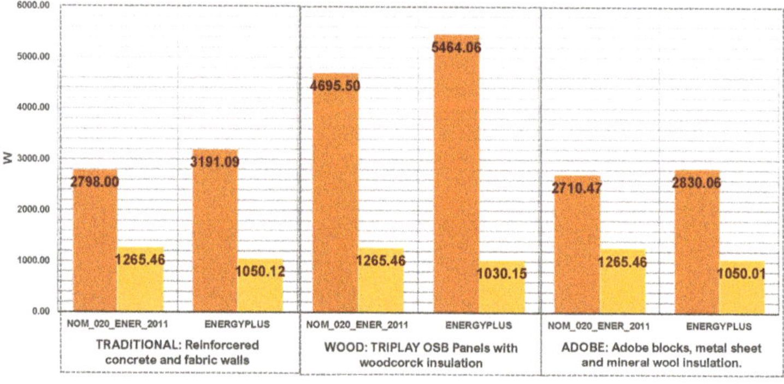

Figure 14. Comparison of the heat gains of the thermal envelope for the three construction systems chosen for testing: traditional reinforced concrete and fabric walls, wood, and adobe with metal sheet. The two graphs with the NOM-020-ENER-2011 label show the calculated heat gains results using the standard method. The two graphs with the Energy Plus label show the heat gains results using the simulation tool software.

It provides a good staring point for new inquiries because this construction system is most environmentally friendly. It produces less waste and has lower primary energy needs for its construction than the traditional reinforced concrete system. The results showed that the three construction systems perform their best in V1, Mexico City, having low heat gains.

This results proved the viability of using vernacular construction techniques for improving the energy efficiency of housing projects. A similar study was developed by Zhai and Previtali. They selected a variety of vernacular techniques (in accordance with climatic zones) and followed a similar methodology to ours. They creates a construction techniques and materials catalogue, and split them into categories by roof, wall, and floor. After comparing and cataloguing the materials and systems, several simulation were run with BEopt software (developed by the U.S. Department of Energy). The computer optimization tool was able to find a combination of vernacular construction techniques that exceeded both the IECC (International Energy Conservation Code) reference case and the observed vernacular case, revealing the potential room for improvement in building codes and vernacular architecture [43].

The simulation also helped to prove the efficiency of the proposed bio-climatic strategies. This helps the architects and energy managers of a project by proving the viability of these design strategies with data. The ventilated air gaps in the warmest climatic zones help to increase the energy savings.This finding agrees with Oropeza-Perez et al. who stated, "Through a sensitivity analysis, it is found that the efficiency of natural ventilation under warm conditions is affected by the following inputs in this order: climate conditions, windows opening schedule, materials of construction, built area, and number of occupants. The potential for saving energy by using natural ventilation is more when the dwelling materials of construction have high heat capacity and the dwelling is located in a hot-dry climate. In a hot-humid climate, low heat capacity materials and natural ventilation help to lower the indoor temperature" [44].

We highlight the differences between the NOM-020-ENER-2011 and Energy Plus results. The Mexican standard results are similar, with very little difference between climatic zones. Conversely, the Energy Plus results have a larger difference between them. This shows how the Mexican standard only considered reference values and standardized coefficients for climate and thermal envelope characteristics, whereas Energy Plus considers more specific variables, like internal gains, shadow elements, and location. In other words, Energy Plus results provide precise information about a project, with its particular characteristics; however, the building must have limited complexity, and the results are basic. As concluded by the authors of the Science Direct article Optimization Tools for Building Energy Model Calibration: "On the one hand, parametric analysis results are exhaustive and show the entire spectrum of results for a given problem, providing a complete picture of the possibilities to consider. On the other, this straightforward 'brute force' approach proved to be quite resource-demanding both regarding calculation time and computational capacity, preventing its implementation when a complex building simulation model is analyzed" [45].

4. Discussion

4.1. Benefits of Using Energy Plus vs. Standard Method

The results produced by the Mexican standard report and Energy Plus report show a variety of differences. The Mexican standard report considers the limited scope of heat gains and verification of standard compliance. Instead, Energy Plus offers a wide variety of possible results (depending on the simulation parameters selected) from energy demands to primary energy consumption (depending also on the level of detail in the model and definition of the project).

Energy Plus results and capabilities are already being used for building energy performance in several softwares and tools. For example, Giancola et al. recorded a map of energy saving potential with a geographic information system (GIS) by applying refurbishment measures and using simulation results [46]. Another direct application for the proposed verification method includes building project management using Building Information Models (BIM). BIM management is a field recently popularized in the construction industry that involves applying a precise team work flow in different departments (structure, cost, engineering, installations, and architectural project) for correctly managing a project. By merging this BIM software with energy simulation software, the manager can analyze and integrate the results, so the pertinent design and engineering decisions can be made on time [47]. Auto-desk developed an energy simulation complement, Revit, for BIM software [48]. With the starting point provided by this research, an energy simulation complement for Mexico could be developed for a BIM management program.

Connecting the Energy Plus simulation tool with the Mexican standard could facilitate applicability to popular architectural and engineer tools on the market. In Spain, the authorities took advantage of simulation tools' practicality to make the energy certification process fast and easy for the construction industry. Therefore, Mexican authorities should follow the same path by starting to include the use of simulation tools for applying the energy standards.

Using the energy simulation engine, the user can obtain a wide range of results, but the results depends on the scope of analysis. Several options of simulation software are available on the market depending on the design process stage in which the user is involved (conception, project, or execution), and the complexity of the work (basic, intermediate, or advanced). Each simulation tool evolves from detailing and requirements, depending on if the user needs a basic program of intuitive interface tools or a specialized study with a wide scope (analyzing different project phases and application magnitudes) [49]. With Energy Plus, the user can enter other important building variables excluded by the standard, like occupation calendar, location characteristics, and HVAC equipment design. The new results allow the user to analyze other variables in addition to those considered in the Mexican standard results of thermal envelope heat gains, such as energy demands, primary energy used for building procedures, an estimated consumption of electricity and sanitary hot water, total hours where the set-point for comfort is not met, and the precise elements that have heat gains or losses (windows, walls, internal gains, soil, ceiling, etc.). Our results pave the way for a parametric analysis tool in the quest for a complete energy verification model. "The parametric analysis tool demonstrates the potential of parametric analysis, in finding optimal building envelope solutions in terms of operational energy, embodied CO_2eq emissions and embodied energy. In the future, the parametric analysis tool may be used for setting energy performance goals and benchmarks, optimizing renewable energy and passive systems, integrating architectural features, minimizing changes during construction and integrating building systems" [50].

4.2. Aims Fulfillment: Why Improved Instead of Optimized?

The aims of the testing methodology were fulfilled by designing an energy model for verification of building compliance with the standard. Notably, the principal aim of the proposed model is to verify regulation compliance in this study. The testing method uses improved models and not an optimized model. The reasons for this are as follows.

Mexican social housing differs from the European or American context. The Mexican government does not provide houses directly for the population. Instead, the government provides loans and financial aid for workers. The principal source of housing acquisition for workers and the poor is self-construction. Due to the economic informality and high poverty rates, those requiring social housing experience problems acquiring a loan, mortgage, or financial aids, so the workers save money and build their own houses [11]. Considering the social housing situation, the proposed model prioritizes energy certification with affordable construction plans. An optimized model that complies with the net-zero carbon buildings 2030 target is a distant and inapplicable aim for Mexican social housing. There is no local suppliers for the required equipment, like thermal-break profiles or high-tech materials, so they need to be imported at high prices. Another problem is the lack of public support for self-construction improvements. To increase economic savings during the construction process, self-constructors prefer to use traditional techniques, and avoid investing in specialized studies, like structural or energy performance analyses. This situation provoked an increase in informal social housing and urban planning problems.

So, to adapt the proposed model for application to the self-construction reality, we compared the energy performance results between the three most-used traditional construction systems. The simulation showed promising results, like the good efficiency of the adobe system for thermal envelope heat gains. The adobe system is well-built by the national labor force and works for the majority of the climatic zones. Adobe is a good material choice for implementing thermal mass strategy, and is more affordable than the reinforced concrete system.

The reason for preferring improved models over an optimized model is for testing the proposed model on a more accurate social housing situation. It is preferable to test and verify compliance with the regulation using affordable and simple bio-climatic strategies (for achieving the energy standard compliance and obtaining an energy efficiency label), rather than implementing complex strategies, high-tech materials, and specialized procedures for achieving optimized results. This does not mean that an NZEB housing model is impossible: the implementation should be step by step. First, the improved models should be promoted for increasing the awareness of the Mexican population about the benefits of energy efficiency strategies. If the market and the authorities increase the application of the Mexican energy standard, the demand for better techniques and materials will increase too. Mexico's authorities are beginning to implement sustainable practices for construction, though there is still a long way to go to achieving the same level of energy efficiency awareness and legislation as in the European Union.

How does the research is aligned with the vision set by Mexico Green Building Council? The Mexico Green Building Council recognizes the NOM_020_ENER_2011 as a mandatory normative, although its applicability has not been strictly monitored by authorities. The research helps the realization of their vision, by focusing on make energy saving building strategies available for a larger segment of population. The Green Council works for bringing educational tools for sustainable building practices, and the proposed model will help as a tool for testing and certifying the norm compliance with social housing models [51].

The proposed model implements technical data from the traditional construction systems available in Mexico. With the climatic zoning designed with Mexican meteorologic information, the model served for testing the building systems energetic performance, their compliance with the national normative, and the possible energy savings by applying several strategies. The model covers several energy requirements for EDGE and LEED certifications. Not all requirements, but is a good starting point for implementing Green Building Council principles of energy performance for self-construction and Mexican social housing. the Mexico Green Building Council encourages the NOM_020_ENER_2011 learning and knowledge for building professionals because of its mandatory use not only in building energy performance, but also because of its utility for materials and equipment certification (home appliances, insulation materials, electric materials, etc.) [52].

How does the proposed model help the AEC (Architecture, Engineering and Construction) Industry for achieving savings for energy consumption? The model design strategies for improvement are oriented for energy savings in operation phase. It did not consider other life cycle stages, like materials extraction or building. But it is possible to make a comparative for energy consumption reduction with an Energy plus report. The normative applied on temperate climatic zones is not functional.

As shown in Figure 15, the normative applied on temperate climatic zones is not functional. The improved model that complied with the energy normative, had higher energy consumption rate than the basic model, resulting on a 1% increment on energy consumption, and a 10% increment of primary energy. But considering that only 10 of 31 Mexican states are classified as tempered climates, the proposed model is applicable and functional for 60% of national territory.

For cold climatic zones (Figure 15), the proposed model showed more efficiency than temperate climates. The improved model that complied with the energy normative, reported lower energy consumption rate, resulting on energy savings of 19% compared to basic model results. For primary energy, the improved model reported energy savings of 21% compared to basic model results. A 20% of energy savings tendency for heating demand cases is a competitive value considering that 10 of 31 Mexican states are classified as cold climates.

V1 – México City		
	Energy Consumption (kWh/m²)	Primary Energy (kWh/m²)
Base Model	45.5	97.1
Improved Model	46.0	106.7
	Savings (%)	
	-1.1	-9.9

V3 – Toluca		
	Energy Consumption (kWh/m²)	Primary Energy (kWh/m²)
Base Model	107.5	202.7
Improved Model	87.6	160.4
	Savings (%)	
	18.5	20.9

Z2 – Mexicali		
	Energy Consumption (kWh/m²)	Primary Energy (kWh/m²)
Base Model	159.3	401.9
Improved Model	73.7	181.9
	Savings (%)	
	53.7	54.7

Figure 15. Comparative Results of Energy Consumption and Primary Energy for the three most representative climatic zones, with their correspondent energy savings percentage.

Remembering the statistics published by INEGI census of 2018, 40% of the northern states' families use air conditioning, and their principal energy consumption waste is electricity (used for HVAC equipment operation). With this context, the results listed in Figure 15 are promising. The improved model that complied with the energy normative, reported lower energy consumption rate, resulted on energy saving of 54% compared to basic model results. For primary energy, the improved model reported energy savings of 55% compared to basic model results. For Warm climatic zones, the proposed model reduces by half the building total energy consumption. A 50% energy saving rate tendency, applicable for 11 of 31 Mexican states with warm climatic classification, is enough for implementing the proposed model at least for this type of climatic zone.

In conclusion, the proposed model reported acceptable energy savings for 21 of 31 Mexican states. In other words, the model is effective and applicable for at least 60% of states, and a 60% of total national territory. It has enough potential market for promoting its application and development, with a national mean energy saving rate of 24% for energy consumption and 21% for primary energy.

5. Conclusions

In this study, we pursued four principal aims. The first aim was the creation of new climatic zoning for Mexico (with a local weather database) for delimiting the winter and summer severity for each capital city state. The actual climatic zoning available was developed by UNAM and considers several geographic and meteorological statistics. In this regard, we concluded that the proposed climatic zoning shares similarities with the Spanish climatic zoning, so this new zoning is compatible with the Open Studio and

SGSAVE work schemes. This compatibility will help with the adaptation of SGSAVE software to the Mexican market.

The second aim was the study of the Mexican energy standard (NOM-020-ENER-2011) for its application in the proposed add-on for Open Studio. For the verification method, reference values and directives were analyzed and applied. Fortunately, Open Studio add-on software was found and used for verifying meeting the standard requirements. We concluded that the Mexican standard has the potential to be adapted to and included in the software. For now, the complement offers a simple report of verification of whether the project complies with the standard, and a final list of radiation and conduction heat gains for the proposed and reference building. Analyzing the results obtained from the testing exercise, we concluded that the standard should be updated on the add-on by programming the new directives and reference values. First, it is important to promote the use of the Mexican standard for motivating implementation of the reference values and directives at present. The strict application of the standard by the relevant authorities should involve promoting the benefits of energy demands/consumption simulations.

The third aim was to research and construct a catalogue of the principal and traditional construction systems in Mexico. Using a public census of Mexican housing, we determined the principal construction system used for walls, ceilings, and floors. By identifying the materials that compose each system, a material catalogue was developed and uploaded in a single file of SGSAVE for the fourth aim. We concluded that, like the standard-verification directives, the information can be adapted for the software and programmed for the Mexican SGSAVE version. However, some additional technical information should be researched and included for new materials (like specific heat). The SGSAVE version includes the most-used construction materials in Spain, the entire Saint Gobain catalogue, and some specialized materials available in Europe. For the new version, some vernacular and manual techniques should be studied and included, like bahareque, adobe, and palm-leaves ceiling. In other words, for completing the catalogue, more profound research to obtain vernacular material technical data is needed.

For the fourth aim, a testing exercise of the verification method was performed. A base model was designed to obtain initial results. Then, we improved/specified the base model for the most common climatic zones. We verified the standard, resulting in all improved models complying with the directives, and we compared the energy standard and Energy Plus results. For the final test, the proposed models of two other construction systems (wood and adobe/steel) were simulated and the results were compared with those for the reinforced concrete system. We found that the proposed model worked in all the tested models for achieving its goal of verifying the compliance with the standard. Although the model succeed for verifying the standard, it did not achieve the same energy savings level for all climatic zones; one zone showed a higher energy demand than that of the base model. An implementation and reshaping of the Mexican standard is recommended for ensuring a minimum level of energy savings for all climatic zones.

We also demonstrated the differences between the Mexican standard and Energy Plus reports. Energy Plus provides a more complete document with specified heat gains sources, virtual energy demands by an HVAC ideal loads scheme (for design references on the definitive HVAC equipment design), etc. We fulfilled our four aims and proved the viability of developing a Mexican SGSAVE version with a local verification method of compliance with the energy standard, with some pending research aims to be fulfilled in future works.

By constructing an adequate tool, the energy standard could be improved into a complete environmental regulation. The environmental standard official report would provide not only verification of meeting the standard; it will also serve to indicate the economic value of a property, the reliability of a mortgage credit, and the fulfillment of energy credits toward a private environmental certification. In conclusion, by knowing all the benefits of improving the existing standard with software within a simulation engine, we can verify and prove the differences between the results before and after the changes between methods.

Further Research

After summarizing all the results of the research, some initial aims were completed. However, other issues remain outstanding. All research projects lead to other starting points for new research. These are the pending aims:

1. The final development of an SGSAVE version for the Mexican standard and context: At the moment, the complementary script only offers the final verification results, analyzing the heat gains. The SGSAVE Mexican version should include an energy performance verification, including building performance with HVAC ideal loads, detecting the elements that more strongly influence the heat gains report, etc.
2. Self-construction energy performance guide: A direct method for conducting the verification in practical use for social benefits is the creation of a self-construction manual with public coverage. This manual should promote the use of vernacular construction systems, and provide the energy performance results for their promotion.
3. Economic viability of certified projects: It is important to prove to the market that a certified building meeting the Mexican energy standard is an attractive business, by evaluating possible usable technologies. One of the principal technologies promoted by this guide will be the photovoltaic panel due to the potential the country has for solar energy harvesting. All climatic zones have a high solar energy collection potential (from 1600 to 1900 kWh per year).

These research opportunities should be explored, taking advantage of professional tools like Energy Plus, Open Studio, and SGSAVE. These are the initial steps of constructing a complete energy certification for Mexico, but there is more work to be done. In Critical Features of Energy Simulation for Single Housing, the authors explained: "These results are useful and beneficial for making certifications, comparing with previous references, checking other alternatives, developing individual and global improvements. Some results allow the integration of costs, make comparisons from a base case or consecutive simulations" [49].

Author Contributions: Conceptualization, All authors; Methodology, A.J.G.D.; Validation and Formal Analysis, all authors; Investigation, Resources, and Data Curation, A.J.G.D.; Writing—Original Draft Preparation, A.J.G.D.; Writing—Review & Editing, Visualization, Supervision, all authors. All authors have read and agreed to the published version of the manuscript.

Funding: This research received no external funding.

Institutional Review Board Statement: Not applicable.

Informed Consent Statement: Not applicable.

Data Availability Statement: Not applicable.

Conflicts of Interest: The authors declare no conflict of interest.

Abbreviations

The following abbreviations are used in this manuscript:

ASHRAE	American Society of Heating, Refrigerating and Air-Conditioning Engineers
BCL	Building Component Library
BEM	Building Energy Model
BIM	Building Information Model
BMS	Building Management Systems
CONAFOR	Comision Nacional de Recursos Forestales
CONUEE	Comision Nacional del Uso Eficiente de la Energia

CTE	Código Tecnico Español
EPS	Expanded Polystyrene
EPW	Energy Plus Weather File
HP	Heat Pump
HVAC	Heating Ventilation and Air Conditioning
IEEC	International Energy Conservation Code
INEGI	Instituto Nacional de Estadistica y Geografia
MPC	Model Predictive Control
NZEB	Nearly Zero Energy Building
NSGA-II	Non-dominated sorting genetic algorithm II
OSB	Oriented Strand Board
PM	Project Management
PMV	Predicted Mean Vote
PV	Photovoltaic
PV-T	Photovoltaic–Thermal Module (PV-T)
LEED	Leadership in Energy and Environmental Design
LCADA	Low Carbon Architecture Danish Agency
RES	Renewable Energy Sources
ROI	Return on Investment
VRF	Variable Refrigerant Flow
SC1	Model without any renewable energy
SC2	Model with the renewable energy consumption but with a fixed set-point of 20
SC3	Improved model with variable set-points
SGSAVE	Saint Gobain Software Avanzado de Verificacion energya (Saint Gobain Advance energy Verification Software)

Appendix A. Summary of Results

The construction system proposed for the basic and improved models used in the test exercise is described in detail on the following diagrams:

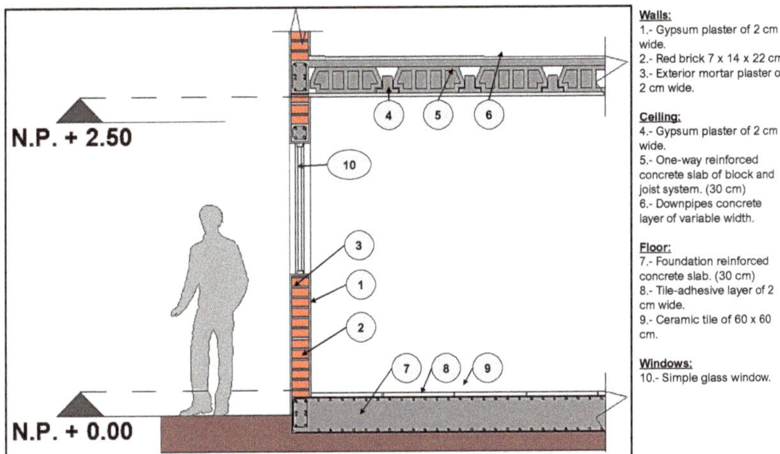

Figure A1. Construction details of the proposed base model.

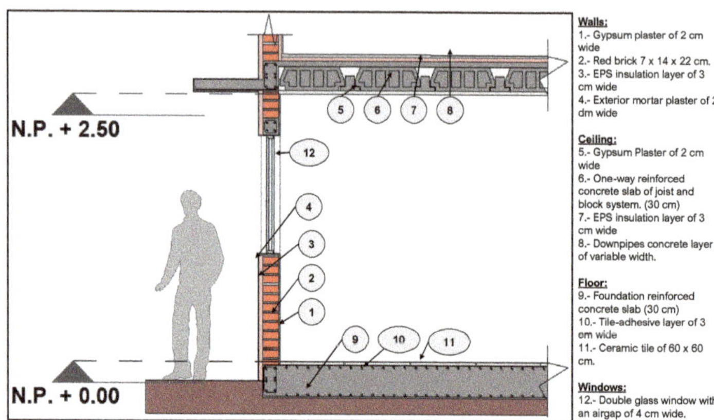

Figure A2. Construction details of the improved model of the V1 climatic zone, Mexico City.

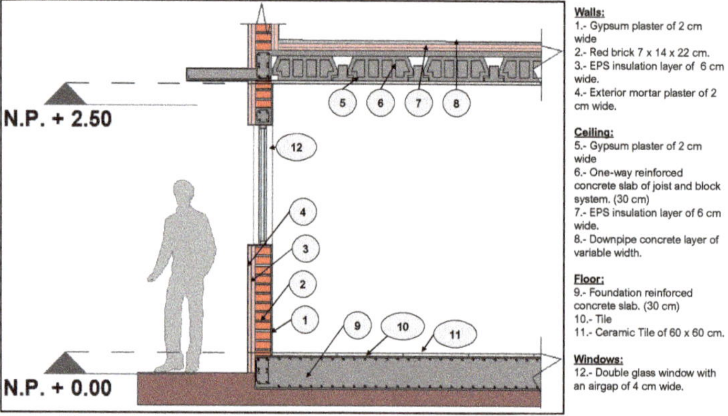

Figure A3. Construction details of the improved model of the V3 climatic zone, Toluca de Lerdo.

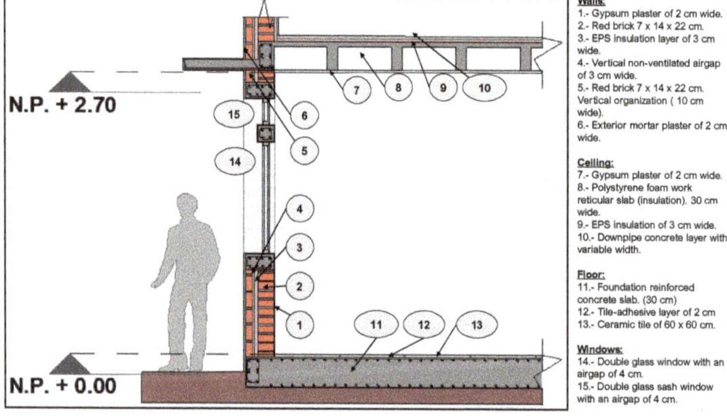

Figure A4. Construction details of the improved model of the W1 climatic zone, Merida.

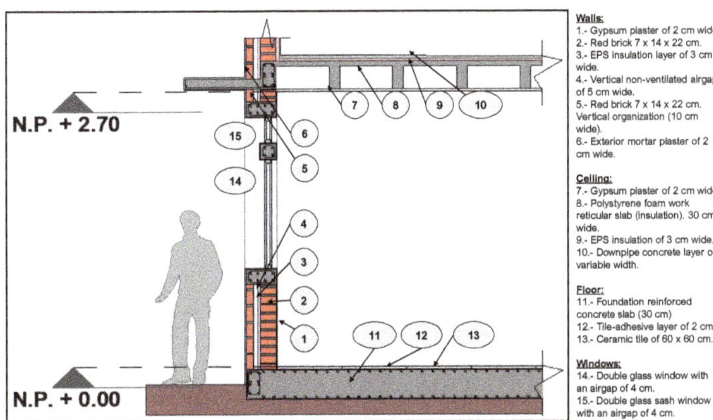

Figure A5. Construction details of the improved model of the X1 climatic zone, Campeche.

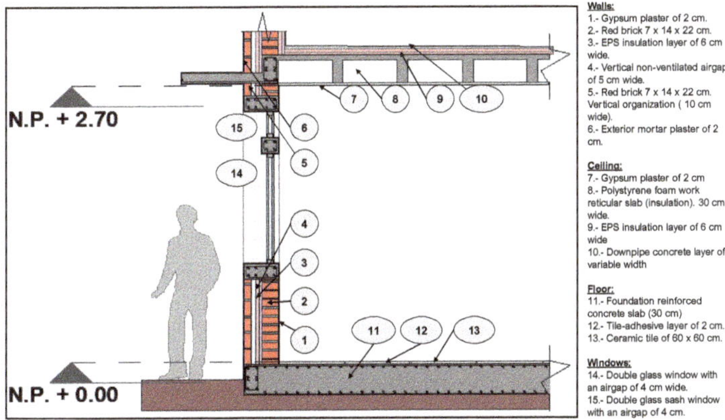

Figure A6. Construction details of the improved model of the Y1 climatic zone, Culiacan.

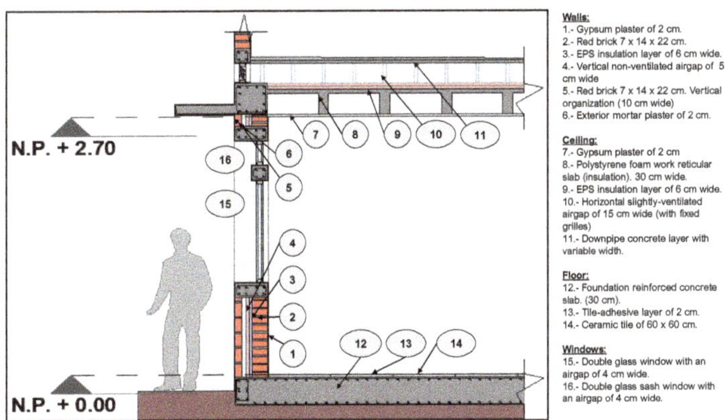

Figure A7. Construction details of the improved model of the Z2 climatic zone, Mexicali.

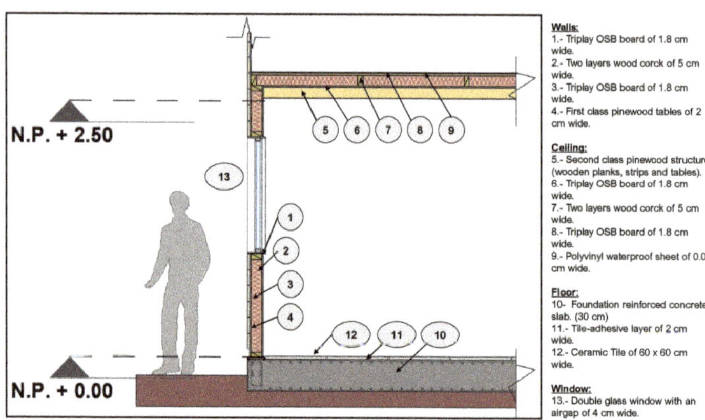

Figure A8. Construction details of the improved model for the wood–cork system.

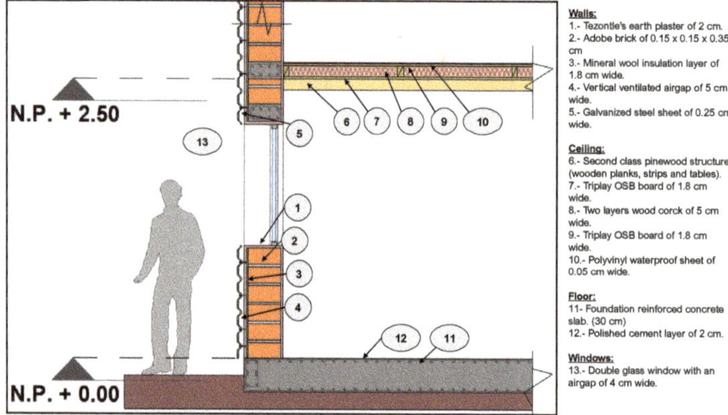

Figure A9. Construction details of the improved model for the adobe (mud and straw) and galvanized steel system.

References

1. Chávez, V. INFONATIV Tiene 100 mil Viviendas Abandonadas y en Litigio por Falta de Pago. El Financiero. 2017. Available online: https://rebrand.ly/elfinanciero_infonavit100viviendas (accessed on 15 May 2020).
2. Griego, D.; Krarti, M.; Hernández-Guerrero, A. Optimization of energy efficiency and thermal comfort measures for residential buildings in Salamanca, Mexico. *Energy Build.* **2012**, *54*, 540–549. [CrossRef]
3. Diario Oficial de la Federación. Norma Oficial Mexicana de Eficiencia Energética en Edificaciones NOM-020-ENER-2011. Mexico. 2011. Available online: https://rebrand.ly/nom020ener2011 (accessed on 15 May 2020).
4. Du, H.; Bandera, C.F.; Chen, L. Nowcasting Methods for Optimising Building Performance. 2019. Available online: http://www.ibpsa.org/proceedings/BS2019/BS2019_210777.pdf (accessed on 11 October 2020)
5. Fernández Bandera, C.; Pachano, J.; Salom, J.; Peppas, A.; Ramos Ruiz, G. Photovoltaic Plant Optimization to Leverage Electric Self Consumption by Harnessing Building Thermal Mass. *Sustainability* **2020**, *12*, 553. [CrossRef]
6. Lucas Segarra, E.; Du, H.; Ramos Ruiz, G.; Fernández Bandera, C. Methodology for the quantification of the impact of weather forecasts in predictive simulation models. *Energies* **2019**, *12*, 1309. [CrossRef]
7. LCADA. Manual Técnico Para la Aplicacion de la NOM-020-ENER-2015. CONUEE. 2015. Available online: https://rebrand.ly/guiacalculoNOM_020_ENER_2011 (accessed on 5 October 2020).
8. Negrete Prieto, R.; Romo Anaya, M. Cuantificando a la clase media en México en la primera década del siglo XXI: Un ejercicio exploratorio. *Real. Datos Y Espacio. Rev. Int. Estad. Y Geogr.* **2014**, *5*, 62–95.
9. *Norma Mexicana NMX-AA-164-SCFI-2013: Edificación Sustentable—Criterios y Requerimientos Ambientales Mínimos*; SEMARNAP: Mexico, 2013; p. 158. Available online: https://biblioteca.semarnat.gob.mx/janium/Documentos/Ciga/agenda/DOFsr/DO3156.pdf (accessed on 10 October 2020).

10. INFONAVIT. Hipoteca Verde. Mexico. 2017. Available online: https://rebrand.ly/INFONAVIThipotecaverde (accessed on 15 May 2020).
11. INEGI. ENCEVI 2018: Primera Encuesta Nacional sobre Consumos de Energéticos en Viviendas Particulares. 2019. Available online: https://rebrand.ly/encuestaENCEVI2018 (accessed on 8 May 2020).
12. Crawley, D.B.; Lawrie, L.K.; Winkelmann, F.C.; Buhl, W.F.; Huang, Y.J.; Pedersen, C.O.; Strand, R.K.; Liesen, R.J.; Fisher, D.E.; Witte, M.J.; et al. EnergyPlus: Creating a new-generation building energy simulation program. *Energy Build.* **2001**, *33*, 319–331. [CrossRef]
13. ASHRAE. ASHRAE fundamentals handbook. In *American Society of Heating Refrigeration and Air-Conditioning Engineers*; ASHRAE: Atlanta, GA, USA, 2001.
14. Segarra, E.L.; Ruiz, G.R.; González, V.G.; Peppas, A.; Bandera, C.F. Impact Assessment for Building Energy Models Using Observed vs. Third-Party Weather Data Sets. *Sustainability* **2020**, *12*, 6788. [CrossRef]
15. Han, J.; Bae, J.; Jang, J.; Baek, J.; Leigh, S.B. The Derivation of Cooling Set-Point Temperature in an HVAC System, Considering Mean Radiant Temperature. *Sustainability* **2019**, *11*, 5417. [CrossRef]
16. Yuan, J.; Emura, K.; Farnham, C. Effects of recent climate change on hourly weather data for HVAC design: A case study of Osaka. *Sustainability* **2018**, *10*, 861. [CrossRef]
17. Zepeda, R.V. *Las Regiones Climáticas de México 1.2. 2*; Universidad Nacional Autonoma de Mexico: Mexico City, Mexico, 2005; Volume 2.
18. Liverman, D.M.; O'Brien, K.L. Global warming and climate change in Mexico. *Glob. Environ. Chang.* **1991**, *1*, 351–364. [CrossRef]
19. Bai, L.; Wang, S. Definition of new thermal climate zones for building energy efficiency response to the climate change during the past decades in China. *Energy* **2019**, *170*, 709–719. [CrossRef]
20. Catálogo de Casas INFONATIV. INFONAVIT Web Page. 2020. Available online: https://rebrand.ly/catalogoinfonavit2020 (accessed on 8 May 2020).
21. EFINOVATIC: Certificación Energética de Edificios. EFINOVATIC Web Page. 2020. Available online: https://www.efinovatic.es/energyPlus/ (accessed on 15 May 2020).
22. Pascual Busain, M.A. *SG SAVE Software v.2.8.0.2*; EFINOVATIC: Pamplona, Spain, 2019.
23. Repository of Free Climate Data for Building Performance Simulation. Climate One Building Web Page. 2017. Available online: http://climate.onebuilding.org/ (accessed on 15 May 2020).
24. *Climate Consultant Software v.6.0.*; U.S. Deparment of Energy: Washington, DC, USA, 2016.
25. Gavilán Casal, A. Análisis Comparativo de la Eficiencia Energética en Edificios Existentes con Diferentes Herramientas de Simulación Energética. Facultad de Ingenieria Industrial. PhD's Thesis, Universidad de Valladolid, Valladolid, Spain, 2015.
26. Forbes. 9 de Cada 10 Mexicanos Quiere Comprar una Casa, 45 no Tiene Recursos. Forbes Economía y Finanzas. 2018. Available online: https://www.forbes.com.mx/9-de-cada-10-mexicanos-quiere-comprar-una-casa-45-no-tiene-recursos/ (accessed on 8 May 2020).
27. INEGI. Encuesta Nacional de Ingresos y Gastos de los Hogares 2018: ENIGH. 2019. Available online: https://rebrand.ly/EnighencuestaINEGI (accessed on 8 May 2020).
28. Gálligo, P.L. *Un Techo Para Vivir: Tecnologías Para Viviendas de Producción Social en América Latina*; Univ. Politèc. de Catalunya: Catalunya, Spain, 2005; Volume 1.
29. Oregon State University. When Is Asbestos Dangerous? Environmental Health and Safety Web Page. 2020. Available online: https://ehs.oregonstate.edu/asb-when (accessed on 28 May 2020).
30. García, A.K. 14 Millones de Viviendas en México no Son Dignas. El Economista. 2020. Available online: https://rebrand.ly/economista14mviviendasnodignas (accessed on 28 May 2020).
31. National Renewable Energy Laboratory. *Openstudio Software v.2.7.0*; USA Department of Energy: Washington, DC, USA, 2019.
32. EUR-LEX. Council Directive 92:75:EEC of 22 September 1992 on the Indication by Labelling and Standard Product Information of the Consumption of Energy and Other Resources by Household Appliances. 1992. Available online: https://eur-lex.europa.eu/legal-content/EN/ALL/?uri=CELEX%3A31992L0075 (accessed on 20 November 2020).
33. Contreras, M. ITOM: Herramienta de Cumplimiento de las NOMs. Youtube Video Platform. 2019. Available online: https://youtu.be/-7GFACJU-wo (accessed on 10 June 2020).
34. Bienvenido-Huertas, D. Analysis of the Impact of the Use Profile of HVAC Systems Established by the Spanish Standard to Assess Residential Building Energy Performance. *Sustainability* **2020**, *12*, 7153. [CrossRef]
35. Código Técnico Español: Documento Básico HE—Ahorro de Energía 2019. Secretaria de Estado de Infraestructuras, Transporte y Vivienda. Ministerio de Fomento, Spain. 2019. Available online: https://rebrand.ly/codigotecnicoespanol2019 (accessed on 10 June 2020).
36. Gumbarević, S.; Burcar Dunović, I.; Milovanović, B.; Gaši, M. Method for Building Information Modeling Supported Project Control of Nearly Zero-Energy Building Delivery. *Energies* **2020**, *13*, 5519. [CrossRef]
37. Heywood, H. *101 Rules of Thumb for Low Energy Architecture*; Routledge: Abingdon-on-Thames, UK, 2019.
38. CONAFOR. Manual de Auto-Construcción de Vivienda con Madera, Mexico. 2000. Available online: https://rebrand.ly/CONAFORmanualviviendamadera (accessed on 10 June 2020).
39. Correa Giraldo, V. El Bahareque, un Sistema Constructivo Sismorresistente y Sustentable Para Soluciones de Vivienda Social en México. Research Gate Web Page. 2014. Available online: https://rebrand.ly/baharequesismoresistente (accessed on 10 June 2020).

40. HOME DEPOT Construction Materials Digital Catalogue. Home Depot Web Page. 2020. Available online: https://www.homedepot.com.mx/materiales-de-construccion (accessed on 15 June 2020).
41. Turley, C.; Jacoby, M.; Pavlak, G.; Henze, G. Development and Evaluation of Occupancy-Aware HVAC Control for Residential Building Energy Efficiency and Occupant Comfort. *Energies* **2020**, *13*, 5396. [CrossRef]
42. Ahn, J. Improvement of the Performance Balance between Thermal Comfort and Energy Use for a Building Space in the Mid-Spring Season. *Sustainability* **2020**, *12*, 9667. [CrossRef]
43. Zhai, Z.J.; Previtali, J.M. Ancient vernacular architecture: Characteristics categorization and energy performance evaluation. *Energy Build.* **2010**, *42*, 357–365. [CrossRef]
44. Oropeza-Perez, I.; Østergaard, P.A. Energy saving potential of utilizing natural ventilation under warm conditions—A case study of Mexico. *Appl. Energy* **2014**, *130*, 20–32. [CrossRef]
45. Lara, R.A.; Naboni, E.; Pernigotto, G.; Cappelletti, F.; Zhang, Y.; Barzon, F.; Gasparella, A.; Romagnoni, P. Optimization tools for building energy model calibration. *Energy Procedia* **2017**, *111*, 1060–1069. [CrossRef]
46. Soutullo, S.; Giancola, E.; Sánchez, M.N.; Ferrer, J.A.; García, D.; Súarez, M.J.; Prieto, J.I.; Antuña-Yudego, E.; Carús, J.L.; Fernández, M.Á.; et al. Methodology for Quantifying the Energy Saving Potentials Combining Building Retrofitting, Solar Thermal Energy and Geothermal Resources. *Energies* **2020**, *13*, 5970. [CrossRef]
47. Fregonara, E. Methodologies for supporting sustainability in energy and buildings. The contribution of Project Economic Evaluation. *Energy Procedia* **2017**, *111*, 2–11. [CrossRef]
48. Autodesk. About Energy Analysis for Autodesk® Revit®. Autodesk Knowledge Network Webpage. 2017. Available online: https://rebrand.ly/jhg2x (accessed on 15 May 2020).
49. García-Alvarado, R.; González, A.; Bustamante, W.; Bobadilla, A.; Muñoz, C. Características relevantes de la simulación energética de viviendas unifamiliares. *Inf. Constr.* **2014**, *66*, 005. [CrossRef]
50. Lolli, N.; Fufa, S.M.; Inman, M. A parametric tool for the assessment of operational energy use, embodied energy and embodied material emissions in building. *Energy Procedia* **2017**, *111*, 21–30. [CrossRef]
51. Mexico, G.B.C. Mexico Green Building Council Webpage: Aims and Projects. 2021. Available online: https://www.gbci.org/mexico (accessed on 20 January 2021).
52. Sánchez, C.G. ¿Qué es y Que se Obtiene Con la Edificación Sustentable? CMIC Web Page. 2013. Available online: https://rb.gy/hsqjcw (accessed on 20 January 2021).

sustainability

Article

Determining the 2019 Carbon Footprint of a School of Design, Innovation and Technology

Guillermo Filippone [1,*]**, Rocío Sancho** [2] **and Sebastián Labella** [3]

[1] Department of Product Design, Escuela Universitaria de Diseño, Innovación y Tecnología ESNE, 28016 Madrid, Spain

[2] Department of Interior Design, Escuela Universitaria de Diseño, Innovación y Tecnología ESNE, 28016 Madrid, Spain; rocio.sancho@esne.es

[3] CarbonFeel, Funciona Fundation, 28007 Madrid, Spain; sebastian.labella@carbonfeel.org

* Correspondence: guillermo.filippone@esne.es

Abstract: As a contribution to the fight against climate change, ESNE's 2018/19 carbon footprint has been evaluated using the CarbonFeel methodology, based on ISO 14069 standards. In the scenario studied, greenhouse gas (GHG) emissions produced by direct and indirect emissions have been included. For comparative purposes, a second scenario has been analyzed in which fossil fuels used for heating are replaced by electrical energy from renewable sources. A decrease of 28% in GHG emissions has been verified, which could even reach 40% if the energy for thermal conditioning was replaced by renewables.

Keywords: climate change; global warming; carbon footprint; GHG emissions; climate emergency

Citation: Filippone, G.; Sancho, R.; Labella, S. Determining the 2019 Carbon Footprint of a School of Design, Innovation and Technology. *Sustainability* **2021**, *13*, 1750. https://doi.org/10.3390/su13041750

Academic Editor: Roberto Alonso González Lezcano

Received: 31 December 2020

Accepted: 29 January 2021

Published: 6 February 2021

Publisher's Note: MDPI stays neutral with regard to jurisdictional claims in published maps and institutional affiliations.

1. Introduction

The consumption of resources in the last century has experienced an exponential growth in all fields, as indicated in the Special Report on Climate Change and Land of the Intergovernmental Panel on Climate Change (IPCC) [1], boosted in a synergetic cycle by the exponential increase in population, growing up from 1 billion at the beginning of the 20th century to the current 7.7 billion, with perspectives of reaching between 9.5 (the most optimistic scenario), to 12.5 million in 2100 (the most pessimistic) [2]. The IPCC report clearly outlines that the increase in Greenhouse Gas (GHG) emissions (mainly CO_2, N_2O, CH_4, O_3, CFC, H_2O) produced by human activity are responsible for the acceleration of the current climate change. The IPCC have estimated an average increase of 1.5 °C as a safe limit to avoid catastrophic and irreversible global changes for the planet. Above 2 °C, the consequences can have unpredictable effects on life on Earth [3]. Primary energy consumption continues expanding (1.3% last year) [4]. This process has also intensified the generation of solid, gaseous, liquid and radioactive wastes [5]. This path has brought us to the record level of GHG in the history of the planet, raising from the 300 CO_2 ppm maximum historical level to more than 415 ppm nowadays [6].

The European Union (EU), signatory of the Paris Agreement, assumed a leadership role in promoting measures to restrict it to 1.5 °C [7]. European policies have been establishing frameworks for action, first until 2020 and then for 2030 [8]. The European Green Deal presents an action plan to make the EU's economy sustainable [9], and a proposal for the first European Climate Law (EUR-Lex, 2020) establishing the framework to achieve climate neutrality and amend Regulation (EU) 2018/1999 [10]. The Draft Law 121/000019 on Climate Change and Energy Transition submitted in May 2020 in Spain, aims to achieve, by 2050, climate neutrality and an electricity system based exclusively on generation of renewable sources [11].

1.1. Approaches to Environmental Assessment of Buildings

With the building and construction sector being one of the major sources of emissions, since the first initiatives to fight against climate change, several approaches have been proposed for its assessment. The Life Cycle Assessment (LCA) is a methodology for assessing environmental impacts associated with every stage of the life cycles of products, including the final disposition, which is also used for construction [12]. Beyond the contribution of this methodology to the understanding of the polluting effects of a product or construction throughout all stages, it is often difficult to put into practice and too complex to be analyzed by designers in order to make decisions about the improvement of the selection of designs and materials. Ecological Footprint (EF) is another commonly used approach to measure the ecological assets of natural resources of a given activity or population in terms of "Global hectares", tracking six categories of productive surface areas: cropland, grazing land, fishing grounds, built-up land, forest area, and carbon demand on land [13,14]. Both methodologies could be complementary, since the LCA is more detailed in terms of coverage of impact categories and EF takes into account the carrying capacity of the territory [15].

Without leaving aside the validity of these approaches, the climate urgency in terms of global warming makes it appropriate to emphasize the Carbon Footprint (CF) approach for design optimization. The CF derives its name from the EF, but does not share the sense of pressure in terms of use of territory; it expresses the impact on global warming in units of tons of CO_2, taking into consideration not only carbon dioxide emissions, but also other gases with greenhouse potential effects in relation to CO_2 (GHG). Several definitions of CF can be found in the literature. The Global Footprint Network interprets CF as "the fossil fuel footprint part of the EF or the demand on CO_2 land" [16]. A more comprehensive definition it is provided by Wiedmann and Minx: "The carbon footprint is a measure of the exclusive total amount of carbon dioxide emissions that is directly and indirectly caused by an activity or is accumulated over the life stages of a product" [17]. Although there is no standard methodology for evaluation, a variety of literature supporting the use of CF in construction impact evaluations across the world can be found [18] for new buildings and rehabilitation [19]. CF can therefore be assessed by different methods and different functional units if they meet the requirements of the definition [20]. Schools, universities, or any building with educational purposes are also potential GHG emitters [21,22]. Determining their CFs can contribute to the elaboration of a plan for to reduce their emissions [23], as well as to improve the design of new infrastructures, as some research shows [24].

1.2. Contribution of Construction to GHGs

In Spain, about 9% of GHG emissions in 2018/2019 was associated with construction, as described in the 2020 Report on Greenhouse Gas Emissions Series 1990–2018 of the Ministry for Ecological Transition and the Demographic Challenge [25] (p. 105). From a global perspective, the UN Global Status Report 2017 establishes that buildings and construction together account for 36% of global final energy use and 39% of energy-related carbon dioxide emissions when upstream power generation is included [26]. For this reason, ESNE aims to shape policies for the reduction in GHG emissions, energy saving and reduction, and optimized waste management. In order to face this challenge, it is essential to identify the initial GHG emissions. Subsequently, it is proposed to register the contribution to its emissions in the Spanish Inventory System (SEI). This is described in the cited report [13].

In this project, to calculate ESNE's 2018/19 CF, the methodology developed by CarbonFeel has been used. It is framed in the SchoolFeel program of CarbonFeel to support the fight against climate change at the level of schools and universities [27]. In an effort to sensitize students, the idea is to promote the understanding of the phenomenon and of individual influence, the motivation to act collaboratively, and finally, to provide them with the necessary knowledge to incorporate carbon accounting in their daily activities. In this way, these future professionals and managers of design and industrial activities will be able to be proactive by promoting measures to reduce, compensate and mitigate their effects.

2. Materials and Methods

2.1. Methodology

The aforementioned tools allowed us to carry out and analyze the carbon footprint inventory of products, processes and organizations, based on different standards: ISO 14067 for products, ISO 14069 for corporate footprint or Global Protocol for Greenhouse Gas Emissions (GPC) at scale community for cities. It also enables ecological footprint—Global Footprint Network—and hydric footprint—Water Footprint Network—analyses. The CarbonFeel Initiative has been promoted by the NGO Funciona Foundation for International Collaboration, which facilitates its free use by students of educational institutions that are members of the ResearchFeel alliance for research and teaching. It is a set of solutions that provides a calculation tool called BookFeel, a methodological guide (ProjectFeel) that provides a series of deliverables that ensure total transparency of the calculation [28]. It is structured using the semantic language Footprint Electronic Exchange Language (FEEL) based on the standard XML Schema (XSD) of the World Wide Web Consortium (W3C) and proposes the use of primary data in order to avoid controversial questions about the use of secondary data in the calculation of carbon footprint.

The results are expressed in kilograms of CO_2 equivalent [$kgCO_2$-e]. This equivalence is calculated according to the greenhouse effect potential of the main GHGs: carbon dioxide (CO_2), methane (CH_4), nitrogen oxide (N_2O), hydrofluorocarbons (HFCs), perfluorocarbons (PFCs), sulfur hexafluoride (SF_6) and nitrogen trifluoride (NF_3), all referring to the CO_2 effect. The methodology and calculation algorithm of emissions follows the guidelines given by the "Guide for the calculation of the carbon footprint" of the Spanish Office of Climate Change [29]:

$$Emissions\ [gCO_2\text{-}e] = Activity\ Data \times Emission\ Factor$$

The activity data may be the mass of material, consumptions or other methods to evaluate each item. Data have to be configured according to the use conditions, such as the chapter it affects, type of impact, units or year of calculation. It takes into account not only the GHGs emitted, but also the GHG sink effect—e.g., gas absorption due to own, promoted or managed forested areas by the university—when present.

However, the CF of a facility (building, school, hotel, etc.), service or product provides necessary but not sufficient information to determine the efficiency against GHG emissions. In an increasingly demanding society in terms of knowledge about the carbon footprint produced by services or products consumed and the impact this has on global warming, the education sector can be a driving force in responding to this demand. One of the objectives of this methodology is to define parameters of comparability, not only for an installation or product over time, but on a comparative level between similar ones in such a way that it facilitates decision-making on the choice of study centers as well as contributing to the reduction in these effects. The following activity rates are proposed for weighting CF, depending on the number of users, hours of use and surface area in order to facilitate the comprehension of the client's or user's contribution to GHG emissions: $kgCO_2$-e/m^2, $kgCO_2$-e/student or $kgCO_2$-e/student·hour [30–33]. The use of the $kgCO_2$-e per student assessment is justified as it is an objective measure for comparison and will serve as a verification ratio of the improvements adopted.

BookFeel is structured on different levels for the scenario configuration:

- SCOPE STRUCTURE—this describes a hierarchical set of chapters that groups the different emission sources, according to each protocol, structured in three levels [28,34,35]:

  Scope 1, combustion direct emissions—changes in land use . . . ;
  Scope 2, indirect energy emissions—main electric power purchased;

Scope 3, other indirect emissions—footprint acquired with purchased products, services or waste generated by the activity.

Each chapter is associated with an algorithm that must be configured according to its scope of application.

- Analytic structure—in a corporate footprint, this could be constituted by the production centers, and these by sections or departments.

Table 1 give an example of BookFeel calculation factors, algorithms, conversion factors and sources. All data, factors and algorithms sources and URLs are explicitly referenced.

Table 1. Examples of algorithms and emission factors.

Factor	Algorithms
P986	CO_2-e Emissions = E000 × F186 × 0.000001 Total emissions in tCO_2-e E000 = consumption of material/service/object of study (UF, Functional unit) 0.000001 = Conversion factor gCO_2 to tCO_2 F186. Scale factor per general functional unit Value = 8.1000 (gCO_2/UF) Source: Winnipeg Sewage Treatment Program South End Plant https://www.winnipeg.ca/finance/findata/matmgt/documents//2012/682-2012//682-2012_Appendix_H-WSTP_South_End_Plant_Process_Selection_Report/PSR_rev%20final.pdf Comments: Appendix 7 Material Plastic Fiber GRP. Functional unit: gr Year: 2012

2.2. GHG Scenario of ESNE

The study was conducted during the 2018/19 academic year. The starting situation will be called Scenario 1. ESNE's facilities include two buildings for educational and research purposes, equipped with classrooms, offices, workshops and laboratories, cafeteria and garage. It takes up 4000 m^2, and is located in a residential area in the north of the city of Madrid, in Chamartin District, Alfonso XIII avenue. It is used daily by approximately 1500 students, using 900,000 student·hours, calculated as follows:

student-hour = students enrolled × total number of hours per day × number of days

As regards the modern part of the facilities, the air conditioning is based on cold-heat equipment with heat pump, which coexists with heating by hot water radiators from a diesel boiler in the former area. For lighting, different types of LED lamps or panels and energy-saving lamps are combined. The exterior is a ventilated façade with brick walls with projected insulation. The enclosures and windows are double glazed with an insulating chamber.

To elaborate the inventory of the carbon emissions of ESNE, the followings scope structures have been taken into account: Scope 1, fuels burned, and fugitive emissions (refrigerant gas); Scope 2, electricity consumed; Scope 3, materials for teaching, paper, business and field trips and workers commuting. The analytic structure includes the following items: a single campus, GHG emissions caused by person for their daily commute to work, business and field trips, and materials acquired for academic, teaching and research purposes.

Some inputs have been neglected because they are not directly under control of ESNE. Student transportation is a highly variable component that does not depend on ESNE decisions. The university campus is located in the center of Madrid. There is not a parking area available for students, therefore, 95% of them use public transport to access the campus. The university promotes the use of collective transport, walking or cycling. The consulted literature does not indicate that there is much potential for improvement in this regard [36]. This is not the case for teachers and staff, since the working hours and the availability of a parking area strongly influences the choice of the mode of transport.

Cafeteria services, provided by an external company, have not been included either, but may be included in future specific studies; as an aged building, the materials used in its construction have not been included either. The tables below describe the emission sources that have been included.

2.3. Calculations Criteria

For emissions due to air travel, the calculation methodology and factors given by the UN agency International Civil Aviation Organization (ICAO) have been used [37]. In substance, this methodology applies the best publicly available industry data for aircraft types, route specific data, passenger load factors and cargo carried. The CO_2 emissions per passenger was assessed taking into a total of 565,000 km per year of business and field trips flights, by means of these four basic steps:

1. Estimation of the aircraft fuel burn.
2. Calculation of the passengers' fuel burn based on a passenger/freight factor.
3. Seat occupied = total seats $\times$ load factor
4. CO_2 emissions/passenger = (passengers' fuel burn $\times$ 3.16)/seat occupied

It is interesting to note that for flights over 3000 km, the CO_2 emissions per passenger in the premium cabin are twice as high as the corresponding emissions per passenger in the economy cabin, as seen in other studies [38].

As regards car, bus and train travel, calculation was based on these criteria:

- 2018 Guidelines Defra Conversion Factors [39].

  By car: average diesel car 0.178 kg CO_2 person/km, one seat occupied; average petrol car 0.184 kgCO_2 person/km, one seat occupied.
  By bus: regular diesel bus, 0.023 kg CO_2 person/km.

- For train and metro, a Renfe/SNCF methodology based on the Ecopassenger calculator was considered [40]:

  By train/Metro: 0.025 kg CO_2 person/km (regular Spanish electric mix).

It should be noted that by using the railroad electric mix with green certificates instead of the regular Spanish electric mix, GHG emission drops by half [41].

For daily employees' commutes, the values shown in Table 2, obtained from an internal survey, express the number of kilometers traveled by employees to work per year. Results are presented in rounded numbers.

Table 2. Employee travel to work (Scenario 1).

Mode of Transport	Km/Year
—Without consumption	**44,040**
Walking	43,320
Bicycle	720
—Railway transport	**304,920**
Metro	99,960
RENFE suburban trains	204,760
Tramway	200
—Road transport	**314,440**
Citybus	31,960
Private electric vehicle. (car/bicycle/motorbike)	35,880
Private car diesel	127,040
Private car gasoline	61,080
Private motorbike	58,480
Total	**663,400**

In Table 3, business and field trips values are shown. It describes annual travel for educational and business reasons of students, lecturers and managers, by road, air and railway modes of transport:

Table 3. Business and field trips (Scenario 1).

Mode of Transport	km/Year
—Road	**57,060**
Bus	47,480
Diesel car [1]	9580
—Train	**37,400**
—Air	**565,000**
Total	**695,460**

[1] Average occupation 3 passengers per car.

Table 4 summarizes the main emission sources, including trips and employees commuting, organized in the corresponding three scopes.

Table 4. Main emission sources inventory (Scenario 1).

Scope	Chapter	Data	Units
1	Fossil fuels (Diesel C)	11,000	L
	Leakage Refrigerant gas (R-410A)	4.3 [1]	kg
2	Electricity	241,572 [1]	kWh
3	Materials		
	Textiles	1439	kg
	Wood	100	kg
	Cardboard + paper + books	11,850	kg
	IT equipment	524	kg
	Water	1628 [1]	m^3
	HPDE 3D printer	4.2 [2]	kg
	Furniture	452	kg
	Business and field trips	695,000	km/year
	Land	94,460	km/year
	Employee commuting	663,400	km/year
	Waste		
	Cardboard + paper	6500	kg
	Light packaging	2100	kg
	Remaining fraction	4200	kg

[1] Invoices supplier company. [2] High density polyethylene. [3] Calculated results.

3. Results

The results of the calculations obtained are presented in Table 5. The total gives a figure of about 255,548 kgCO$_2$-e.

Electrical energy consumption, producing 72,471 kg CO$_2$-e (Scope 2), stands out as the main source of emissions. The 31,548 kg produced by Diesel C stationary combustion for heating (Scope 1) represents the second most significant emission. The WTT transmission and distribution losses (18,802 kg CO$_2$-e) also represents an important source (Scope 3). This source is not manageable as it depends on the electrical system. For purchased products, from 34,645 kg CO$_2$-e of materials, 10,075 kg are related to paper consumption (included photocopying) and 11,388 kg to textiles used for fashion practices. As regards commuting, road private combustion modes correspond to 28,290 kg CO$_2$-e, significantly higher compared to urban bus (2564 kg CO$_2$-e) and rail transport (14,626 kg CO$_2$-e). Concerning business and field trips, flights represent a significant part of GHG emissions: 21% of Scenario 3 and 12% of total ESNE emissions were international flights (28,992 kg

CO_2-e), representing the main part of the total of 29,855 kg CO_2-e. Private car road transport (6382 kg CO_2-e) is also significant compared to buses (991 kg CO_2-e), even more taking in account the km·person relation. AVE transportation means the less affecting mode (1153 kg CO_2-e).

Table 5. ESNE carbon footprint 2018/19 (Scenario 1).

Scope	CF (kg CO_2-e)
1. Direct emissions and absorptions	40,526
Stationary combustion	31,548
Refrigerant leakage (R-410A)	8978
2. Indirect energy emissions (Electricity consumed)	72,471
3. Other indirect emissions	142,549
Energy not included in direct and indirect	
WTT transmission and distribution losses [1]	18,802
Products purchased	34,645
Water (natural)	643
Wood/cork/basketry/rubber/plastic products	158
Furniture	2501
Paper, books and cardboard	11,254
Computer, electronic and optical products	8699
Textile products	11,387
Employee commuting	47,391
Train/Metro/Tram	14,626
Road transport	32,765
Urban bus	2564
Private car	
Diesel	16,755
Petrol	11,523
Electric (incl. bike and skateboard)	1923
Business and field trips	38,382
International and national flights	29,855
Road transport	7374
By bus	991
Private car	6382
Train (AVE)	1153
Waste	3328
Cardboard + paper	366
Light packaging	252
Remaining fraction	2710
Total	**255,548**

[1] WTT (Well to tank): additional emissions (related to electricity).

4. Discussion

Based on the obtained results, some clues can be found to outline alternatives for reducing greenhouse emissions. The contribution of electricity to emissions is noteworthy, being the highest negative contributor and coinciding with other results of other reviewed studies [23]. This source represents a difficult optimization; today, education is strongly linked to technological progress and the use of tools that require considerable electrical expenditure, and electrical devices (PCs, lighting, air conditioning) are already of maximum efficiency. As regards air conditioning, the insulation of the enclosure has recently been improved; a ventilated façade covers the entire surface and enclosures have double-glass windows with air chambers. The replacement of the aluminum carpentry by a more efficient one with thermal break could be assumed, but this option would represent an important investment. However, a better solution is within reach: the availability of certified renewable energy in Spain means that the best option is to replace the supplier with one with certified green energy. Taking into account this option, the purchased electricity will have no GHG emissions, reducing the total to 183,078 kg CO_2-e (Scenario 2).

In this case, the GHG emissions of Scenario 2, compared with the initial situation (Scenario 1), represents a saving of 28% (Figure 1). This is the easiest and most immediate way to reduce the CF of ESNE, and can be reached with noninvestments, probably even with discounts from suppliers, since electricity from renewable sources in Spain usually has a lower annual cost than the regular mix.

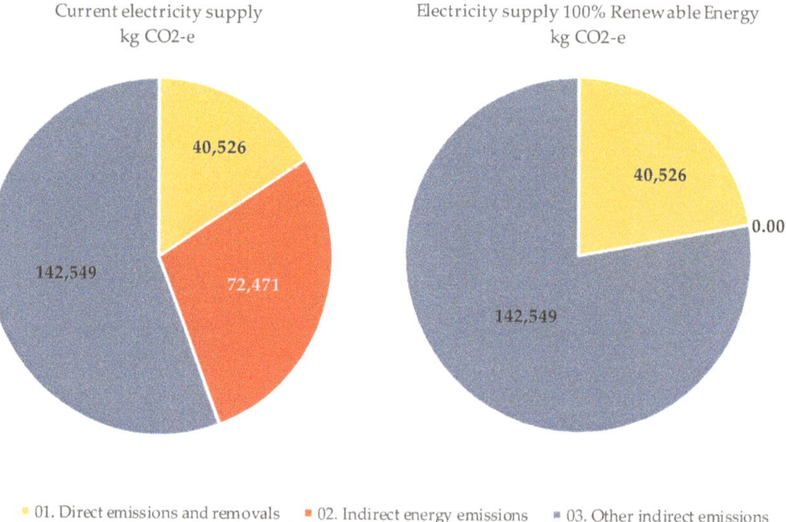

Figure 1. Electricity supply carbon footprint distribution scopes (kg CO_2-e).

The change between scenarios can be reflected by means of the indexes described in Section 2.1. Table 6 shows the activity rates in both scenarios (255,548 kg CO_2-e for Scenario 1, and 183,078 for Scenario 2), considering 1500 students enrolled and a total of 900,000 h/year, as indicated in Section 2.2:

Table 6. Activity rates of Scenarios 1 and 2.

Ratios	Kg CO_2-e/m^2	Kg CO_2-e/Student	Kg CO_2-e/Student·h
Built-up area: 4000 m^2 Students: 1500			
Scenario 1	64	170	0.28
Scenario 2	46	122	0.20
Reduction		−28%	

The emission ratio per person decreases from 0.28 to 0.20 kg CO_2-e/student·hour, and the hourly emission is 420 to 300 kg CO_2-e/hour for the total number of students. Each student enrolled in a full academic year of 600 teaching hours per year, meaning a total emission of 170 kg CO_2-e in the present situation, and 120 kg CO_2-e for Scenario 2. This figure should be updated annually, taking into account the improvements made to correct the resulting carbon footprint. Educational centers could inform students and future students of these data in an exercise of transparency or even promotion.

The next option in importance to decrease GHG emissions may be found in the 31,548 kg CO_2-e from the combustion of heating oil. To maintain the current water radiator system, this fuel could be replaced by natural gas, but this solution is expensive, since it requires the replacement of the current low efficiency boiler, unfeasible to adapt for use with natural gas. However, replacing direct combustion heating with heat pump air conditioners powered by green electricity would be a better solution. By means of a quick calculation, the additional electricity consumption needed can be estimated considering: a boiler combustion efficiency of 80%, a lower calorific value of 1028 kWh/L for Diesel C [42] and an efficiency of 60% for Split heating equipment [43]. This would result in an approximate electricity consumption of 90,000 kWh. This would increase annual electricity consumption to 331,000 kWh, 37% higher, but would eliminate the consumption of 11,000 L of a fossil fuel, its GHG emissions and highly polluting smoke. The result would be a decrease in CF to 155,500 kg CO_2-e, 60% of the initial amount.

In order to further CF improve, the next option should be to decrease GHG emissions from transportation. Employees produce 28,290 kg CO_2-e a year to go to work every with private fuel vehicles. According to the results obtained in the survey conducted using 116 employees, 37 use this mode of transport. This represents about 0.79 kg CO_2-e/km·person (or 764 kg CO_2-e/year per person). In comparison, about 80 people using public transport produce 17,190 kg CO_2-e, a ratio of about 0.05 kg CO_2-e/km·person (183 kg CO_2-e/year per person), only 6% in terms of km·person (24% in a year·person basis). It should be noted that several electric (six in the present day) and hybrid cars are continuously being incorporated into the workforce. Measures such as time optimization, including the reduction in attendance days, could significantly improve this balance.

Business and field trips give another perspective. While there are significant improvements to be made, long-distance air travel has few solutions. In this account, there are 73 person flights in Europe and long trips, 18 person traveling to Beijin and 10 to Miami (USA). It should be noted that international flights cause a significant amount of 409 kg CO_2-e/year, but, taking in account the distance (kg CO_2-e/km·person), the emission ratio is as low as intercity bus transportation. Once again, the lower emissions are due to high-speed railway transportation, less than 5% of private diesel cars (Table 7).

Table 7. Carbon Footprint (CF) ratios per year of business and field trips (Scenario 1).

Ratios	kg CO_2-e/km	kg CO_2-e/Person
—Travel to work		
Private combustion vehicles	0.788	764
Public transport	0.051	183
—Study and business trips		
Private diesel cars	0.667	236
Bus	0.054	16
Flights	0.053	409
Train AVE	0.031	29

5. Conclusions

The results obtained and literature review make it possible to draw conclusions based on a proposal for reducing carbon footprints.

If the calculation of the student-hour activity rate is applied, accompanied by calculation rules agreed by the sector, a register could be developed to allow comparability and to help mitigate global warming caused by educational activities and infrastructure.

The impact of Scope 2 is the highest of the factors studied, referring to the University's electricity expenditure. This incidence could be eliminated if the production of electrical energy was supplied by a company with a 100% renewable energy source, where the contribution of kg CO_2-e emissions disappears, in addition to replacing old and inefficient installations with systems that use less energy.

Another action to be taken is the rehabilitation of old buildings in order to improve energy efficiency. In this study, the emission of greenhouse gases due to heating is 31,548 kg CO_2-e. The insulation of the building envelope is an action to be taken into account to reduce the thermal transmittance and the heating energy consumption by around 90% if it is combined with adequate ventilation and a more efficient heating system, reaching values of 3155 kg CO_2-e [44].

On the other hand, the carbon footprint produced by ESNE could be mitigated with the contribution of green spaces responsible for absorbing carbon, with an annual action of extending the trees in an institution, as was considered by Diponegoro University on its university campus [22] or Trisakti University in Jakarta [45], and studied at Suranaree University of Technology in Thailand, where the green area captured 40% of the total emissions produced by the university [46]. In the event that no land is available on campus, one option to consider would be the creation of green façades or vertical gardens, which also contribute to the insulation of the building's façade envelope [47].

University education should include sustainability and sustainable development in training actions [48], instructing students in sustainable development in all areas, as well as informing them of the impact that their way of life has on the planet, efficiency in the use of electricity and water, reduction in the use of paper (10), and contribute to raising awareness of the three Rs method "reduce, reuse and recycle", in addition to the alternatives presented, in order to improve the impact of kg CO_2-e [49].

Another point to be dealt with would be the study of the optimization of working time in attendance, trying to reduce attendance as much as possible, to avoid trips that are not essential, even considering the possibility of limiting working days to four.

Author Contributions: Conceptualization, G.F.; Formal analysis, S.L.; Investigation, R.S.; Writing–original draft, G.F. All authors have read and agreed to the published version of the manuscript.

Funding: This research received no external funding

Institutional Review Board Statement: Not applicable

Informed Consent Statement: Not applicable

Data Availability Statement: The data described in the research are made public in this document, based on the research carried out by the authors.

Conflicts of Interest: The authors declare no conflict of interest.

References

1. IPCC. Special Report on Climate Change and Land (SRCCL). 2019. Available online: https://www.ipcc.ch/site/assets/uploads/sites/4/2020/07/03_Technical-Summary-TS_V2.pdf (accessed on 23 September 2020).
2. UN Population. 2019. Available online: https://www.un.org/en/sections/issues-depth/population/#:~{}:text=The%20world\T1\textquoterights%20population%20is%20expected,nearly%2011%20billion%20around%202100 (accessed on 23 September 2020).
3. IPCC 2019. IPCC 2019 Refinement to the 2006 IPCC Guidelines for National Greenhouse Gas Inventories. 2019. Available online: https://report.ipcc.ch/sr15/pdf/sr15_spm_fig1.pdf (accessed on 28 September 2020).
4. bp Statistical Review of World Energy. 2020. Available online: https://www.bp.com/content/dam/bp/business-sites/en/global/corporate/pdfs/energy-economics/statistical-review/bp-stats-review-2020-full-report.pdf (accessed on 21 September 2020).
5. What a Waste 2.0: A Global Snapshot of Solid Waste Management to 2050, the World Bank Open Knowledge Repository. 2020. Available online: https://openknowledge.worldbank.org/handle/10986/30317 (accessed on 17 October 2020).
6. NASA. Global Climate Change. 2020. Available online: https://climate.nasa.gov/vital-signs/carbon-dioxide/ (accessed on 17 October 2020).
7. UNFCCC. Paris Agreement. 2015. Available online: https://unfccc.int/sites/default/files/english_paris_agreement.pdf (accessed on 23 July 2020).
8. UE 2030 Climate and Energy Policy Framework, European Council. 2014. Available online: http://data.consilium.europa.eu/doc/document/ST-169-2014-INIT/en/pdf (accessed on 19 October 2020).
9. European Green Deal, UE. 2020. Available online: https://ec.europa.eu/info/strategy/priorities-2019-2024/european-green-deal_en (accessed on 20 October 2020).
10. European Commission. Regulation of the European Parliament and of the Council. 2020. Available online: https://eur-lex.europa.eu/legal-content/EN/TXT/PDF/?uri=CELEX:52020PC0080&from=EN (accessed on 20 July 2020).
11. BOCG. Boletín Oficial de las Cortes Generales, 121/000019 Proyecto de Ley de Cambio Climático y Transición Energética. 2020. Available online: http://www.congreso.es/public_oficiales/L14/CONG/BOCG/A/BOCG-14-A-19-1.PDF (accessed on 20 July 2020).
12. Ding, G. *Life Cycle Assessment (LCA) of Sustainable Building Materials: An Overview*; Woodhead Publishing Limited: Sawston, UK; Cambridge, UK, 2014.
13. Wackernagel, M.; Kitzes, J. Ecological Footprint. 2008. Available online: https://www.sciencedirect.com/science/article/pii/B9780080454054006200 (accessed on 21 September 2020).

14. González-Vallejo, P.; Marrero, M.; Solís-Guzmán, J. The ecological footprint of dwelling construction in Spain. *Ecol. Indic.* **2015**, *52*, 75–84. [CrossRef]
15. Castellani, V.; Sala, S. Ecological Footprint and Life Cycle Assessment in the sustainability assessment of tourism activities. *Ecol. Indic.* **2012**, *16*, 135–147. [CrossRef]
16. Wackernagel, M.; Monfreda, C.; Moran, D.; Wermer, P.; Goldfinger, S.; Deumling, D.; Murray, M. National Footprint and Biocapacity Accounts 2005: The underlying calculation method. *Land Use Policy* **2004**, *21*. Available online: https://www.researchgate.net/publication/228649734_National_Footprint_and_Biocapacity_Accounts_2005_The_underlying_calculation_method (accessed on 22 July 2020).
17. Wiedmann, T.; Minx, J. A Definition of Carbon Footprint. C. C. Pertsova. *Ecol. Econ. Res. Trends* **2008**, *1*, 1–11.
18. Fenner, A.; Kibert, C.; Woo, J.; Morque, S.; Razkenari, M.; Hakim, H.; Lu, X. The carbon footprint of buildings: A review of methodologies and applications. *Renew. Sustain. Energy Rev.* **2018**, *94*, 1142–1152. [CrossRef]
19. Camacho, C.; Pereira, J.; Gomez, M.; Arrero, M.M. Huella de carbono como instrumento de decisión en la rehabilitación energética. Películas de control solar frente a la sustitución de ventanas. *Rev. Hábitat Sustentable* **2018**, *8*, 20–31. [CrossRef]
20. Gao, T.; Liu, Q.; Wang, J. A comparative study of carbon footprint and assessment standards. *Int. J. Low-Carbon Technol.* **2013**, *9*, 237–243. [CrossRef]
21. Sangwan, K.; Bhakar, V.; Arora, V.; Solanki, P. Measuring carbon footprint of an Indian university using life cycle assessment. In Proceedings of the 25th CIRP Life Cycle Engineering (LCE) Conference, Procedia CIRP 69, Copenhagen, Denmark, 30 April–2 May 2018.
22. Syafrudin, S.; Zaman, B.; Budihardjo, M.; Yumaroh, S.; Gita, D.; Lantip, D. Carbon Footprint of Academic Activities: A Case Study in Diponegoro University. In *IOP Conference Series: Earth and Environmental Science*; IOP Publishing: Tembalang, Indonesia, 2020.
23. Zakaria, R.; Aly, S.H.; Hustim, M.; Oja, A.D. A Study of Assessment and Mapping of Carbon Footprints to Campus Activities in Hasanuddin University Faculty of Engineering. In *IOP Conf. Series: Materials Science and Engineering, Proceedings of The 3rd EPI International Conference on Science and Engineering (EICSE2019), 24-25 September 2019, South Sulawesi, Indonesia*; IOP Publishing: Tembalang, Indonesia, 2020.
24. Paguigan, G.; Jacinto, D. Carbon Footprint Inventory of Buildings in Isabela State University: Benchmark for Future Design Alternatives. In Proceedings of the 4th International Research Conference on Higher Education, Cabagan, Isabela, 27–29 August 2018.
25. MITECO. Informe de Inventario Nacional Gases de Efecto Invernadero, Serie 1990–2018. 2020. Available online: https://www.miteco.gob.es/es/calidad-y-evaluacion-ambiental/temas/sistema-espanol-de-inventario-sei-/es-2020-nir_tcm30-508122.pdf (accessed on 12 November 2020).
26. UN-Environment. Towards a Zero-Emission, Efficient, and Resilient Buildings and Construction Sector, Global Status Report 2017. 2017. Available online: https://www.worldgbc.org/sites/default/files/UNEP%20188_GABC_en%20%28web%29.pdf (accessed on 15 September 2020).
27. CarbonFeel. SchoolFeel. 2020. Available online: http://www.carbonfeel.org/Carbonfeel_2/SchoolFeel.html (accessed on 12 July 2020).
28. CarbonFeel. BookFeel. 2020. Available online: http://www.carbonfeel.org/Carbonfeel_2/BookFeel.html (accessed on 7 July 2020).
29. MITECO. Ministerio Para la Transición Ecológica, Guía Para el Cálculo de la Huella de Carbono y Para la Elaboración de un Plan de Mejora de una Organización. 2020. Available online: https://www.miteco.gob.es/es/cambio-climatico/temas/mitigacion-politicas-y-medidas/guia_huella_carbono_tcm30-479093.pdf (accessed on 20 July 2020).
30. Lo-Iacono-Ferreira, V.; Torregrosa-López, J.; Capuz-Rizo, S. The use of carbon footprint as a key performance indicator in higher education institutions. In Proceedings of the 22nd International Congress on Project Management and Engineering, Madrid, Spain, 11–13 July 2018.
31. Güereca, L.; Torres, N.; Noyola, A. Carbon Footprint as a basis for a cleaner research institute in Mexico. *J. Clean. Prod.* **2013**, *47*, 396–403. [CrossRef]
32. Yañez, P.; Sinha, A.; Vásquez, M. Carbon Footprint Estimation in a University Campus: Evaluation and Insights. *Sustainability* **2020**, *12*, 181. [CrossRef]
33. More, A.; Patil, M.; Shinde, N.; Waghere, S.; Dharwal, K.; Bhave, S. Carbon Footprint of International University Dr. D. Y. Patil, Akurdi, Pune. *Int. J. Res. Appl. Sci. Eng. Technol.* **2018**, *6*, 1952–1954. [CrossRef]
34. Budihardjo, M.; Syafrudin, S.; Putri, S.; Prinaningrum, A.; Willentiana, K. Quantifying Carbon Footprint of Diponegoro University: Non-Academic Sector. In *IOP Conference Series: Earth and Environmental Science*; IOP Publishing: Tembalang, Indonesia, 2020.
35. Khandelwal, M.; Jain, V.; Sharma, A.; Bansal, S. Students' Attitude toward Carbon Footprints of a Leading Private University in India. *Manag. Econ. Res. J.* **2019**, *5*, 7. [CrossRef]
36. Lim, M.; Hayder, G. Performance and Reduction of Carbon Footprint for a Sustainable Campus. *Int. J. Eng. Adv. Technol.* **2019**, *9*, 1.
37. ICAO. International Civil Aviation Organization. Available online: https://www.icao.int/environmental-protection/CarbonOffset/Documents/Methodology%20ICAO%20Carbon%20Calculator_v11-2018.pdf (accessed on 21 October 2020).
38. Ciers, J.; Mandic, A.; Toth, L.D.; Op't Veld, G. Carbon Footprint of Academic Air Travel: A Case Study in Switzerland. *Sustainability* **2019**, *11*, 80. [CrossRef]
39. Defra Factors. Greenhouse Gas Reporting: Conversion Factors 2018. 2018. Available online: https://www.gov.uk/government/publications/greenhouse-gas-reporting-conversion-factors-2018 (accessed on 10 October 2020).

40. Hüttermann, R. *EcoPassenger. Environmental Methodology and Data*; Ifeu: Heidelberg/Hannover, Germany, 2016.
41. Renf/SCNF. EcoPassenger. 2016. Available online: http://ecopassenger.org/bin/query.exe/hn?ld=uic-eco&L=vs_uic& (accessed on 2 December 2020).
42. IDAE-Instituto Para la Diversificación y Ahorro de la Energía. Guía Técnica de Diseño de Centrales de Calor Eficientes. 2010. Available online: https://www.idae.es/uploads/documentos/documentos_11_Guia_tecnica_de_diseno_de_centrales_de_calor_eficientes_e53f312e.pdf (accessed on 23 September 2020).
43. IDEA. Prestaciones Medias Estacionales de las Bombas de Calor Para Producción de Calor en Edificios. 2014. Available online: https://energia.gob.es/desarrollo/EficienciaEnergetica/RITE/Reconocidos/Reconocidos/Otros%20documentos/Prestaciones_Medias_Estacionales.pdf (accessed on 26 September 2020).
44. Ull, J.B.; Gupta, A.; Mumovic, D.; Kimpian, J. Life cycle cost and carbon footprint of energy efficient refurbishments to 20th century UK school buildings. *Int. J. Sustain. Built Environ.* **2014**, *3*, 1–17.
45. Rahma, N.; Rosnarti, D.; Purnomo, A. The carbon footprint of Trisakti University's campus in Jakarta, Indonesia. In *IOP Conference Series: Earth and Environmental Science*; IOP Publishing: Tembalang, Indonesia, 2020; Volume 452.
46. Karuchit, S.; Puttipiriyangkul, W.; Karuchit, T. Carbon Footprint Reduction from Energy-Saving Measure and Green Area of Suranaree University of Technology. *Int. J. Environ. Sci. Dev.* **2020**, *11*, 4. [CrossRef]
47. Patarlageanu, S.; Negrei, C.; Dinu, M.; Chiocaru, R. Reducing the Carbon Footprint of the Bucharest University of Economic Studies through Green Facades in an Economically Efficient Manner. *Sustainability* **2020**, *12*, 3779. [CrossRef]
48. Lozano, R. Incorporation and institutionalization of SD into universities: Breaking through barriers to change. *J. Clean. Prod.* **2006**, *9*, 787–796. [CrossRef]
49. Qafisheh, N.; Sarr, M.; Hussain, U.; Awadh, S. Carbon Footprint of ADU Students: Reasons and Solutions. *Environ. Pollut.* **2017**, *6*, 7. [CrossRef]

 *sustainability*

Article

Novel Use of Green Hydrogen Fuel Cell-Based Combined Heat and Power Systems to Reduce Primary Energy Intake and Greenhouse Emissions in the Building Sector

Jordi Renau [1,*], Víctor García [1], Luis Domenech [1], Pedro Verdejo [1], Antonio Real [1], Alberto Giménez [1], Fernando Sánchez [1], Antonio Lozano [2] and Félix Barreras [2]

[1] Technical School of Design, Architecture and Engineering (ESET), Cardenal Herrera CEU University (UCHCEU)—CEU Universities, C/San Bartolomé 55, 46115 Alfara del Patriarca, Valencia, Spain; vicgarpe@uchceu.es (V.G.); luis.domenech@uchceu.es (L.D.); pverdejo@uchceu.es (P.V.); antonio.real@uchceu.es (A.R.); algisan@uchceu.es (A.G.); fernando.sanchez@uchceu.es (F.S.)

[2] LIFTEC, CSIC-University of Zaragoza, C/María de Luna 10, 50018 Zaragoza, Spain; a.lozano@csic.es (A.L.); felix@litec.csic.es (F.B.)

* Correspondence: jordi.renau@uchceu.es

Abstract: Achieving European climate neutrality by 2050 requires further efforts not only from the industry and society, but also from policymakers. The use of high-efficiency cogeneration facilities will help to reduce both primary energy consumption and CO_2 emissions because of the increase in overall efficiency. Fuel cell-based cogeneration technologies are relevant solutions to these points for small- and microscale units. In this research, an innovative and new fuel cell-based cogeneration plant is studied, and its performance is compared with other cogeneration technologies to evaluate the potential reduction degree in energy consumption and CO_2 emissions. Four energy consumption profile datasets have been generated from real consumption data of different dwellings located in the Mediterranean coast of Spain to perform numerical simulations in different energy scenarios according to the fuel used in the cogeneration. Results show that the fuel cell-based cogeneration systems reduce primary energy consumption and CO_2 emissions in buildings, to a degree that depends on the heat-to-power ratio of the consumer. Primary energy consumption varies from 40% to 90% of the original primary energy consumption, when hydrogen is produced from natural gas reforming process, and from 5% to 40% of the original primary energy consumption if the cogeneration is fueled with hydrogen obtained from renewable energy sources. Similar reduction degrees are achieved in CO_2 emissions.

Keywords: hydrogen; PEM fuel cells; cogeneration; building sustainability; energy saving

Citation: Renau, J.; García, V.; Domenech, L.; Verdejo, P.; Real, A.; Giménez, A.; Sánchez, F.; Lozano, A.; Barreras, F. Novel Use of Green Hydrogen Fuel Cell-Based Combined Heat and Power Systems to Reduce Primary Energy Intake and Greenhouse Emissions in the Building Sector. *Sustainability* **2021**, *13*, 1776. https://doi.org/10.3390/su13041776

Academic Editor: András Reith

Received: 30 December 2020

Accepted: 2 February 2021

Published: 7 February 2021

Publisher's Note: MDPI stays neutral with regard to jurisdictional claims in published maps and institutional affiliations.

1. Introduction

Europe aims to achieve climate neutrality by 2050, which means net-zero greenhouse gas emissions. Entire society and economic sectors must join this task to reach the final objective, from industry to mobility, building, agriculture, etc. Building sector in Europe consumed 40% of the final energy in 2018, with just household being 26% of the final energy, similar to the industry sector [1]. According to the Spanish "Instituto para la Diversificación y Ahorro de la Energía (IDAE)" report [2], space heating and sanitary hot water are responsible of 58% to 75% of the final energy consumption in flats and single-family houses in Spain, respectively. Single-family houses account for the biggest energy share supplied from renewable energy, which is close to 40% in the Mediterranean area. However, this is not enough to meet the 2050 EU objective, because 47% of the energy supply still comes from fossil fuels. The Energy Performance of Buildings Directive (EPBD) (2010/31/EU) [3] is the legislative framework "to achieve a high energy efficiency and decarbonize building stock by 2050". EPBD states that from 31 December 2020 all new edifications must be nearly zero-energy buildings (nZEB). These are "buildings with a very high energy performance and the low amount of energy that these buildings require comes mostly from renewable sources". Primary energy consumption analysis is how the EPBD evaluates the building energy efficiency due to the variety of energy sources that are used. "The concept of primary energy attempts to provide a simple metric for all forms of energy that are supplied to, transmitted through, a defined boundary" [4]. The EPBD leaves the member states to determine the methodology to calculate the primary energy factor and CO_2 emissions for each end-use energy source depending on the energy supply grid circumstances. Improving energy efficiency and the share of renewable sources are also main targets of the Spanish Government, as they are reflected in the Integrated National Energy and Climate Plan 2021–2030 [5]. According to this plan, high-efficiency renewable cogeneration facilities are going to be part of the comprehensive strategy for energy efficiency in cities.

2. Background

Cogeneration, also known as combined heat and power system (CHP), shows the ability to decrease primary energy consumption and reduce greenhouse gas emissions due to the increase in building energy efficiency [6,7]. Thermal energy demand can be supplied by a heater or a boiler, but the use of the CHP technology could provide the same thermal energy consumption and a fraction or the whole electrical demand depending on the CHP technology used. CHP technologies must be easily scalable for low power ratios to make them suitable for residential applications. CHP systems for buildings can be classified depending on the rated thermal power as micro- (1–5 kW) and small-scale ($\leq$50 kW) units [8]. Because CHP systems produce the energy at the point of use, they can be referred as decentralized energy sources [9]. This advantage of decentralized generation includes an improved energy efficiency, which means an optimized fuel utilization that results in decreased CO_2 emissions and primary energy consumption and a reduction in the national transmission losses, that account for 2%–11% of the losses in the European transmission network [10]. However, benefits of the distributed generation are only achievable if there is a proper energy management between generation and consumption [10]. Proper management would require of an energy storage system. Electrical energy storage (EES) improves the self-consumption ratio for small CHP units [11] and thermal energy storage (TES) eliminates system oversize and also optimizes the use of produced energy [12]. All the CHP units considered in this paper will integrate both EES and TES systems.

Figure 1 shows the different CHP technologies suitable to be installed in a residential building [13]. The zones have been delimited using the technical parameters obtained from the "Cogen Challenge Project" document [14], where micro-scale and small-scale cogeneration technologies are analyzed. Each technology is represented as a colored fuzzy area, delimited by four straight lines, two thick solid lines and two dotted ones. Thermal/electrical efficiency is indicated in the vertical axis and thermal/electrical rated power in the horizontal one. The represented surface covers the power range for the applicability in buildings, from single-family houses to blocks of apartments. The figure can be read as follows. Solid lines represent two possible CHP configurations in each technology. The right-side solid line of each area indicates the most common or typical "small-scale" unit of the technology and the left-side one corresponds to the smallest CHP unit possible as indicated in [14]. Solid lines can be understood as an operating point of the real CHP unit, where the upper extreme of the line is for the CHP unit thermal characteristic and the lower one corresponds to the electrical one.

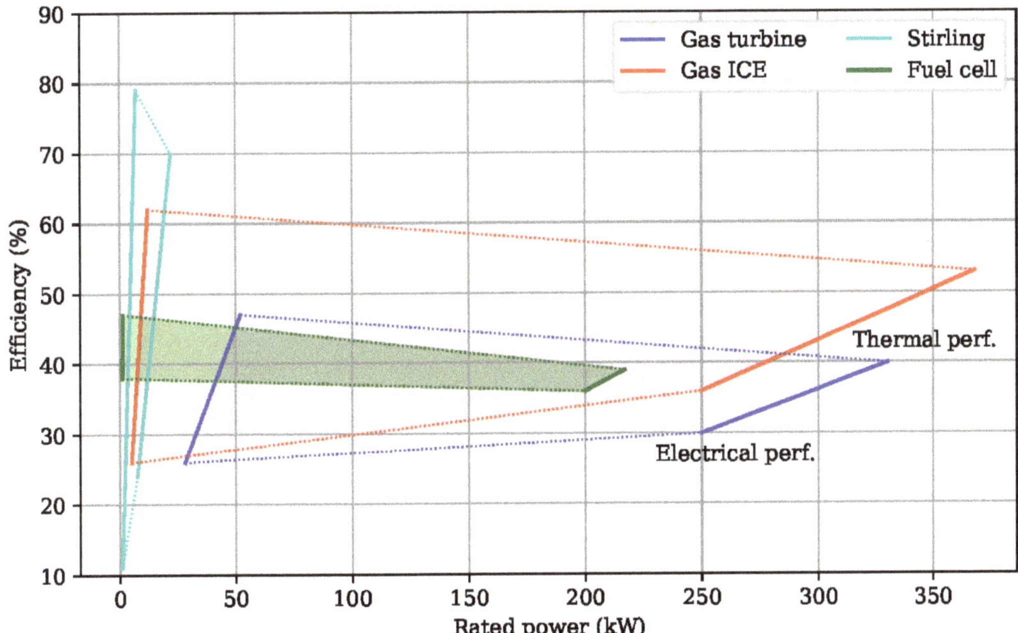

Figure 1. Comparison of the technical parameters in different gas-fueled combined heat and power technologies.

For example, the solid line of the right side in the blue area means the "typical small-scale gas turbine CHP unit" as stated in [14], which means a rated electrical power of 250 kW and around 330 kW for its rated thermal power. The corresponding conversion efficiencies, around 30% and 40% for the electrical and thermal energies respectively, are read in the vertical axis. On the left side of the same fuzzy area, the electrical and thermal rated power of the smallest gas turbine CHP unit analyzed are 30 kW (26%) and 50 kW (47%), respectively. Both units are connected with the dotted lines that create the fuzzy area, which can be considered as an operational chart for the technology in the small-scale use. In other words, this area can be understood as the operating range for each technology. According to Figure 1 the following conclusions can be extracted:

- Stirling CHP technology is suitable for any kind of fuel. Its performance is similar to that of a boiler due to its high thermal energy conversion efficiency with the additional benefit of the electrical production that can compensate some of the building intakes. Nevertheless, the heat-to-power (HtP) ratio, which is an important selection parameter [15], is too high for residential uses meaning that the energy production is unbalanced with respect to the thermal energy demand. Stirling technology is included in Figure 1 chart due to its scientific interest, but it is not going to be considered in the present analysis.

- Gas turbine and internal combustion engine (Gas ICE) are mature technologies that can be scaled from small to large sizes. Both can consume natural gas, which is a fuel widely available in the residential building sector, but its consumption should be minimized due to environmental constraints. The use of pure hydrogen into ICEs and turbines has several technical problems that is now under research and still need to be improved [16]. In this paper both options are going to be analyzed.

- Fuel cell-based CHP (FC-CHP) is the most promising technology due to its balanced heat-to-power ratio, better adapted to the residential building energy profiles, which are more electricity demanding [17]. Fuel cell presents the highest electrical conversion efficiency. Fuel cells are easily scaled from few watts or kilowatts to hundreds of kilowatts keeping a constant energy conversion efficiency when they are fueled from pure hydrogen [18–21]. When this pure hydrogen comes from a green production process, the energy obtained can also be considered as green or carbon-free.

FC-CHP are classified as a function of the fuel cell technology used in the power unit. The most common technologies in commercial units are based on polymer exchange membrane fuel cells (PEMFC), The most successful examples of these systems can be found in Japan and Europe [22,23]. PEMFC can be classified into low- (up to 80 °C) and high-temperature (from 120 °C to 180 °C) devices. They only differ in the working temperature required by the polymer used as solid electrolyte membrane. Low-temperature PEM fuel cell-based CHP systems are the most common. In this paper both PEM technologies are considered, paying special attention to a high-temperature PEM fuel cell-based micro-CHP system specifically conceived in the framework of the MICAPEM project that is been integrated into an existent nearly-zero energy house, developed and built for the international Solar Decathlon 2012 contest [24,25]. The use of high-temperature PEMFC is promising due to the improved chemical kinetics in the electrodes, better tolerance to CO impurities in the fuel, simplification of the water management because it is produced in vapor phase and simpler and compact heat recovery system because of the higher enthalpy of the thermal energy [26,27]. A majority of the significant studies in the literature involving a high-temperature PEM fuel cell-based CHP system are theoretical works [28–31] Only one report on tests in an experimental facility has been found [32].

The objective of this research paper is to expose, using numerical simulations, how fuel cell-based CHP systems can drive a potential reduction of primary energy consumption and CO_2 emissions in the building sector. Numerical simulations are performed using preliminary results from the characterization of the high-temperature PEM fuel cell prototype built and tested to be installed in a demonstrative scale CHP facility. Once installed, the CHP technology will be evaluated and a novel oil-based refrigeration system for HT-PEMFC will also be tested, as explained in Section 3.3. In the same project framework, a hydrogen electrolyzer integrated with the solar system is also being installed to link with the green hydrogen source requirement objective.

3. Methods and Materials

Numerical simulations were performed using the electrical and gas energy consumptions from four real dwellings in the east coast (Mediterranean area) of Spain. Weather in this region can be classified as a "Csa climate" with hot, dry summers and cool, wet winters, according to Köppen climate international classification [33].

Using the information from actual energy invoices, four daily consumption datasets have been created. Four 10-apartment building consumption profiles were determined considering simultaneously factors from the single dwelling datasets. The selected dwellings are described as follows:

- Id 1: 140 m^2 two-story terraced house, four inhabitants. The gas consumers are the boiler, used for heating and on-demand sanitary hot water, and the kitchen cooktop.
- Id 2: 75 m^2 flat, four inhabitants. Natural gas is consumed only in the on-demand water heater. Electrical induction cooktop.
- Id 3: 90 m^2 flat, four inhabitants. Gas-powered boiler for heating and hot water production. Electrical induction cooktop.
- Id 4: 90 m^2 flat, three inhabitants. Natural gas-powered on-demand water heater. Electrical cooktop and individual electrical oil heaters in each bedroom.

Data obtained from the gas and electrical utility invoices are one-month aggregated information that must be statistically treated to create the useful datasets to simulate the daily energy demand of the users. The numerical treatment has consisted of normal randomized daily energy consumption estimation using the daily seasonal average consumption values and its seasonal data standard deviation (Figure 2). This process was applied to both energy invoices (electricity and gas), taking the billing date into account to correct the consumption data of the different energy suppliers.

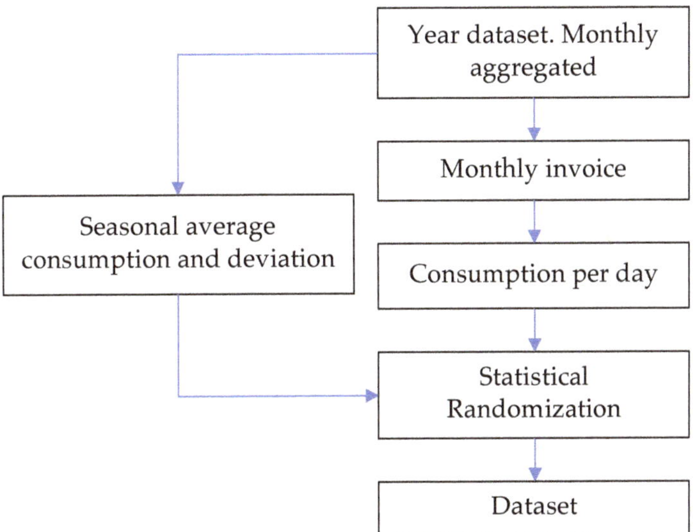

Figure 2. Statistical analysis from the annual energy invoices to randomize a daily energy consumption dataset for each building using seasonal values (average and standard deviation).

The results of the treatment of the numerical data can be observed in Figure 3, where the vertical axis is the daily electrical consumption and the horizontal one represents the daily thermal energy one. Darker dots are the daily consumption calculated from the utility invoices and the "*x*" markers are the randomized values obtained from the numerical treatment.

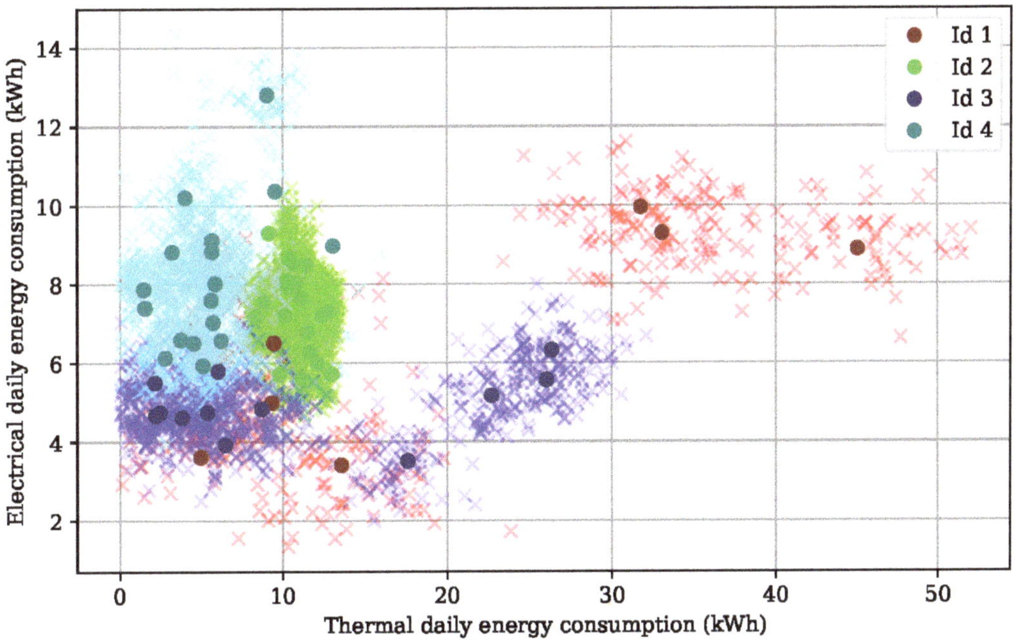

Figure 3. Result of the numerical treatment of the utility invoices to create an energy consumption dataset for each dwelling.

Electrical and thermal data in Figure 3 are correlated values, where two main tendencies can be observed. Dwellings with gas-powered heating facilities (Id 1 and Id 3) show a greater thermal energy demand for heating seasons. This results in two different "x-clouds": one energy demand cloud at the right-top side of the chart (Id 1: from 30 to 50 kWh$_{th.}$, Id 3: from 15 to 30kWh$_{th.}$), where the heating demand can be detected; and a second cloud below 15 kWh$_{th.}$ that is overlapped with the two less-thermal demanding dwellings (Id 2 and Id 4). In the case of dwellings Id 2 and Id 4, the seasonal variation is notorious in the vertical axis due to the increase in the electrical energy demand during the heating season caused by the use of electrical heaters. When the thermal demand is limited to hot water, the energy consumption is function of the number of inhabitants as can be observed for the daily energy thermal values for Id 2 and Id 4, respectively. Heat-to-power (HtP) is calculated as the thermal demand over the electrical demand. This ratio is a season-dependent value, and normally a year-based calculation is provided. The results obtained with the datasets sorted in decreasing order are 3.4 (Id 1), 2.3 (Id 3), 1.5 (Id 2) and 0.7 (Id 4).

Considering each dwelling dataset, four 10-dwelling buildings were created using a randomized factor to simulate a centralized CHP. Heat-to-power ratios for the building datasets are similar in value and order.

3.1. Simulation Algorithm

Cogeneration facilities are designed to supply the user's thermal energy demand and to provide electricity as a secondary energy source [15]. Industrial-scale CHPs use the heat directly, but residential-scale CHPs require a thermal energy storage system (TES) to manage the energy demand avoiding system oversizing [12,34]. Because of this, two simplifications have been assumed for the simulation analysis. The first simplification is that thermal energy demand is supplied daily from a thermal energy buffer (Figure 4), that can be charged with the CHP unit and discharged by the user without any time dependence. The second simplification is that the electrical energy is also managed using an electrical energy storage system (EES) with the suitable capacity to manage the daily consumption. EES ensures the generation–consumption correlation [24]. Electrical energy surplus is daily exported to the electrical grid.

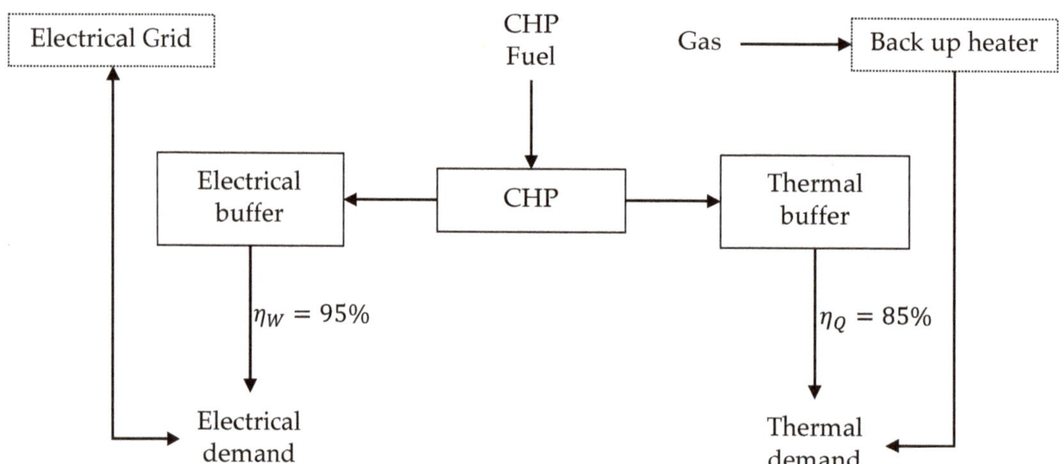

Figure 4. Cogeneration functioning block diagram for the simulation algorithm.

Figure 4 shows the functioning block diagram of the CHP system for the simulation algorithm programmed in a Python [35] script that is graphically described in Figure 5. The code is used to evaluate a day-by-day energy balance from the user dataset for each individual dwelling and the 10-dwelling buildings.

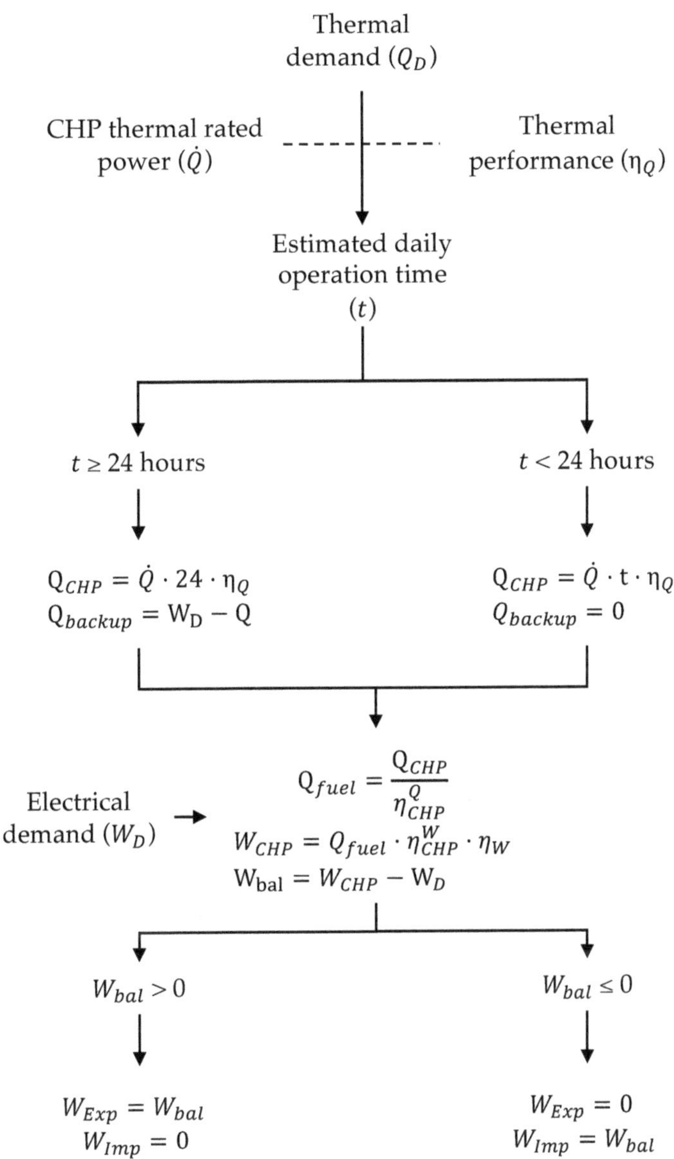

Figure 5. Python script code algorithm for day-to-day energy analysis decision diagram.

The calculation process starts with the evaluation of the total time that the CHP requires to produce the daily thermal demand, which will depend on the rated thermal power of the CHP technology used. This time can be calculated with the equation:

$$t = \frac{Q_D}{\dot{Q}_{CHP} \cdot \eta_Q}, \tag{1}$$

where Q_D is the daily thermal energy demand, $\dot{Q}_{CHP}$ is the rated CHP thermal power and η_Q the thermal efficiency of the energy conversions required to feed the demand. Energy conversions are due to the energy extraction process of the thermal energy from the buffer used to match the energy production and demand (Figure 4). The same efficiency (85%) is considered for all the CHP technologies. The result of Equation (1) is the time that the CHP requires to produce the daily thermal energy, where two different situations can occur. If the total time calculated is lower than 24 h, the CHP technology is capable to provide the entire daily thermal demand and no back-up energy system will be required. In case of a CHP technology that is not suitably sized the total estimated time can be greater than 24 h, and the maximum achievable energy will be the obtained from the CHP unit working at the rated power the entire day. In this case, it is considered that the shortage of thermal energy will be compensated by a back-up system, e.g., a boiler, using natural gas as fuel.

The fuel consumption to produce the thermal energy with the CHP can be determined as:

$$Q_{fuel} = \frac{Q_{CHP}}{\eta_{CHP}^{Q}}, \tag{2}$$

where Q_{CHP} stands for the thermal energy produced by the CHP unit, and η_{CHP}^{Q} is the energy performance for the thermal energy flow in the CHP system. The back-up energy (Q_{backup}) equals the lack of thermal energy because the value comes from the boiler consumption as can be read in Figure 5.

The electrical energy produced with the CHP unit depends on the CHP gas consumption and the electrical efficiency of the unit, η_{CHP}^{W}, as:

$$W_{CHP} = Q_{fuel} \cdot \eta_{CHP}^{W} \cdot \eta_W, \tag{3}$$

where η_W is additional electrical efficiency due to the energy conversions. The electrical energy produced will be used to provide the daily electrical energy demand (W_D). The electrical energy balance is determined as:

$$W_{bal} = W_{CHP} - W_D. \tag{4}$$

Attending to the sign value reported from this equation, electrical energy will be imported from the grid when it is negative and exported or sold to the grid if it is positive.

3.2. Primary Energy Factors

Primary energy (PE) is a concept used to compare different kinds of energy sources, but the scale used in the calculations is relevant [4,36]. In this research, the primary energy factors published by the Spanish Government in 2016 [37] are used to determine the building performance required for legalization. Corresponding values are summarized in Table 1 for both utilities considered, namely, the national electrical grid and the natural gas supply facility. The PE factor depends on the energy carrier and relates the primary energy consumed to provide one kWh to the end-user, in this case the final energy consumed by the residential users. PE factors are also divided into renewable and non-renewable. As can be observed in Table 1, renewable factors are smaller that non-renewable ones, but they are values above zero. This means that a certain amount of energy is required to serve the renewable source, i.e., maintenance tasks.

Table 1. Primary energy factors and CO_2 conversion factor established by the Spanish Government for the electrical and natural gas utilities.

	Primary Energy kWh$_{primary}$/kWh$_{final}$ **No Renewable Source**	**Primary Energy** kWh$_{primary}$/kWh$_{final}$ **Renewable Source**	**CO_2 Emission Factor** kg CO_2/kWh$_{final}$
National electrical utility	1.954	0.414	0.331
Natural gas utility	1.190	0.005	0.252

Not only the PE, but also the CO_2 emissions will be compared to determine the benefits of the cogeneration technologies. The PE reductions are measured from the initial situation, and they can be evaluated using the equation:

$$PE = \left(Q_D \cdot f_g^{PE} + W_D \cdot f_e^{PE}\right), \tag{5}$$

where f^{PE} stands for the primary energy factor (subscripts g and e represent gas and electricity, respectively), Q_D is the total thermal energy demanded and W_D stands for the total electrical energy. The PE consumption with the use of a CHP system will depend on the technology. A general case is shown in Equation (6):

$$PE_{CHP} = \left(\left(Q_{fuel} + Q_{backup}\right) \cdot f_g^{PE} + W_{Imp} \cdot f_e^{PE}\right) - W_{Exp} \cdot f_e^{PE}, \tag{6}$$

where exported energy (W_{Exp}) is considered as a primary energy decrement due to its decreasing effect in primary energy consumption. When hydrogen or any other fuel obtained from renewable energy sources is used, the gas terms (Q_{fuel} and Q_{backup}) in Equation (6) can be neglected. Carbon emissions can be calculated using the same equations, just replacing the primary energy factors with the CO_2 emission factor (f^{CO_2}).

$$CO_2 = \left(Q_D \cdot f_g^{CO_2} + W_D \cdot f_e^{CO_2}\right) \text{ and} \tag{7}$$

$$CO_2^{CHP} = \left(\left(Q_{fuel} + Q_{backup}\right) \cdot f_g^{CO_2} + W_{Imp} \cdot f_e^{CO_2}\right) - W_{Exp} \cdot f_e^{CO_2}. \tag{8}$$

3.3. Fuel Cell Stack and Its Cooling System Design

The power unit of the CHP in the present research consists of a prototype of high-temperature PEM fuel cell and its novel cooling system that were designed and developed specifically for this project. The 40-cells high-temperature PEM fuel cell stack (HT-PEMFC) is formed by 41 JP-945 graphite bipolar plates 280 mm high × 195 mm wide × 5 mm thick manufactured by Mersen, as well as two stainless steel end plates where all the connectors for the reactant gases, H_2 and O_2/air, are placed. The flowfield geometry in both anode and cathode sides consisted of straight parallel channels with a land-to-channel ratio of 1, as recommended by the MEA manufacturer. The anode side was formed by 47 channels 1 mm wide, 1.5 mm deep, and a total length of 210 mm, while the cathode side is formed by 87 channels with a width of 1 mm and a depth of 2 mm, and a total length of 120 mm. With this design, pressure losses were minimized to 5.87 Pa in the anode and 2.6 Pa in the cathode, ensuring both the homogenous distribution of the reactant gases over the electrodes and the correct water management. Commercial high-temperature membrane-electrode assemblies (MEAs) G1018 Dapozol-110, manufactured by Danish Power System (DPS) with a rectangular active area of 163.5 cm^2, were used [38]. The MEAs are formed by phosphoric acid doped PBI polymeric membranes, with a nominal thickness of 650 ± 50 μm, gas diffusion layers of non-woven carbon paper and a platinum load of 1.5 mg cm^{-2} in both electrodes. The nominal thickness of the electrodes is 250 μm, including the GDL, the microporous layer and the catalyst layer. To obtain the best results, a minimum compression rate of 13% is advised, as well as a recommended working temperature ranging from 150 °C to 180 °C. Figure 6a shows the manufactured prototype developed by the PEMFC research team from LIFTEC-CSIC in Zaragoza (Spain), which has an ample expertise in this field [39,40]. Figure 6b shows the electrical and thermal performance of the HT-PEMFC stack. The vertical axes represent the voltage (left axis and red curve) and the power (right axis and green curves), and the horizontal values are the current produced by the electrochemical device. Solid green line corresponds to the electrical power, and the dashed green line is the estimated thermal power.

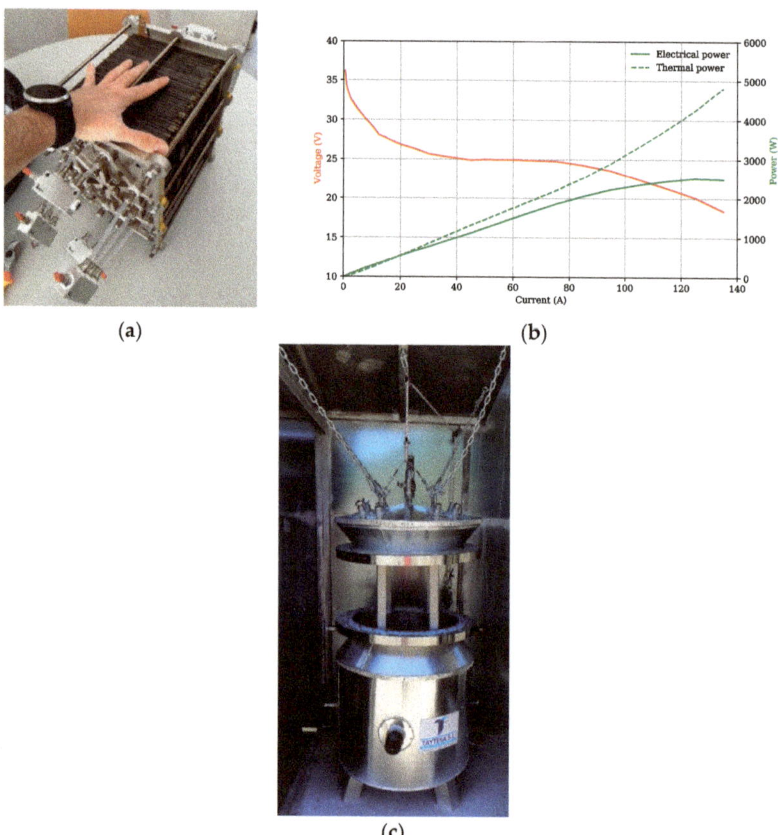

Figure 6. Fuel cell system developed for the MICAPEM project. (**a**) Fuel cell stack designed and manufactured by LIFTEC-CSIC laboratory. (**b**) Performance curves obtained in the test bench. (**c**) High-temperature PEM fuel cell cooling system.

The fuel cell rated operating point is set in 1.6 kW and 1.9 kW for electrical and thermal power respectively, which ensures a long lifetime of the MEAs [41]. Despite this, the fuel cell is capable to achieve a maximum electrical and thermal power of 2.5 kW and 4.8 kW, respectively with excellent performance. Both operating points are considered in the primary energy analysis.

High-temperature PEM fuel cells work in a temperature range from 120 °C to 180 °C that has to be controlled to avoid fast degradation. Figure 6c shows the manufactured novel cooling system specially designed to preheat the stack before starting, and to maintain the required temperature during the HT-PEMFC operation. The novelty of the system is the use of an isothermal oil bath with a dielectric oil that helps not only to keep the fuel cell temperature in the suitable range, but also as an energy buffer system extracting heat from the oil. Preliminary results show that for such system the heat extraction efficiency is higher than the value selected for the algorithm (85%), but no re-calculations have been performed with this higher efficiency.

4. Results

Simulations with the above discussed algorithm were performed using the datasets for single dwellings and 10-dwelling buildings. Five different CHPs were considered, which are summarized in Table 2, where "Gas ICE", "Gas turbine" and "Fuel cell" are the smallest CHP commercial units as shown in Figure 1, from [14]. "MICAPEM$_{rated}$" and "MICAPEM$_{max}$" is the fuel cell-based CHP system with the manufactured fuel cell operating at both rated and maximum power points, respectively. The simulations for single-dwellings and buildings used the same CHP unit characteristics from Table 2 to help in the results comparison.

Table 2. Combined heat and power smaller units and MICAPEM project fuel cell-based characteristics.

	Thermal Power (kW)	Thermal Efficiency (%)	Electrical Power (kW)	Electrical Efficiency (%)
Gas ICE	12	62	5	26
Gas turbine	52	47	28	26
Fuel cell	1.2	47	1	38
MICAPEM$_{rated}$	1.9	49	1.6	46
MICAPEM$_{max}$	4.8	59	2.5	37

Calculations were performed considering three possible scenarios for the primary energy and carbon dioxide emissions reductions. These scenarios are classified as a function of the fuel used to power the CHP installation and the back-up heater, if it is necessary.

- Natural gas scenario: In this case, all of the thermal systems, namely, CHP and backup heater, were fueled with natural gas. Gas ICE and gas turbine can use this fuel directly, but not the fuel cell-based systems, which require pure hydrogen. So, a natural gas reforming process was considered. The efficiency for such process depends on the gas volume managed, and it is fixed in 87% for the small-scale reforming system required in this study [42].
- Green gas scenario: In this scenario, all of the systems were powered with carbon-free fuels, like green hydrogen. Thus, hydrogen production did not require non-renewable primary energy consumption. In fact, today actual environmental impact of hydrogen production is not zero because the distribution infrastructure is not fully developed yet [43–47]. Nevertheless, in the present research it is considered as a carbon-free fuel because it is locally produced with a hydrogen electrolyzer included in the CHP infrastructure of the project. Same thermal and electrical efficiency was considered for gas ICE and gas turbine fueled with green source gas.
- Expected scenario is the most probable situation. Here, natural gas is considered as the fuel for gas ICE, gas turbine, and backup heaters in all the cases, but the fuel cell-based power system is fueled with pure green hydrogen.

Results are graphically displayed in Figures 7–12. Vertical axes values vary depending on the variable analyzed, but the horizontal axis is the same in all the Figures, namely, the CHP technologies. In each technology the four individual dwellings or the 10-dweling buildings are shown for an easier comparison.

Each Figure caption groups the simulation results for the dwellings (Id 1, Id 2, Id 3, and Id 4) and buildings datasets (B-Id). In case of Figures 7 and 8, where PE and CO$_2$ reductions are shown, graph grouping includes the results for the three analyzed scenarios. Figures 9–12 are only function of the energy demand and CHP technology and independent from the fuel scenario.

4.1. Primary Energy Consumption

Final primary energy (PE) consumption is calculated according to Equation (6) where the exported electricity has a positive effect because of the decrease in PE consumption (negative in the equation). Figure 7 groups the graphs with the simulation results for the PE reduction in the vertical axis as the percentage value of the PE consumed with the CHP operative over the PE consumption calculated for the dataset (without an operative CHP installation). Figure 7a1,a2 resume the simulation results under the "Natural gas scenario" for single dwellings and 10-dwelling buildings, respectively. Similarly, Figure 7b1,b2 summarize the results for the same dwellings when the "green gas scenario" is simulated, while Figure 7c1,c2 correspond to the results for the "expected scenario".

4.2. Carbon Emissions Results

Figure 8a–c show the percentage of CO_2 emissions over current emissions. Values are determined by Equations (7) and (8). These plots are similar to PE reduction graphs due to the linear relationship between the two variables, but it should be noted the relative value in each scenario.

4.3. Cogeneration and Energy Demand Rates

The graphs on this section are independent of the scenarios because the energy balances depend on the consumption and the CHP parameters but not on the fuel. Figure 9a,b show the importation and exportation of electrical energy from/to the utility over the electrical demand of each individual and building dwellings respectively.

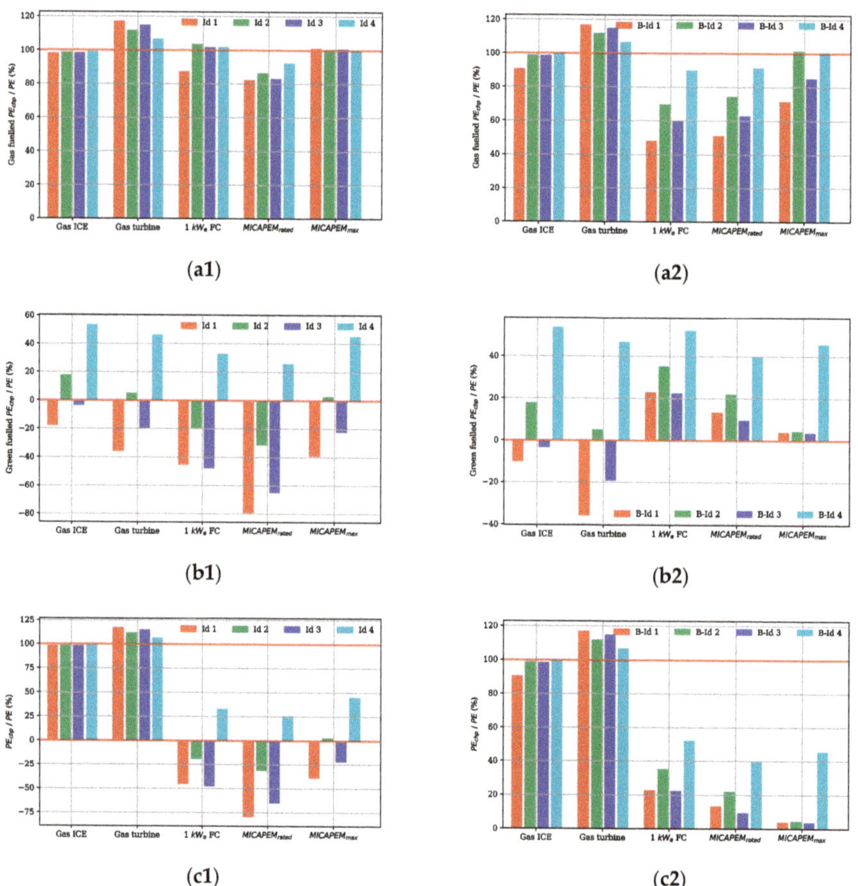

Figure 7. Primary energy consumption reduction percentage over the initial PE per technology and dataset. (**a**) Natural gas scenario simulation results; (**b**) green gas scenario; (**c**) expected scenario. Plots on the left, labeled with (**1**), correspond to the single unit dwellings. Plots on the right, labeled with (**2**), are for the 10-dwelling buildings datasets.

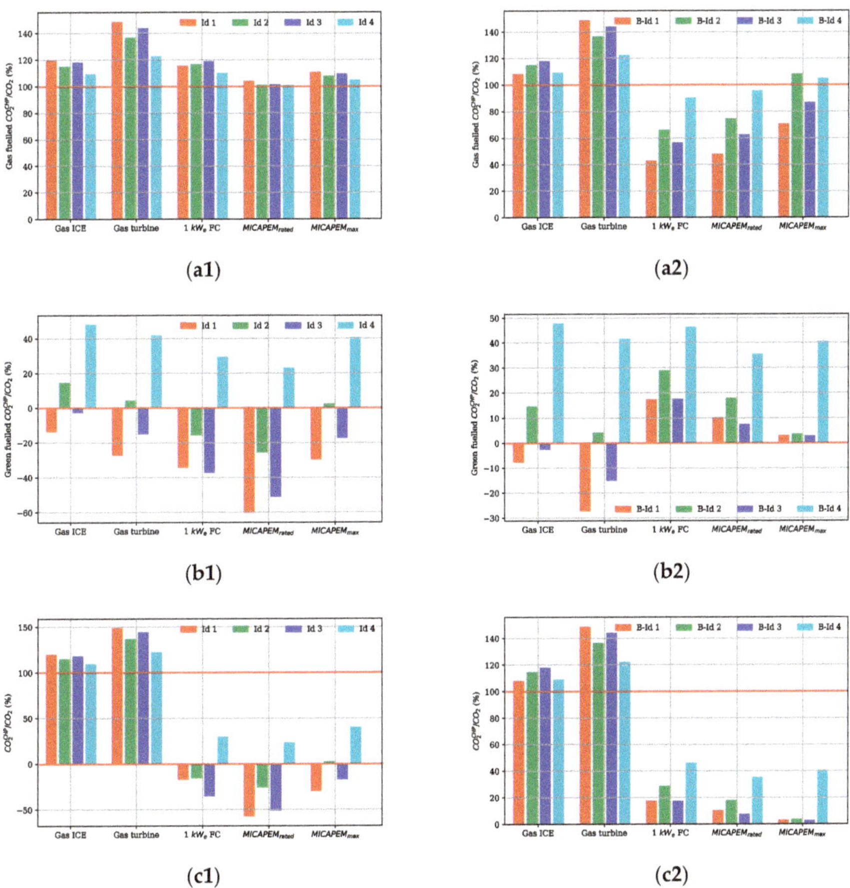

Figure 8. Relative CO_2 emission over the current situation of the dwellings (**a**) using natural gas-fueled CHP; (**b**) considering all the systems powered with green fuel as hydrogen (carbon-free); (**c**) expected scenario, each CHP system with the corresponding fuel and back-up heaters powered with natural gas. Plots on the left, labeled with (**1**), correspond to the single unit dwellings. Plots on the right, labeled with (**2**), are for the 10-dwelling buildings datasets.

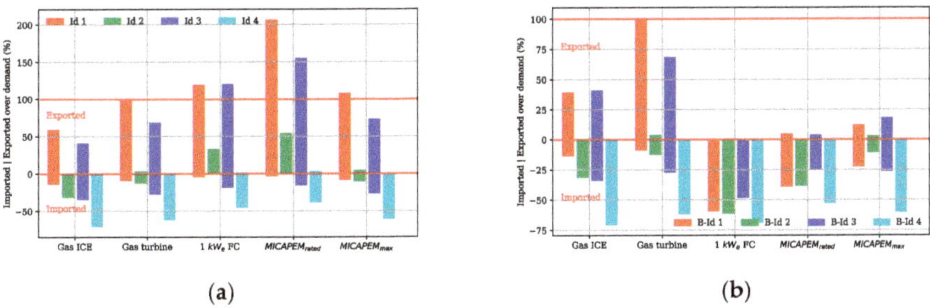

Figure 9. Electrical energy import/export over the electrical demand Results of the electrical energy exchange with the grid over the demanded energy for (**a**) the single-dwelling simulation; (**b**) the 10-dwelling buildings simulation.

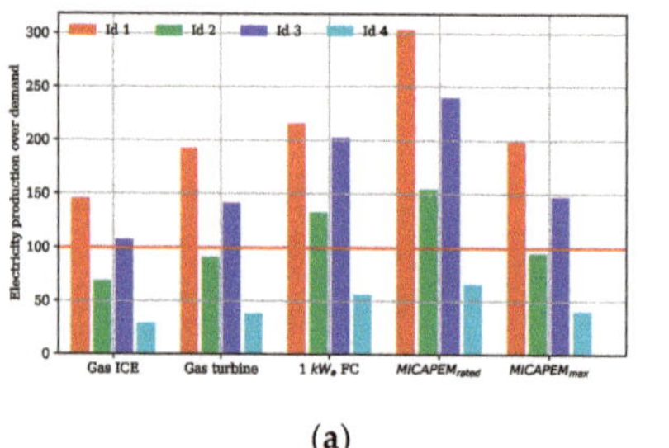

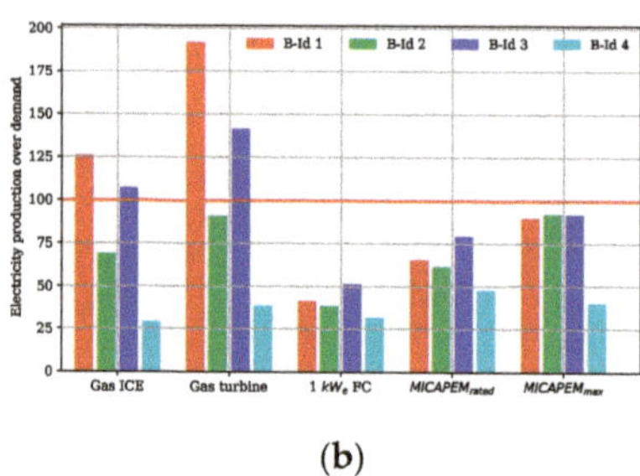

(a)　　　　　　　　　　(b)

Figure 10. Results for the relative electrical production over the electrical energy demanded. (**a**) Individual dwellings. (**b**) 10-dwelling buildings.

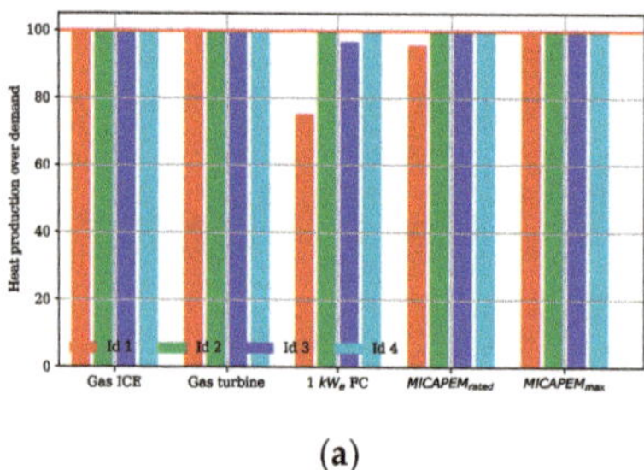

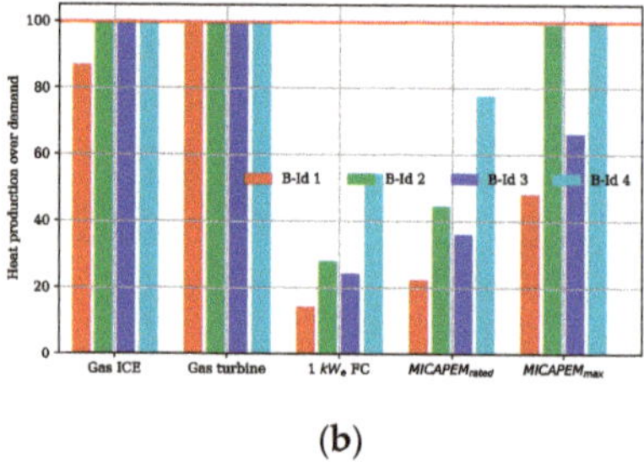

(a)　　　　　　　　　　(b)

Figure 11. Relative thermal energy produced with the CHP vs. thermal demand. (**a**) Individual dwellings. (**b**) 10-dwelling buildings.

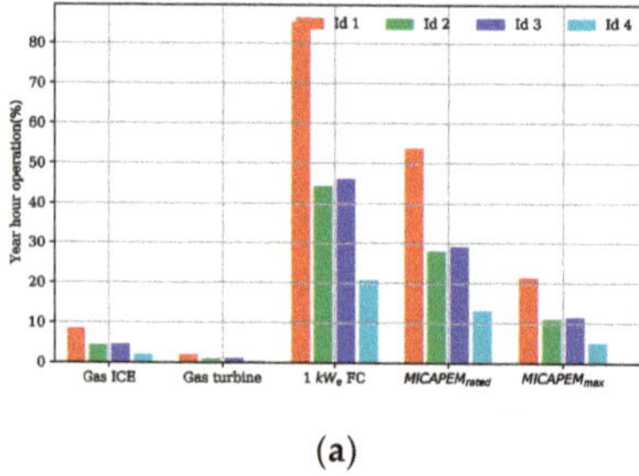

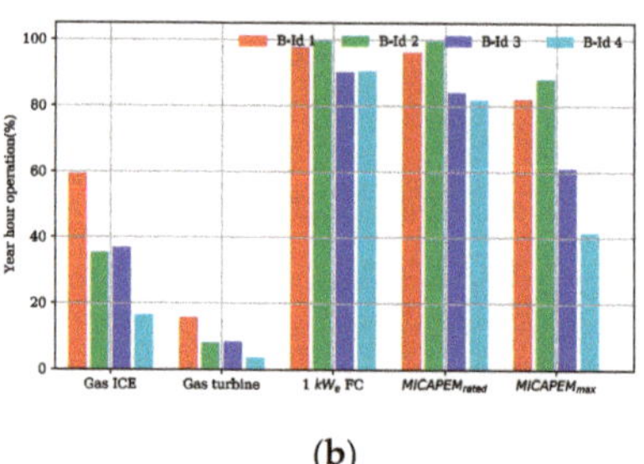

(a)　　　　　　　　　　(b)

Figure 12. Operating time over the total year-hour rate. (**a**) Results for individual dwelling simulation. (**b**) Results for 10-dwelling building simulation.

In Figure 10a,b, the results for the relative electrical production over the electrical energy demanded are depicted. The plots show the capacity of the CHP system to meet the electrical energy demand of the consumers. Values above 100% mean that the system exceeds the electrical demand resulting in an energy surplus.

Figure 11a,b show the percentage of thermal energy produced with the CHP against the thermal demand. When the system is well-sized the value is 100% because the entire thermal demand is met with the CHP production. When the system is under-sized, the thermal production has to be compensated with some thermal energy produced with the back-up boiler. There is not a thermal energy surplus because the system control was set to feed the thermal demand without exceeding it.

The use of CHP systems has to be economically viable. This viability depends not only on the facility cost, but also on the rate of use. Figure 12a,b show the percentage of the year-hour that each technology will operate in each dwelling and building respectively.

The time is calculated with Equation (1) as a function of the thermal demand and the production capacity of the CHP system. As can be observed in Figure 12a for the individual dwellings simulation, where the fuel-cell-based CHP systems present a higher duty cycle, the smaller the rated power, the higher the duty cycle. A similar behavior is observed in the results for the 10-dwelling building simulation (Figure 12b).

5. Discussion

The simulation results for the "natural gas scenario" show some interesting conclusions. In this scenario, natural gas is used as the only fuel for all CHP systems. Fuel cell-based technologies reform the gas to obtain pure hydrogen, which results in an additional gas consumption. Even so, the fuel cell-based CHP systems have a significant impact into PE reduction as can be observed in Figure 7a1,a2. It is noteworthy that the smallest fuel cell CHP ("1 kW$_e$ FC") system presents the greatest PE reductions for the 10-dwelling building simulations. PE consumptions are from ca. 50% for B-Id 1 to ca. 90% for B-Id 4. On the contrary, the same facility does not show such good behavior in the single dwelling simulation. This is due to the better efficiency of the back-up heater for the energy conversion of natural gas into heat. In other words, the "1 kW$_e$ FC" system used in the high-rise buildings means a longer operation time because of the lower proportion of heat demand fed from the CHP (see Figures 11b and 12b for operating time and thermal demand respectively). Thus, more energy from the back-up heater is needed to cover the heating demand and this means that lower energy from the natural gas is required. Even so, the use of the CHP decreases the consumed energy from the electrical grid, which benefits the PE reduction. This can also be observed in the CO_2 emissions in Figure 8a1,a2, where only fuel cell-based CHP systems show an effective greenhouse gases reduction because of the impact of the reduction in electrical energy importation. It is easy to see in Figure 7a2 that the higher the HtP ratio the better the PE reduction.

"MICAPEM$_{rated}$" fuel cell CHP was designed according to a standard dwelling consumption, showing the best PE reduction in all scenarios. Focusing on the first scenario (Figure 7a1) the PE consumption is around 80% for the user with the higher HtP ratio (Id 1) and around 90% for the lower HtP ratio (Id 4). However, CO_2 emissions are maintained because of the increment in natural gas consumption to meet the thermal demand (Figure 8a1). Electrical energy surplus achieves a maximum value, doubling the energy demand, as can be observed in Figure 9a. In the same Figure, Id 4 and B-Id 4 always import electricity from grid because the datasets are based on a high electrified user with a low HtP ratio, ca. 0.7.

The results for the "green gas scenario" are also interesting and as expected, the use of carbon-free fuel has a significant impact on the PE reduction in buildings. The higher the energy exported; the higher PE reduction is achieved. Main differences for fuel-cell-based systems in Figure 7b1,b2 are the PE reduction rate sign. On the one hand, in Figure 7b1 it is negative because the CHP unit is able to cover the entire thermal demand (see also Figure 11a) yielding an exported electrical energy surplus. On the other hand, 10-dwelling building results in Figure 7b2 show that the reduction is not so high because neither the thermal nor the electrical demand are fully covered by the CHP unit and despite the fact that green gas can be used in the backup heater, the electricity has to be imported from the commercial grid. Decreasing the energy importation rather than creating an energy surplus is the preferred situation for the actual electrical systems because energy consumers can be managed more easily than small energy producers (self-consumption without net balance). The case Id 4 is an exception not only in the carbon-free fuel scenario, but also in the other ones, because it is a highly electrified consumer with a low thermal energy demand. In fact, this means that the total operation time of the units is lower than others, so electricity has to be imported. Nevertheless, the PE reduction is significant for Id 4 due to the cogeneration.

The "expected scenario" is the most probable scenario, where only hydrogen is used to power the fuel cell-based CHP systems and natural gas for back-up heaters and the other CHP technologies. This means that the fuel consumption term in Equation (6) can be neglected for fuel cell-based system exclusively. Figure 7c1 shows that the most powerful thermal CHP units ("gas ICE" and "gas turbine") result in a worst PE reduction for all dwellings due to the low operating time (see Figure 12a) and, consequently, the lowest electrical energy production (see Figure 9a). The negative value in the PE ratio for fuel-cell-based CHP units is due to the electrical energy surplus. Figure 7a1 shows, again, an optimal design point in the "MICAPEM$_{rated}$" characteristics. For the 10-dwelling buildings, the higher the power the higher the PE reduction due to the ability to provide the energy demands from a low consumption of carbon-free fuel for PE production. The contrary can be observed in Figure 7a2 for the "natural gas scenario", where a greater fuel consumption is penalized.

Figure 8a1,a2 show the CO_2 reduction results for the "gas-fueled scenario" for both individual and building dwellings, respectively. The use of CHP in dwellings does not have a carbon emissions reduction due to the higher gas consumption because of the efficiency reduction compared to the use of a boiler. Nevertheless, the 10-dwelling building simulation shows that CO_2 emissions are lower for the fuel cell-based system due to the electrical generation and the increase in global efficiency. Similarly, in both the "green fuel scenario" and the "expected scenario", the reduction of CO_2 emission shows the same behavior compared to the PE reduction.

The size of the CHP system is an important design and a critical economical parameter. The size of the CHP is directly related to the ability to meet the energy demand, but indirectly related to the operation time (and the economic viability). Figure 11 shows the share of thermal energy demand produced with the CHP system and Figure 12 the share of operation hours per year. The total operating time for the smaller units of the "*gas ICE*" and the "gas turbine" CHP units applied to low thermal demand consumers like a single dwelling (Figure 12a), makes its use unviable (less than 5%-year hours). "Gas turbine" units still are unviable for typical buildings in Spain (less than 20%-year hours). Contrary, fuel cell-based technologies, due to their lower power appear to be a better solution for CHP systems in the building sector, ca. 80%-year hours in the best cases. The low thermal power is not a handicap because fuel-cell-based CHP systems are fully scalable.

6. Conclusions

When the primary energy (PE) consumption in Spanish buildings is calculated with the official factors summarized in Table 1, which depend on the energy carrier and its energy source, it was demonstrated that the electrical energy carrier is 1.6 times more demanding than the natural gas from non-renewable sources. Even so, there is a tendency to electrify the consumptions because it is an energy carrier that can be more easily decarbonized. Some industrial heating systems are electrified due to the availability of powerful transmission lines. Despite of this, building centralized heating systems are not normally electrified because of the limitations of electrical grids in the cities.

Fuel cell-based CHP systems are a good solution to provide the energy demand for heating and hot water in buildings, showing a decrease in both PE consumption and CO_2 emissions, even if the hydrogen is obtained from natural gas reforming. However, this PE reduction is directly related to the thermal energy conversion efficiency of the CHP and the boiler because in the best situation a 50% PE reduction can be achieved with an energy production of ca. 20% of thermal demand and ca. 50% of the electrical demand. When the fuel cell-based CHP systems are powered with carbon-free hydrogen, the PE reduction is higher when the system is able to meet a big share of the energy demand, which corresponds to a better fit of the heat-to-power ratio between production and demand. The use of micro-CHP units integrated into smart grids can help to reduce not only the thermal and electrical demand of the user but also the electrical demand from the nearby with the proper energy management.

The economic viability of fuel cell-based CHP units is similar to other technologies such as solar thermal systems that are projected to supply a maximum share of the thermal demand, ensuring the higher possible operating time. Based on the results of Figure 12, the optimal situation corresponds to centralized systems where the total operating time is above 80% of the year-hours and the thermal demand can be fully supplied.

Author Contributions: The first author, J.R., has taken lead on all the steps of the research and writing process. The rest of authors have contributed as follow: research methodology, F.S., A.L. and F.B.; simulation, V.G., L.D. and A.R.; validation, P.V., A.G.; writing—original draft preparation, F.S., A.L. and F.B.; supervision, F.S. and A.L.; writing—review and editing, F.S., A.L. and F.B. All authors have read and agreed to the published version of the manuscript.

Funding: This research was funded by the Secretariat of State for Research of the Spanish Ministry of Economy and Competitiveness (DPI2015-69286-C3-1-R), the Spanish Ministry of Science and Innovation (RTI2018-096001-B-C33), and the Aragon Government (LMP246_18).

Institutional Review Board Statement: Not applicable.

Informed Consent Statement: Not applicable.

Acknowledgments: This work has been partially funded by the Secretariat of State for Research of the Spanish Ministry of Economy and Competitiveness under the project MICAPEM (ref.: DPI2015-69286-C3-1-R) and by the Spanish Ministry of Science and Innovation under the project DOVELAR (ref.: RTI2018-096001-B-C33). LIFTEC research team would also acknowledge the funded provided by the Aragon Government under the project LMP246_18.

Conflicts of Interest: The authors declare no conflict of interest.

Nomenclature

Abbreviations

CHP	Combined heat and power system
EES	Electrical energy storage system
EPBD	Energy Performance of Buildings Directive (2010/31/EU)
HtP	Heat-to-power energy ratio
PE	Primary energy
PEMFC	Polymer exchange membrane fuel cell
TES	Thermal energy storage system

Variables

CO_2^{CHP}	CO_2 emissions from the combined heat and power unit (kg)
f^{CO_2}	CO_2 emissions conversion factor, kg of CO_2 per kWh of end-use energy
f^{PE}	Primary energy conversion factor, kWh of primary energy per kWh of end-use energy
$\dot{Q}$	Rated thermal power of the combined heat and power unit (kW)
Q_{backup}	Back-up heater energy flow (kWh)
Q_D	Daily thermal energy demand (kWh)
Q_{fuel}	Fuel energy flow in the combined heat and power unit (kWh)
Q_{CHP}	Thermal energy produced by the combined heat and power unit (kWh)
t	Estimated daily operation time of the combined heat and power unit (h)
W_{bal}	Electrical energy balance (kWh)
W_{CHP}	Electrical energy produced by the combined heat and power unit (kWh)
W_D	Daily electrical energy demand (kWh)
W_{Exp}	Net electrical exported energy to the grid (kWh)
W_{Imp}	Net electrical imported energy from the grid (kWh)
η_Q	Thermal efficiency to supply the energy demand from the generation
η_{CHP}^{Q}	Thermal energy performance for the combined heat and power unit
η_{CHP}^{W}	Electrical energy performance for the combined heat and power unit
η_W	Electrical efficiency to supply the demand from the generation

Subscripts

e	electricity
g	gas

References

1. Energy Statistics—An Overview—Statistics Explained. Available online: https://ec.europa.eu/eurostat/statistics-explained/index.php/Energy_statistics_-_an_overview#Final_energy_consumption (accessed on 27 December 2020).
2. IDEA. SECH PROJECT-SPAHOUSEC Analyses of the Energy Consumption of the Household Sector in Spain IDAE General Secretary Planning and Studies Department 16th of June of 2011. 2014. Available online: https://ec.europa.eu/eurostat/cros/system/files/SECH_Spain.pdf (accessed on 27 December 2020).
3. EPBD. Energy Performance of Buildings Directive 2010/31/EU (recast). *Off. J. Eur. Union* **2010**, *153*, 13–35.
4. Hitchin, R.; Thomsen, K.E.; Wittchen, K.B. Primary Energy Factors and Members States Energy Regulations—Primary Factors and the EPBD, no. 692447. 2018, p. 4. Available online: https://www.epbd-ca.eu/wp-content/uploads/2018/04/05-CCT1-Factsheet-PEF.pdf (accessed on 19 December 2020).
5. Integrated National Energy and Climate Plan 2021–2030. 2020. Available online: https://ec.europa.eu/energy/sites/ener/files/documents/es_final_necp_main_en.pdf (accessed on 27 December 2020).
6. Amber, K.P.; Day, T.; Ratyal, N.I.; Ahmad, R.; Amar, M. The Significance of a Building's Energy Consumption Profiles for the Optimum Sizing of a Combined Heat and Power (CHP) System—A Case Study for a Student Residence Hall. *Sustainability* **2018**, *10*, 2069. [CrossRef]
7. Atănăsoae, P. Technical and Economic Assessment of Micro-Cogeneration Systems for Residential Applications. *Sustainability* **2020**, *12*, 1074. [CrossRef]
8. Brett, D.J.L.; Brandon, N.J.; Hawkes, A.; Staffell, I. Fuel cell systems for small and micro combined heat and power (CHP) applications. In *Small and Micro Combined Heat and Power (CHP) Systems*; Elsevier: Amsterdam, The Netherlands, 2011; pp. 233–261.
9. Elmer, T.; Worall, M.; Wu, S.; Riffat, S.B. Fuel cell technology for domestic built environment applications: State-of-the-art review. *Renew. Sustain. Energy Rev.* **2015**, *42*, 913–931. [CrossRef]
10. CEER. CEER Report on Power Losses 2017. Available online: https://www.ceer.eu/documents/104400/-/-/6f455336-d8c8-aa6f- (accessed on 12 January 2021).
11. Uchman, W.; Kotowicz, J.; Li, K.F. Evaluation of a micro-cogeneration unit with integrated electrical energy storage for residential application. *Appl. Energy* **2021**, *282*, 116196. [CrossRef]
12. Pérez-Iribarren, E.; González-Pino, I.; Azkorra-Larrinaga, Z.; Gómez-Arriarán, I. Optimal design and operation of thermal energy storage systems in micro-cogeneration plants. *Appl. Energy* **2020**, *265*, 114769. [CrossRef]
13. Onovwiona, H.; Ugursal, V. Residential cogeneration systems: Review of the current technology. *Renew. Sustain. Energy Rev.* **2006**, *10*, 389–431. [CrossRef]
14. Cogen Challenge. Pick the Right Cogeneration Technology. A Technology Checklist of Small-Scale Cogeneration. 2006. Available online: https://ec.europa.eu/energy/intelligent/projects/sites/iee-projects/files/projects/documents/cogen_challenge_technology_checklist.pdf (accessed on 23 April 2019).
15. Bhatia, S.C. Cogeneration. In *Advanced Renewable Energy Systems*; Elsevier: Amsterdam, The Netherlands, 2014; pp. 490–508.
16. Verhelst, S.; Wallner, T. Hydrogen-fueled internal combustion engines. *Prog. Energy Combust. Sci.* **2009**, *35*, 490–527. [CrossRef]
17. Hawkes, A.; Staffell, I.; Brett, D.; Brandon, N. Fuel cells for micro-combined heat and power generation. *Energy Environ. Sci.* **2009**, *2*, 729–744. [CrossRef]
18. Fiskum, R.J.; Hadder, G.R.; Chen, F.C. Fuel cells in residential and commercial building applications. In Proceedings of the Intersociety Energy Conversion Engineering Conference, Atlanta, GA, USA, 8–13 August 1993.
19. Arsalis, A. A comprehensive review of fuel cell-based micro-combined-heat-and-power systems. *Renew. Sustain. Energy Rev.* **2019**, *105*, 391–414. [CrossRef]
20. Barelli, L.; Bidini, G.; Gallorini, F.; Ottaviano, A. An energetic–exergetic analysis of a residential CHP system based on PEM fuel cell. *Appl. Energy* **2011**, *88*, 4334–4342. [CrossRef]
21. Olabi, A.; Wilberforce, T.; Sayed, E.T.; Elsaid, K.; Abdelkareem, M.A. Prospects of Fuel Cell Combined Heat and Power Systems. *Energies* **2020**, *13*, 4104. [CrossRef]
22. European Project. Ene-Field. Available online: http://enefield.eu/category/about/ (accessed on 24 December 2020).
23. Ene-Farm Project. Japan LP Gas Association. Available online: https://www.j-lpgas.gr.jp/en/appliances/ (accessed on 24 December 2020).
24. Renau, J.; Domenech, L.; García, V.; Real-Fernández, A.; Montés, N.; Sanchez, F.P. Proposal of a nearly zero energy building electrical power generator with an optimal temporary generation–consumption correlation. *Energy Build.* **2014**, *83*, 140–148. [CrossRef]
25. Real-Fernández, A.; Garcia, V.G.; Domenech, L.; Renau, J.; Montés, N.; Sanchez, F.P. Improvement of a heat pump based HVAC system with PCM thermal storage for cold accumulation and heat dissipation. *Energy Build.* **2014**, *83*, 108–116. [CrossRef]
26. Jensen, J.O.; Li, Q.; Pan, C.; Vestbø, A.P.; Mortensen, K.; Petersen, H.N.; Sørensen, C.L.; Clausen, T.N.; Schramm, J.; Bjerrum, N.J. High temperature PEMFC and the possible utilization of the excess heat for fuel processing. *Int. J. Hydrogen Energy* **2007**, *32*, 1567–1571. [CrossRef]
27. Zhang, J.; Xie, Z.; Zhang, J.; Tang, Y.; Song, C.; Navessin, T.; Shi, Z.; Song, D.; Wang, H.; Wilkinson, D.P.; et al. High temperature PEM fuel cells. *J. Power Sources* **2006**, *160*, 872–891. [CrossRef]
28. Arsalis, A.; Nielsen, M.P.; Kær, S.K. Modeling and off-design performance of a 1kWe HT-PEMFC (high temperature-proton exchange membrane fuel cell)-based residential micro-CHP (combined-heat-and-power) system for Danish single-family households. *Energy* **2011**, *36*, 993–1002. [CrossRef]

29. Yang, Y.; Zhang, H.; Yan, P.; Jermsittiparsert, K. Multi-objective optimization for efficient modeling and improvement of the high temperature PEM fuel cell based Micro-CHP system. *Int. J. Hydrogen Energy* **2020**, *45*, 6970–6981. [CrossRef]
30. Arsalis, A.; Nielsen, M.P.; Kær, S.K. Modeling and optimization of a 1 kWe HT-PEMFC-based micro-CHP residential system. *Int. J. Hydrogen Energy* **2012**, *37*, 2470–2481. [CrossRef]
31. Boulmrharj, S.; Khaidar, M.; Bakhouya, M.; Ouladsine, R.; Siniti, M.; Zine-Dine, K. Performance Assessment of a Hybrid System with Hydrogen Storage and Fuel Cell for Cogeneration in Buildings. *Sustainability* **2020**, *12*, 4832. [CrossRef]
32. Jo, A.; Oh, K.; Lee, J.; Han, D.; Kim, D.; Kim, J.; Kim, B.; Kim, J.; Park, D.; Kim, M.; et al. Modeling and analysis of a 5 kWe HT-PEMFC system for residential heat and power generation. *Int. J. Hydrogen Energy* **2017**, *42*, 1698–1714. [CrossRef]
33. Arnfield, J. *Köppen Climate Classification*; Encyclopedia Britannica: Chicago, IL, USA, 2017.
34. Yu, D.; Meng, Y.; Yan, G.; Mu, G.; Li, D.; Le Blond, S. Sizing Combined Heat and Power Units and Domestic Building Energy Cost Optimisation. *Energies* **2017**, *10*, 771. [CrossRef]
35. Python Software Foundation. Available online: https://www.python.org/ (accessed on 13 January 2021).
36. Lightfoot, H.D. What engineers and scientists should know about scales for measuring primary energy: Why they are necessary and how to use them. In Proceedings of the 2nd Climate Change Technology Conference, Hamilton, ON, Canada, 12–15 May 2009; p. 10.
37. Gobierno de España. Factores de Emisión de CO_2 y Coeficientes de Paso a Energía Primaria de Diferentes Fuentes de Energía Final Consumidas en el Sector de Edificios en España. 2016. Available online: https://energia.gob.es/desarrollo/EficienciaEnergetica/RITE/Reconocidos/Reconocidos/Otrosdocumentos/Factores_emision_CO2.pdf (accessed on 19 December 2020).
38. Chandan, A.; Hattenberger, M.; El-Kharouf, A.; Du, S.; Dhir, A.; Self, V.; Pollet, B.G.; Ingram, A.; Bujalski, W. High temperature (HT) polymer electrolyte membrane fuel cells (PEMFC)—A review. *J. Power Sources* **2013**, *231*, 264–278. [CrossRef]
39. Renau, J.; Barroso, J.; Lozano, A.; Nueno, A.; Sánchez, F.; Martín, J.; Barreras, F. Design and manufacture of a high-temperature PEMFC and its cooling system to power a lightweight UAV for a high altitude mission. *Int. J. Hydrogen Energy* **2016**, *41*, 19702–19712. [CrossRef]
40. Barreras, F.; Lozano, A.; Roda, V.; Barroso, J.; Martín, J. Optimal design and operational tests of a high-temperature PEM fuel cell for a combined heat and power unit. *Int. J. Hydrogen Energy* **2014**, *39*, 5388–5398. [CrossRef]
41. Alegre, C.; Lozano, A.; Manso, Á.P.; Álvarez-Manuel, L.; Marzo, F.F.; Barreras, F. Single cell induced starvation in a high temperature proton exchange membrane fuel cell stack. *Appl. Energy* **2019**, *250*, 1176–1189. [CrossRef]
42. Bhushan, B. Handbook Technology. Available online: https://www.engineering-airliquide.com/es/technology-handbook (accessed on 16 October 2020).
43. IEA. *The Future of Hydrogen*; IEA: Paris, France, 2019.
44. Abdin, Z.; Zafaranloo, A.; Rafiee, A.; Mérida, W.; Lipiński, W.; Khalilpour, K.R. Hydrogen as an energy vector. *Renew. Sustain. Energy Rev.* **2020**, *120*, 109620. [CrossRef]
45. IREMA. Hydrogen: A Renewable Energy Perspective. 2019. Available online: https://www.irena.org (accessed on 16 October 2020).
46. Veneri, O. Hydrogen as Future Energy Carrier. In *Green Energy and Technology*; Springer: London, UK, 2011; pp. 33–70. [CrossRef]
47. Winter, C.-J. Hydrogen energy—Abundant, efficient, clean: A debate over the energy-system-of-change. *Int. J. Hydrogen Energy* **2009**, *34*, S1–S52. [CrossRef]

 *sustainability*

Article

Indirect Economic Effects of Vertical Indoor Green in the Context of Reduced Sick Leave in Offices

Jutta Hollands * and Azra Korjenic

Research Unit of Ecological Building Technologies, Institute of Material Technology, Building Physics and Building Ecology, Faculty of Civil Engineering, Vienna University of Technology, A-1040 Vienna, Austria; azra.korjenic@tuwien.ac.at
* Correspondence: jutta.hollands@tuwien.ac.at

Abstract: Low indoor humidity has been shown to influence the transmission of respiratory diseases via air. A certain proportion of sick leave in offices is therefore attributable to dryness of air. An improvement in these conditions thus means a reduction in sick leave, which is accompanied by cost savings for companies. Vertical indoor greening has a verifiable positive effect on air humidity, especially in winter months. In this article, the correlation between improved air humidity in greened rooms and reduction of sick leave due to improved air humidity was described. The resulting indirect economic effect was determined by comparing the costs for construction, green care, and technical maintenance of indoor greenery with savings due to lower sick leave. Based on long-term measurement data on air humidity and temperature, and actual cost values for three buildings, located in Vienna, Austria, with 6 greened and 3 reference rooms without greenery, the correlation of the method was derived and finally formulated in a generalized way using dimensioning factors. Only considering the influence on air humidity, profitability of 6.6 m^2 vertical greening installed in an example office with six workplaces equipped with technical ventilation and saving of two sick days already results after about 4.5 years.

Keywords: hygrothermal comfort; indoor green; vertical greenery; indoor air quality; cost-benefit-ratio; sick leave; absenteeism; alternative quantification method

 **check for updates**

Citation: Hollands, J.; Korjenic, A. Indirect Economic Effects of Vertical Indoor Green in the Context of Reduced Sick Leave in Offices. *Sustainability* **2021**, *13*, 2256. https://doi.org/10.3390/su13042256

Academic Editor: Roberto Alonso González Lezcano
Received: 2 February 2021
Accepted: 17 February 2021
Published: 19 February 2021

Publisher's Note: MDPI stays neutral with regard to jurisdictional claims in published maps and institutional affiliations.

1. Introduction

People spend about 90% of their time indoors [1]. A large proportion of this time is spent in offices. In Vienna, the share of office workers in 2001 was 28.6% of all employees, and the trend is rising [2]. Moreover, in Germany, a rise in office working places can be observed, as a study shows: In 2020, 71% of all employees in Germany worked at least partly in an office, which means 32 million people, whereas in 2015, it was only about 52% (22.5 million) [3]. Austrian law assumes a normal working time of 8 h per day or 40 h per week [4]. For occupations that are mainly performed in offices, this thus accounts for a share of around 24% of the total weekly time. Due to this amount of time spent indoors, indoor air quality is also increasingly becoming the focus of numerous studies. In many cases, the quality of indoor air is rated as insufficient [5–8]. In addition to the detection of pollutants in indoor air, the temperature and climatic conditions are also the focus of investigations. Temperature and climatic conditions are perceived as the biggest disturbances in office work environment, directly followed by noise pollution [9,10]. Air humidity especially plays a very important role. The occurrence of the following health effects in working spaces is associated with too low humidity: Drying of mucous membranes, colds, eye complaints, skin complaints, and electrostatic charging and discharging [11]. Several studies have shown that the perceived indoor air quality is enhanced by indoor air pollutants, the protective mucous layer in the respiratory tract, and tear films. This results in complaints and diseases of the respiratory tract and eyes [5].

297

The question of the development of diseases, depending on the relative humidity, was already raised in the 1960s [12]. In this context, a connection between the survival of pathogens and relative humidity was established. Diseases or irritations of the skin, eyes, and upper respiratory tract are often associated with low relative humidity indoors during the cold season [5,13]. Dry, cold respiratory air favors infections of the upper respiratory tract such as colds and throat infections in particular [14,15]. This is probably due to a higher stability of virus particles at low humidity and low temperatures. This has already been shown for rhinoviruses [16], influenza A viruses [17], and numerous other viruses [16], which are typical pathogens of the common cold. Due to the increased stability, the transmission of these viruses is particularly favored. Studies have already been conducted to examine the effects of prolonged exposure to low humidity on perceived indoor air quality, sensory irritation symptoms in the eyes and respiratory tract, work performance, sleep quality, virus survival, and voice disorders. Results showed that an improvement in indoor humidity can have a positive effect on perceived indoor air quality, eye symptoms, and possibly work performance in the office environment [5,10,18,19]. However, effects on increased diseases are not only attributed to the higher stability of the viruses depending on the humidity, but are also caused by the influence on the host. Thus, due to low humidity, the host defense changes as well as tissue repair is reduced as Kudo et al. [20] showed in their study on mice. Lowen et al. [21] summarize as a result of their study with guinea pigs as model host the mechanisms of influenza virus transmission as a function of humidity at three levels: Level of host concerning the mucociliary clearance and the associated defense potential, level of particle concerning the stability of the influenza virions, and level of vehicle in the form of respiratory droplets. They state that there is a possibility of reducing influenza virus spread by "maintaining room air at warm temperature (>20 °C) and either intermediate (50%) or high (80%) RHs" [21]. These studies on animals will aid in understanding the ways and types of transmission between human populations [21].

However, it is important that the relative humidity does not reach too high values, as this allows selected viruses to survive, as well as the growth of mold spores and fungi. The relative humidity must therefore be within a certain defined range in order to achieve a positive health effect. This optimal range where overall health risks may be minimized regarding relevant biological and chemical interactions has already been defined in 1985 by Sterling et al. [22], with a relative humidity between 40–60%. This optimal comfortable range between 40–60% is also pointed out by Arundel et al. [23] as a result of their study. In this study, different studies from schools, offices, and barracks were summarized, which deal with the "indirect health effects of relative humidity in indoor environments" with the clear statement that absenteeism or respiratory infections were found to be lower among people working or living in environments with mid-range versus low or high relative humidity [23]. Other studies also came to the result of an optimal range of relative air humidity concerning the viability of bacteria and the viability of viruses [24], the virus stability and transmission rates [25], and the reduction of human stress levels in comparison to drier conditions [26].

Furthermore, temperature is also attributed an important role in the spread and the toll of influenza. Shaman et al. [27] therefore investigated the relationship between absolute humidity and influenza survival and transmission, with the result that this relationship has even stronger significance than when considering the dependence of relative humidity. The consideration of hygrothermal comfort as a function of not only humidity but also air temperature is therefore crucial. This connection has also been pointed out by Wolkoff [28] in his review article concerning indoor air humidity and air quality and their influence on health. He gives an overview of numerous studies conducted in schools, offices, hospitals, and factories investigating the influence of air humidity on ocular surface, sleep quality, and the airways, but also its influence on the survival of influenza virus with the conclusion that not only relative air humidity plays a decisive role, but everything that is connected to it such as air pollutants. Due to the complexity of this, more attention should be paid to the term of absolute humidity, as already done by [27].

These effects of low humidity in office rooms in the winter period inevitably manifest themselves in higher absences due to illness. Employees in Austria spent an average of 13.1 days on sick leave in 2018, compared to 12.5 days in 2017. Short absences due to illness (1–3 days) are very common and accounted for about 40% of all recorded sick leave cases in 2018. However, they are not recorded, which means that the actual sickness rate is higher. The most frequent causes of sickness are mainly diseases of the musculoskeletal and respiratory systems [29]. Together, these illnesses cause about 50% of all sick leave cases and 43% of all sick leave days. The overall economic costs of sick leave and accidents are made up of several components that can be measured with varying degrees of accuracy. While the direct payments made by companies and social insurance agencies in the form of continued pay and sick pay can be estimated relatively accurately. However, there is little evidence of the indirect economic costs or the medical treatment costs incurred in the health care system. In 2017, continued salary payments in Austria accounted for 2.9 billion euros, and a further 725 million euros were spent on sick pay. The directly attributable sick leave costs thus amount to 1% of Austria's GDP. Sickness-related absences from the workplace also lead to losses in added value and possibly to other operational costs (productivity losses, costs for replacement employees, follow-up costs of accidents at work, etc.) that exceed the direct costs of continued remuneration of the sick employee. These costs are difficult to quantify, as they vary greatly depending on the economic cycle, the industry, and the size of the company. Under highly simplified assumptions, it can be estimated that, in addition to the cost of salary replacement, sickness-related absenteeism generates indirect business and economic costs of 0.8% to 1.7% of GDP. In addition to these direct and indirect sick leave costs, there are also costs to the health care system in the form of medical care, hospitals, medication, etc. The above-mentioned cost factors are directly related to sickness absence; a decrease in sickness-related absenteeism has a correspondingly positive effect on these factors [30]. Not least because of the high costs involved, companies worldwide are striving to reduce absenteeism. Since the subject matter and the reasons for absences are very different and complex, different approaches to their reduction are also pursued. These include organizational measures related to the scope of duties, but also the upgrading of the workplace and the creation of a positive working environment in the offices with the aim of health promotion [31].

Milton et al. [32] investigated the connection between sick leave and indoor air quality among office workers in the USA. They established the link between the cost of sick leave and the currently recommended air exchanges, which, based on the length of sick leave attributable to air quality and the labor costs of an employee, can save about USD 400 per employee per year by improving indoor air quality through air exchange with the outside. In the mentioned article, ventilation is considered the main factor in improving indoor air quality. In any case, air exchange is the best way to prevent the spread of viruses and pathogens that are transmitted through the air. Moreover, the relative humidity influenced by humidifiers is included in this study with the knowledge that, in any case, too high humidity should be avoided, as this can not only lead to a higher survival rate of certain viruses, but also allows the development of mold spores and fungi. As also highlighted in Arundel et al. [23], maintaining a relative humidity between 40 and 60% should therefore be ensured. In their article, authors clearly state, supported by various epidemiological studies, that there is a significant correlation between absentee rates and relative humidity indoors. This correlation has also been investigated by Reiman et al. [33] in their study on humidity as a non-pharmaceutical intervention for influenza A in different classrooms. Comparing humified rooms to control rooms, they observed a significant reduction of the total number of influenza A virus positive samples. Taylor et al. [34] point out that there is a connection between low indoor relative humidity and reduced outdoor air ventilation and sick leave and productivity. Mendell et al. [35] suggest that health benefits for indoor workers by improving the building environments can lead to high economic benefits. One of the other measures they advise is the influence of temperature and humidity of air.

Indoor greening has many advantages. In addition to the aesthetic enhancement of the room, it can not only contribute to a reduction of the reverberation time and thus to better speech intelligibility, but also influences the air quality in a room. This has already been proven in numerous studies and investigations [36–38].

In particular, vertical indoor greening in the form of wall greenery has a great effect, since a large area of vegetation can be created on a small floor surface. Among other things, vertical indoor greening has a positive effect on hygrothermal comfort. Particularly in winter, this is a great advantage due to the health effects of too low humidity. This has already been shown by means of measurement data from [36] and international studies such as [39–41].

Further, Reimherr and Kötter [42] examined the effects of indoor greening in offices on health, well-being, and work performance in the context of a research project. Through their surveys, they found out that with about 55%, the psychological and psychosomatic effects have the greatest health-promoting effect, followed by the advantages of air humidification (30%). Furthermore, the reduction of dust and noise as well as the reduction of pollutants are also cited. Similar results were obtained by Fjeld et al. [43] through a survey addressing neuropsychological symptoms, mucous membrane symptoms, and skin symptoms through indoor air conditions among office workers. The situation with and without plants in the office was compared, and it was found that complaints regarding cough and fatigue were reduced by 37% and 30% through plants present. They though clearly suggest that foliage plants in offices can lead to an improvement in health and a reduction in symptoms of discomfort. Studies by Smith and Pitt [44] also show that plants can be a low maintenance tool to improve indoor air quality. Their in situ measurements show that plants can not only influence the humidity in offices, but can also influence other air pollutants such as VOCs.

Vertical indoor greening also has the advantage that very little to no floor space is lost in the room, and yet plants can be available in large numbers in the room. In comparison to individual plants in pots or troughs, however, wall plantings are associated with higher costs for installation as well as for the upkeep and maintenance of the technical system.

When making decisions about investments in buildings, costs and benefits are always weighed against each other. Cost–benefit analyses are therefore used to compare the monetary advantages and disadvantages. In a cost–benefit analysis, the value of a project is thus quantified in monetary terms with the aim of the support of social decision making on a rational basis. A plan is worthy of realization if, compared to doing nothing, the sum of its advantages is greater than the sum of its disadvantages [45], or as defined by Cambridge Dictionary, "the process of comparing the costs involved in doing something to the advantage or profit that it may bring" [46]. However, such cost–benefit evaluations are very complex for indoor and outdoor greening of buildings. This is not least due to the fact that the positive effects of the living, nevertheless technical, system of the vertical green are varied and not only the investor profits, e.g., in the form of energy saving, but also substantial positive effects on the health as well as also on the cityscape, which are so far difficult to quantify and/or in a further step to monetarize, as already explained in detail in [47]. A classical cost–benefit analysis is therefore not the correct instrument to illustrate the effects holistically for building greenery. Alternative assessment and evaluation concepts are therefore necessary.

In this article, the costs of an investment and operation of vertical indoor greening are to be examined and analyzed in relation to the benefits in the form of reduced sickness-related downtime in office buildings due to improved humidity thanks to the vertical indoor green. These comparisons and the conclusions drawn from them are based on the following context: Particularly in winter, interiors often have too low humidity. This has health effects for the people who stay in these rooms—this also applies to offices and the people who work in them and who are on sick leave because of these health consequences.

Vertical greenery improves the hygrothermal comfort in indoor spaces and, especially in winter, can contribute to an increase in air humidity to a comfortable level and thus also influence the associated health consequences. Sick leave due to health consequences and the associated absence cause costs for the company, which are reduced accordingly when sick days are reduced. Indoor greenery can contribute to this reduction, but it also causes costs for installation and maintenance. These costs for greenery and possible savings by reduced sick days are compared, and a method of quantifying and monetizing the effects of vertical greening is shown.

2. Methodology and Approach

In the context of the investigations for this article, a comparison was made between the cost savings due to less absence through illness and the costs for vertical greening in the interior of office spaces.

The investigations of this article are based, on the one hand, on the measurements of relative air humidity and air temperature in greened and non-greened interiors of three Viennese school buildings, which were equipped with different vertical greening systems within the scope of research projects. All project results can be found in [48,49]. These projects provided extensive long-term measurement data. The evaluations of the hygrothermal comfort for the classrooms in summer and winter have already been published [36]. In addition, recommendations for the dimensioning of vertical indoor greening in classrooms in relation to hygrothermal comfort were developed on the basis of formulas applied [49,50]. For the present study, the hygrothermal measurement data are filtered again and evaluated accordingly. This allows statements about the percentage of improvement of the hygrothermal comfort and thus the improvement of the indoor air quality based on this parameter. The three school buildings investigated differ in their construction method and in the way they are ventilated: A non-insulated old building in brick construction without technical ventilation system, a new building in reinforced concrete construction with a thermal insulation composite system and ventilation system, and one without a technical ventilation system. It is therefore possible to make statements for three different structural situations for these locations. They will be referred to as Building A, B, and C in the following. As an example for building C, Figure 1 shows the three different rooms as they exist in each of the three buildings: A reference room, a green room with the trough system, and a green room with the fleece system. This figure also contains the calculation results obtained. Figure 2 shows the greening with the fleece system as an example from building B.

The measured data of air humidity and air temperature were collected per building in two greened classrooms and one non-greened comparable classroom, which served as reference rooms, over several years in a measuring interval of 5 min. More details on used measurement instruments as well as measurement settings can be equivalently found in [36]. Two different vertical greening systems were used at each of the three locations: A fleece system and a trough system. These two systems were described in detail in [48,49] and shown in Figure 1 in a sketch. The plants used in the vertical greenings were selected within the framework of the research projects by the project partners with many years of expertise in vegetation technology as well as a landscape gardener involved in the project, so that the plant selection is optimally designed for use in vertical indoor greenings. The selection of the plants is attributed a high value, but this should not be the focus of the present investigations, but should always be accompanied by an expert. In order to be able to make statements about the effect of the greening also for office rooms, these measuring data from classrooms were filtered in such a way that only times in winter period when the rooms were not used for teaching were used for the present analysis, so that there is no influence of the presence of the students. In addition, all measured data were checked for plausibility, and data gaps and outliers were processed accordingly. In a further step, it was determined how many workplaces could be arranged in the respective classrooms in accordance with the Austrian workplace regulations. In this way, the vertical green area per workplace in the respective room under consideration was highlighted.

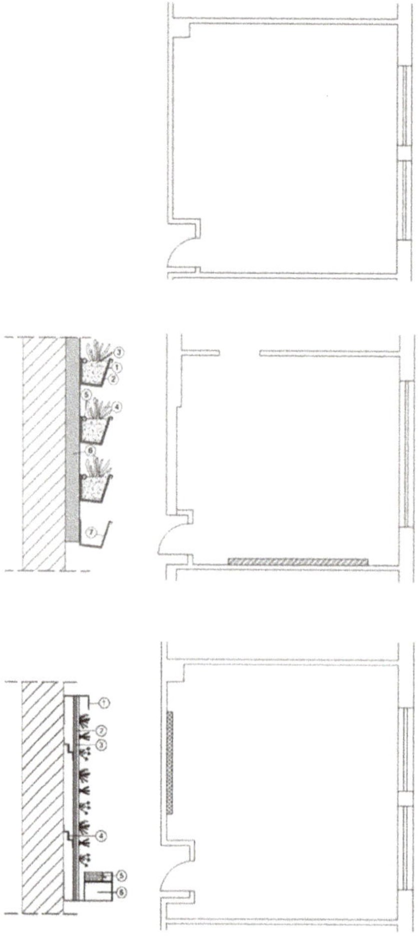

Reference Room – Not greened		
room size	64	m²
greenery size	-	m²
possible workstations	12	
m2-greenery per workstation	-	m²
installation costs for greenery	-	€/m²
costs for technical maintenance & green care	-	€/(m²·a)
improvement of hygrothermal comfort	-	%

Greened Room – Trough System		
room size	59	m²
greenery size	11.4	m²
possible workstations	11	
m2-greenery per workstation	1.0	m²
installation costs for greenery	1254	€/m²
costs for technical maintenance & green care	250	€/(m²·a)
improvement of hygrothermal comfort	408	%

Greened Room – Fleece System		
room size	63	m²
greenery size	5.6	m²
possible workstations	11	
m2-greenery per workstation	0.5	m²
installation costs for greenery	2375	€/m²
costs for technical maintenance & green care	400	€/(m²·a)
improvement of hygrothermal comfort	419	%

Figure 1. Overview of buildings and rooms with according calculation results.

Figure 2. Fleece system in Building B.

The hygrothermal comfort indoors and its criteria have already been examined in detail. In addition to relative humidity and air temperature, detailed analyses also take into account factors such as physical conditions, the activity of the persons, and their clothing. In order to enable statements as general as possible and in accordance with the available data, the definition of hygrothermal comfort according to Frank [51] is used in the present investigations and the measured values are analyzed according to these defined areas. This method is the same as that used in [36] for the analysis. Figure 3 shows these areas. Thus, measured values within the red framed area are in the comfortable range, which means that both the measured air temperature and the relative humidity are in a range that is comfortable for persons present. If a combination of relative humidity and air temperature is within the green framed area but outside of the red area, these measurements are called "still comfortable". Outside of this green area, the existing conditions are considered "not comfortable". This means that the temperature is either too cold or too warm, and the air is too dry or too humid.

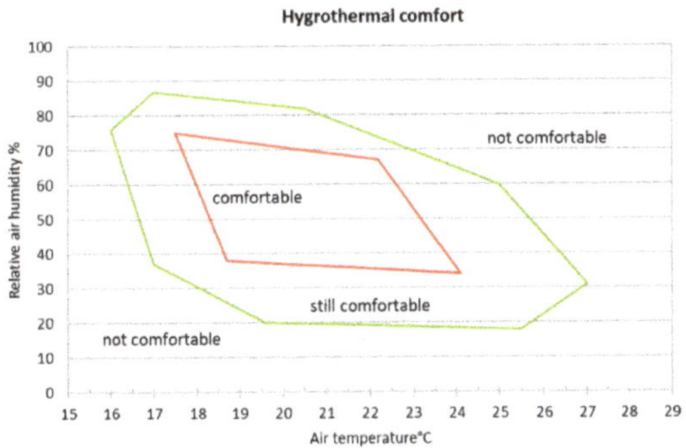

Figure 3. Hygrothermal comfort according to Frank [51].

The costs for the installation and construction of the greenery for the three locations were summarized and calculated on the basis of the actual costs incurred. Moreover, the costs for operation as well as green care and technical maintenance were collected and presented in values per year for the three locations and the two different greening systems used. Due to the locations as well as the different functioning of the greening systems, these vary. In summary, the costs of the greening systems could be calculated per workplace and year for each greening system used at the three locations.

In a further step, these costs are compared to the costs for the absence due to sickness of one person per day, which were determined based on the explanations in Section 1 and the average annual income of employed persons (including apprentices) and the working hours in hours per year according to Statistik Austria [52].

By improving the hygrothermal comfort in the greened rooms compared to the non-greened rooms, a reduction of the number of sick days is then possible on the basis of the correlations explained in Section 1, which allows a statement about the positive monetary effect of vertical greening in the office space on the saving of salary costs for employees due to fewer sick days.

In a final step, different initial situations addressing the connection between vertical indoor greening and reduced sick leave are considered. These should show in which way the method described in the article can be applied or which statements can be formulated based on the considerations. Situations such as the profitability of greening after a certain number of years or with a certain reduction of sick days per employee are considered before finally a generalization of assumptions based on dimensioning factors is carried out. Due to the compactness of the considerations, the approach is briefly described in the chapter of the actual calculation.

Cost-benefit analyses and also the method of evaluation presented in the context of this article have their limitations and only provide a decision support based on a comparison and do not represent an actual decision. They contain both value estimations as well as uncertainties and contain beyond that in principle no examination of legal defaults.

3. Results and Discussion

Following the procedure explained in Section 2, the next subchapters describe the considerations, calculations, and analyses performed and present the results of these.

3.1. Number of Workstations in Monitored Rooms

The monitored rooms are located in three different school buildings in Vienna, as already mentioned under 2. The nine rooms are six classrooms with greenery and three reference rooms without greenery. As can be seen in table in Section 3.5, all nine rooms have different room sizes and volumes—the room size varies between about 52 m^2 and 84 m^2; the room volume between 193 m^3 and 259 m^3.

According to the Workplace Ordinance applicable to Vienna, which among other things regulates the necessary size of a workplace in offices, "at least 8.0 m^2 for one employee must be provided plus at least 5.0 m^2 for each additional employee" per room in accordance with *§24 (1) AStV. §24 (3) AStV also stipulates that at least 12.0 m^3 of airspace per employee must be available "for work with low physical stress", which also includes normal office activities. [53]

In accordance with these legal requirements, the possible number of hypothetical workplaces in the nine monitored rooms was determined. The results are summarized in table in Section 3.5. The calculation was based on both the existing floor space and the air volume, and it turned out that for all nine rooms, the floor space was decisive. The number of workstations in the considered rooms ranges between 8 and 15.

3.2. Costs for Greening Systems

The construction as well as the green care and technical maintenance of vertical greening systems comes with costs. These are divided into investment costs, which are incurred once when the greening system is set up, and ongoing costs for plant care and technical maintenance of the system. The calculations also include costs for electricity and water consumption for lighting and irrigation of the green areas. They are divided into costs for green care and technical maintenance including fertilizer, plant material, as well as water and electricity consumption.

The costs considered in the research of this article are based on the real consumption of electricity and water measured in one of the schools and on the costs for green care and maintenance for the company that took over the maintenance after the end of the research project. The necessary lighting of the vertical indoor greenery is provided by LED strips or spotlights. The irrigation is done by a connection to the house water pipe and a micro-drip system according to the needs of each of the two different greening systems.

Due to the different functionalities of the two greening systems under consideration, the water consumption and the costs for green care and maintenance also differ. While in the trough system the plants are placed in technical substrate, comparable to a conventional flower pot, in the fleece system three different fleece layers are used for protection as well as for water distribution and storage. Since the plants are inserted bare-rooted into the fleece system and the fleece serves as a substrate substitute, the water consumption is significantly higher than with the trough system. Since this characteristic makes the system less resilient, the costs for care and maintenance are also higher than for the trough system. A more detailed analysis and explanation of the costs are included in [47].

Due to the comparability of the systems at the three locations, these real costs per m^2 known for one location are also used for the other two locations. Table in Section 3.5. contains the corresponding values for the six greened rooms.

The construction costs are incurred once only and are therefore allocated over the considered time period as shown in table in Section 3.5. This is done using a straight-line depreciation of the installation.

3.3. Costs for Sick Leave

The question to be answered in the following subchapter is: What does an hour and resulting from this a day of sick leave in an office cost? The winter period is considered in particular, since the frequency of sicknesses caused by low air humidity is highest during this period, and the effect on hygrothermal comfort due to indoor greenery is most significant.

Based on the conditions explained in Section 1 and on data from Statistik Austria from 2017, the average annual income of employed persons in Austria is 38,828 euros. If these costs are divided by the usual annual hourly rate of 1720 h per year, the average labor costs per hour per person are 22.57 euros. This annual hourly rate factor already includes the annual vacation days and public holidays. This results in labor costs of 180.60 euros for a regular 8-h workday. Table 1 contains the results of these calculations.

Table 1. Average labor costs of employed persons in Austria 2017 [54].

Average Annual Income of Employed Persons (Including Apprentices)	in EUR	38,828.00
Working hours per year	in h	1720
Working hours per day	in h	8
Average labor costs per day per person	in EUR	180.60
Average labor costs per hour per person	in EUR	22.57

The company incurs direct costs in the form of wage costs per person and working day of 180.60 euros. These costs represent only the direct costs as explained in Section 1. Not included are costs for, e.g., overtime of colleagues to compensate the workload or costs that arise from the delay of projects. Moreover, indirect costs for the health service as well as insurance are not considered. The actual costs for one sick day per employee are therefore significantly higher. Due to the existing data situation and the difficulties in the determination in particular of the indirect costs in the context of these present investigations, only the pure wage costs which must be further paid are taken into account.

Accordingly, the answer to the above question of the cost of one day of absence of an office employee can be answered: There are direct costs for continued payment of wages in the amount of 180.60 euros per day. Determining the indirect costs of absence is very complex and is therefore not quantified in this article.

3.4. Improvement of Hygrothermal Comfort

The measured data were evaluated according to the filter criteria and definition ranges for hygrothermal comfort (Figure 2) explained in Section 2. The results of these evaluations are shown in Table 2, and for one of the buildings as an example in Figure 4. It is clearly visible that the point cloud of the measurement data of the greened rooms (green measurement points) has clearly shifted into the comfortable area compared to the non-greened reference room (blue measurement points).

Based on the evaluations, it was calculated to what extent the greened rooms improve in comparison to the non-greened reference room. This percentage improvement is shown in Table 2. It can be clearly seen that all six greened rooms offer significantly higher hygrothermal comfort in comparison to the non-greened rooms and therefore have a healthier indoor climate. This also means that at no time was relative humidity too high, as is illustrated in Figure 4 for Building A as an example, so there is no risk of mold.

Due to the applied filter criteria explained in Section 2, these evaluations of the measured data obtained in school classes can also be applied to office rooms and, in particular, statements can be made about the winter period, which is important for sick leave due to respiratory diseases. The analysis of the monetary connection between the improved air quality and the days of sickness is presented in the next subchapter.

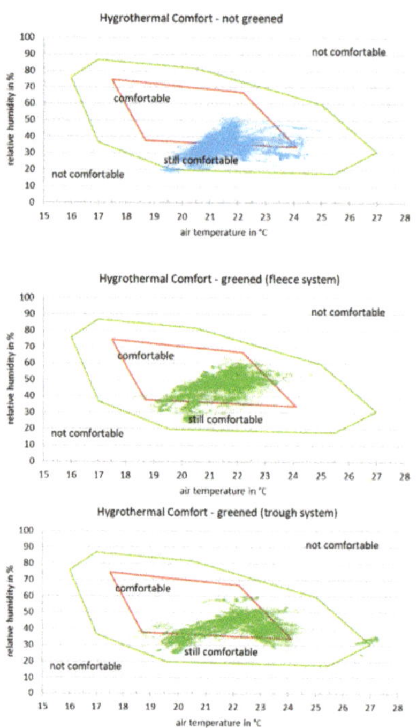

Figure 4. Hygrothermal comfort in not greened reference room and two greened rooms (fleece and trough system) for winter period represented by measured air temperature and relative air humidity with applied filter criteria (Building A).

Table 2. Evaluation of the measured data on hygrothermal comfort in the greened rooms compared to the reference rooms.

Building	Room	Comfortable		Still Comfortable		Not Comfortable		Total		Difference Greened to Not Greened (Comfortable)	Improvement Greened to Not Greened (Comfortable)
	not greened	22,229	34%	42,916	65%	510	1%	65,655	100%	-	-
A	greened (trough)	47,561	72%	17,039	26%	1055	2%	65,655	100%	39%	214%
	greened (fleece)	55,153	84%	10,445	16%	57	0%	65,655	100%	50%	248%
	not greened	2659	19%	10,872	77%	659	5%	14,190	100%	-	-
B	greened (trough)	10,499	74%	2370	17%	1321	9%	14,190	100%	55%	395%
	greened (fleece)	9466	67%	4678	33%	46	0%	14,190	100%	48%	356%
	not greened	2566	17%	11,440	76%	963	6%	14,969	100%	-	-
C	greened (trough)	10,476	70%	4486	30%	7	0%	14,969	100%	53%	408%
	greened (fleece)	10,752	72%	4127	28%	90	1%	14,969	100%	55%	419%

Sustainability **2021**, *13*, 2256

3.5. Connection between Vertical Indoor Greening and Reduced Sick Leave

The results explained so far and the calculation results described below are summarized in Table 3. This table also shows the procedure explained in Section 2. In the following subchapters, the connection between the improved indoor conditions thanks to greening and the associated costs for installation as well as green care and technical maintenance and the possible savings due to reduced sick days are studied by different approaches.

Table 3. Overview of the calculation results for all considered rooms and greenery systems.

Building	Room Size	Room Height	Room Volume	Greenery System	Greenery Size	Number of Possible Workstations According to AStV by m² [53] (3.1)	Number of Possible Workstations According to AStV** by m3 [53] (3.1)	m²-Greenery per Workstation	Installation Costs for Greenery Total [49] (3.2.)	Installation Costs for Greenery per m² (3.2.)	Costs for Technical Maintenance and Green Care per m² and Year [49] (3.2)	Costs per Sick Day per Person [54] (3.3.)	Improvement of Hygrothermal Comfort (Comfortable) Greened to Not Greened (3.4.)	Installation Costs with Linear Depreciation on a Years per m² (3.5.1.)	Total Greenery Costs Yearly with Linear Depreciation on x Years per m² (3.5.1.)	Yearly Total Greenery Costs per Workstation	From ... Days Less Sick Leave per Person It Will Pay Off to Have Greenery. (3.5.3.)	After ... Years, Greening Pays Off with a Reduction in Sick Leave Per Person by d Day/Year (3.5.2)
	(m²)	(m)	(m³)		(m²)			(m²)	(EUR)	(EUR/m²)	(EUR/m²)	(EUR)	(%)	(EUR/m²)	(EUR/m²)	(EUR)	(d)	(a)
A	67	4	242	trough	17	12	20	1.4	21,500	1265	247	181	214	181	428	606	3.4	6.3
A	54	4	200	fleece	6.5	10	16	0.7	8200	1262	394	181	248	180	574	373	2.1	2.2
A	52	4	193	none	0	9	16	0.0	-	-	-	181	-	-	-	-	-	-
B	74	3	236	trough	9	14	19	0.6	9300	1033	250	181	395	148	398	256	1.4	1.4
B	82	3	259	fleece	5.6	15	21	0.4	14,500	2589	400	181	356	370	770	287	1.6	2.0
B	84	3	240	none	0	16	19	0.0	-	-	-	181	-	-	-	-	-	-
C	59	3	189	trough	11.4	11	15	1.0	14,300	1254	250	181	408	179	429	445	2.5	3.5
C	63	3	202	fleece	5.6	11	16	0.5	13,300	2375	400	181	419	339	739	376	2.1	2.8
C	64	3	206	none	0	12	17	0.0	-	-	-	181	-	-	-	-	-	-
					decisive										a = 7			d = 3.5

For each of the considerations and the initial situations described below, the green area per workplace in the different rooms was used. It is dependent on the size of the installed vegetation as well as the number of possible workstations in the room under consideration. This results in costs for the greening per workstation. The respective values are shown in the Table 3.

It is to be pointed out again expressly that in the following considerations, only the improvement of the air humidity is used as reason, however numerous further reasons speak for indoor greenery, which were not considered in the context of the present investigations due to so far lacking data. In Section 4, these connections are explained prospectively.

3.5.1. Initial Situation: Profitability after Seven Years Using Linear Depreciation

For this first consideration, it is assumed that the greening system as a technical system is depreciated on a linear basis over seven years—the installation costs are therefore spread over 7 years. The costs for green care and technical maintenance are considered as annual costs.

Assuming a usage period of 7 years, the following statement can be made depending on the green space considered: From d days less sick leave per person, it will pay off to have greenery. The number of d days varies between 1.4 and 3.4 (Table 3).

3.5.2. Initial Situation: Profitability after A Certain Number of Years

In a further step, the following statement shall be made: "With a reduction of d sick days, the construction of the greenery is already paid off after x years".

To answer this, the annual costs for the greening per workstation are set in connection with the number of workstations in the room and the size of the greening and with the annual savings with reduced sick leave by a certain number of days at about 181 euros per day each. This can be calculated by the following Equation (1) and respectively Equation (2).

$$ws_{green} \cdot x \left(\frac{C_0}{x} + C_{care} \right) = C_{sick} \cdot d \cdot x \qquad (1)$$

$$x = \frac{ws_{green} \cdot C_0}{C_{sick} \cdot d - ws_{green} \cdot C_{care}} \qquad (2)$$

x—number of years; C_{sick}—costs per sick day per person; d—number of sick days; C_0—installation costs for greenery per m^2; C_{care}—costs for technical maintenance and green care per m^2 and year; ws_{green}—m^2-greenery per workstation.

3.5.3. Initial Situation: Reduction of Sick Days in Number of Days

Studies have shown that in greened offices with correspondingly improved humidity, sickness-related days of absence decreased by up to 3.5 days per employee [55]. The calculations based on the costs to be attributed to the greening and the saved wage costs calculated with Equation (2) show that the greening systems installed in the rooms are rewarded after only 1.4 to 6.3 years.

This large difference between the considered rooms or rather the exception with 6.3 years is especially due to the size of the greening with troughs in building A.

3.5.4. Initial Situation: Reduction of Sick Days in Percent

Another approach that has been followed is based on the correlation that statistically speaking, when humidity improves into a comfortable area, there is a certain percentage decrease in sickness absence due to respiratory diseases. As already explained in Section 1, this correlation has already been scientifically proven with regard to the transmission and survival of viruses at different levels of humidity. In addition, in an experiment described under [56], 30–40% fewer symptoms related to symptoms of the mucous membranes were detected. Based on their research, Fjeld et al. [43] found that plants reduce dry throat symptoms by 25% and coughing symptoms by as much as 37% due to the increased humidity in the room.

The reduction of these complaints results in reduced sick leave due to these symptoms. Here, an assumption of a reduction of 25% is made.

Based on the average number of sick days in Austria due to respiratory diseases, which were explained in Section 1, this results in a reduction of sick days by about 2 days.

This results in a profitability of the installed greening systems after between 3.3 and 12.7 years, depending on the greening system and area under consideration. Moreover, in this case, the large spread of values is due to the large differences in the size of the rooms and the installed greening systems.

3.5.5. Generalization of Assumptions Based on Dimensioning Factors

The calculations carried out so far are based on data obtained in the course of research projects in implemented projects. Within the framework of the research project, a formula for the dimensioning of vertical indoor greenery to achieve the optimum level of comfort in relation to humidity (relative humidity 45%) in the interior could also be developed. The dimensioning differentiates between technical and manual ventilation and between good and bad user behavior with regard to ventilation or the circulation of the ventilation system, i.e., the existing air exchange. Using these dimensioning factors, the necessary green area is calculated in m^2 depending on the floor area of the room (technical ventilation, circulation 1x per hour: 0.2; technical ventilation, circulation <1x per hour: 0.1; manual ventilation, good user behavior 0.08) [48,50]. Figure 5 shows the surface of an office depending on workplaces and the resulting surface of greenery for the different options of ventilation.

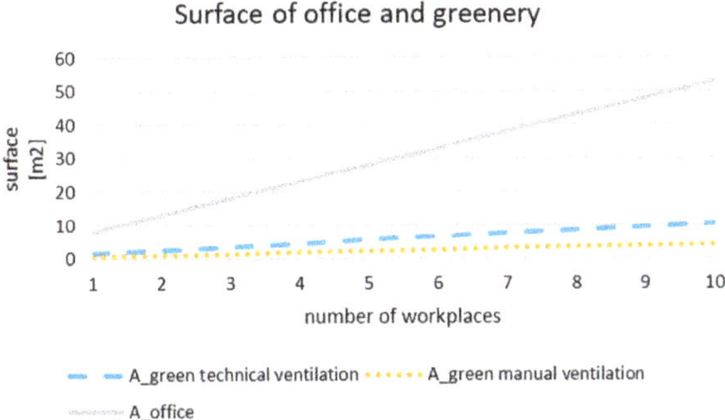

Figure 5. Surface of office (A_{office}) depending on workplaces (y) and resulting surface of greenery (A_{green}) with technical and manual ventilation according to dimensioning factor (g).

If the sizes of the greenery installed in the three buildings in Vienna are checked on the basis of these dimensioning factors, it is noticeable that they are significantly larger than those calculated after the dimensioning. However, this does not represent a contradiction, but the developed formula is in a way based on these research results. If the systems are too large in relation to the room size, there is a risk that the humidity in the room is too high due to the greenery. However, this fear could be excluded on the basis of the long-term measurement data. In other words: These effects can also be achieved with smaller surfaces only in relation to air humidity. However, if other effects are also considered, such as the influence on the room acoustics or aesthetic aspects, other greening areas may well prove to be useful. However, it is always necessary to pay attention to the increase in humidity and to select an optimum of effects. Last but not at least, these effects also depend on the choice of plants.

Furthermore, this generalization now includes costs for installation as well as green care and technical maintenance based on current manufacturer information. These costs amount to 800 euros per m^2 for the installation and 150 euros per m^2 per year for the maintenance of the green care and technical maintenance.

Based on the calculations carried out for the model buildings, the following Equation (3) applies for a general dimensioning, whereby the following conditions apply ((4) to (7)). For an office space, the dimensioning factor thus results in a calculation according to Equation (8) for the number of years after which the installation of greenery based on reduced sick leave has paid off. Figure 6 visualizes these Equations for better understanding.

$$A_{green} \cdot C_{green,\,annual} = C_{sick,office,annual} \tag{3}$$

$$A_{green} = A_{office} \cdot g \tag{4}$$

$$A_{office,min} = 8 + 5\,(y - 1) \tag{5}$$

$$C_{green,\,annual} = \left(\frac{C_0}{x} + C_{care} \right) \tag{6}$$

$$C_{sick,office,annual} = C_{sick} \cdot y \cdot d \tag{7}$$

$$x = \frac{C_0}{\dfrac{C_{sick} \cdot y \cdot d}{A_{office} \cdot g} - C_{care}} \tag{8}$$

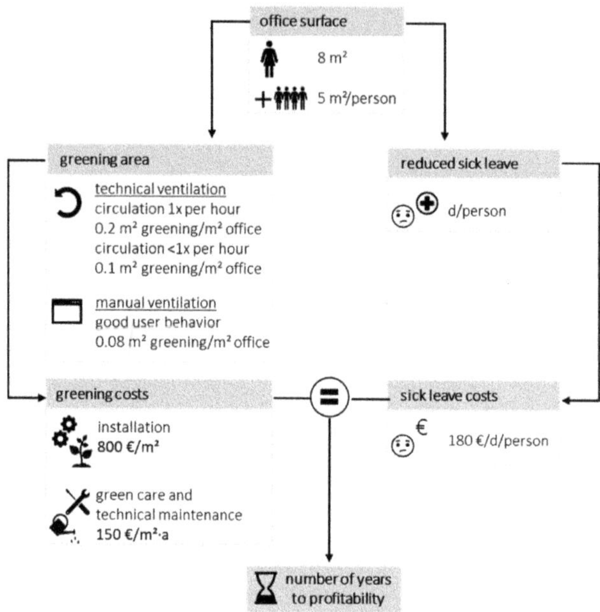

Figure 6. Overview of application of Equations (3)–(8) to calculate number of years to profitability of greenery through reduced sick leave.

A_{green}—surface greenery; $C_{green,annual}$—total annual costs for greening per m²; $C_{sick,office,annual}$—annual costs for sick leave per office space; A_{office}—surface of office; y—number of work places; g—dimensioning factor for greenery; $A_{office,min}$—minimum size of an office for x employees; C_0—installation costs for greenery per m²; x—number of years; C_{care}—costs for technical maintenance and green care per m² and year; C_{sick}—costs per sick day per person; d—number of reduced sick days.

For an office space with 6 workplaces and the minimum size of 33 m² specified by AStV and assuming technical ventilation and a circulation of 1x per hour (dimensioning factor 0.2) as well as the costs for the greenery according to the above-mentioned manufacturer's specifications, it follows that, assuming a reduction in sick leave by 2 days per person, the greenery will be profitable after about 4.5 years. For the same space with manual ventilation, the greening will be profitable already after about one year, since the size of the greened area of about 3 m² is significantly smaller than in the first example with 6.6 m². Figure 7 illustrates the number of years to profitability comparing the costs for greenery and the according benefits for reduced sick leave for different cases such as different sizes of offices (3 or 6 persons) and different reduction of sick leave (2 or 3 days) as well as different types of ventilation (manual and technical). Profitability is given as soon as the line of costs intersects with the line of benefits.

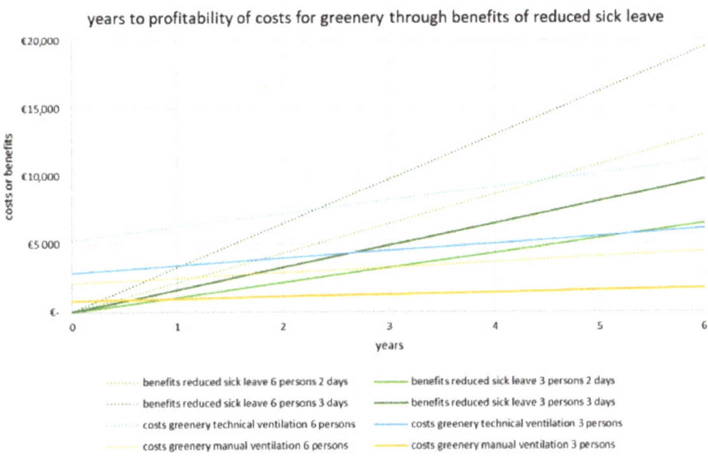

Figure 7. Years to profitability of costs for greenery through benefits of reduced sick leave for different cases.

4. Conclusions and Outlook

The calculations and analyses carried out have shown that an economic consideration of indoor greening in relation to its effects on sick leave in offices is definitely worthwhile due to the improvement of humidity in the room and its impact on human health.

The positive effects of vertical indoor greening on the relative humidity of the indoor air have already been proven worldwide and can also be confirmed for the six greened sample rooms under consideration on the basis of the analyzed measurement data. On this basis, the effects on an office space and the workplaces and employees located in it were derived.

The effects on sick leave were assumed in these studies based on expert literature and only the effect of improved humidity in the form of hygrothermal comfort was considered. Other known positive effects of greening have not been considered so far. In the office environment, these include in particular the improvement of the reverberation time and thus the influence on the room acoustics as well as the possible enhancement of the working environment through greening. It can therefore be assumed that there may be further positive effects on working life and employee satisfaction. In particular, unspecific clinical patterns related to sick building syndrome should be further considered in this context.

A comparison of these effects on the humidity of greenery with conventional air humidification systems or extended possibilities of building technology systems could also be made. However, such systems also cause costs for technical maintenance and, in addition, they only pursue the one benefit of the change in air humidity and, as technical systems, do not achieve any further advantages as is the case with greening.

It should be pointed out that this article is not a financial report, but presents a method to demonstrate and quantify the effects that have not been considered in investment decisions about vertical greening up to now.

Furthermore, it is to be differentiated in connection with the presented method between the macro economical costs of sick leave and the costs for one company, as was already explained in the introductory chapters. So far, only wage costs were considered, and no macroeconomic total calculation were aimed at, which would include also costs of insurance, hospitals, etc. The calculated profitability is therefore a conservative consideration.

Additionally, it must be taken into account that the calculations are subject to uncertainties and, especially when using living, technical systems and when considering the effects on humans, a generalization is not always exactly possible. Uncertainties exist, for example, with regard to personnel costs, prices of greenery, actual reduction of sick leave, as well as in the generalization of sick days and the assumption that the air quality in the office space considered was not optimal before. With regard to a practical application of this method by a specific company, however, it should be pointed out that the formulas presented, with their in-house data for personnel costs and sick leave, provide direct results that are subject to greater certainty.

In a further step, the extension of this presented method to other areas such as outdoor greening is possible. However, this expansion is more complex due to the fact that the effects and responsibilities cannot be clearly assigned. In the presented example of office space and the saving of wage costs, it can be assumed that the person who invests in the greening is also the one who benefits from the savings. Due to the different levels on which vertical greenery is effective in outdoor areas, as shown in [47], the analysis is therefore also much more complex. One instrument that should also be considered in this context is the Cost-Efficiency-Analysis, because it allows the intersection between buildings and urban planning to be made visible, which is also the focus of the greening of buildings.

Author Contributions: Conceptualization, J.H. and A.K.; methodology, J.H. and A.K.; validation, J.H. and A.K.; formal analysis, J.H.; investigation, J.H. and A.K.; writing—original draft preparation, J.H.; writing—review and editing, J.H. and A.K.; visualization, J.H.; supervision, A.K. Both authors have read and agreed to the published version of the manuscript.

Funding: Open Access Funding by TU Wien.

Institutional Review Board Statement: Not applicable.

Informed Consent Statement: Not applicable.

Data Availability Statement: Not applicable.

Acknowledgments: The authors acknowledge the TU Wien Bibliothek for financial support humans through its Open Access Funding Program.

Conflicts of Interest: The authors declare no conflict of interest.

References

1. European Commission. Indoor Air Pollution: New EU Research Reveals Higher Risks than Previously Thought. 2003. Available online: https://ec.europa.eu/commission/presscorner/detail/en/IP_03_1278 (accessed on 26 August 2020).
2. Binder, B.; Gielge, J.; Peer, C. Beiträge zur Stadtentwicklung—Bürobeschäftigte in Wien. Wien. 2006. Available online: https://www.wien.gv.at/stadtentwicklung/studien/pdf/b008039.pdf (accessed on 26 August 2020).
3. German Business Interior Association IBA. IBA Publications—IBA-Studie 2019/20. 2020. Available online: https://iba.online/en/service/magazines/ (accessed on 23 November 2020).
4. Federal Ministry Republic of Austria. Working Time Act. Available online: https://www.ris.bka.gv.at/Dokumente/Erv/ERV_1969_461/ERV_1969_461.html (accessed on 23 November 2020).
5. Carrer, P.; Wolkoff, P. Assessment of indoor air quality problems in office-like environments: Role of occupational health services. *Int. J. Environ. Res. Public Health* **2018**, *15*, 741. [CrossRef]
6. Wyon, D.P. The effects of indoor air quality on performance, behaviour and productivity. *Pollut. Atmos.* **2005**, *14*, 35–41.
7. Śmiełowska, M.; Marć, M.; Zabiegała, B. Indoor air quality in public utility environments—A review. *Environ. Sci. Pollut. Res.* **2017**, *24*, 11166–11176. [CrossRef] [PubMed]
8. Hutter, H.P.; Moshammer, H.; Wallner, P.; Damberger, B.; Tappler, P.; Kundi, M. Health complaints and annoyances after moving into a new office building: A multidisciplinary approach including analysis of questionnaires, air and house dust samples. *Int. J. Hyg. Environ. Health* **2006**, *209*, 65–68. [CrossRef] [PubMed]
9. Statista. Welche der Folgenden Faktoren Empfinden Sie am Arbeitsplatz als Störend? Available online: https://de.statista.com/statistik/daten/studie/184025/umfrage/meinung-zu-stoerfaktoren-am-arbeitsplatz-in-bueros/ (accessed on 26 August 2020).
10. Bluyssen, P.M.; Roda, C.; Mandin, C.; Fossati, S.; Carrer, P.; de Kluizenaar, Y.; Mihucz, V.G.; de Oliveira Fernandes, E.; Bartzis, J. Self-reported health and comfort in 'modern' office buildings: First results from the European OFFICAIR study. *Indoor Air* **2016**, *26*, 298–317. [CrossRef] [PubMed]
11. Gossauer, E. *Nutzerzufriedenheit in Bürogebäuden: Eine Feldstudie*; Fraunhofer IRB: Stuttgart, Germany, 2007. [CrossRef]
12. Hemmes, J.H.; Winkler, K.C.; Kool, S.M. Virus Survival as a Seasonal Factor in Influenza and Poliomyelitis. *Nature* **1960**, *188*, 430–431. [CrossRef] [PubMed]
13. Von Hahn, N. Trockene Luft und ihre Auswirkungen auf die Gesundheit—Ergebnisse einer Literaturstudie. *Gefahrst. Reinhalt. Luft* **2007**, *67*, 103–107.

14. Mäkinen, T.M.; Juvonen, R.; Jokelainen, J.; Harju, T.H.; Peitso, A.; Bloigu, A.; Silvennoinen-Kassinen, S.; Leinonen, M.; Hassi, J. Cold temperature and low humidity are associated with increased occurrence of respiratory tract infections. *Respir. Med.* **2009**, *103*, 456–462. [CrossRef]

15. Noti, J.D.; Blachere, F.M.; McMillen, C.M.; Lindsley, W.G.; Kashon, M.L.; Slaughter, D.R.; Beezhold, D.H. High Humidity Leads to Loss of Infectious Influenza Virus from Simulated Coughs. *PLoS ONE* **2013**, *8*, e57485. [CrossRef]

16. Gardinassi, L.G.; Simas, P.V.M.; Salomão, J.B.; Durigon, E.L.; Trevisan, D.M.Z.; Cordeiro, J.A.; Souza, F.P.D. Seasonality of viral respiratory infections in southeast of Brazil: The influence of temperature and air humidity. *Braz. J. Microbiol.* **2012**, *43*, 98–108. [CrossRef]

17. Lowen, A.C.; Steel, J. Roles of Humidity and Temperature in Shaping Influenza Seasonality. *J. Virol.* **2014**, *88*, 7692–7695. [CrossRef] [PubMed]

18. Brightman, H.S.; Milton, D.K.; Wypij, D.; Burge, H.A.; Spengler, J.D. Evaluating building-related symptoms using the US EPA BASE study results. *Indoor Air* **2008**, *18*, 335–345. [CrossRef]

19. Reinikainen, L.M.; Jaakkola, J.J.K. Effects of temperature and humidification in the office environment. *Arch. Environ. Health* **2001**, *56*, 365–368. [CrossRef]

20. Kudo, E.; Song, E.; Yockey, L.J.; Rakib, T.; Wong, P.W.; Homer, R.J.; Iwasaki, A. Low ambient humidity impairs barrier function and innate resistance against influenza infection. *Proc. Natl. Acad. Sci. USA* **2019**, *166*, 10905–10910. [CrossRef]

21. Lowen, A.C.; Mubareka, S.; Steel, J.; Palese, P. Influenza virus transmission is dependent on relative humidity and temperature. *PLoS Pathog.* **2007**, *3*, 1470–1476. [CrossRef]

22. Sterling, E.M.; Arundel, A.; Sterling, T.D. Criteria for Human Exposure to Humidity in Occupied Buildings. *ASHRAE Trans.* **1985**, *91*, 611–622.

23. Arundel, A.V.; Sterling, E.M.; Biggin, J.H.; Sterling, T.D. Indirect health effects of relative humidity in indoor environments. *Environ. Health Perspect.* **1986**, *65*, 351–361. [PubMed]

24. Lin, K.; Marr, L.C. Humidity-Dependent Decay of Viruses, but Not Bacteria, in Aerosols and Droplets Follows Disinfection Kinetics. *Environ. Sci. Technol.* **2020**, *54*, 1024–1032. [CrossRef] [PubMed]

25. Audi, A.; AlIbrahim, M.; Kaddoura, M.; Hijazi, G.; Yassine, H.M.; Zaraket, H. Seasonality of Respiratory Viral Infections: Will COVID-19 Follow Suit? *Front. Public Health* **2020**, *8*, 576. [CrossRef]

26. Razjouyan, J.; Lee, H.; Gilligan, B.; Lindberg, C.; Nguyen, H.; Canada, K.; Burton, A.; Sharafkhaneh, A.; Srinivasan, K.; Currim, F.; et al. Wellbuilt for wellbeing: Controlling relative humidity in the workplace matters for our health. *Indoor Air* **2020**, *30*, 167–179. [CrossRef]

27. Shaman, J.; Kohn, M. Absolute humidity modulates influenza survival, transmission, and seasonality. *Proc. Natl. Acad. Sci. USA* **2009**, *106*, 3243–3248. [CrossRef] [PubMed]

28. Wolkoff, P. Indoor air humidity, air quality, and health—An overview. *Int. J. Hyg. Environ. Health* **2018**, *221*, 376–390. [CrossRef]

29. Statista. Krankheit und Beruf. 2017. Available online: https://de.statista.com/statistik/studie/id/6697/dokument/krankheit-und-beruf-statista-dossier/ (accessed on 30 August 2020).

30. Leoni, T. Fehlzeitenreport 2019 Krankheits- und unfallbedingte Fehlzeiten in Österreich, Die Flexible Arbeitswelt: Arbeitszeit und Gesundheit Krankheits- und unfallbedingte Fehlzeiten in Österreich Die flexible Arbeitswelt: Arbeitszeit und Gesundheit. Wien. 2019. Available online: https://www.wifo.ac.at/publikationen/studien?detail-view=yes&publikation_id=62103 (accessed on 30 August 2020).

31. Shrivastava, S.R.; Shrivastava, P.S.; Ramasamy, J. A comprehensive approach to reduce sickness absenteeism. *J. Inj. Violence Res.* **2015**, *7*, 43–44.

32. Milton, D.K. Risk of sick leave associated with outdoor air supply rate, humidification, and occupant complaints. *Indoor Air* **2000**, *10*, 212–221. [CrossRef]

33. Reiman, J.M.; Das, B.; Sindberg, G.M.; Urban, M.D.; Hammerlund, M.E.M.; Lee, H.B.; Spring, K.M.; Lyman-Gingerich, J.; Generous, A.R.; Koep, T.H.; et al. Humidity as a non-pharmaceutical intervention for influenza A. *bioRxiv* **2018**, *13*, 1–15. [CrossRef] [PubMed]

34. Taylor, S.H.; Scofield, C.M.; Graef, P.T. Improving IEQ to reduce transmission of airborne pathogens in cold climates. *ASHRAE J.* **2020**, *62*, 30–47.

35. Mendell, M.J.; Fisk, W.J.; Kreiss, K.; Levin, H.; Alexander, D.; Cain, W.S.; Girman, J.R.; Hines, C.J.; Jensen, P.A.; Milton, D.T.; et al. Improving the health of workers in indoor environments: Priority research needs for a National Occupational Research Agenda. *Am. J. Public Health* **2002**, *92*, 1430–1440. [CrossRef] [PubMed]

36. Tudiwer, D.; Korjenic, A. The effect of an indoor living wall system on humidity, mould spores and CO2-concentration. *Energy Build.* **2017**, *146*, 73–86. [CrossRef]

37. Bornehag, C.G.; Sundell, J.; Sigsgaard, T. Dampness in buildings and health (DBH): Report from an ongoing epidemiological investigation on the association between indoor environmental factors and health effects among children in Sweden. *Indoor Air Suppl.* **2004**, *14* (Suppl. 7), 59–66. [CrossRef]

38. Burchett, M. *Towards Improving Indoor Air Quality with Pot –Plants—A Multifactorial Investigation*; University of Technology Sydney: Sydney, Australia, 2009; pp. 1–44. ISBN 0 7341 2171 7.

39. Tarran, J.; Torpy, F.; Burchett, M. Use of living pot-plants to cleanse indoor air—Research review. In Proceedings of the Sixth International Conference on Indoor Air Quality, Ventilation and Energy Conservation in Buildings—Sustainable Built Environment, Sendai, Japan, 28–31 October 2007; Volume III, pp. 249–256.
40. Brilli, F.; Fares, S.; Ghirardo, A.; de Visser, P.; Calatayud, V.; Muñoz, A.; Isabella, A.; Federico, S.; Alessandro, A.; Vincenzo, V.; et al. Plants for Sustainable Improvement of Indoor Air Quality. *Trends Plant Sci.* **2018**, *23*, 507–512. [CrossRef]
41. Moya, T.A.; van den Dobbelsteen, A.; Ottelé, M.; Bluyssen, P.M. A review of green systems within the indoor environment. *Indoor Built Environ.* **2019**, *28*, 298–309. [CrossRef]
42. Reimherr, P.; Kötter, E. *Auswirkungen von Innenraumbegrünungen in Büros auf Gesundheitszustand, Wohlbefinden und Arbeitsleistung*; Bayerische Landesanstalt für Weinbau und Gartenbau, Sachgebiet Zierpflanzenbau: Würzburg/Veitshöchheim, Germany, 1999.
43. Fjeld, T.; Veiersted, B.; Sandvik, L.; Riise, G.; Levy, F. The Effect of Indoor Foliage Plants on Health and Discomfort Symptoms among Office Workers. *Indoor Built Environ.* **2003**, *7*, 204–209. [CrossRef]
44. Smith, A.; Pitt, M. Healthy workplaces: Plantscaping for indoor environmental quality. *Facilities* **2011**, *29*, 169–187. [CrossRef]
45. Boardman, A.E.; Greenberg, D.H.; Vining, A.R.; Weimer, D.L. *Cost-Benefit Analysis: Concepts and Practice*; Cambridge University Press: Cambridge, UK, 2017.
46. Cambridge University Press. Cambridge Dictionary—Cost-Benefit Analysis. 2020. Available online: https://dictionary.cambridge.org/de/worterbuch/englisch/cost-benefit-analysis (accessed on 23 November 2020).
47. Hollands, J.; Korjenic, A. Ansätze zur ökonomischen Bewertung vertikaler Begrünungssysteme. *Bauphysik* **2019**, *41*, 38–54. [CrossRef]
48. Korjenic, A.; Tudiwer, D.; Hollands, J.; Fischer, H.; Mitterböck, M.; Gonaus, T.; Salonen, T.; Blaha, A.; Pitha, U.; Weiss, O.; et al. GRÜNEzukunftSCHULEN Grüne—Publizierbarer Endbericht. Wien. 2020. Available online: http://www.grueneschulen.at/endbericht/ (accessed on 26 August 2020).
49. Korjenic, A.; Tudiwer, D.; Penaranda Moren, M.S.; Hollands, J.; Salonen, T.; Mitterböck, M.; Pitha, U.; Zluwa, I.; Stangl, R.; Kräftner, J.; et al. GrünPlusSchule@Ballungszentrum. Wien. 2018. Available online: https://nachhaltigwirtschaften.at/de/sdz/projekte/gruenplusschule-ballungszentrum-hocheffiziente-fassaden-und-dachbegruenung-mit-photovoltaik-kombination-optimale-loesung-fuer-die-energieeffizienz-in-gesamtoekologischer-betrachtung.php (accessed on 26 August 2020).
50. Weiss, O.; Fichtenbauer, L. Evapotranspiration of Indoor Greenery—As Part of the Research Project, GREENfutureSCHOOLS. In Proceedings of the 16th International Conference on Urban Health (ICUH 2019), Xiamen, China, 4–8 November 2019.
51. Frank, W. Raumklima und Thermische Behaglichkeit. *Ber. Bauforsch.* **1975**, *104*, 36.
52. Statistik Austria. Tabellenteil zum, Allgemeinen Einkommensbericht 2018. Bericht über die durchschnittlichen Einkommen. 2018. Available online: https://www.statistik.at/web_de/statistiken/menschen_und_gesellschaft/soziales/personen-einkommen/allgemeiner_einkommensbericht/index.html (accessed on 30 August 2020).
53. Arbeitsstättenverordnung—AStV: Verordnung der Bundesministerin für Arbeit, Gesundheit und Soziales, mit der Anforderungen an Arbeitsstätten und an Gebäuden auf Baustellen festgelegt und die Bauarbeiterschutzverordnung Geändert Wird (23.10.2020). Available online: https://www.ris.bka.gv.at/GeltendeFassung.wxe?Abfrage=Bundesnormen&Gesetzesnummer=10009098 (accessed on 23 October 2020).
54. Statistics Austria. General Income Report 2018. Available online: https://www.statistik.at/web_en/statistics/PeopleSociety/social_statistics/personal_income/general_income_report/index.html (accessed on 23 November 2020).
55. Bundesanstalt für Arbeitsschutz und Arbeitsmedizin. *Wohlbefinden im Büro*; Bundesanstalt für Arbeitsschutz und Arbeitsmedizin: Dortmund, Germany, 2008; p. 40.
56. Plants, I.I. Health and Well Being and the Benefits of Office Plants. 2018. Available online: https://www.ieqindoorplants.com.au/benefits-of-office-plants/ (accessed on 26 August 2020).

 *sustainability*

Review

Towards a Sustainable Indoor Lighting Design: Effects of Artificial Light on the Emotional State of Adolescents in the Classroom

David Baeza Moyano [1], Mónica San Juan Fernández [2] and Roberto Alonso González Lezcano [3,*]

[1] Department of Chemistry and Biochemistry, Universidad San Pablo CEU, Campus Montepríncipe, Boadilla del Monte, 28925 Madrid, Spain; baezams@ceu.es
[2] Faculty of Languages and Education. Campus de Princesa-Madrid. Universidad Antonio de Nebrija, 28015 Madrid, Spain; monica.sanjuan.f@gmail.com
[3] Architecture and Design Department, Escuela Politécnica Superior, Universidad San Pablo CEU, CEU Universities, Campus Montepríncipe, Boadilla del Monte, 28668 Madrid, Spain
* Correspondence: rgonzalezcano@ceu.es

Received: 29 April 2020; Accepted: 21 May 2020; Published: 22 May 2020

Abstract: In recent years, articles have been published on the non-visual effects of light, specifically the light emitted by the new luminaires with light emitting diodes (LEDs) and by the screens of televisions, computer equipment, and mobile phones. Professionals from the world of optometry have raised the possibility that the blue part of the visible light from sources that emit artificial light could have pernicious effects on the retina. The aim of this work is to analyze the articles published on this subject, and to use existing information to elucidate the spectral composition and irradiance of new LED luminaires for use in the home and in public spaces such as educational centers, as well as considering the consequences of the light emitted by laptops for teenagers. The results of this research show that the amount of blue light emitted by electronic equipment is lower than that emitted by modern luminaires and thousands of times less than solar irradiance. On the other hand, the latest research warns that these small amounts of light received at night can have pernicious non-visual effects on adolescents. The creation of new LED luminaires for interior lighting, including in educational centers, where the intensity of blue light can be increased without any specific legislation for its control, makes regulatory developments imperative due to the possible repercussions on adolescents with unknown and unpredictable consequences.

Keywords: natural lighting; visual comfort; artificial lighting; indoor lighting design; chronodisruption; circadian rhythms; daylighting; sustainable lighting design; LED luminaires; indoor environment quality; classroom lighting

---

1. Introduction

In recent years, an unprecedented technological revolution has taken place. Many of the luminaires that humans use are being replaced by new light emitting diode (LED) light sources. Unlike the luminaires that have been used until now, which had fixed and known emission spectra, the spectral composition and intensity in each range of the visible spectrum of these new LED luminaires can be made to order, even including fixed or remote electronic control systems, and can be varied as desired at any time.

It is a scientific fact that human beings have genetically evolved to perform their activities during the day and rest during the night. The evolution of the human species has occurred under periodic and relatively stable cycles of light and darkness, next to the main cycle, which is linked to the rotation of the Earth on its own axis (light–dark cycle, 24 h) and the rotation of the earth around the sun (seasons, changed daylengths as a function of latitude, annual cycle, lunar cycle). Natural selection has favored the presence of biological clocks through all lines of evolution, allowing animals to optimally benefit from these cycles [1–4].

Maintaining proper lighting cycles, with appropriate levels of light during the day and darkness at night is beneficial to the synchronization of the circadian system. It is not only the amount or type of light we are exposed to that is important, but also the time of day when that light is received. Biological rhythms persist in constant environments; the origin of such rhythms is endogenous and depends on an internal clock. They are not, by default, driven by external signals but do have the ability to become driven by several of these signals.

The increasing use of electronic equipment late at night from an early age has aroused curiosity and fear about various related consequences, especially for children and adolescents [5]. Therefore, in this review, we examine all relevant scientific and regulatory information that exists or is under development. Numerous scientific studies demonstrate the influence of light on our biorhythms, and therefore, on the moods of adolescents [6,7].

The evolution in lighting with the introduction of LED lights that can emit any wavelength at any intensity has raised concerns about what combinations of light can be received and emitted by electronic equipment and disturbing doubts about such light's consequences for the health of adolescents.

The main objective of this review is to study the different sources of artificial light that adolescents may receive and the physiological and psychological consequences of their absorption. From this general objective, the following specific objectives were developed:

a. determine the nature of the light emitted from electronic equipment that can be used by adolescents from the perspective of the possible risks of blue light on the retina and its effects on circadian cycles and mood,

b. evaluate the new sources of indoor lighting that can be installed in educational establishments today and their possible influence on mood,

c. study the current regulations (or those under development) related to new sources of artificial light, and

d. propose control or preventive measures to improve the health of adolescents.

The light used in the personal and educational contexts of adolescents can be harmful, so there must be an awareness of the need to implement ethical and healthy regulations related to lighting to optimize the health and performance of young people in school contexts [1,3]. It is essential to carry out a bibliographic review and a review of the current regulations in all proposed areas to successfully legislate and set limits on the emissions of new LED lights according to their uses.

Research and practice in the domain of indoor lighting design have become a relevant topic in the search for indoor environment quality. A keyword search of articles and abstracts related to these research areas was carried out through the Web of Science (Figure 1). Indoor lighting design was the keyword.

Figure 1. Keywords for indoor-lighting-design-related issues.

2. Methodology and Sources of Information

The aim is to study certain processes that occur in nature without intervening or manipulating the possible variables. For this purpose, an analysis of the physiological and psychological consequences of the absorption of light through the human visual system will be carried out through an exhaustive bibliographic and documentary review based on the proposed objectives. In this review, we undertake a study of the existing literature in the scientific community concerning the effect that artificial lighting can have on adolescents, including an exhaustive study of the literature on lighting in classrooms, the use of natural lighting, quality requirements of the interior environment in classrooms, and how lighting influences the visual health of adolescents.

The Federation of National Manufacturers Association for Luminaires and Electrotechnical Components in the European Union (CELMA) published measurements of several LED luminaires with an illuminance of 500 lx showed that none of these luminaires had blue emissions in group 2. EN 62471 classifies light sources into four risk groups: 0, 1, 2, and 3 (where 0 = no risk and 3 = high risk). The sun is classified as high risk, or group 3 [8]. Further measurements by the French Agency for Food, Environment and Health and Safety at Work (ANSES) in 2010 showed that all measured LED sources were in groups 0 and 1 according to EN 62471 UNE EN 62471:2009, the "Photobiological safety of lamps and appliances using lamps". This risk is measured according to two different methods, depending on the product being measured [8].

La EN 12464-1, the European standard for indoor lighting, has detailed the recommended values of illuminance and uniformity for classrooms (6.2 Educational Buildings) with light levels between 300 lx, and 500 lx (EN 12464-1, 2002) [9]. Figure 2 shows the blue light retinal exposure risk weighted irradiance for small sources (E_B).

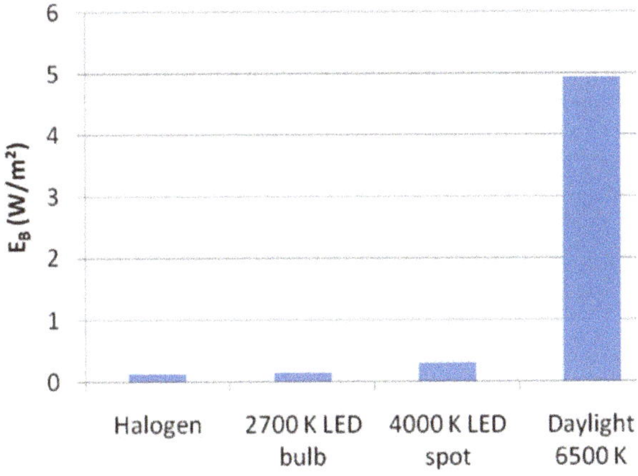

Figure 2. Comparison of the irradiance values of some lamp types at 500 lx (typical for indoor lighting) with daylight at 5000 lx (typical for outdoor lighting) [10].

Light emitting diodes (LEDs) are light sources in which light is created within solid-state materials. The light emission is obtained by the interaction of an electric field with the solid material. The physical process is called "electroluminescence". This phenomenon was discovered as early as 1907, and the first practical product based on it was created in 1962. The semiconductor material used in LEDs is selected so that it emits in the visible or ultraviolet range. Different materials produce light with different wavelengths (different colors). They have the characteristic of having a very narrow range of emissions.

The general lighting levels used in interiors via LED luminaires with cold color temperatures of 4000 K, 6500 K, and even 9000 K (whose blue components very high) are never enough to cause damage to the retina, as long as one does not look at the naked light source. The possible photobiological risk of blue light can be assessed using the criteria set out in EN 62471 [10], which have been indicated above.

From three billion years ago until 130 years ago, living organisms had one part of the day of sunlight, and other part of the day, darkness. From 130 years ago until today, thanks to artificial light, humans have experienced short periods of sleep, dim light inside the buildings where we work, short periods of time receiving natural light, bright light in the bathroom during the night, and dim light in the bedroom while we sleep [11,12].

Before the invention of electric light, humans had a "first sleep", after which they would arise and engage in tasks or go and visit neighbors. Then, they would have a "second sleep". This whole process took about 12 h. Today, we have only one sleep that lasts 7 or 8 h at most [13,14].

The study of the non-visual effects of light entering our eyes is so topical that the 2017 Nobel Prize in Physiology and Medicine was awarded jointly to Jeffrey C. Hall, Michael Rosbash, and Michael W. Young [15] for their discoveries on the molecular mechanisms of circadian rhythm control. Below are some terms that have emerged in this new branch of science.

Chronobiology is the branch of science that deals with the physiological variations of living beings called circadian rhythms, which occur approximately every 24 h, as well as the study of the processes of synchronization with the surrounding environment The variation of endogenous circadian rhythms that are determined from individual differences and psychological factors is described as a chronotype. Twenty-five percent of people are active from noon onwards, go to bed late, and then get up late; these people are commonly called "owls" [11].

Smolensky and colleagues define chronodisruption as an alteration in the internal temporal order of physiological, biological, and behavioral rhythms. If chronodisruption becomes chronic, the asynchrony, advancement, or retardation of peripheral clocks may occur [16].

3. Possible Effects of Artificial Light Absorption on Adolescents

3.1. Studies on the Non-Visual Effects of Light Absorption through Our Visual System

Stevens and Zhu [17] observed that the sun is our primary source of light during the day and that for millions of years sunlight has shaped mammals' endogenous circadian rhythms including when to wake up, body temperature, metabolism, oscillations of gene expression, and hormone production throughout our bodies. Electric light, in contrast, is dim and alters all aspects of our internal circadian rhythms. Its intensity and spectral content are often not adequate during the day for proper circadian resetting and are too great at night for "true darkness" to be detected.

In 1975, the scientist Richard J. Wurtman wrote an extensive article talking about the effects of light on our internal organs, such as the ovaries, and tissues, such as the breasts. This publication was visionary because of the knowledge of the scientific community at that time [18].

Guido et al. [19] established that the retina contains a biological pacemaker that influences the entire circadian system. Glickman et al. [20] stated that an illuminance of 2500 lx is necessary to suppress nocturnal melatonin in humans, but it was later determined that under certain conditions, such as below 1 lx, melatonin can be suppressed in humans.

There is a growing trend in the scientific field to a consensus that exposure to light influences many psychological processes through at least two pathways unrelated to the phenomenon of vision [21]. The best-known pathway is related to the regulation of melatonin secretions by the pineal gland [22]. This pathway controls circadian rhythms. Exposure to light at night, particularly at short wavelengths, suppresses melatonin and influences insomnia. The other pathway acts on the level of alertness by activating a mechanism separate from that of melatonin suppression, during which cortisol is secreted [21].

Given the large amount of data that is gradually appearing on the positive and negative effects of artificial light related to health, Erren et al. defined photohygiene as exposure to light under the optimal conditions of periodicity, quality, and quantity [23].

A study carried out by Cho et al. [24] stated that sleeping with the lights on causes acute negative effects on the structure and quality of sleep. A later study showed that these negative effects could affect aspects related to memory by producing corneal levels of less than 10 lx. The experiment was conducted with a 5779 K LED source with a diffuser [25].

The World Health Organization's (WHO's) International Agency for Research on Cancer (IARC) has classified "shift work involving circadian disruption" as probably carcinogenic to humans (group 2a). Continuous light contributes to acute confusional syndrome in adult intensive care units, where environmental constancy is the norm [26].

Dim light melatonin onset (DLMO) refers to the onset of melatonin secretion under low light conditions. The human body is programmed for this moment to occur when the sun is setting. Graham and Wong [27,28] noted in their study that intense blue light cancels the night peak 10–20 min after exposure, which returns to its initial value after 40 min once the stimulus is removed.

In relation to the time of day, exposure to light can generate clock advances or delays as defined by the phase response curve (PRC), so bright light at the beginning of the biological night (from the start of melatonin elevation to the time when the minimum body temperature is produced) generates a phase delay, while in the morning (from the minimum body temperature to 8 h later), it produces an advance. Epidemiological, clinical, and experimental studies with animal models show that the chronodisruption produced by artificial light during the night may be related to pathologies such as the increased incidence of metabolic syndrome [29], cardiovascular disease [29], and cognitive and affective disorders [30].

Exposure to light at night (LAN), reduced or inadequate light intensity during the day, or decreased contrast in the light–dark cycle all contribute to chronodisruption (CD). The cognitive decline, affectivity, behavioral and sleep disturbances, and limitations in the daily activities of elderly patients with senile dementia and their caregivers have been associated with alterations in circadian cycles [31]. Artificial light at night (ALAN) is drawing the attention of researchers and environmentalists for its ever-increasing evidence on its capacity for "desynchronization" of organismal physiology [32]. Obesity is a common disorder with many complications. Although chronodisruption plays a role in obesity, few epidemiological studies have investigated the association between artificial light at night (ALAN) and obesity. Since sleep health is related to both obesity and ALAN, Koo et al. [33] investigated the association between outdoor ALAN and obesity after adjusting for sleep health and the association between outdoor ALAN and sleep health.

Simón and Sánchez [34] noted that about 20% of people in today's society spend most of their time during the day indoors under dim lighting, with low physical activity and irregular, short sleep cycles. The authors suggest that these factors could contribute to the prevalence of chronodisruption possibly facilitating pathologies such as cancer, intestinal conditions, metabolic syndrome, cardiovascular diseases, mood disorder, and cognitive impairment. Chronodisruption may also adversely affect melatonin and cortisol levels. Cortisol is a regulator of stress-related functions [34]. In 1998, Sterling and Eyer [35] defined the role of allostasis as "maintaining stability through change" and noted that cortisol secretion and stress are a "body's adaptation to an unknown situation that must be transient and therefore blocked or stopped". The authors comment that night shift workers have a high risk of circadian disruption and, therefore, hormonal alterations. Similarly, Mirick et al. [36] suggest that a low level of urinary sulfatoximelatonin is related to working at night, which results in higher levels of cortisol.

Stevens and Zhu [17] state that light is a regulator of psychology and behavior and that its effects have evolved over millennia throughout which illumination has provided reliable information about the time of day. The authors claim that the advent of electric light has now altered this relationship with patterns of light exposure reflecting personal tastes and social pressures. It is important, therefore, that non-visual light effects be incorporated into lighting design. For example, one might ask to what extent existing architectural lighting replicates the biological effects of natural light, how lighting could be used to minimize the harmful effects of shift work while promoting alertness and safety, or how light therapy could be optimized. The lighting industry and scientists have begun research in this direction. They argue that we must first determine how light impacts human behavior and psychology. There are two different techniques for measuring light and there are different scientific criteria to determine which of the two is the most suitable: radiometry (quantitative analysis) and photometry (qualitative analysis).

Several studies have observed that humans are adopting increasingly nocturnal lifestyles, both for work and for leisure, which has resulted in the night becoming excessively illuminated, while we spend most of the day in poorly lit interiors. This results in an increasing gap between our habits and the natural synchronizers of the circadian system. Chronodisruption or circadian disruption (CD) is the physiological price of exposure to light at night [16,17].

It is well established that light affects both visual and non-visual systems. Little attention has been given to testing the effects of light on building occupants' non-visual responses, and, consequently, lighting specifiers have been offered little guidance on the design and application of lighting for non-visual effects. A study conducted by the Lighting Research Center, by Figueiro et al. (2019) [37], helped to fill that gap through field-testing of light exposures from a novel luminaire designed to promote entrainment and alertness throughout the day in actual office environments. The data supported the inference that light exposures, when properly applied, can promote circadian entrainment and increase alertness.

Recent research has shown that exposure to bright white or high blue light stimulates alertness, but these effects are not seen in tasks that demand a high cognitive level. Individual and psychological differences have been taken into account to explain the variability in the cognitive effects of light. Sensitivity to light depends on individual differences in the PER3 gene clock involved in sleep–wake regulation, age, the cognitive domain, and task difficulty [38]. Some authors claim that exposure to bright daylight indoors can result in positive vitality, alertness, and help promote a healthy, active day. These studies reveal that bright light induces improvements in alertness when healthy participants are experimentally deprived of sleep or light prior to exposure to this indoor light [39].

Experimental studies have shown that the magnitude and duration of non-visual light effects depend on previous light doses [40]. Exposure to bright indoor light would induce weaker non-visual effects in spring than in autumn and winter [39].

Angel Correa et al. [38] observed that bright white light or light enriched in the range of blue increases alertness but that this is not effective for high cognitive level tasks, such as sustained attention to task response (SART). The authors observed that the results varied greatly depending on the individuals and their previous states of alertness, with higher results among those who previously had a better state of alertness or vigilance.

Wright Jr. et al. [41] conducted an experiment in which they recruited eight participants (two women and six men) aged around 30 years old whose circadian cycles were previously studied at their work and home. The participants then spent a week camping in the mountains without electricity. Among the eight subjects, there was a wide range of chronotypes (i.e., larks and owls) and sleep times. The beginning of melatonin release is approximately 2 h after sunset. Due to the habits of modern life, this release now occurs later. After a week of camping, the beginning of melatonin secretion occurred closer to sunset and elimination occurred at dawn, aligning larks and owls more closely to the duration of natural light. In their conclusions they state that "increased exposure to sunlight may help reduce the health consequences of circadian disruption".

3.2. Analysis of the Possible Negative Effects of Blue Light from Electronic Devices on the Eyes of Adolescents

The first type of cancer to be associated with the suppression of melatonin production was breast cancer in a 1987 publication by Stevens [42]. Stevens claims that the level of ambient light in the room during the sleep period is associated with other diseases such as diabetes, obesity, depression, and affective disorders, supported by multiple publications, such as Obayashi et al. (2013) [43] and McFadden et al. (2014) [44]. The authors conclude by stating that, although the light that reaches us from the sun contains a high irradiance at all visible wavelengths, its maximum is 480 nm, which is perceived by humans as a beautiful blue (e.g., a clear day in the middle of the morning). This is the most optimal wavelength to signal to an organism that it is day rather than night [43].

Experimental and epidemiologic studies suggest that light at night (LAN) exposure disrupts circadian rhythm, and this disruption may increase breast cancer risk. Ritonia et al. (2020) [45] found no association between residential outdoor LAN and breast cancer for either measure of LAN. The authors found no association when considering interactions for menopausal status and past/current night work status. Ritonia et al. [45] are consistent with studies reporting that outdoor LAN has a small effect or no effect on breast cancer risk.

Konis et al. [46] conducted a 12-week study of 77 patients in eight dementia care facilities with the assumption that increased exposure to indoor lighting with natural light could reduce depression and other neuropsychiatric symptoms. The authors stated that they found positive results, but that these results should not be interpreted as a definitive prescription to replace existing treatments and that further studies should be carried out on the subject.

Lai et al. [47] concluded from their studies on mice that blue light is harmful; moreover, epidemiological studies on humans were found to be inconclusive regarding positive or negative effects of the use of blue-light-blocking intraocular lenses (IOLs). In an experiment conducted with 18 groups of four rats that were exposed to light between 460 and 480 nm with intensities of 0.6 and 1.5 W/m^2, the authors concluded that the safe amount of time that these animals can be exposed to these irradiances is 12 h with an intensity of 0.6 W/m^2 and for 4 h with an intensity of 1.5 W/m^2 [48].

Roehlecke et al. [49] irradiated a culture of mouse retinal cells using blue light with a peak emission at 405 nm and an intensity of 10 W/m^2 and stated in their conclusions that oxidative stress was produced and reactive oxygen species (ROS) were generated. Torkaz et al. [50] (2013) stated that the retina is especially exposed to oxidative stress due to high oxygen pressure, UV exposure, and blue light promoting the generation of ROS. Nakanishi et al. [51] stated that cultured ROS cells exposed to blue LEDs with an emission peak at 470 nm and an intensity of 48 W/m^2 showed a significant increase in ROS occurrence.

Chamorro et al. [52] radiated human retinal pigment epithelial cells) (HRPEpiC) using blue light emitting LEDs with peak emissions at 468 nm, using green light with peak emission at 525 nm, and using red light with peak emissions at 616 nm, as well as white light, in 12 h light–dark cycles. The authors observed ROS, cell DNA damage, and apoptosis (delayed cell death).

3.3. Measurements of Light Emitted by Luminaires and Electronic Equipment Used by Adolescents

There is a growing number of articles on the subject of light emitted by luminaires and equipment. In an article published in 2013, Hassoy et al. [53] noted that mobile phones are widely used by children and adolescents. The authors provide data from several countries including 76% phone use in Hungary, 79% in Sweden, and 94% in Germany.

In 2014 in Barcelona, the Spanish testing laboratory accredited by the Empresa Nacional de Acreditación (ENAC) FUTTEC S.L carried out, with a double monochromator spectrometer calibrated by the Centro Superior de Investigaciones Científicas (CSIC), different irradiance measurements on the blue ranges of mobile phones, laptops, and indoor lights in W/m^2 compared to the same ranges within the blue range of UVI 6 and UVI 9.

The measurements were made at the distances that the electronic devices are located from our eyes. Figure 3 shows a graph of the ultraviolet index (UVI) measurements (around five) given by the Department of Astronomy and Meteorology of the Faculty of Physics at the University of Barcelona. The maximum irradiance was observed around 480 nm.

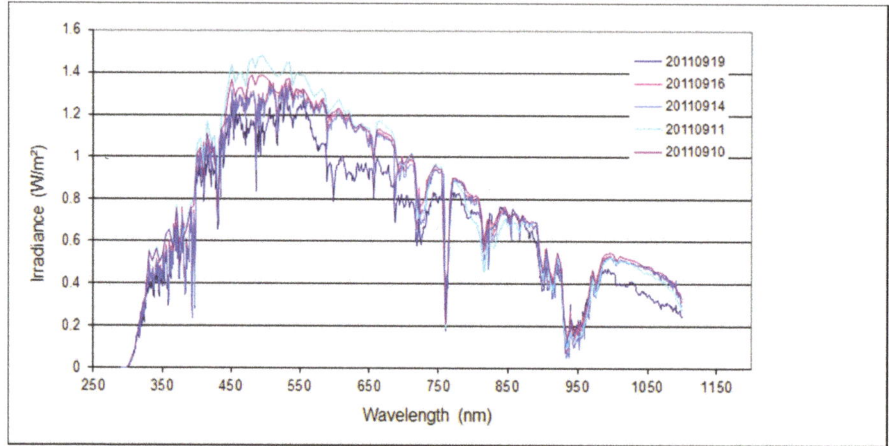

Figure 3. Several solar spectral distributions provided by the Faculty of Physics of the University of Barcelona [54].

These results were made public at the Congress of the General Council of the Official Associations of Pharmacists of Spain (CGCOF). Below are the results of the irradiance measurements of 10 nm between 400 nm and 600 nm performed by the FUTTEC test laboratory including:

- irradiance at solar noon on a spring day with a UVI of 9 measured by the Astronomy Department of the University of Physics of the University of Barcelona;
- irradiance at solar noon on a summer day with a UVI of 6 measured by the Astronomy department of the University of Physics of the University of Barcelona;
- a laptop screen with a blue background in a totally dark room. A laptop screen with a blue background in a lit room at working distance;
- emissions of a cold white light luminaire (6500 K) in an interior corridor at the distance from the ceiling indicated by regulations;
- display of a desktop computer at a working distance with a blue background in an illuminated office;
- irradiance in a work room of the PHILIPS company with the irradiance of a 2700 K LED light;
- a 4000 K LED light; and
- irradiance of a Samsung mobile phone screen with a blue background in total darkness.

In Figure 4, the green line represents measurements on a computer screen in a workroom, where the windows facing the street are very close. This is why the measured irradiance is superior to the irradiances of smartphones, computers, mobile phones, and LEDs, and similar in shape to the graphics of sunlight.

Escofet and Bará [55] show the analysis of two computer screens and two smartphones. The irradiance results from the computers are around 450 nm of 4×10^{-3} W/m^2, and those from the two smartphones around 450 nm are 6–7×10^{-4} W/m^2. In their article, they show the absorption of different filters to be placed in front of the screens to reduce the irradiance in the blue range and recommend their use.

In 2018, new measurements were made in an optics laboratory on an optical bench in total darkness with a double monochromator spectrometer from FUTTEC S.L. calibrated by the CSIC. The measured electronic equipment belonged to students studying optics who were asked to give their electronic equipment a blue background with maximum luminosity. The measurements were taken at the distance at which they said they would put the equipment in place to use it. A nanometer by nanometer scan was performed from 350 to 500 nm to analyze the area in the blue range, as well as the irradiance peak. Table 1 shows the irradiance and peak light results from these measurements.

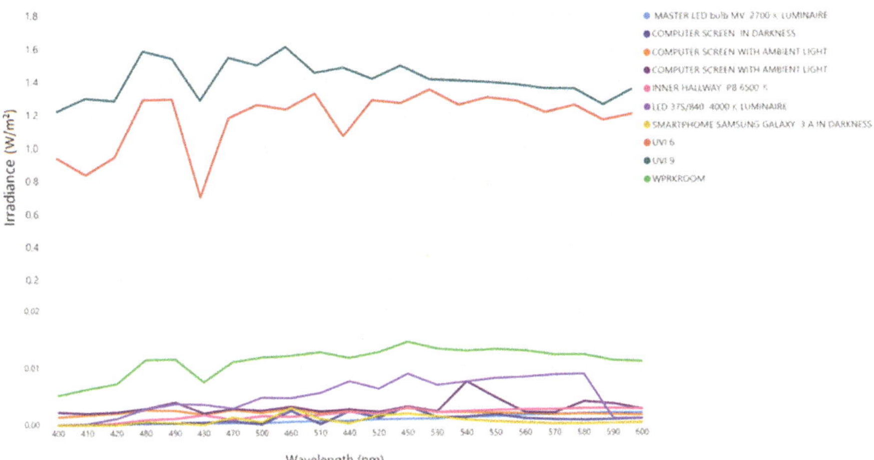

Figure 4. Comparison of the irradiance of artificial light sources with that of the sun. The 400–500 nm irradiance of electronic equipment for personal use, indoor lighting, and solar ultraviolet index (UVI) (modified vertical axis scale). Authors' own.

Table 1. Peak emission and irradiance of electronic devices. Results of the measurement of the peak wavelength and irradiance in the blue range of electronic equipment. Authors' own.

Model	Irradiance Peak Wavelength	Irradiance W/m^2
Laptop: HP Pavilion	445 nm	0.003597388
iPhone 6-S	448 nm	0.001446128
iPad sixth generation	448 nm	0.001565119
Samsung Galaxy J7	455 nm	0.001976249
Samsung S9	452 nm	0.000369329
iPhone 8 Plus	449 nm	0.000465317
iPad second generation	451 nm	0.003354588

3.4. Studies on the Effects of Artificial Light Sources on Adolescents

Several authors [56–59], have referred for years to the problem of mobile phone addiction, which they define in different ways as mobile phone dependence, the problematic use of mobile phones or cell phones, mobile phone abuse, and nomophobia (a fear of being without one's phone). Symptoms include preoccupation with the device, excessive use with loss of control, use of the device in inappropriate or dangerous situations, adverse effects on relationships, symptoms from lack of use (feeling angry, tense, depressed when the mobile phone is not accessible, constantly worrying that the battery will run out, etc.), and tolerance symptoms (the need for a better mobile phone, more software, or more hours of use). These symptoms are similar to those of substance dependence, which is why some researchers consider mobile phone dependence (MPD) to be an independent diagnosis.

A study was carried out in 2015 that surveyed 415 secondary school adolescents, among whom 251 were boys (60.48%) and 164 were girls (39.51%). The average age of the participants was 13.99 ± 0.8 years. The average time spent each day with a mobile phone was 131.77 ± 119.9 min. In this way, the prevalence of mobile phone dependence was demonstrated. Indian boy and girl adolescents comprised 33.5% and 39.6% of the study, respectively, and it was conducted using the International Classification of Diseases 10th edition (ICD-10), classification criteria for mental and behavioral disorder dependence syndromes [60].

LeBourgeois et al. [61] published an article on the effects of machine-emitted light on physiological circadian and warning sleep timers. They claimed that a high percentage of young people and adolescents have insufficient sleep periods. They referred to a 2004 article entitled "National Sleep Foundation. Sleep in America poll: Teens and Sleep", in which a large majority of studies were shown to find an adverse association between electronics consumption and sleep health, as well as a short time in bed and reduced sleep time. U.S. population data show that approximately 30% of preschool children and between 50% and 90% of school-age children and adolescents do not sleep as much as they need to. Data from a study of 454 teenagers revealed that more than 60% go to bed with their mobile phones and more than 45% use their mobile phone as an alarm or as a light. In addition, recent studies of 2000 students in grades four through six indicated that sleeping close to what is defined as "a small screen" was associated with increased fatigue.

Additionally, Crowley et al.'s [62] recent findings indicate that prepubescent children compared to postpubescent adolescents have greater melatonin suppression under low (15 lx), moderate (150 lx), and bright (5000 lx) light exposure in the hours prior to bedtime. Turner and Mainster [63] stated that children are more sensitive to light than adults based on their eye structures featuring a larger pupil size and a higher lens transmission.

Another study was conducted on 9846 adolescents between the ages of 16 and 19, taking into account the type and frequency of electronic equipment used in bed and the hours spent in front of screens during rest time. Sleep variables were calculated based on time in bed, duration of sleep, latency time before sleep, and state upon waking after sleep. The results of the study confirm that the adolescents spent a great deal of time during the day and in bed using electronic equipment. The use of electronic devices during the day and night was associated with an increased risk of a short sleep duration, a long sleep onset latency, and increased sleep deficits. The conclusion of the study is that the frequent use of electronic equipment is as common in adolescents during the day as it is at night. These results demonstrate a negative relationship between the use of technology and sleep, suggesting that recommendations on healthy use could involve a restriction of device usage [64].

A total of 746 surveys were conducted by Feizhou Zheng et al. [65] with children in Chongqing, China between October 2011 and May 2012 with data such as mobile phone use, if they feel well, and other possible confounding factors. Fatigue was significantly associated with years of mobile phone use and daily call duration. Headache was also significantly associated with daily call duration. There was no significant association between mobile phone use and other physical symptoms in the children and no consistent association with fatigue. The Independent Expert Group on Mobile Phones (IEGMP) reported several possible reasons for such sensitivity, including the information below.

- Children are more vulnerable to potentially harmful agents than adults because their nervous systems are developing.
- For anatomical reasons such as the smallness of their heads, thinner skulls, and increased nerve conductivity of their brains, children can absorb more energy from mobile phones than adults.
- Due to their early and extensive exposure to these devices, children will tend to accumulate more harmful health effects.

Yoshimura et al. [66] conducted a study on the mobile phone use of 23 nursing students. The measurements of the mobile devices showed the peak light of the mobiles at 453 nm, and the values of illuminance produced by each of the mobile devices were 25.3 to 42.6 lx (simulating their use in a seated position) and 50.5 to 80.4 lx (lying down).

Bae [67], in an editorial in a Korean journal of medicine in 2017, referred to several studies on the mobile phone dependency of adolescents. Min referred to an article by Lin et al. [68], which stated that dependence on smartphones is a problem that has spread throughout the world and, based on the study, proposed diagnostic criteria for smartphone addiction. Min also referred to a study by Elhai et al. [69] that related the use of smartphones to anxiety and psychopathological depression. In the editorial, Min states that the studies being published allude to various physical and psychological problems, including ophthalmological, orthopedic, and sleep disorders due to the use of smartphones. This supports the claim that adolescents are a group at higher risk for addiction because of their developing brains, as stated in the National Information Society Agency (KR) study, "Survey on smartphone overuse" (2017) and in Long J et al. [70].

On 5 October 2017, the Spanish National Statistics Institute (INE) published the "Survey on Equipment and Use of Information and Communication Technologies in Households. Year 2017". The statistics show that the proportion of children (from 10 to 15 years old) who use information and communications technology (ICT) is, in general, very high. Thus, the use of computers among children is very widespread (92.4%) (even more so for the use of the internet (95.1%)). In 2016, the number of minors using the internet exceeded the number of those using a computer. The differences by sex were not very significant. By age, the results suggested that the use of computers and the internet was a majority practice among children aged ≤ 10 years old. In turn, the use of mobile phones was found to increase significantly (Table 2) after the age of 10, reaching 94.0% in the population aged 15 years old [71].

Table 2. Percentage of underage information and communications technology (ICT) users among children aged 10–15 years. Percentage of underage ICT users by gender and age [71].

	Computer Use	Internet Use	Cellular Availability
Total	92.4	95.1	69.1
Gender			
Men	91.1	94.9	68.2
Women	83.8	95.2	70.0
Age			
10	88.4	88.8	25.0
11	89.3	91.0	45.2
12	95.8	95.8	75.0
13	93.6	96.8	83.2
14	95.1	98.9	92.8
15	92.5	99.2	94.0

4. Lighting in the Classrooms

4.1. Classroom Lighting and Energy Saving

Mareno and Labarca (2015) [72] have suggested that the traditional daylight analysis methods provide very limited information compared to the new dynamic methods that integrate factors such as sky type and local light conditions. These methods reinforce the evaluation of applied daylight strategies in favor of integrated design. This is shown in the design of classrooms by the method of dynamic simulation, which provides quantitative information on the proposed strategies, taking into account the levels of illumination for the visual tasks of the students.

The methods and metrics that incorporate climatic variables have been studied in order to refine the climate data needed to apply dynamic methodologies. Climate-based daylight modeling (CBDM) allows the integration of different variations of light in the simulations in relation to the local climate, generating for a specific moment a series of predictions which, in general, are for each hour of a complete year. At the same time, passive design, as an architectural principle, seeks to provide comfort conditions within buildings by optimizing the design through the integration of environmental factors of the site, thus minimizing the use of active means for that purpose. Complementarily, design for energy efficiency seeks a specific purpose for it, providing the best environmental conditions to achieve visual comfort and good lighting quality, and using the least amount of energy possible.

From the perspective of architectural design, a poor conception of lighting strategies can have negative implications, such as increased use of alternative energies, or can directly affect users, creating situations of visual discomfort. Dynamic metrics based on climate data (CBDM) have created a new perspective in the study of natural light by responding to the local climate, which accounts for daily and seasonal variations of daylight when combined with weather data. These have displaced traditional daylight metrics, like the daylight factor (DF), whose limitation is that it does not consider the dynamic aspects of light: the latitude, the different seasons, the different times of the day, the variations of the skies, and the orientation of the building [73].

The metrics developed and used in dynamic daylight performance measurements are based on the time interval in which the basal levels of illuminance and luminance are reached within a building. These time intervals typically extend throughout the year, based on external data such as annual solar radiation and depending on the location of the building [10]. Among the metrics whose analyses are based on the time variable, we have daylight autonomy (DA), useful daylight illuminance (UDI), and continuous daylight autonomy (DAcon), which have been given a more comprehensive evaluation based on climate files, the building's orientation, and the time that the building is occupied. Daylight autonomy (DA) sets an illumination value to guarantee autonomy to work only with daylight; however, we can have an exceptional autonomy without guaranteeing visual comfort, since by not setting an upper illumination limit, we can have too much daylight at certain times of the year.

The useful life of daylighting defines a range of illuminances that can be said to constitute useful levels of illumination [74]. What is new about this metric is that upper and lower limiting illuminance values are incorporated, while compliance intervals are set by integrating the concept of target ranges. The new series defined in 2012 were the useful daylight illuminance (UDI) "fell-short" (UDI-f), when the illuminance is less than 100 lx; the UDI supplementary (UDI-s), when the illuminance is greater than 100 lx and less than 300 lx; the UDI autonomous (UDI-a), when the illuminance is greater than 300 lx and less than 3000 lx; the UDI combined (UDI-c), when the illuminance is greater than 100 lx and less than 3000 lx; and, finally, the UDI exceeded (or UDI-e), when the illuminance is greater than 3000 lx [75].

Al-Khatatbeh (2017) [76] stated that in classrooms, light levels are directly related to energy consumption due to the use of artificial light. Therefore, this study aimed to improve visual comfort and energy efficiency in existing classrooms by investigating various adaptation methods for passive daylighting techniques in north-facing classrooms at the Jordan University of Science and Technology (JUST). The data from this research were obtained using computer simulations and real measurements. The combination of the office window and the south-facing anidolic ceiling provided about 62% of the light needed for the classrooms, and reduced the energy consumption required for lighting and heating by 16.3%. According to Yener [77], classroom lighting should be adequate for activities such as reading and writing on the blackboard and at desks. Kruger and Dorigo [78] found that each country has its own classroom lighting standards, but they all fall within the 300–500 lx range. Many are based on guidelines published by the Illuminating Engineering Society of North America (IESNA) and the European Standards (CEN).

Currently, the tools that are available to evaluate green buildings include the evaluation of Indoor Environmental Quality (IEQ) and obtaining the health of the occupants, and one of the elements that can be evaluated is visual comfort. Adaptive re-use is one of the well-known strategies to improve the sustainability of existing buildings in order to reduce material, transport, and energy consumption, as well as pollution levels [79]. Changes in building function have degraded the level of many buildings' IEQs, including their indoor lighting performance. The performance of indoor lighting is generally measured by the lighting level (E), daylight factor (DF), and the uniformity ratio. The illuminance level may vary in each room, with their different functions.

DF = Ei/Eo × 100(%), where Ei = indoor illuminance, Eo = outdoor illuminance (varies from 20–130 klx), and DF = daylight factor.

Based on previous research, Susan and Prihatmanti (2016) [80] stated that lighting in educational institutions is a critical factor, because poor lighting is not only detrimental to the visual comfort of the occupants, but could also lead to eye fatigue.

4.2. Problems with Poor Lighting in the Classroom

Winterbottom and Wilkins [81] measured, in 90 secondary school classrooms, with the variable of flicker, the lighting of the desks and the blackboards. The results showed that 80% of the classrooms were illuminated with fluorescents, which can cause headaches and impair visual task performance. While the illuminance (from excessive day and artificial lighting) was in excess of the recommended levels in 88% of the classrooms, in 84% it exceeded the levels to a degree that visual comfort would decrease. The lighting was not always adequately controlled, depending on the class design and infrastructure. The ambient light needed for close work at desks reduced the contrast of images. Venetian blinds in 23% of the classrooms had spatial features suitable for inducing glare. These findings provided information on small-scale reports linking student performance, behavior, and learning to classroom lighting. In rooms where there is no uniform lighting, when the immediate task area is brighter than the surrounding area, the effects of glare can be significant. Compared to the recommendations of the Chartered Institution of Building Services (2004) [82], the lighting varied from inadequate to excessive. It is important to note that the data for this study were collected in the summer months, when daily lighting may have been higher than the average for the year.

Winterbottom and Wilkins [81] stated in their article that there are numerous studies that have also noted changes in behavior under particular forms of lighting. Schreiber (1996) [83] suggested that children are more relaxed and interested in activities when classroom brightness is reduced; while Shapiro, Roth, and Marcus (2001) [84] found that maladaptive behaviors were less frequent under indirect full-spectrum fluorescents. Rittner and Robbin (2002) [85] indicated that daylight helps students retain and learn information. Some authors have emphasized the importance of daylighting, but with the need for integrative systems of natural and artificial light.

CIBSE (2004) [86] provides lighting design recommendations for different types of classrooms, in the range of 300 lx to 500 lx; the adoption of such values helps to restrict glare to reasonable levels (it is worth noting that a new installation with new lamps and clean surfaces can provide 25% more lighting than the designed lighting, and only half as much as the initial lighting is present when the lamps are old and dirt has accumulated).

Ho et al. [87] conducted a study to minimize classroom lighting costs in Taiwan by taking advantage of natural light. In their article, they state that there is ample evidence of the damage caused to children's vision as a result of poor lighting conditions in classrooms. Optimal shading of classroom windows is important to improve daylight illumination in the subtropics.

Guan and Yan (2016) [88] state that daylight varies greatly due to the movement of the sun, changing seasons, and various climatic conditions. Customized static light assessments, known as static daylight assessments, representing simulations of only one time of the year or one time of the day, are inadequate for evaluating the dynamics of daylight variability. Using the graphic tool Temporal Map to display annual daylight data, this study compared different passive architectural design strategies under climatic conditions of five representative cities and selected the most appropriate schematic design for each city, which in turn was integrated with the Chinese academic calendar to obtain an improvement in occupational time. This modified map connected design work with human activity, making the daylight evaluation more accurate and efficient. In addition, the prevalence rate of myopia is extremely high among young Chinese people according to current government statistics [89]: up to 40.89%, 67.33%, and 79.2% in primary, junior, and high school, respectively, and remarkably, 84.72% in university. In fact, the database, derived from the national adolescent health survey conducted every five years, shows that nearsightedness ratios have been steadily rising in recent decades [89–91]. Meanwhile, high levels of light have been shown to have preventive effects for myopia [92,93]. Given the potential energy savings and health benefits for students, improving the quality of light in such environments must be a priority.

4.3. Effects of Classroom Lighting on Student Performance

Improving student progress is vital to a nation's competitiveness. Scientific research shows how the physical environment of classrooms influences student progress. The structural facilities of buildings have a profound influence on learning. Inadequate light, noise, poor air quality, and poor heating in classrooms are factors known to be relevant to poor student progress. Students exposed to more natural light (i.e., daylight) in their classrooms perform better than students exposed to less natural light [94]. Cheryan et al. [95] conducted a study of more than 2000 classrooms in California, Washington, and Colorado, in which they found students who were exposed to a large amount of daylight in their classes had better reading and math test scores than students who were exposed to less daylight in their classes (2–26% higher, depending on school district), even after statistical control of the student population that included race and socioeconomic status [96]. According to the National Center for Education Statistics (Alexander and Lewis, 2014) [97], 16% of schools with permanent buildings and 28% of schools with temporary buildings (i.e., portable) have unsatisfactory or very unsatisfactory natural lighting. Although the incorporation of natural light can be beneficial, it should be done with care to avoid visual discomfort [95].

Choi et al. (2014) [98] investigated the relationship between indoor environmental quality (IEQ) in university classrooms as a whole and student outcomes, including satisfaction with IEQ, perception of learning, and course satisfaction. The results were collected from the students.

Lighting conditions have always been an important IEQ criterion, including the sources of natural and artificial ambient and task lighting. Each of these elements has a unique role in user experiences within the built environment. Exposure to various types of light may be associated with psychological responses to human performance. Studies conducted in elementary school settings found a positive and significant correlation between the presence of daylight and student performance across three different school districts. Daylight received through skylights has a positive effect on students in their classrooms. Subsequent studies comparing classrooms with a large amount of daylighting with classrooms with less daylighting showed a 21% increase in student performance [96,99,100].

López-Chao et al. (2020) [101] state that empirical research has shown the influence of architectural spatial variables on student performance. Their article explored the relationship between the learning space and mathematics and artistic activities in 583 primary school students in Galicia (Spain). For this study, the Indoor Physical Environment Perception scale was adapted and validated, and utilized in 27 classrooms. The result of this exploratory factor analysis evidenced that the learning space has three structural categories: workspace comfort, natural environment, and building comfort. Sick building syndrome (SBS) shows that poor quality environments harm the health of users. Specifically, students perform better in brighter classrooms. Similarly, young children can differentiate their lighting needs according to the task at hand, while visual comfort is a key element for artistic activities, especially for drawing.

Heschong [96] included a focus on solar lighting as a way to isolate daylight as a source of illumination, and separated the effects of illumination from other qualities associated with light entering through windows. In this project, the author established a statistically convincing connection between daylight and student performance, and between lighting and retail performance. The author analyzed test score results for over 21,000 students from three districts located in Orange County, California; Seattle, Washington; and Fort Collins, Colorado. The author reviewed architectural plans, aerial photographs, and maintenance records, and visited a sample of the schools in each district to classify the daylighting conditions in over 2000 classrooms. Each classroom was assigned a series of codes on a simple 05 scale indicating the size and tint of its windows, the presence and type of any skylighting, and the overall amount of daylighting expected.

Kuller and Lindsten [102] are private investigators who followed the health, behavior, and hormone levels of 88 8-year-old students in four classes over the course of a school year. The four classes had very different daylight and artificial lighting conditions: two had natural light while two did not, and two were illuminated with warm white light emitting fluorescents (3000 K) while two had very cold white light emitting fluorescents (5500 K). The researchers found a significant correlation between daylight level patterns, hormone levels, and student behavior, and concluded that the practice of having no windows in classrooms should be abolished.

Wilkins (2002) [103] analyzed the performance of primary school classroom lighting given the impact that natural light has on the educational experience of students. Boyce [104] and Dudek [105,106] showed that natural light increases productivity, positively affects human performance, and has biological effects on the production of the hormone cortisol, regulating light-dark cycles, and the ability of students to concentrate. Ultimately, this is a relevant environmental aspect to be studied in order to understand the results of daylighting of different environments.

5. Existing and Developing Regulations Related to the Non-visual Effects of Light Absorbed through Our Eyes

In 2011, the European Commission published the Green Paper "Let's light up the future: Accelerating the deployment of innovative lighting technologies". This document raises concerns about blue light's potential risks to vision. The Scientific Committee on Emerging and Newly Identified Health Risks (SCENIHR) released a document in 2012 entitled "Health effects of artificial lights", which explored the scientific evidence on the potential impacts of artificial light radiating into the visible spectrum on human health [107].

In 2013, the International Commission on Illumination (CIE) published the "Report on the first international workshop on circadian and neurophysiological photometry". This document features multiple references to the article mentioned in this paper by Lucas et al. [108]. This CIE paper describes the five photoreceptors located in the human eye (the three types of cones, the rods, and the ipRGCs) and notes that they all influence the stimulation of melatonin synthesis, although the paper concludes that much remains to be determined about their mechanisms of action. The importance of pupil contraction in the non-visual effects of light absorption is also mentioned [109].

In 2016, the CIE [110] published the "Research Roadmap for healthful interior lighting applications", which refers to multiple studies with statements such as the following.

- The use of lamps with an illuminance of 10,000 lx for 30 min has been tried with positive results for the treatment of people who have been diagnosed as suffering from seasonal affective syndrome (SAD).
- Bright white light can have an activation effect that is linked to increased alertness, reduced drowsiness, and increased levels of vigilance both during the day and at night. During the day, high exposure to white light can modulate the brain's cognitive functions such as logical reasoning and creativity.
- People choose places with higher light intensity when they feel low in vitality and alertness. During the day, periods of exposure to bright light (more than 1000 lx) can lead to better social and emotional interaction. This may be due to a direct effect of bright light on serotonin metabolism.

In 2017, the CIE [111] held a working meeting: the CIE stakeholder workshop for temporary light modulation standards for lighting systems. Current lighting systems vary widely in that their light output shows temporary variations or flickering (temporal light modulation, TLM). Temporal light modulation (TLM) is known to affect human visual perception, neurobiology, and behavior, sometimes adversely. Many organizations have developed standards and regulations, and certifying organizations have been active in research focused on these issues. Some researchers are studying these effects, but such activities are currently uncoordinated and at risk of being inefficient. The CIE, with support from Energy Efficiency Canada, the national association of electricity manufacturers, Philips Lighting, BC Hydro, and Research Canada, agreed to hold a stakeholder workshop in Ottawa, Canada on 8–9 February 2017. The objective of this meeting was to create a roadmap for research, offer recommendations, and develop regulatory activities related to the temporary variations of light from lighting systems to accelerate the international regulatory development process in an efficient manner and avoid the overlap and duplication of such efforts.

In 2018, the CIE [112] published a document titled "ED/IS 026/E:2018 CIE System for metrology of optical radiation for light responses influenced by intrinsically-photosensitive retinal ganglional cells". The introduction of this paper states that light is the main synchronizer of human biological clocks and may result in the acute suppression of nocturnal melatonin release. While the functioning of cones and rods in the vision process are well known, the function of spectral sensitivity and a quantitative and qualitative analysis of melatonin-based photoreception remain unexplored.

The aim of this international standard is to define the role of spectral sensitivity, quantify it, and measure it to describe visible radiation and UVA according to its ability to stimulate each of the five types of photoreceptors, i.e., how much each contributes to the melatonin content of intrinsically photosensitive retinal ganglion cells (ipRGCs), and thus their mediation on the non-visual effects of light on humans. This international standard is applicable to visible optical radiation of 380–780 nm.

This international standard does not give information on particular lighting applications or a quantitative prediction of the response of ipRGCs; it also, does not mention health or safety issues, such as the results of light treatments, flickers, or photobiological health.

Experts around the world are collaborating on the development of this standard which states that light is the main synchronizer of human biological clocks and can lead to an acute suppression of melatonin release at night.

6. Results and Discussions

Throughout this review, we have examined works by scientists who are experts in different branches of knowledge related to humans. These studies confirm that human beings are designed to carry out all their physiological and cognitive activities during the day and to rest at night. Our civilization has changed the natural light–dark cycle that humans experience. Researchers have expressed concern over electric lighting as a potential disruptor of the natural light–dark cycle [16].

Some articles [10,59,113] have stated that LED lights are dangerous because they emit much more blue light than other types of lights and the sun. In studies on the development of LEDs, several examples of technical publications have refuted the theory that new light sources radiate more blue light than traditional sources (rather the opposite), including the sun. New LED-type light sources do not necessarily produce a higher percentage of blue light than traditional light sources (mercury vapor lamps, halogens, etc).

In addition, under normal working conditions, in accordance with current lighting legislation, around 500 lx should be received at one's workplace, regardless of the type of luminaire used.

If we compare the illuminance emitted by lights against those of solar irradiance, we find that a cloudy day in the northern hemisphere will produce about 3000 lx; about 50,000 lx on a sunny summer day in Spain; and about 100,000 lx in the tropics. Thus, we can conclude that the possible risk of blue light on the retina under general lighting indoors is low to very low, even if one uses cold color temperatures as general lighting.

The neurons responsible for the melatonin, serotonin balance (ipRGCs), are sensitive to light from UVA to red, with their peak in blue. By absorbing light at the beginning of the day, melatonin is eliminated and replaced by serotonin [59]. Within the range of the visible spectrum, the most abundant energy we receive from the sun is blue, and its peak is approximately 480 nm. This is, therefore, the peak of the maximum sensitivity of melatonin. Beings who are active during the day have evolved by adapting to sunlight [114].

Since the discovery of these neurons 25 years ago, the scientific world has not stopped discovering evidence of the synchronization of our organs, and even our individual cells, with the light and dark cycles of nature. These discoveries are considered so important for the scientific world that the 2017 Nobel Prize in Physiology and Medicine was awarded jointly to Jeffrey C. Hall, Michael Rosbash, and Michael W. Young for their discoveries of the molecular mechanisms underlying the control of circadian rhythms.

Artificial light has facilitated great advances in many fields, but it can also pose a risk as it can alter the natural cycles of sleep and wakefulness, in many cases reducing the hours of sleep by people being able to continue to carry out activities during the night period. An alteration of sleep can pose a risk to health, leading to chronodisruption. Chronodisruption is an alteration of the internal temporal order of physiological, biological, and behavioral rhythms. If it becomes chronic, asynchrony, advancement, or retardation of the peripheral clocks may occur [17].

Epidemiological studies show that chronodisruption is associated with an increased incidence of metabolic syndrome, cardiovascular disease, cognitive and emotional disorders, premature aging, and some cancers such as breast, prostate, and colorectal cancer, as well as the worsening of pre-existing pathologies.

The scientific community has shown, with multiple studies, that humans need biorhythms to receive sufficient amounts of light in the blue range during the day but in a different proportion depending on the time of day. Erren and Reiter define correct lighting as photohygiene. This factor is already considered so important for health that the International Agency for Research on Cancer (IARC) of the World Health Organization (WHO) included shift work as a cancer-inducing factor in 2010.

Many articles recommend that the reception of artificial light should be as close as possible to daylight. It should be more intense with a higher proportion of blue, with a maximum value at mid-morning, which should then decrease in intensity and quantity, mainly in its proportion of blue. For more than ten years, correct or incorrect lighting has been associated with the good or bad physical and psychological conditions of people [17].

Increasingly more scientists are advocating the study and development of luminaires that are more in line with sunlight [21], and there is a call for interior lighting that not only differs throughout the day, but also differs depending on the seasons based on the response to their cumulative absorption, i.e., there should be more intense irradiance in public spaces (including schools) in autumn and winter and less in spring and summer [39,40].

Experiments have been carried out on patients with dementia and other neuropsychiatric pathologies where researchers improved the symptoms of their illnesses by receiving light closer to daylight. This light was more intense and featured a greater amount of light in the blue range than that which they received in the centers where they reside [31].

Notably, Glickman et al. [20] observed that, although it was initially thought that an illuminance of 2500 lx is necessary to suppress nocturnal melatonin in humans, under certain conditions, values as low as 10 lx [114], and possibly 1 lx or less, can suppress melatonin in humans. This would support the recommendation to sleep in complete darkness since, if we have any lights on, that very small amount of light could pass through our eyelids.

The current majority of studies states that the minimum amount of light that we receive from 9:00–10:00 p.m. will delay the transition from serotonin to melatonin and, therefore, delay the process of rest that is essential for health. Moreover, the sudden reception of low-intensity light delays the process since this light is not received for at least 40 min [27]. If these disorders persist over time, chronodisruption and undesirable cortisol levels may occur. Scientists have related this disorder to metabolic syndrome, cancer, obesity, diabetes, cardiovascular disease, and cognitive and affective disorders.

The accredited test laboratory FUTTEC measured the amount of blue light emitted by various electronic equipment used by children and adolescents in 2014 and 2019. The results show that the irradiance in the blue range of these devices (at the distance they are commonly used) is between 1000 and 10,000 times lower than what we can receive at the same time from the sun on a spring day with a UVI of 6. Compared to a day with a UVI of 9, this amount would be up to 30% higher on a spring day.

Based on these results, it is not considered plausible that this equipment is harmful to the retina, as our eyes can survive 1000 to 10,000 times more light than they receive from these devices from the sun.

Yoshimura et al. [115], in their study on the use of mobile phones among 23 nursing students, found the peak of light of the mobile phones at 453 nm, and the illuminance in the seated measurements were from 25.3 to 42.6 lx; lying down, these values ranged from 50.5 to 80.4 lx. The authors logically note that this amount of light is not likely harmful to the eye (as per other studies) but can affect circadian cycles and sleep quality.

Escofet and Bará [55] studied the emissions of two computer screens and two smartphones with irradiance results from the computers of 4×10^{-3} W/m^2 around 450 nm and $6\text{–}7 \times 10^{-4}$ W/m^2 from the two smartphones. The authors recommend the use of filters to protect from such irradiation. This irradiance, however, is between thousands and tens of thousands of times lower than the irradiance we receive from the sun.

An important issue for optometrists is that if we remove the part of the spectrum that we can most comfortably accommodate at close distances, as explained in the introduction to this paper, we would have to make more accommodation efforts and would require greater light intensity to do the same job. Would that not be more counterproductive than removing a small percentage of blue?

The results of the measurements and scientific publications show that the composition of light used for indoor lighting can influence the balance between serotonin and melatonin and therefore, our circadian cycles.

- Analysis of the possible negative effects of blue light from electronic devices on the eyes of adolescents

Various news media have reaffirmed the results (according to scientific studies) that suggest the damaging effects of blue light emitted by sources of artificial light on the retina (both LED lights and computers and mobile phones). These results are supported by published articles that correlate a greater exposure to light emitted by LEDs with a greater absorption of light in the blue range which, as a consequence, can accelerate the development of cataracts, pose a greater risk of damage to the retina, and, therefore, increase the prevalence of age-related macular degeneration (AMD). Many companies in the optical sector offer filters that absorb much of the blue light emitted by electronic devices to protect children and adolescents from this hypothetical risk.

After observing the prevalence of this concern, a search was made for scientific articles that corroborate this concern. The aim was to find an article showing the results of the irradiation of cell cultures or exposure to guinea pigs via the light actually emitted by this electronic equipment.

Notably, all the studies to date that note the possible damage from blue light in living tissue, especially the retina, have been carried out on cultures or living animals using monochromatic LED sources (emitting only in the most oxidizing part of the blue spectrum), with significantly higher irradiance levels than those actually emitted by electronic equipment.

Moreover, the experimental animals used include mice and rabbits, which are nocturnal animals whose visual systems are much more sensitive to light than those of humans, without any kind of algorithm or objective method to transpose these results into the possible effects on humans.

As a result, we believe that it is not possible to directly associate ICT's emissions with the possible real risks in the retinas of adolescents.

- Studies on the effects of artificial light sources on adolescents

There is an increasing number of publications by researchers (mainly from Asian countries) that relate low indoor lighting and natural light to the increase in myopia among children and adolescents worldwide, which in some countries, such as Singapore, involves more than 90% of children. The specific causes are unknown. No article has provided conclusive results [112]. This pandemic remains a challenge for the scientific community.

The dependence and addiction due to the indiscriminate use of mobile phones by children and adolescents all over the world already have attention within the scientific community, (e.g., nomophobia). This is a different problem from chronodisruption described above.

There is talk of a greater risk in children than in adults due to physiological reasons such as increased nerve conductivity and unwanted effects such as headache, dizziness, sleep problems, insomnia, dependence syndrome, and mental or behavioral disorders [56–59,67,115].

According to the results of the INE 2017, the percentage of minors who use a mobile phone in Spain reached 94% by the age of 15, making it a health priority to provide relevant information and control the use of ICT by adolescents [71].

- Existing and developing regulations for the non-visual effects of light absorbed through our eyes

The authorities in the European Community, through the CIE, and internationally through the International Organization for Standardization (ISO), have been sensitive to the concerns of the scientific community, and groups with experts in different fields have been increasingly established to study the non-visual effects of light.

In the development of this study, the work of the European Community was broken down as a Green Paper published in 2011. When this document was published, the scientific community was not yet aware of the technological revolution of the new LED lighting systems to come, especially from the perspective of their possible influence on people's states of mind.

European bodies studying health risks, such as the SCENIHR and the CIE from 2012 onwards, have gradually entered into the exciting study of this new field of science.

Concern over ICT irradiance is growing. This concern is reflected by the proposal produced at the end of 2018 for a group of experts from the CIE and ISO to jointly create a new document (integrative lighting) to use this research for the development of standards that define the composition of light in each area in which it will be used.

Lighting in the Classroom

- Classroom lighting and energy saving

The problems generated by the excessive consumption of fossil fuels and their effect on the fragile balance of our planet are well known, and a reduction in expenditure due to a decrease in energy consumption would allow countries to use these funds for the health or education of their inhabitants. Therefore, the reduction of energy consumption must be a priority for all public administrations, both for the savings it would entail and for the reduction of emissions of polluting gases [79,82,88,89].

In recent years, publications have appeared on studies and proposals for new parameters for cataloging buildings (IEQ), for specific data analysis of the parameters to be studied (CDBM), and on different parameters for obtaining more specific data, taking into account variables such as the amount of light at each time of day and season (DF, DA, UDI, DAcon, UDI-e, UDI-f, UDI-s, UDI-a, UDI-c) [72–75,79].

- Problems with poor lighting in classrooms

Studies on classroom lighting in this century and last century have reflected the widespread concern around obtaining the best possible lighting for children. We found an abundance of publications that discuss the unwanted effects on children of having classrooms that are either under-lit or over-lit [79,81,82]. The problems of inadequate illumination can be divided into temporary physiological problems of the visual system, psychological problems, and permanent problems of the visual system. For many years, there have been publications on possible temporary physiological disorders of the visual system, such as glare, headache, and/or fixation problems [81,82,85,87]. Some authors have realized, from the results of their studies, that poor lighting, whether natural or artificial, has notable negative effects on children's performance and mood [84,85].

The current pandemic of child myopia is an issue of global concern, but it is especially of concern in Asian countries, where it is most prevalent. One of the theories proposed by scientists who study this problem is that this increase in child myopia may be due to the short amount of time that children are outdoors receiving natural light. The recommendation for indoor lighting is between 300 and 500 lx, while outdoor light even on a cloudy day is tens of thousands of lx [90–94,116].

Figueira et al. [116] show that 1-h and 2-h exposure to light from self-luminous devices significantly suppressed melatonin by approximately 23% and 38%, respectively. The authors' previous studies suggest that adolescents may be more sensitive to light than other populations.

- Effects of classroom lighting on student performance

In 1992, the researchers Kuller and Lindsten [102] published an article in which they associated insufficient light levels with inadequate levels of hormones, leading to negative effects on children's behavior, and recommended that there be no classrooms built without natural lighting. At the time of their study, the relationship between light and the pineal gland and biorhythms was still unknown. Later studies found a relationship between correct artificial lighting or a combination of natural and artificial light and better performance of students in the classroom [98–101].

The influence of light on people's moods remains a subject of study today, throughout the scientific world. Official and scientific bodies involved in the world of lighting, such as the International Committee on Illumination (CIE) and the Lighting Research Center, have been researching and publishing studies for decades relating to the possible effect of lighting on the moods of students in the classroom, and the influence of this on their learning ability and their performance [102,109–112,116].

7. Conclusions and Recommendations

The functioning of human biorhythms responds to the proportion and intensity of the light we receive, whether natural or artificial. We are light-based beings designed to live with the rhythm of sunlight. Published articles that refer to the possible damage of blue light emitted by electronic equipment or LED lights do not scientifically prove that such equipment can damage the retinas of its users, including teenagers.

According to the research conducted so far, the light received from luminaires at night and from the electronic equipment used by adolescents, however low, is believed to have a negative influence on the balance of circadian cycles and on the quality of sleep.

Numerous studies around the world have come to the same conclusion. There is a growing pandemic of myopia in children and adolescents in certain areas of Asia, such as Singapore and South Korea, that exceeds 80%. This pandemic is suspected to be related to the use of electronic devices, but so far, its direct cause and possible solutions remain unknown.

The new generation of LED lights whose blue light intensity can be increased without increasing the illuminance, makes it possible for the professionals responsible for educational centers to recognize and alter the composition of their centers' lights since this composition can influence the "mood" of the affected people, including adolescents.

No publications were found on the composition of light in adolescent training centers or the psychological responses to it. For all these reasons, there is growing concern among international bodies, both at the European and the global level. These bodies have created groups of experts to urgently develop regulations to control the composition of the lights in new luminaires.

Based on the conclusions of this review, the following control or preventive measures are suggested. These measures could be promoted by administrations, educational centers, and families.

- Develop a "manual for the good use of electronic equipment by adolescents".
- Public and private managers should train the professionals in charge of the maintenance of public buildings, in this case schools, so that they can acquire the necessary knowledge for the correct acquisition and assembly of the lights to be placed in their educational centers.

- Make this circumstance known to the public bodies responsible for consumer protection, so they can inform people, Thus, when consumers acquire new luminaires for their homes, they can consider the proportion of blue light those luminaires contain due to their non-visual effects.

7.1. Limitations of Study and Foresight

The difficulties of this study stem from the scarcity of existing research in this field. The effects of continuously increasing or decreasing the percentage of blue light in a systemic way are unknown for human beings and, therefore, for children and adolescents. This is because, until a few years ago, it was not possible to make luminaires with alterable light components.

There is no possibility to control the luminaires that are being made in real time for two reasons: (1) any *company or person can buy LEDs, group them, and use a programming system with a remote control to change the proportion and composition of their light as desired* and (2) as there is no legislation on the non-visual effects of light, the proportion of blue light cannot be legally limited.

Due to these limitations, no concrete recommendations can be given to safely improve the health of adolescents. The scientific study of the effects of non-ionizing light on living beings is known as photobiology. This field does not currently exist as a separate subject, but is a new branch of knowledge that involves dermatologists, ophthalmologists, physicists, opticians, and chemists, each in their own fields. The aim of photobiologists is to integrate the specialty into all degrees of training in the health field (pharmacy, medicine, veterinary medicine, architecture, and engineering related to animal and plant biology).

Given its relevance to healthy human functioning, this content should be integrated into the curriculum from the earliest stages of education, with greater emphasis on the secondary and high school stages and the professional branches related to the field. In this sense, students should be aware of the impact of healthy ethical light in different contexts:

- greater precision and safety for patients who have light applied to their light treatments,
- improving the quality of life of tabulated animals by receiving light much more similar to that of the sun;
- a higher quality and optimization of greenhouse crops by irradiating them with light that is the most similar in each case to that of genetically designed crops; and
- finally, improving the health of all humans, including adolescents, by providing much healthier light inside buildings. The ideal lighting conditions involve the provision of natural light according to the time of day and the season, so that artificial light processes can radiate with light that is as similar as possible to that received when going out onto the street.

Therefore, the establishment of natural or artificial lighting according to natural parameters should also be a priority issue in the design and construction of new educational centers, as well as in the lighting of existing centers, to facilitate the transformation of toxic light into a new ethical–healthy model. In this sense, it is also necessary to train senior educational managers (central and autonomous community administration staff, as well as the directors or presidents of educational centers).

Additionally, this research offers a framework for study and reflection that will allow national and international governments to regulate lights in order to promote naturalization processes and new models that are applicable to the domestic, health, professional, and educational fields, thereby facilitating the incorporation of new measures and training that will make it possible to overcome the lack of existing regulations and training.

7.2. Lighting in Classrooms

The lighting of school classrooms has changed from being mostly mercury vapor lamps that emitted a peak in the blue range, another in the green range, and another in the red range but with a wide spectral distribution, to being LED lighting, for which the composition of the light and the proportion of each part of the visible spectrum that the students now receive is different to that received from fluorescent lamps. LEDs do not emit a wide range of blue (blue LEDs do) but, as their emission is centered around 460 nm, even though the luminaires have a quantity of lx similar to that which would be obtained if a fluorescent lamp were measured with a lux meter, the effect of the proportion of blue on the melatoninergic receptors could be very different. Do those responsible for the maintenance of educational facilities understand the quantity of blue from the new LED luminaires they are installing and, therefore, the light that the students receive with the new LED luminaires, and how it differs from that which they received before?

Author Contributions: Conceptualization, D.B.M. and M.S.J.F.; methodology, D.B.M. and M.S.J.F.; formal analysis, D.B.M. and M.S.J.F.; investigation, D.B.M. and M.S.J.F.; resources, D.B.M.; data curation, D.B.M. and R.A.G.L.; writing—original draft preparation, D.B.M., M.S.J.F., and R.A.G.L.; writing—review and editing, D.B.M. and R.A.G.L.; visualization, D.B.M. and R.A.G.L.; supervision, M.S.J.F. and R.A.G.L.; project administration, D.B.M.; funding acquisition, D.B.M. and R.A.G.L. All authors have read and agreed to the published version of the manuscript.

Funding: We are especially grateful to FUTTEC and the Universidad San Pablo CEU for providing us with the necessary resources to carry out this research. The authors wish to thank CEU San Pablo University Foundation for the funds dedicated to the Project CEU-Banco Santander (Ref: MVP19V14) provided by CEU San Pablo University and financed by Banco Santander.

Conflicts of Interest: The authors declare no conflicts of interest.

References

1. Hysing, M.; Pallesen, S.; Stormark, K.M.; Jakobsen, R.; Lundervold, A.J.; Sivertsen, B. Sleep and use of electronic devices in adolescence: Results from a large population-based study. *BMJ Open* **2015**, *5*, e006748. [CrossRef] [PubMed]

2. Millodot, M.; Sivak, J. Influence of accommodation on the chromatic aberration of the eye. *Br. J. Physiol. Opt.* **1973**, *28*, 149–174.

3. Shen, J.K.; Dong, A.; Hackett, S.F.; Bell, W.R.; Green, W.R.; Campochiaro, P.A. Oxidative damage in age-related macular degeneration. *Histol. Histopathol.* **2007**, *22*, 1301–1308. [PubMed]

4. Grubisic, M.; Haim, A.; Bhusal, P.; Dominoni, D.M.; Gabriel, K.M.A.; Jechow, A.; Kupprat, F.; Lerner, A.; Marchant, P.; Riley, W.; et al. Light Pollution, Circadian Photoreception, and Melatonin in Vertebrates. *Sustainability* **2019**, *11*, 6400. [CrossRef]

5. Moore, L.A. Ocular protection from solar ultraviolet radiation in sport: Factors to consider when prescribing. *S. Afr. Optom.* **2003**, *62*, 72–79.

6. Hassoy, H.; Durusoy, R.; Karababa, A.O. Adolescents' risk perceptions on mobile phones and their base stations, their trust to authorities and incivility in using mobile phones: A cross-sectional survey on 2240 high school students in Izmir, Turkey. *Environ. Health* **2013**, *12*, 10. [CrossRef] [PubMed]

7. Alvarez, A.A.; Wildsoet, C.F. Quantifying light exposure patterns in young adult students. *J. Mod. Opt.* **2013**, *60*, 1200–1208. [CrossRef]

8. Comité Europeo de Normalización. *Seguridad Fotobiológica de Lámparas y de los Aparatos que Utilizan Lámparas*; EN62471; CEN: Bruselas, Belgium, 2012.

9. European Committee for Standardization. *Light and Lighting—Lighting of Work Places—Part 1: Indoor Work Places*; EN 12464-1; CEN: Brussels, Belgium, 2002.

10. CELMA. Optical Safety of LED Lighting. Federation of National Manufacturers Association for Luminaires and Electrotechnical Components for Luminaires in the European Union. Available online: https: //moodle.polymtl.ca/pluginfile.php/354010/mod_resource/content/2/CELMA-ELC_LED_WG%28SM% 29011_ELC_CELMA_position_paper_optical_safety_LED_lighting_Final_1st_Edition_July2011.pdf (accessed on 19 March 2019).

11. Sánchez-Muniz, F. Cronodisrupción y desequilibrio entre cortisol y melatonina ¿Una antesala probable de las patologías crónicas degenerativas más prevalentes? *J. Negat. No Posit. Results* **2017**, *2*, 619–633. [CrossRef]

12. Morley, M.W.; Goldberg, P.; Uliyanov, V.A.; Kozlikin, M.B.; Shunkov, M.V.; Derevianko, A.P.; Jacobs, Z.; Roberts, R.G. Hominin and animal activities in the microstratigraphic record from Denisova Cave (Altai Mountains, Russia). *Sci. Rep.* **2019**, *9*. [CrossRef]

13. Rutger, A.W. The circadian system of man—Results of experiments under temporal isolation. *Comp. Biochem. Physiol. Part A Physiol.* **1980**, *67*, 532. [CrossRef]
14. Zeitzer, J.M. Control of Sleep and Wakefulness in Health and Disease. In *Progress in Molecular Biology and Translational Science*; Elsevier BV: Amsterdam, The Netherlands, 2013; Volume 119, pp. 137–154.
15. NobelPrize.org. *The Nobel Prize in Physiology or Medicine 2017*; NobelPrize.org: Stockholm, Sweden, 2019.
16. Smolensky, M.H.; Hermida, R.C.; Reinberg, A.; Sackett-Lundeen, L.; Portaluppi, F. Circadian disruption: New clinical perspective of disease pathology and basis for chronotherapeutic intervention. *Chrono. Int.* **2016**, *33*, 1101–1119. [CrossRef] [PubMed]
17. Stevens, R.G.; Zhu, Y. Electric light, particularly at night, disrupts human circadian rhythmicity: Is that a problem? *Philos. Trans. R. Soc. B Boil. Sci.* **2015**, *370*, 20140120. [CrossRef] [PubMed]
18. Wurtman, R.J. The Effects of Light on the Human Body. *Sci. Am.* **1975**, *233*, 68–77. [CrossRef]
19. Guido, M.E.; Garbarino-Pico, E.; Contín, M.A.; Valdez, D.J.; Nieto, P.S.; Verra, D.M.; Acosta-Rodríguez, V.; De Zavalía, N.; Rosenstein, R.E. Inner retinal circadian clocks and non-visual photoreceptors: Novel players in the circadian system. *Prog. Neurobiol.* **2010**, *92*, 484–504. [CrossRef] [PubMed]
20. Glickman, G.; Levin, R.; Brainard, G.C. Ocular input for human melatonin regulation: Relevance to breast cancer. *Neuroendocrinology.* **2002**, *23*, 17–22.
21. Cajochen, C. Alerting effects of light. *Sleep Med. Rev.* **2007**, *11*, 453–464. [CrossRef]
22. CIE. Report on the First International Workshop on Circadian and Neurophysiological Photometry, 2013. 2015. Available online: http://files.cie.co.at/785_CIE_TN_003-2015.pdf (accessed on 19 March 2020).
23. Erren, T.C.; Reiter, R.J. Light Hygiene: Time to make preventive use of insights—old and new—into the nexus of the drug light, melatonin, clocks, chronodisruption and public health. *Med. Hypotheses* **2009**, *73*, 537–541. [CrossRef]
24. Cho, C.-H.; Lee, H.-J.; Yoon, H.-K.; Kang, S.-G.; Bok, K.-N.; Jung, K.-Y.; Kim, L.; Lee, E.-I. Exposure to dim artificial light at night increases REM sleep and awakenings in humans. *Chrono Int.* **2015**, *33*, 1–7. [CrossRef]
25. Sancar, A.; Lindsey-Boltz, L.; Kang, T.-H.; Reardon, J.T.; Lee, J.H.; Ozturk, N. Circadian clock control of the cellular response to DNA damage. *FEBS Lett.* **2010**, *584*, 2618–2625. [CrossRef]
26. Madrid Pérez, J.A.; Rol de Lama, M.Á. Ritmos, relojes y relojeros. Una introducción a la Cronobiología. *Eubacteria* **2015**, *33*, 1–7.
27. Schroeder, M.M.; Harrison, K.R.; Jaeckel, E.R.; Berger, H.N.; Zhao, X.; Flannery, M.P.; Pierre, E.C.S.; Pateqi, N.; Jachimska, A.; Chervenak, A.P.; et al. The Roles of Rods, Cones, and Melanopsin in Photoresponses of M4 Intrinsically Photosensitive Retinal Ganglion Cells (ipRGCs) and Optokinetic Visual Behavior. *Front. Cell. Neurosci.* **2018**, *12*. [CrossRef] [PubMed]
28. Wong, K.Y.; Dunn, F.A.; Berson, D. Photoreceptor Adaptation in Intrinsically Photosensitive Retinal Ganglion Cells. *Neuron* **2005**, *48*, 1001–1010. [CrossRef] [PubMed]
29. Garaulet, M.; Madrid, J.A. Chronobiology, genetics and metabolic syndrome. *Curr. Opin. Lipidol.* **2009**, *20*, 127–134. [CrossRef] [PubMed]
30. Pandi-Perumal, S.R.; Srinivasan, V.; Spence, D.W.; Moscovitch, A.; Hardeland, R.; Brown, G.M.; Cardinali, D.P. Ramelteon: A review of its therapeutic potential in sleep disorders. *Adv. Ther.* **2009**, *26*, 613–626. [CrossRef]
31. Der Lek, R.F.R.-V.; Swaab, D.F.; Twisk, J.; Hol, E.; Hoogendijk, W.J.; Van Someren, E.J. Effect of Bright Light and Melatonin on Cognitive and Noncognitive Function in Elderly Residents of Group Care Facilities. *JAMA* **2008**, *299*, 2642. [CrossRef]
32. Khan, Z.; Labala, R.K.; Yumnamcha, T.; Devi, S.D.; Mondal, G.; Devi, H.S.; Rajiv, C.; Bharali, R.; Chattoraj, A. Artificial Light at Night (ALAN), an alarm to ovarian physiology: A study of possible chronodisruption on zebrafish (Danio rerio). *Sci. Total Environ.* **2018**, *628–629*, 1407–1421. [CrossRef]
33. Koo, Y.S.; Song, J.-Y.; Joo, E.Y.; Lee, H.-J.; Lee, E.; Lee, S.K.; Jung, K.-Y. Outdoor artificial light at night, obesity, and sleep health: Cross-sectional analysis in the KoGES study. *Chrono Int.* **2016**, *33*, 301–314. [CrossRef]
34. Martín, C.S.; Sánchez-Muniz, F.J. Chronodisruption and cortisol and melatonin imbalance, a probable prelude of most prevalent pathologies? *J. Negat. No Posit. Results* **2017**, *2*, 619–633. [CrossRef]
35. Abbas, K. Handbook of life stress, cognition and health. *Behav. Res. Ther.* **1990**, *28*, 104. [CrossRef]
36. Mirick, D.K.; Davis, S. Melatonin as a Biomarker of Circadian Dysregulation. *Cancer Epidemiol. Biomark. Prev.* **2008**, *17*, 3306–3313. [CrossRef] [PubMed]
37. Figueiro, M.; Steverson, B.; Heerwagen, J.; Yucel, R.; Roohan, C.; Sahin, L.; Kampschroer, K.; Rea, M. Light, entrainment and alertness: A case study in offices. *Light. Res. Technol.* **2019**. [CrossRef]

38. Correa, Á.; Barba, A.; Padilla, F. Light Effects on Behavioural Performance Depend on the Individual State of Vigilance. *PLoS ONE* **2016**, *11*, e0164945. [CrossRef] [PubMed]

39. Smolders, K.C.; De Kort, Y. Bright light and mental fatigue: Effects on alertness, vitality, performance and physiological arousal. *J. Environ. Psychol.* **2014**, *39*, 77–91. [CrossRef]

40. Chang, A.-M.; Scheer, F.A.J.L.; Czeisler, C.A.; Aeschbach, D. Direct Effects of Light on Alertness, Vigilance, and the Waking Electroencephalogram in Humans Depend on Prior Light History. *Sleep* **2013**, *36*, 1239–1246. [CrossRef]

41. Wright, K.P.; McHill, A.W.; Birks, B.R.; Griffin, B.R.; Rusterholz, T.; Chinoy, E.D. Entrainment of the human circadian clock to the natural light-dark cycle. *Curr. Boil.* **2013**, *23*, 1554–1558. [CrossRef]

42. Stevens, R.G. ELECTRIC POWER USE AND BREAST CANCER: A HYPOTHESIS. *Am. J. Epidemiol.* **1987**, *125*, 556–561. [CrossRef]

43. Obayashi, K.; Saeki, K.; Iwamoto, J.; Ikada, Y.; Kurumatani, N. Exposure to light at night and risk of depression in the elderly. *J. Affect. Disord.* **2013**, *151*, 331–336. [CrossRef]

44. McFadden, E.; Jones, M.E.; Schoemaker, M.J.; Ashworth, A.; Swerdlow, A.J. The Relationship Between Obesity and Exposure to Light at Night: Cross-Sectional Analyses of Over 100,000 Women in the Breakthrough Generations Study. *Am. J. Epidemiol.* **2014**, *180*, 245–250. [CrossRef]

45. Ritonja, J.; McIsaac, M.A.; Sanders, E.; Kyba, C.C.M.; Grundy, A.; Cordina-Duverger, E.; Spinelli, J.J.; Aronson, K.J. Outdoor light at night at residences and breast cancer risk in Canada. *Eur. J. Epidemiol.* **2020**, 1–11. [CrossRef]

46. Konis, K.; Mack, W.J.; Schneider, E.L. Pilot study to examine the effects of indoor daylight exposure on depression and other neuropsychiatric symptoms in people living with dementia in long-term care communities. *Clin. Interv. Aging* **2018**, *13*, 1071–1077. [CrossRef] [PubMed]

47. Lai, E.; Levine, B.; Ciralsky, J. Ultraviolet-blocking intraocular lenses. *Curr. Opin. Ophthalmol.* **2014**, *25*, 35–39. [CrossRef] [PubMed]

48. Meng, Z.-J.; Chen, X.; Zhang, J.; Li, Y.; Wang, W. Influence of 460-480 nm wavelength light at three different irradiance on retina tissue of SD rats. *Chin. J. Ophthalmol.* **2013**, *49*, 438–446.

49. Roehlecke, C.; Schümann, U.; Ader, M.; Brunssen, C.; Bramke, S.; Morawietz, H.; Funk, R.H.W. Stress Reaction in Outer Segments of Photoreceptors after Blue Light Irradiation. *PLoS ONE* **2013**, *8*, e71570. [CrossRef]

50. Tokarz, P.; Kaarniranta, K.; Blasiak, J. Role of antioxidant enzymes and small molecular weight antioxidants in the pathogenesis of age-related macular degeneration (AMD). *Biogerontology* **2013**, *14*, 461–482. [CrossRef]

51. Nakanishi-Ueda, T.; Majima, H.J.; Watanabe, K.; Ueda, T.; Indo, H.P.; Suenaga, S.; Hisamitsu, T.; Ozawa, T.; Yasuhara, H.; Koide, R. Blue LED light exposure develops intracellular reactive oxygen species, lipid peroxidation, and subsequent cellular injuries in cultured bovine retinal pigment epithelial cells. *Free Radic. Res.* **2013**, *47*, 774–780. [CrossRef]

52. Chamorro, E.; De Luna, J.M.; Vázquez, D.; Bonnin-Arias, C.; Pérez-Carrasco, M.J.; Sánchez-Ramos, C. Effects of Light-emitting Diode Radiations on Human Retinal Pigment Epithelial CellsIn Vitro. *Photochem. Photobiol.* **2012**, *89*, 468–473. [CrossRef]

53. Simsek, H.; Hassoy, H.; Oztoprak, D.; Yilmaz, T. Medical students' risk perceptions on decreased attention, physical and social risks in using mobile phones and the factors related with their risk perceptions. *Int. J. Environ. Health Res.* **2019**, *29*, 255–265. [CrossRef]

54. Sola, Y.; Lorente, J. Contribution of UVA irradiance to the erythema and photoaging effects in solar and sunbed exposures. *J. Photochem. Photobiol. B Boil.* **2015**, *143*, 5–11. [CrossRef]

55. Escofet, J.; Bará, S. Reducing the circadian input from self-luminous devices using hardware filters and software applications. *Light. Res. Technol.* **2015**, *49*, 481–496. [CrossRef]

56. Lopez-Fernandez, O.; Honrubia-Serrano, L.; Freixa-Blanxart, M.; Gibson, W. Prevalence of Problematic Mobile Phone Use in British Adolescents. *Cyberpsychol. Behav. Soc. Netw.* **2014**, *17*, 91–98. [CrossRef] [PubMed]

57. King, A.L.S.; Valença, A.M.; Silva, A.C.; Sancassiani, F.; Machado, S.; Nardi, A.E. "Nomophobia": Impact of Cell Phone Use Interfering with Symptoms and Emotions of Individuals with Panic Disorder Compared with a Control Group. *Clin. Pract. Epidemiology Ment. Health* **2014**, *10*, 28–35. [CrossRef] [PubMed]

58. Bragazzi, N.L.; Del Puente, G. A proposal for including nomophobia in the new DSM-V. *Psychol. Res. Behav. Manag.* **2014**, *7*, 155–160. [CrossRef] [PubMed]

59. Pickard, G.E.; Sollars, P.J. Intrinsically Photosensitive Retinal Ganglion Cells. *Rev. Physiol. Biochem. Pharmacol.* **2011**, *162*, 59–90. [CrossRef]

60. Nikhita, C.S.; Jadhav, P.R.; Ajinkya, S. Prevalence of Mobile Phone Dependence in Secondary School Adolescents. *J. Clin. Diagn. Res.* **2015**, *9*, VC06–VC09. [CrossRef]

61. LeBourgeois, M.K.; Hale, L.; Chang, A.-M.; Akacem, L.; Montgomery-Downs, H.E.; Buxton, O.M. Digital Media and Sleep in Childhood and Adolescence. *Pediatrics* **2017**, *140*, S92–S96. [CrossRef]

62. Crowley, S.J.; Cain, S.W.; Burns, A.C.; Acebo, C.; Carskadon, M.A. Increased Sensitivity of the Circadian System to Light in Early/Mid-Puberty. *J. Clin. Endocrinol. Metab.* **2015**, *100*, 4067–4073. [CrossRef]

63. Turner, P.L.; A Mainster, M. Circadian photoreception: Ageing and the eye's important role in systemic health. *Br. J. Ophthalmol.* **2008**, *92*, 1439–1444. [CrossRef]

64. Dube, N.; Khan, K.; Loehr, S.; Chu, Y.; Veugelers, P. The use of entertainment and communication technologies before sleep could affect sleep and weight status: A population-based study among children. *Int. J. Behav. Nutr. Phys. Act.* **2017**, *14*. [CrossRef]

65. Zheng, F.; Gao, P.; He, M.; Li, M.; Tan, J.; Chen, D.; Zhou, Z.; Yu, Z.; Zhang, L. Association between mobile phone use and self-reported well-being in children: A questionnaire-based cross-sectional study in Chongqing, China. *BMJ Open* **2015**, *5*, e007302. [CrossRef]

66. Yoshimura, M.; Kitazawa, M.; Maeda, Y.; Mimura, M.; Tsubota, K.; Kishimoto, T. Smartphone viewing distance and sleep: An experimental study utilizing motion capture technology. *Nat. Sci. Sleep* **2017**, *9*, 59–65. [CrossRef] [PubMed]

67. Bae, S.M. Smartphone Addiction of Adolescents, Not a Smart Choice. *J. Korean Med. Sci.* **2017**, *32*, 1563–1564. [CrossRef] [PubMed]

68. Lin, Y.-H.; Chiang, C.-L.; Lin, P.-H.; Chang, L.-R.; Ko, C.-H.; Lee, Y.-H.; Lin, S.-H. Proposed Diagnostic Criteria for Smartphone Addiction. *PLoS ONE* **2016**, *11*, e0163010. [CrossRef] [PubMed]

69. Elahi, H.; Wang, G.; Li, X. Smartphone Bloatware: An Overlooked Privacy Problem. In *Computer Vision*; Springer Science and Business Media LLC: Berlin, Germany, 2017; Volume 10656, pp. 169–185.

70. Long, J.; Liu, T.; Liao, Y.; Qi, C.; He, H.; Chen, S.-B.; Billieux, J. Prevalence and correlates of problematic smartphone use in a large random sample of Chinese undergraduates. *BMC Psychiatry* **2016**, *16*, 408. [CrossRef] [PubMed]

71. INE. Encuesta sobre Equipamiento y Uso de Tecnologías de Información y Comunicación en los Hogares. Año 2017. Instituto Nacional de Estadística. Available online: https://www.ine.esprensatich_2017.pdf (accessed on 30 April 2019).

72. Moreno, M.P.; Labarca, C.Y. Methodology for Assessing Daylighting Design Strategies in Classroom with a Climate-Based Method. *Sustainability* **2015**, *7*, 880–897. [CrossRef]

73. Reinhart, C.F.; Mardaljevic, J.; Rogers, Z. Dynamic Daylight Performance Metrics for Sustainable Building Design. *LEUKOS* **2006**, *3*, 7–31. [CrossRef]

74. Nabil, A.; Mardaljevic, J. Useful daylight illuminances: A replacement for daylight factors. *Energy Build.* **2006**, *38*, 905–913. [CrossRef]

75. Mardaljevic, J.; Andersen, M.; Roy, N.; Christoffersen, J. *Daylighting Metrics: Is There a Relation Between Useful Daylight Illuminance and Daylight Glare Probability?* Ibpsa-England Bso12: Loughborough, UK, 2012; pp. 189–196.

76. Al-Khatatbeh, B.J.; Ma'Bdeh, S.N. Improving visual comfort and energy efficiency in existing classrooms using passive daylighting techniques. *Energy Procedia* **2017**, *136*, 102–108. [CrossRef]

77. Yener, A.K. Daylight Analysis in Classrooms with Solar Control. *Architect. Sci. Rev.* **2002**, *45*, 311–316. [CrossRef]

78. Krüger, E.L.; Dorigo, A.L. Daylighting analysis in a public school in Curitiba, Brazil. *Renew. Energy* **2008**, *33*, 1695–1702. [CrossRef]

79. Bullen, P.A. Adaptive reuse and sustainability of commercial buildings. *Facilities* **2007**, *25*, 20–31. [CrossRef]

80. Maria, Y.S.; Prihatmanti, R. Daylight Characterisation of Classrooms in Heritage School Buildings. *Plan. Malays. J.* **2017**, *15*, 209–220. [CrossRef]

81. Winterbottom, M.; Wilkins, A. Lighting and discomfort in the classroom. *J. Environ. Psychol.* **2009**, *29*, 63–75. [CrossRef]

82. CIBSE GUIDE F. Energy efficiency in buildings. 2004. Available online: https://epdf.pub/energy-efficiency-in-buildings-cibse-guide-f.html (accessed on 15 January 2020).

83. Schreiber, T.; Schmitz, A. Improved Surrogate Data for Nonlinearity Tests. *Phys. Rev. Lett.* **1996**, *77*, 635–638. [CrossRef] [PubMed]
84. Shapiro, M.; Roth, D.; Marcus, A. The effect of Lighting on the Behavior of Children Who Are Developmentally Disabled. *J. Int. Spec. Needs Educ.* **2001**, *4*, 19–23. Available online: http://login.ezproxy.library.ualberta.ca/login?url=http://search.ebscohost.com/login.aspx?direct=true&db=eric&AN=EJ657365&site=eds-live&scope=site (accessed on 3 March 2020).
85. Rittner, H.; Robbin, M. Color and light in learning. *Sch. Plan. Manag.* **2002**, *41*, 57–58.
86. Mulligan, H. Energy Efficiency in Commercial Buildings. *Facilities* **1993**, *11*, 18–21. [CrossRef]
87. Ho, M.-C.; Chiang, C.-M.; Chou, P.-C.; Chang, K.-F.; Lee, C.-Y. Optimal sun-shading design for enhanced daylight illumination of subtropical classrooms. *Energy Build.* **2008**, *40*, 1844–1855. [CrossRef]
88. Guan, Y.; Yan, Y. Daylighting Design in Classroom Based on Yearly-Graphic Analysis. *Sustainability* **2016**, *8*, 604. [CrossRef]
89. Ministry of Education of the People's Republic of China. *Reports on National Students' Constitution and Health of People's Republic of China in 2010*; Ministry of Education of the People's Republic of China: Beijing, China, 2011.
90. Ministry of Education of the People's Republic of China. *Reports on National Students' Constitution and Health of People's Republic of China in 2005*; Ministry of Education of the People's Republic of China: Beijing, China, 2006.
91. Ministry of Education of the People's Republic of China. *Reports on National Students' Constitution and Health of People's Republic of China in 2000*; Ministry of Education of the People's Republic of China: Beijing, China, 2001.
92. Wang, Y.; Ding, H.; Stell, W.K.; Liu, L.; Li, S.; Liu, H.; Zhong, X.-W. Exposure to Sunlight Reduces the Risk of Myopia in Rhesus Monkeys. *PLoS ONE* **2015**, *10*, e0127863. [CrossRef]
93. Hua, W.-J.; Jin, J.-X.; Wu, X.; Yang, J.-W.; Jiang, X.; Gao, G.-P.; Tao, F. Elevated light levels in schools have a protective effect on myopia. *Ophthalmic Physiol. Opt.* **2015**, *35*, 252–262. [CrossRef] [PubMed]
94. Edwards, L.; Torcellini, P. Literature Review of the Effects of Natural Light on Building Occupants. *Lit. Rev. Eff. Nat. Light Build. Occup.* **2002**, *55*. [CrossRef]
95. Cheryan, S.; Ziegler, S.A.; Plaut, V.C.; Meltzoff, A.N. Designing Classrooms to Maximize Student Achievement. *Policy Insights Behav. Brain Sci.* **2014**, *1*, 4–12. [CrossRef]
96. Heschong, L. *Daylighting in Schools: An Investigation into the Relationship between Daylighting and Human Performance. Detailed Report*; HMG-R-9803; Pacific Gas and Electric Company: San Francisco, CA, USA, 1999; Available online: http://eric.ed.gov/?id=ED444337 (accessed on 10 March 2020).
97. Alexander, D.; Lewis, L.R. *Condition of America's Public School Facilities: 2012–13. First Look*; NCES: Washington, D.C., USA, 2014; pp. 2014–2022.
98. Choi, S.; Guerin, D.; Kim, H.-Y.; Brigham, J.K.; Bauer, T. Indoor Environmental Quality of Classrooms and Student Outcomes: A Path Analysis Approach. *J. Learn. Spaces* **2014**, *2*, 1–14.
99. Kelting, S.; Montoya, M. Green Building Policy, School Performance, and Educational Leaders' Perspectives in USA. In Proceedings of the Construction Challenges in the New Decade, Kuala Lumpur, Malaysia, 5–7 July 2011.
100. Kelting, S.; Montoya, M. Green Building Policy and School Performance. *ICSDC 2011* **2012**, 112–118. [CrossRef]
101. Lopez-Chao, V.; Lorenzo, A.A.; Saorin, J.L.; De La Torre-Cantero, J.; Melián-Díaz, D. Classroom Indoor Environment Assessment through Architectural Analysis for the Design of Efficient Schools. *Sustainability* **2020**, *12*, 2020. [CrossRef]
102. Kuller, R.; Lindsten, C. Health and behavior of children in classrooms with and without windows. *J. Environ. Psychol.* **1992**, *12*, 305–317. [CrossRef]
103. Wilkins, A. Coloured overlays and their effects on reading speed: A review. *Ophthalmic Physiol. Opt.* **2002**, *22*, 448–454. [CrossRef]
104. Boyce, P. Human Factors in Lighting, Second Edition. In *Human Factors in Lighting*, 2nd ed.; Informa UK Limited: Colchester, UK, 2003.
105. Dudek, M. *Schools and Kindergartens*; Walter de Gruyter GmbH: Berlin, Germany, 2014.
106. Mark, D. *A Design Manual Schools and Kindergartens*; Birkhäuser: Basel, Switzerland, 2007.

107. SCENIHR. *Health Effects of Artificial Light*; Scientific Committe on Emerging and Newly Identified Health: Bruselas, Belgium, 2012.
108. Lucas, R.J.; Peirson, S.N.; Berson, D.M.; Brown, T.M.; Cooper, H.M.; Czeisler, C.A.; Figueiro, M.G.; Gamlin, P.D.; Lockley, S.W.; O'Hagan, J.B.; et al. Measuring and using light in the melanopsin age. *Trends Neurosci.* **2013**, *37*, 1–9. [CrossRef]
109. CIE. Iluminemos el futuro. Acelerando el despliegue de tecnologías de iluminación innovadoras. In *LIBRO VERDE*; CIE: Bruselas, Belgium, 2013.
110. CIE. *Research Roadmap for Healthful Interior Lighting Applications*; International Commission Illumination: Viena, Austria, 2016.
111. CIE. *Stakeholder Workshop for Temporal Light Modulation Standars for Lghting Systems*; International Commission Illumination: Viena, Austria, 2017.
112. CIE. *System for Metrology of Optical Radiation for Light Responses Influenced by Intrinsically-Photosensitive Retinal Ganglion Cells*; International Commission Illumination: Viena, Austria, 2018.
113. Bullough, J.D.; Bierman, A.; Rea, M.S. Evaluating the blue-light hazard from solid state lighting. *Int. J. Occup. Saf. Ergon.* **2017**, *25*, 311–320. [CrossRef]
114. Dharani, R.; Lee, C.-F.; Theng, Z.X.; Drury, V.B.; Ngo, C.; Sandar, M.; Wong, T.-Y.; A Finkelstein, E.; Saw, S.-M. Comparison of measurements of time outdoors and light levels as risk factors for myopia in young Singapore children. *Eye* **2012**, *26*, 911–918. [CrossRef] [PubMed]
115. Figueiro, M.; Overington, D. Self-luminous devices and melatonin suppression in adolescents. *Light. Res. Technol.* **2016**, *48*, 966–975. [CrossRef]
116. Figueiro, M.G.; Wood, B.; Plitnick, B.; Rea, M.S. The impact of light from computer monitors on melatonin levels in college students. *Neuroendocrinol. Lett.* **2011**, *32*, 158–163. [PubMed]

Review

Requirements for the Construction of New Desalination Plants into a Framework of Sustainability

Francisco Berenguel-Felices *, Antonio Lara-Galera * and María Belén Muñoz-Medina

Escuela Técnica Superior de Ingenieros de Caminos, Canales y Puertos, Universidad Politécnica de Madrid, 28040 Madrid, Spain; mariabelen.munoz@upm.es
* Correspondence: franciscoberenguel@ciccp.es (F.B.-F.); alargal@ciccp.es (A.L.G.)

Received: 4 May 2020; Accepted: 18 June 2020; Published: 23 June 2020

Abstract: Population growth has increased in the last two centuries. In the driest countries, water supply alternatives are scarce, and desalination is an alternative to guarantee water supply. The question is what conditions must be met by the new desalination plants to achieve the objectives of sustainability. The present study is an analysis of the social, economic, and environmental variables that are critical in the development of desalination plants: technology used, energy sources, correction of the generated environmental impacts, and the most appropriate contractual model for its development. These attributes justify at the time of writing why reverse osmosis is the safest and most efficient technology among those available and those that are under investigation. It is proposed to incorporate renewable energy production sources, although it is still necessary to continue depending on the significant contribution of the traditional energy sources. The need will also be demonstrated to adopt corrective measures to mitigate against the impact produced on the environment by energy production and to implement monitoring plans to confirm the validity of these corrective measures. Finally, turnkey contracts are proposed because osmosis technology is complex, although technology should be justified by means of a decision support system. One of the determining factors is proposed in this present analysis.

Keywords: sustainable development; desalination; reverse osmosis; renewable energies; environmental impacts; decision support systems; types of contract

---

1. Introduction

"Sustainable development is the one which satisfies the present needs without threatening the capacity of future generations to meet their own needs". This definition by Brundtland [1] clearly shows the idea of limits imposed on technology and "social organization due to the capacity of the environment to satisfy present and future needs", as Jonker and Harmsen mention [2].

Population growth has increased dramatically throughout the world in the last two centuries. Global population is expected to go from the present 7.7 billion inhabitants to some 10 billion people by the year 2050. One of the biggest challenges to face in the near future is how to guarantee drinking water supply all over the world. This is an important challenge for humanity, since drinking water availability will have to increase significantly.

Apart from this important population growth, another effect that has been observed in the past few decades is the rural exodus to the big cities. This migration is causing demographic imbalance, depopulating certain areas and causing big population increases in others. In the driest countries, the most densely populated areas are actually on coastal regions due to very well-known social and economic reasons.

Siegel [3] says that according to the American government, 40 out of 50 states and in 60% of the USA surface, there will shortly be an alarming difference between the water available and the increasing demand for this resource. He also explains how Israel, with 60% desert land, can be an example for all other countries, not only because of solving their water problem, but also for having sufficient to provide Palestinians and Jordanians with the resource. Even Iran depends on a similar water system and China knows enough about the Israeli water system to be able to manage its own water needs.

The situation in Cape Town, South Africa, as described by Bates Ramírez [4], should serve as an example, since a critical water situation arose when four million inhabitants ran out of water supply. This was the first time that such a large number of people were ever put at such a serious risk of lack of water. The dangerous situation in South Africa was a wakeup call for other cities in similar circumstances, such as Mexico City, Sao Paulo, and Cairo, who all face water shortages.

As the global population grows and climate change increases temperatures, water will become even more scarce.

Oceans have roughly 97.5% of the planet's water, and the 2.5% remaining water deposit is divided up in glaciers, ice, phreatic layers, rivers, lakes, and atmosphere.

According to the Madrid Complutense University [5], polar ice caps and glaciers hold 69.3% of fresh water, groundwater, 30.3%, lakes, 0.26% and rivers, just 0.006%. The remaining fresh water is found in living beings (0.003%) on the planet, including the atmosphere. Therefore, of the total freshwater reserve on the planet, we only have a volume of 127,679,000 cubic hectometers in rivers and lakes.

In AQUASTAT [6], three types of water withdrawal are distinguished: agricultural (including irrigation, livestock, and aquaculture), municipal (including domestic), and industrial water withdrawal. A fourth type of anthropogenic water use is the water that evaporates from artificial lakes or reservoirs associated with dams. It is worth highlighting the consensus on the use of water by humans, allocating 12% for domestic use, 19% for industry, and 69% for agricultural use. These numbers, however, are strongly biased by a few countries which have very high-water withdrawals. Table 1 shows the water withdrawal ratios by continent.

Table 1. Water withdrawal percentages by continent.

	Agricultural	Municipal	Industrial
World	69	19	12
Europe	21	57	22
Americas	51	34	15
Oceania	60	15	25
Asia	81	10	9
Africa	82	5	13

Source: AQUASTAT.

The above shows not only the scarcity of the water resources and the difficulty to guarantee water supply, but also that there are not many water supply alternatives. Water desalination is particularly suitable for cities situated near the coast or with brackish water.

According to data published by the International Desalination Association (IDA) and Global Water Intelligence (GWI) in the Water Security Handbook 2019–2020 [7], over 17,000 desalination plants have been contracted, reaching a total of 107 Hm^3/day of cumulative installed desalination capacity in 2019. Desalination is operational in 174 countries. There are more than 300 million people around the world who rely on desalinated water for some or all their daily needs, with 146 Hm^3/day of cumulative installed reuse capacity in 2019.

The various desalination technologies are differentiated by cost, product quality, and energy consumed. Most plants desalinate through a thermal process or using membranes.

Thermal desalination methods use heat to evaporate saltwater and they condense it again, now without salt. They basically imitate the natural water cycle of evaporation and rainfall.

Urrutia [8] explained that thermal processes that have been used since the appearance of desalination in the 1950s; they are mainly used in oil exporting countries today.

Torres Corral [9] pointed out that from these beginnings until the 1980s, desalination was mainly done by distillation processes, building dual water and electric power plants, as long as the market for the power plant was viable.

In the following section we will show how the cost of the selected desalination technology is the most significant cost in the final price of desalinated water. Torres justified how the various oil crises have significantly affected applied technology. The first crisis in 1973 led to optimizing the process. The second crisis in 1979 caused a shift to the use of the reverse osmosis process, by developing membranes which removed over 99% of salt, with mechanical resistance capable of withstanding 70 Kg/cm^2 to overcome the osmotic pressure.

Reverse osmosis is based on the natural osmosis which occurs in cell membranes in living organisms, in which water diffusion moves from an area with low concentration of solutes to another with a higher concentration. The system used to desalinate is the opposite (hence the term "reverse")—the saltwater propelled to break the osmotic pressure goes through a semi-permeable membrane, which retains water with higher saline concentration (brine) and allows water for human consumption to pass.

According to Stover [10], in the 1990s and 2000s, the innovation in the desalination industry focused on reducing energy consumption, improving the performance and reliability of the reverse osmosis membranes and the innovation of energy recovery devices. It is also worth highlighting the improvement in the processes, such as the use of a second layer of reverse osmosis for the retained water on the first stage (brine), increasing fresh water compared to raw water and decreasing residual brine. With these measures, not only did the energy used for desalination fall by half, but it was also possible to build fairly big reverse osmosis desalination plants.

Scott [11] showed that the fact that desalination costs have decreased in recent years is due to the progressive incorporation of membrane processes by those countries where energy is expensive, thus being able to replace thermal processes.

However, even in these oil exporting countries, the tendency is changing due to oil prices, as Ibáñez Mengual [12] stated, since the evaporating processes are associated with a thermal power plant. When there is an imbalance between supply and demand for electric energy, this is reflected in desalinated water production decreasing. This explains why at the time of writing desalination projects in the Middle East used 50% evaporation technology and 50% reverse osmosis technology, with a tendency to increase this latter technology.

Another relevant question to consider is the cost. There are several factors that contribute to costs: the type of technology used, the type of water to desalinate, the quality of the water that is demanded, the cost of energy, etc. Usually, the cost is divided into three blocks: investment costs, fixed operating costs and variable operating costs.

Voutchkov [13] estimated a cost share for the membrane desalination technology: approximately 35% costs are for energy; the recovery part roughly 30%, then personnel (5%), taxes (8.5%), and industrial profit (6%). The remaining percentage to reach 100% would be made up by membrane replacement, chemical products, maintenance, and other costs. It is clear from these figures how important the cost of energy is. Voutchkov reported the worldwide evolution of energy consumption, which has gone from 22 kWh/m^3 in the early 1970s to a consumption of the order of 2.8 kWh/m^3 (pure desalination) nowadays. It was during the 1990s when the greatest technological advances occurred. The impact of these advances can be seen on the production price of the cubic meter of desalinated water, which in the 1980s was around more than \$2/m^3, and currently it is hovering around \$0.60/m^3 of desalinated water.

However, desalination is not free from controversy and criticism related to the impact in the environment caused by the desalination plants. Latteman and Höpner [14] warned that the Persian Gulf has always had intense desalination activity, but that other regional centers were prominently emerging, such as the Mediterranean Sea, the Red Sea, the coastal waters of California, China, and Australia. Kämpf and Clarke [15] claimed in 2012 that the brine discharge was non-compliant with the Environmental Impact Assessment (EIA). The monitoring process in South Australia was flawed, and in 2015 the current license was modified based on the results of an independent review of the monitoring performed for the desalination plant operations. The Environment Protection Authority (EPA) has set strict compliance limits and monitoring requirements for the environmental license for the plant. Fuentes-Bargues [16] published a study on the environmental impact assessment process in Spain for seawater desalination projects with 12 years' worth of data, identifying brine discharge as the main impact. However, Shemer and Semiat [17] defended the use of reverse osmosis desalination and claimed that brine discharge has minimal impacts. Recently, Saracco [18] warned of the risk of brine contamination and pointed out how substantial damage to the marine ecosystem can be observed in the Persian Gulf area, that requires corrective action. Most of the cited authors agree that the main effect of seawater desalination plants is the discharge of the brine and its impact on the marine environment. They also agree that the solution is for the environmental authorities of each country to implement environmental impact studies that establish strict compliance limits and monitoring requirements to verify that the measures adopted are adequate.

The desalination process has required technology development these past decades at all the stages of the technology to reduce cost and negative impact and still meet the needs of society. This is more pressing where there are not enough freshwater resources to supplement quality desalinated water.

Finally, it is worth mentioning that to achieve more efficiency in the process, more complex and expensive technologies have been developed, resulting in increasing the size of the desalination plants to decrease operation and maintenance costs. Thus, in 2018, each of the ten biggest desalination plants in the world produced more than half a million cubic meters a day, with the largest of them reaching a million cubic meters daily. The following factors have to be taken into account before justifying the decision to implement a project: technical complexity, efficiency, size, and cost, forcing a search for the best-suited method of procurement to develop the project, and achieving its objectives.

The objective of the present paper is to study the requirements that new desalination plants need to meet to be compatible with sustainability requirements. Four areas are considered:

- The desalination technology, the necessary energy, and its production;
- The environmental impact and measures to neutralize it;
- The management of the construction and smooth running of the plants.

2. Desalination Technology

The main desalination technologies are divided into evaporation (distillation) and membranes. To evaporate, heat and electricity are necessary, while membranes only need electrical energy and have a considerably low consumption.

Torres [9] explained that in the distillation processes there are several systems depending on the use of the condensation heat of steam.

The most relevant are:

- Multi-stage evaporation or multi-stage flash distillation, known by its acronym MSF;
- Multi-effect evaporation (MED);
- Mechanical Vapor Compression (MVC).

A multiple-effect evaporator is an apparatus for efficiently using the heat from steam to evaporate water. In a multiple-effect evaporator, water is boiled in a sequence of vessels, each held at a lower pressure than the previous one. MSF is a water desalination process that distills sea water by flashing a portion of the water into steam in multiple stages that are essentially countercurrent heat exchangers.

The MED is similar to the previous process but operating at a lower pressure. In the case of multiple-effect evaporation plants, the exhaust vapors from the product are used to heat the downstream-arranged evaporation effect so that the steam consumption is reduced accordingly.

MVC refers to a distillation process where the evaporation of sea or saline water is obtained by the application of heat delivered by compressed vapor. This system is the most thermodynamically efficient process of single-purpose thermal desalination plants.

Reverse osmosis (RO) is a water purification process that uses a partially permeable membrane to remove ions, unwanted molecules, and larger particles from drinking water. In reverse osmosis, an applied pressure is used to overcome osmotic pressure. A part of the inlet water is desalinated, producing a certain amount of water with a high concentration of salt called brine.

Voutchkov [13] estimated a cost share, that approximately 35% goes to energy. Table 2 shows the energy consumption of the desalination technologies described.

Table 2. Desalination technologies' power requirements. Own elaboration.

Desalination Technology	Energy Requirement (kWh/m^3)
MSF	21–58
MED	15–58
MVC	7–12
RO	3–5

Desalination by nanofiltration has a similar principle to the one used for reverse osmosis. The main difference with the latter is the characteristics of the semipermeable membranes used in this technique, which offer a higher percentage of rejection of some ions from salts, which can operate at lower pressures (Parlar et al. [19]). Nanofiltration is a process that has been used in recent years but basically limited to some stages of the drinking water purification, such as softening, discoloration, and elimination of micro contaminants.

Desalination by electrodialysis consists of the passage of ions under the effect of a continuous electric current through a series of selective cationic and anionic permeable membranes, which allows the electrochemical separation of ions. The membranes, separated from each other by a few millimeters, are placed between two electrodes so that the incoming water circulates. The membranes let the ions in, by being transferred through them from a low concentrate to a higher concentrate (Lee and Kang, [20]).

Electrodialysis has proved very viable, especially in brackish water desalination, in effluent treatment, and in industrial processes. It is suitable for connecting directly to photovoltaic panels, taking advantage of the use of solar energy, and it is particularly recommended in areas with isolated saline aquifers where the connection to the electrical network is difficult and expensive.

Asociación Española de Desalación y Reutilización, AEDYR [21] states that nowadays the most globally used technique to desalinate water is reverse osmosis, which reaches almost 70% of the total available technologies; followed by MSF (18%), MED (7%), nanofiltration (3%), and finally electrodialysis (2%).

The question is whether there is room for improvement in desalination technology, although advances in this field will undoubtedly continue to occur. Inside the reverse osmosis system, the key component is the membranes. The ones that are used at present are the result of more than 50 years of research in polymers. In the USA, MIT researchers are experimenting with graphene membranes, which require less pressure and therefore, less energy. Other researchers have studied the use of carbon nanotube membranes. Unfortunately, these technologies have not yet been developed for industrial use.

Jeff Urban [22] from Berkeley Laboratory described the line of open investigation to develop desalination by direct osmosis through a highly concentrated extraction solution to extract water from sea water. In a Berkeley Laboratory, they are developing extraction solutions, in gel form, which would extract water effectively and would then separate spontaneously from this water thanks to the application of low amounts of heat. This line is still in the research stage and the first steps would be taken by giving direct osmosis a complementary role in the brackish water treatment.

Another open line of investigation is desalination by solar energy. Many areas with water scarcity usually have a decent insolation level that can be used as solar energy. De Luis López and Gómez Benítez [23] mentioned small installations (up to 15 m^3/day) of solar stills to provide drinking water in Greece. The Freeport Plant, in the Gulf of Mexico, is more important, with a multiple stage system (LTV, Long Tube Vertical Multiple Effect Distillation), which guarantees a relatively good output through a progressive evaporation process at a constant decreasing pressure, producing 4000 m^3/day of desalinated water. It is a small installation if we compare it with the big desalination plants that have been built in recent years. Some current investigations also revealed a model which has a manifold, an evaporation tower, and a condensation tower, but with no conclusive results yet.

Subramani and Jacangelo [24] made a critical review of new emerging desalination technologies, considering membranes that incorporate nanoparticles, carbon, or graphene nanotubes; they also analyzed alternative technologies like the ones based on the deionization and on microbial desalination cells. From all these options only nanocomposite membranes have been commercialized.

Estevan and García [25] stated that in Spain, desalination has evolved very positively since the first facilities were launched in the early 1970s, that were designed by thermal type processes (MSF, MED and MVC). These facilities were large energy users with specific consumption which could exceed 30–40 kilowatts/hour per cubic meter of desalinated water. In the 1980s the first reverse osmosis installations were introduced and coexisted with the evaporation technologies, mainly MVC, and with important energy reduction consumption: 15 kWh/m^3 for vapor compression plants and 8–10 kWh/m^3 for those of reverse osmosis. The evolution of specific consumption in the field of desalination by reverse osmosis, through successive technological innovations in energy recovery systems, reduced to 3 kWh/m^3, contributing very significantly to the huge increase of production capacity. The graph of the evolution of the installed capacity/specific consumption ratio in Spain done by Centro de Estudios y Experimentación de Obras Públicas (CEDEX) is attached below in Figure 1.

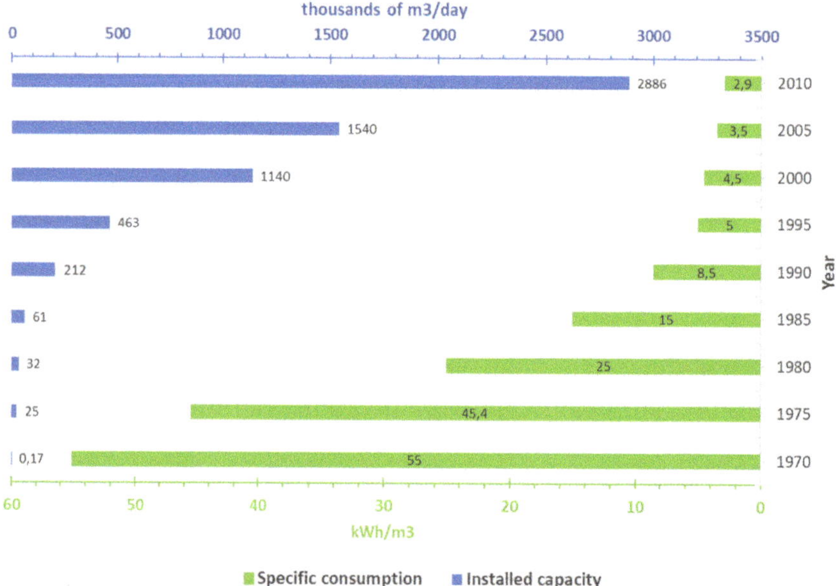

Figure 1. Evolution of energy consumption to desalinate in Spain 1970–2010. Source: CEDEX.

Today, reverse osmosis processes have been achieved under 2.7 kWh/m^3, by slowly reducing the consumption through energy recovery systems and by achieving a higher efficiency in the membranes, which are the key elements in the process.

Jia et al [26] analyzed the energy consumption, greenhouse gas (GHG) emissions, and cost of seawater desalination in China. The energy consumption and GHG increased from 81 MWh to 1561 MWh from 2006 to 2016. The unit product cost (UPC) of seawater desalination is shown in the Table 3. They concluded that there was potential for energy consumption, GHG emission, and cost reduction with the application of energy recovery units, the integration of desalination plants and renewable energies or low potential heat, as well as the development of new technologies.

Table 3. Unit product cost of seawater desalination technologies.

Desalination Technology	UPC (USD in 2016)
RO	0.8 to 1.3
MED	2.0
MSF	3.6
ED	3.0

Source: Jia et al.

However, the energy cost is still the most significant in large industrial desalination plants, as well as the consequences regarding the sustainability of generating the necessary energy.

Improvement is still possible, mainly in three aspects:

- Design;
- Equipment and materials, especially highlighting membranes and pumps;
- Energy recovery systems.

The next significant step would be to lower the working pressure, which would reduce energy consumption [20]. Nevertheless, this cost should be put in the context of what it really involves, by comparing it to other activities or development, which are either not considered or are even positive measures, without analyzing them altogether. This is like promoting the electric car without considering the origin of the energy necessary to charge those vehicles.

The Water Corporation [27], regarding energy consumption, estimated that desalination uses more energy than water supply by using traditional methods, such as the gravity feeding of water from a dam. However, the energy used to provide enough desalinated water daily for a family of four is the same quantity as to operate an air conditioner for just an hour.

AEDYR [21] explained the consumption equivalence to desalinate with the following reasoning: if we bear in mind that the energy consumption of an average home in Spain is 13,141 kWh/year, and the daily average consumption per person is 150 liters/day, taking as a reference that the average energy consumption to produce 1 m^3 of desalinated water is 3 kWh/m^3, with the energy consumption of an average home, 80 people can be supplied with desalinated water for a whole year.

For years, there has been progress in renewable energy production plants, mainly solar and wind, associated with big desalination plants. Kalogirou [28] studied seawater desalination using renewable energy sources. Charcosset [29] provided a state-of-the art review on membrane processes associated with renewable energies for seawater and brackish water desalination. Eltawil [30] provided a review of renewable energy technologies integrated with desalination systems.

Petersen et al. [31] and Lindermann [32] studied wind and solar powered desalination plants for the Mediterranean, Middle East, and Gulf countries. Ghermandi and Messalem [33] provided a state-of-the art on renewable powered seawater desalination plants. Palenzuela et al. [34] valued the use of solar power and desalination plants in arid regions.

Initially, renewable energies were not efficient enough to meet the energy demand of large desalination plants. The technological development produced in recent years allows a wind or solar plant to guarantee the demanded electricity supply. For this reason, it is now common to award construction contracts for seawater desalination plants associated with photovoltaic or wind solar plants. In 2019, ACWA Power [35] awarded the Taweelah desalination plant (reverse osmosis) in the United Arab Emirates with a capacity of 909,000 m^3/day, including the construction of a 40 MW solar photovoltaic plant. In 2020, ACWA Power [36] awarded the construction of the 600,000 m^3/day Jubail 3A osmosis plant in Saudi Arabia, associated with a solar plant.

As another example, Southern Seawater Desalination Plant, SSDP, (Figure 2) located in Binningup, Australia, produces up to 100 billion liters of fresh drinking water a year, around 30% of Perth's water supply. It started production in 2011. The plant is owned by Water Corporation, a public company dependent on the Western Australian Government, which is the main provider of drinking water supply and wastewater treatment services to more than two million people throughout more than 2.6 million square kilometers in Western Australia. Water Corporation [37] says that since production commenced in 2011, WC has purchased energy from a wind farm and a solar farm near Geraldton: Mumbida Wind Farm [38] (55 MW) and Greenough Solar Farm [39] (80 Has, 10 MW). Both the wind and solar farms were developed on the back of a long-term energy purchase agreement associated with the Southern Seawater Desalination Plant.

Figure 2. SSDP, general view. Source: Water Corporation.

Stover [10], a member of the Board of Directors of the International Desalination Association, claims that reverse osmosis is still the dominant technology for desalination. Innovation is promoted to increase freshwater performance, to reduce residual brine, and to deal with harder water sources, because innovation stimulates the growth of desalination in industrial and inland brackish water applications.

3. Environmental Impact of Desalination Plants

There used to be a huge controversy about the environmental impact of desalination plants, but measures taken these past few years have improved the situation and people now accept the technology. The actions taken to minimize the environmental impact have been simultaneously studied with the technological developments done to make desalination plants more efficient. Some countries have passed strict environmental regulations for both environment effect investigation and control of compliance with the corrective measures approved, significantly improving this matter.

However, the type of technology used to desalinate, and the environmental regulations of certain countries or regions can still cause significant impacts. Raluy et al [40] studied life cycle assessment of MSF, MED, and RO; Mezher et al [41] analyzed technoeconomic assessment and environmental impacts of desalination technologies such as MSF, MED, RO, and hybrid MSF/MED-RO; Van der Bruggen and Vandecasteele [42] gave an overview of process evolutions in desalination of seawater by distillation versus membrane filtration; Najafi et al [43] developed environmental cost analysis of MSF, MED, MVC, and RO, making a performance comparison between MSF and RO (Table 4).

Table 4. Performance comparison of multi-stage flash distillation (MSF) and reverse osmosis (RO).

Component	MSF	RO
Recovery percentage	10–20%	30–50%
Investment ($/m$^3 \cdot$day)	1000–1500	1500–7000 (Including 10% for membranes)
Chemicals $/m^3	0.03 to 0.05	0.06 to 0.10
Brine Quality	Chemicals and Heat	Chemicals
Robustness	High	Medium (Problems: fouling sensitive and feed water monitoring)
Improvement Potential	Low	High

Source Najafi et al.

In this respect, Saracco [18] warned of the risk of contamination by brine disposal, which is significantly higher using evaporation as compared to osmosis. Saracco also pointed out the substantial damage done to the marine ecosystem in the Persian Gulf area.

Jones et al. [44] reported that according to their estimates, brine production is around 142 Hm^3/day, approximately 50% higher than had been foreseen. Brine production in Saudi Arabia, the United Arab Emirates, Kuwait, and Qatar represents 55% of the global total, where there are not even emissaries with diffusers or prior dilution. They warn of the need to establish brine management strategies to limit negative environmental impacts and to reduce the economic cost of its disposal, thus stimulating new developments in desalination plants to safeguard water supplies for current and future generations.

According to AEDYR [21], Table 5 shows the salinity of different types of water, indicating the level of salt reduction that desalination plants must achieve.

Nevertheless, there are already countries that are working on the correction of environmental impacts. Spain, Australia, and other developed countries have applied legislation that guarantees the correction of environmental impacts through follow-up programs whose reports are made public.

The environmental impacts and the corresponding corrective measures must be studied as much during the construction of a plant as during the desalination plant lifetime.

In Spain, Law 21/2013 on Environmental Assessment [45] includes the Directives of the European Union in this regard. This law requires an Environmental Impact Statement to be made for each project, which must be approved by the Ministry of the Environment. In the case of desalination plants, ACUAMED, a public company under the Ministry of the Environment, supervises compliance with environmental impact measures.

Table 5. Different salinity water.

Water Type	Salinity (gr/l)
Freshwater	Less than 1 gr/l
Sea water average salinity	35
Sea water	35–45
Brackish water	3–25
Red Sea	42–46
Persian Gulf	40–44
Mediterranean Sea	36–39
Caribbean Sea	34–38
Indian Ocean	33–37
Pacific Ocean	33–36
Atlantic Ocean	33–36
Baltic Sea	6–18
Caspian Sea	12
Dead Sea	350–370

Source AEDYR.

Martínez de la Vallina [46], Environment Director of the Public Company ACUAMED in charge of the construction and operation of desalination plants in Spain, presented a technical communication in the National Environment Congress held in 2008, where he explained the adopted measures to minimize and restore the environmental impacts caused by the construction and operation of desalination plants on the coast of the Mediterranean Sea. Among other things, it said:

> The environmental impact assessment procedure tries to establish the minimum thresholds under which alterations to the environment caused by an action would or would not be acceptable, paying attention not only to the characteristics of the action involved, but also to the environmental conditions–broadly understood- of the area on which action might be needed.

> And in this sense, it has to be underlined that the impact of a desalination plant is not at all more than the residual impact of previous larger human actions, such as the urbanization and extensive occupation of thousands of hectares which lack water resources in quantity and quality enough to meet the demand typical for this accelerated building process.

From the requirements established by ACUAMED and the experience accumulated in the projects developed by the authors, the aspects to consider are indicated.

The most relevant points to bear in mind during construction are:

- Starting from the study phase, the representatives of the communities that live near or within the catchment area of a possible plant location should be included in the decision process that may affect these communities;
- The location of the plant and its integration into the environment. This is always difficult because the plant will necessarily always be situated near the coastline;
- The areas affected by work installations, quarries, landfill sites, etc., in order to consider restoration measures;
- The seawater intake area and its connection to the plant;
- The marine and land fauna which might be either temporarily or permanently affected, which would require studies for corrective measures;

- The marine and land flora affected by the works;
- The connections of the plant with the nearby road network and the effects construction vehicles have on it;
- The connections with the electricity grid and its possible environmental impact;
- The piping connections with the general system supply network.

The most relevant points to consider during the operation of the plant are:

- Maintenance of the adopted measures for the environmental integration of the plant;
- Brine discharge control measures to preserve the marine flora and fauna in the area;
- Purification of reject waters resulting from the treatment of drinking water supplied to the network;
- Conservation and maintenance of the adopted measures not to harm or damage the marine and land fauna;
- Conservation and maintenance of the road network due to the deterioration by the vehicles from the plant;
- Adequate surveillance of the plant's connection pipe network.

The environmental protection measures adopted in Australia are now outlined here for (a) the construction of large desalination plants compatible with sustainable development, including their corrective measures and (b) the monitoring measures to make sure the compliance and outcome of the measures are followed during the operation of the plant.

In the case of Australia during the tender stage for the award and construction of a desalination plant, all the interested companies were given the report and recommendations from the Environmental Protection Authority (EPA), Report 1302 [47]. Regarding the specific desalination plant mentioned above, the Southern Seawater Desalination Project, the Water Corporation transferred this report from the Environmental Protection Authority in Perth, Western Australia.

The report must set out the key environmental factors identified in the course of the assessment, and the EPA's recommendations as to whether or not the proposal may be implemented. If the EPA recommends that implementation be allowed, the conditions and procedures to which implementation should be subject.

The EPA decided that the following key environmental factors relevant to the proposal required detailed evaluation in the report:

(a) Water quality and marine biota—impacts from construction and operation of the desalination plant;
(b) Terrestrial fauna—impacts from clearing of habitat;
(c) Terrestrial vegetation and wetlands—impacts from clearing during infrastructure construction;
(d) Greenhouse gas emissions—proposed no net greenhouse gas emissions.

The following principles were considered by the EPA in relation to the proposal:

(a) The precautionary principle;
(b) The principle of inter-generational equity;
(c) The principle of conservation of biological diversity and ecological integrity; and
(d) The principle of waste minimization.

Having considered the proponent's information provided, the EPA has developed a set of conditions that it recommends be imposed if the proposal by the Water Corporation of Western Australia, to construct and operate an 100 GL per annum reverse osmosis seawater desalination plant at Binningup, and associated infrastructure, is approved for implementation. Matters addressed in the conditions include the following:

(1) Water quality and marine biota;
(2) Marine fauna;

(3) Terrestrial fauna;
(4) Terrestrial flora and vegetation;
(5) Wetlands; and
(6) Greenhouse gas emissions.

For the above-mentioned project, EPA required the set of conditions shown in Table 6.

Table 6. Key environmental factors.

Element	Description
General	
Capacity	50 Gigalitres per year initial capacity 100 Gigalitres per year ultimate capacity
Power requirement	50 Megawatts annual average
Power source	100% renewable energy from Western Power Grid
Clearing of vegetation required	Not more than 20 hectares (at plant site)
Rehabilitation	7 hectares minimum
Offset (rehabilitation)	13 hectares minimum
Seawater intake	
Intake volume	Average 722 Megaliters per day
Length (indicative)	Extending from 400 to 600 m offshore
Number	Up to 4 pipes
Diameter	Up to 3 m
Concentrated seawater discharge	
Discharge volume	418 Megaliters per day (average)
Salinity	Up to 65,000 milligrams per liter
Temperature	Not more than 2 °C above/below ambient seawater
pH	6–8
Length (indicative)	Extending not more than 1100 m offshore
Number	Up to 4 pipes
Diameter	Up to 3 m
Diffuser	Located between 600 and 1100 m offshore and up to 450 m in total length
Sludge	
Sludge production	30 tons per day (approximately)
Water Transfer Pipeline	
Length	30 km (approximately)
Diameter	1400 mm
Destination	Harvey Summit Tank Site
Clearing of native vegetation	Not more than 7 hectares (in pipeline corridor)
Rehabilitation	7 hectares minimum
Harvey Summit Tank Site	
Number of tanks	Up to 4
Capacity of each tank	32 Megaliters
Sump size	2 Megaliters (upgradeable to 5 Megaliters)
Clearing of native vegetation	Not more than 0.1 hectares

Source: Water Corporation.

Water Corporation manages two plants, both located near the open sea. Due to the increased energy, the concentrated seawater discharged during the process mixes very quickly with the surrounding seawater. The discharge and admission pipes on the high seas are designed and located to minimize the effects on sensitive marine habitats, such as seagrasses and reef systems. With regard to the effluent treatment system, the wastewater not assimilable to urban waste is collected and sent to a thickener and subsequent mechanical dehydration by centrifugal pumps.

Christie and Bonnélye [48], in a conference paper at the world congress of the International Desalination Association, held in Dubai in 2009, presented the results of two years of monitoring the operation and environmental impact of the first large desalination plant using reverse osmosis constructed in Australia, the Perth Seawater Desalination Plant (PSDP), with a production capacity of 45 GL/year, which was completed in November 2006 and handed to the Water Corporation in April 2007.

The following conclusions of the study were extracted:

The unprecedented marine monitoring program has included computer modelling for diffuser design and validation, rhodamine dye tracer tests, extensive far field dissolved oxygen tests, a water quality monitoring program, diffuser performance monitoring program, WET testing and Macrobenthic surveys. All studies have proven that the PSDP is having negligible impact on the surrounding environment. Impacts on seawater habitat are limited by a validated diffuser design and treatment of suspended solids.

The power consumption of RO plants is decreasing due to increasing technological gains in plant design, membrane design and energy recovery. RO plants can also easily be powered (offset) by renewable energies. Energy recovery systems such as that used at the PSDP (ERI) are now extremely efficient at recovering energy from the brine wastewater (greater than 96% efficiency). Sourcing power from renewable energy (albeit offset) is an important sustainability principal employed by the PSDP, which is also now being applied by other large-scale Australian desalination plants.

In 2018, the Water Corporation [49] published the 2018 Performance Review Report, which included the result of the environmental monitoring of SSDP during the previous year:

- Environmental factors, risks, and impacts;
- Environmental monitoring plan;
- Environmental behavior;
- Value comparison, best available technology, and improvements in environmental management.

It concludes:

Monitoring results indicate that the SSDP is operating effectively and that the Environmental Quality Objectives for the marine environment are being maintained. Water Corporation has demonstrated that the SSDP has met MS792 criteria for salinity and dissolved oxygen and is achieving the required diffuser performance to meet 99% species protection at the LEPA boundary".

"Seagrass Health Monitoring continues and while a decline in seagrass shoot density was recorded post construction, we have seen a recovery and stabilization of seagrass shoot density over the last three-years of monitoring".

"Water Corporation is in the process of developing a strategy for the purchase of renewable energy and/or carbon offsets for the SSDP".

"Water Corporation and SSWA plan to undertake the following:

- *Continuous Seawater intake and effluent discharge water quality and flow monitoring,*
- *Annual in-situ salinity and diffuser performance monitoring,*
- *Complete the third year of the three-year Seagrass Health Monitoring Program, and*
- *Complete the Swell impact on diffuser performance research project".*

Every year, Water Corporation publishes a detailed monitoring report on the corrective measures of each of the two desalination plants it operates, PSDP [50] and SSDP [51]. The results obtained from the monitoring plan are very positive, since they demonstrate that the natural environment is not affected.

The way to protect the environment developed by countries such as Spain or Australia are examples to be followed by other countries which are developing important desalination projects without adopting the necessary environmental protection.

As a conclusion regarding the environmental impact caused by desalination plants, it can be asserted that by adopting the appropriate corrective measures and monitoring their implementation and effects, said environmental impact is perfectly acceptable due to the enormous benefits of guaranteeing drinking water supplies to populations with severe supply shortage. Moreover, the use of desalinated water frees other sources of supply, allowing the rise of phreatic values, the recharge of aquifers (depleted in many cases by over-exploitation), freeing river intakes, which allows rivers to be recovered as a fish habitat, and the recovery of wetlands and natural lakes.

4. Decision Support System and Type of Contract

A significant increase in the size of the plants has been observed due to technological development and the effect of economies of scale. The bidding process, financing, the selected type of contract, the operation, and maintenance are the key factors for the achievement of the objectives set in the development of a desalination plant. This situation makes these plants increasingly complex, making the choice of contract decisive.

At the end of the 20th century, the type of contract for projects in question was part of the selection process and professional advisors in the sector proposed contract models that helped investors find the best contract for each specific project. These decision support methods or systems are known in the international market as "Decision Support Systems" (DSS). There are numerous authors who have proposed different methods, such as Gordon [52]; Bennett, Pothecary, and Robinson [53]; Molenaar [54]; Konchar and Sanvido [55]; Ibbs et al. [56]; Gransberg, Koch, and Molenaar [57]; Hale et al. [58]; Touran et al. [59]; Park and Kwak [60]; Sullivan et al. [61]; and Jiyong, Wang, and Hu [62].

From among the numerous decision support systems, authors have proposed the method of determining factors [63], published in 2020: (1) client, (2) contractor, (3) contract, (4) budget, (5) financing, (6) risks, and (7) technological developments. It offers a procedure which adapts to any kind of project, specially indicated for large-scale infrastructure projects.

This method consists of the quantitative and qualitative evaluation of the mentioned factors. The qualitative evaluation analyzes the client and contractor capabilities, the suitability of the contract and the assignment of responsibilities, the feasibility of budget compliance and financial availability to meet payments, the risks of the development of the project, and the technological innovations implemented. The quantitative evaluation consists of assigning a score from 1 to 5 to every determining factor, with an increasing value the higher the degree of compliance. The method that will indicate the degree of compliance of each determining factor is applied to each contract modality, so that the higher the score, the more suitable the type of chosen contract will be to develop the project.

The international industrial construction market has great dynamism, which leads those involved in the project to look for new contracting formulas which adapt to their needs, in order to carry out the project. It is normally done by combining other types of contract or modifying the existing ones according to said particular needs. Many authors have written about the most used types of contracts and their differential characteristics, such as El-Wardani, Messner, and Horman [64]; Ohrn and Rogers [65]; Hinze [66]; Chamarro [67]; Franz et al. [68]; and Farnsworth [69].

Therefore, updating the list given by Hernández [70], the most used types of contracts are summarized below:

(a) Traditional contract or "Design then bid" or "Design Bid Build" (DBB). This implies the participation of at least three parties: client, engineering, and contractor. From the legal point of view, it is structured on the conclusion of two contracts: one between client-engineering, and another between client and contractor. This results in the division of the project into design and construction phases. The construction will start once the project is completed. The client uses this method when they feel more confident about a completed project.

(b) Accelerated construction process or "fast track construction". This contract allows construction work to begin before the project has been drawn up completely. It implies a fragmentation of the project into different phases. From the legal point of view, the accelerated construction process can be structured on the basis of: (1) separated contracts between the client and each of the parties involved in engineering and construction; (2) a traditional construction contract; (3) a design and build contract (DB), in which case the contractor will be responsible for the design and the construction; or (4) a Construction Management or Project Management contract. From these contractual formulas, the best suited, considering the fragmentation of the project, are the last two, which will be discussed later.

(c) "Project Management". This attribute the functions of supervision, direction, and coordination of the project as a whole to an entity, called the Project Manager, which through their services, tends to get maximum control and a consequent reduction in the time and costs involved in the execution of the work. Compared to the traditional method, it involves the participation in the process of a fourth party, which assumes certain functions which are usually attributed to the client, the engineer and/or the contractor.

(d) "Design and build" (DB) and turnkey contracts imply a progressive expansion of the obligations assumed by the contractor. It involves that the general contractor is tasked with gathering a group of designers and constructors to carry out the work. In the late twentieth century, English/American law had a great influence in the international arena of contracts. Thus, through the design and build contract, the contractor undertakes to conceive and execute the industrial project in accordance with the needs and requirements of the client. Therefore, the benefits derived from this contract are limited to the construction operation itself. Obligations outside the contract, such as the commissioning of the installation or the training of personnel, are not included in its content, as in the case of the turnkey contracts.

Based on this difference in content between the two contracts, it can be claimed that while a design and build contract can never be equated to a turnkey contract, a turnkey contract, however, always includes the obligations derived from the design and build contract—something that is understandable, considering that, in the international arena, the design and build contract has served as the basis for the configuration of international turnkey contracts. Similarly, both contracts have been the benchmark for the creation of other contractual forms, such as the turnkey product contract and the turnkey market contract.

(e) BOT and BOOT projects. The formula BOT (Build-Operate and Transfer) as well as the contracts called BOOT (Build-Own-Operate and Transfer) are different mechanisms used to finance projects (Project Financing) of mainly infrastructures or public works, through which the public sector has been transferring to the private sector. Traditionally, this was an activity in which the public sector took charge drawing on its own financing.

(f) "Engineering" contracts. Through these contracts, a party (the engineering company) subcontracts another party (called client) to develop manage and supervise a project, and when agreed, depending of circumstances, other obligations, including the maintenance and management of the finished work. Despite the fact that in business practice, the engineering contracts may include different content, a common denominator is observed: they all have an obligation for results. Be it simple or global, whether it involves the development of a project and/or its execution. The contractual compliance is borne by a business entity in which the activity of the individual professional is depersonalized.

(g) "Engineering, Procurement, and Construction Management", also known by its acronym EPCM, which means that in this type of Engineering Contract the Contractor will provide the Client with Engineering Services, Purchasing Management, and Construction Management. In the EPCM contract, the Contractor develops engineering, processes acquisition, and manages the work on behalf of the Client, but does not construct the project. The Contractor, thus, becomes a representative, working hand in hand with the Client, and managing the contractual relations with Suppliers and Contractors. In this way, the Client will be ultimately responsible for the acquisitions and approve all contracts. Thus, the construction risks fall on the client.

(h) "Open Book Estimation", OBE. An OBE contract consists of an agreement between Property/Client and Contractor to carry out work in which the costs are reimbursable to the latter, plus a previously agreed margin. Client and Contractor also agree on how to pay for the works. This type of contract has become important in recent years and has been generalized as a previous phase to the award of a turnkey contract for large industrial installations.

(i) "Progressive Design-Build", PDB, is an emerging variation of alternative contracting methods, which allows the client to hire a project and construction contractor without a price commitment until reasonable design details are defined. They have been used for water treatment plants and for airports. The critical issues are: what responsibilities are transferred to the contractor, the need for the client participation during the design and the provision of cost saving measures that do not jeopardize the quality of the project. Gad el al. [71] and Gransberg and Molenaar [72] have studied this new type of contract.

In the area of desalination, the turnkey contract stands out as the most proven and efficient tool for the development of projects, offering the following advantages:

- There is a single contractor responsible for design and work, so that engineering and construction can be developed in parallel, thus shortening deadlines;
- The dialogue is limited to client/contractor;
- The global assumption of responsibility includes not only the quality requirements established by contract, but also the proposed new technological developments, and consequently, changed or modified orders which generate deviations of deadlines and budget are eliminated or reduced.

It is advisable to introduce a phase in the modality of open book estimation, prior to hiring, that allows designs and prices to be adjusted and agreed, which will reduce project risks, contingencies and deviations during the construction, leading to a better final result. This modality has been successfully offering the solutions given in the mentioned PDB.

5. Conclusions

In the present paper, we have defined: the most efficient desalination technology, energy supply sources, corrective measures for environmental impacts, and the most suitable type of contract for the construction of large desalination plants compatible with sustainable development. A discussion was also included for every corresponding point of each of the analyzed factors, their evolution, and present situation.

Considering that the main question is to guarantee the supply of water in quantity, quality, and safety to millions of people, the proposals made below offer sufficiently argued solutions. The technological developments in this field have evolved and still do very fast, so it is reasonable to expect significant improvements in the near future.

However, today, the recommendations and conclusions are the following:

1. The technology to adopt for desalination is reverse osmosis, considering the several stages and energy recovery measures and opportunities available along the process. In addition, adopting the latest generation membranes is essential to achieve the best efficiency;
2. Regarding the consumption and production of the energy necessary for desalination, the proposed technology, including all the measures to improve efficiency, offers the safest possible means for desalination. Consumption is getting below 2.7 kWh/m^3 through reverse osmosis. It is recommended to associate with the construction of these large desalination plants renewable

energy production parks, fundamentally solar or wind farms; even if it is utopian to think of a plant today producing hundreds of thousands of cubic meters a day, through the exclusive supply of renewable energies;

3. From the environmental point of view, it is essential that the construction of any desalination plant, wherever it may be, should include an environmental impact study during the construction of said plant, and also include a monitoring plan that guarantees the corrective measures and the possibility of adopting new ones if impacts on the environment were detected. This monitoring must be guaranteed by an independent body from the plant operating company and must be published regularly with the supporting documentation of the results obtained during the follow-up.

4. As for the development management of these large infrastructures, it is recommended to use one of the decision support systems that justifies the chosen contractual modality for the project and construction. Today, the best contract to achieve the objective of big, complex, and expensive projects is the turnkey contract. The advantages are mainly that turnkey contracts shorten deadlines when combining design and construction and avoid or even reduce extra costs because of the closed price formula. It is, however, advisable to have an Open Book Estimation (OBE) phase before finally agreeing the binding contract.

5. A significant cost reduction in RO is possible in the short term if the working pressure can be reduced without the membranes losing efficiency.

Author Contributions: Author Contributions: Conceptualization, F.B.-F.; Investigation, F.B.-F.; Methodology, F.B.-F., A.L.-G.; Writing—original draft and editing, F.B.-F.; Formal analysis, F.B.-F.; Writing—review, A.L.-G.; Supervision, A.L.-G., M.B.M.-M.; Validation, M.B.M.-M. All authors have read and agreed to the published version of the manuscript.

Funding: This research received no external funding.

Conflicts of Interest: The authors declare no conflict of interest.

References

1. Brundtland Commission. The World Commission on Environment and Development. In *Our Common Future*; Hauff, V., Ed.; Oxford University Press: Oxford, New York, NY, USA, 1987; ISBN 9780192820808.

2. Jonker, G.; Harmsen, J. *Engineering for Sustainability. A Practical Guide for Sustainable Design*; Elsevier BV: Amsterdam, The Netherlands, 2012; ISBN 9780444538468.

3. Siegel, S.M. *Let There Be Water. Israel's Solution for a Water-Starved World*; Thomas Dunne Books: New York, NY, USA, 2015; ISBN 978-1250073952.

4. Bates Ramírez, V. Inching towards Abundant Water: New Progress in Desalination Tech, Singularity Hub. Available online: https://singularityhub.com/2019/06/18/inching-towards-abundant-water-new-progress-in-desalination-tech/ (accessed on 30 January 2020).

5. Universidad Complutense de Madrid. El Agua en la Tierra. Available online: https://webs.ucm.es/info/diciex/proyectos/agua/El_agua_en_la_tierra.html (accessed on 19 June 2020).

6. AQUASTAT FAO's Global Information System on Water and Agriculture. Available online: http://www.fao.org/aquastat/en/overview/methodology/water-use (accessed on 20 May 2020).

7. International Desalination Association. Desalination by the Numbers. Available online: https://idadesal.org/ (accessed on 20 May 2020).

8. Urrutia, F. Evolución Global de la Capacidad Instalada de Plantas Desalinizadoras. *Ingeniería y Territorio* **2005**, *72*, 68.

9. Torres Corral, M. *Avances Técnicos en la Desalación de Aguas*; Ambienta, Ministerio de Medio Ambiente: Madrid, Spain, 2004; pp. 18–19.

10. Stover, R.L. Innovation Leads the Way to Solving Desalination Challenges. Water Online. Available online: https://www.wateronline.com/doc/innovation-leads-the-way-to-solving-desalination-challenges-0001 (accessed on 23 May 2020).

11. Scott, K. *Handbook of Industrial Membranes*; Elsevier Advanced Technology: Oxford, UK, 1995.

12. Ibáñez Mengual, J.A. *Desalación de Aguas. Aspectos Tecnológicos, Medioambientales, Jurídicos y Económicos*; Editorial Fundación Instituto Euromediterráneo del Agua: Espinardo, Murcia, Spain, 2009; ISBN 978-84-936326-6-3.

13. Voutchkov, N. Desalination-Past, Present and Future. Available online: https://iwa-network.org/desalination-past-present-future/ (accessed on 19 May 2020).

14. Lattemann, S.; Höpner, T. Environmental impact and impact assessment of seawater desalination. *Desalination* **2008**, *220*, 1–15. [CrossRef]

15. Kämpf, J.; Clarke, B. How robust is the environmental impact assessment process in South Australia? Behind the scenes of the Adelaide seawater desalination project. *Mar. Policy* **2013**, *38*, 500–506.

16. Fuentes-Bargues, J.L. Analysis of the Process. of environmental impact assessment for seawater desalination plants in Spain. *Desalination ELSEVIER* **2014**, *347*, 166–174. [CrossRef]

17. Shemer, H.; Semiat, R. Sustainable RO desalination—Energy demand and environmental impact. *Desalination* **2017**, *242*, 10–16. Available online: https://doi.org/10.1016/j.desal.2017.09.021 (accessed on 19 May 2020). [CrossRef]

18. Saracco, R. Desalination Plants Ask for Tech Evolution. Available online: https://cmte.ieee.org/futuredirections/2019/01/19/desalination-plants-ask-for-tech-evolution/ (accessed on 20 May 2020).

19. Parlar, I.; Hacıfazlıoğlu, M.; Kabay, N.; Pek, M.; Yüksel, M. Performance comparison of reverse osmosis (RO) with integrated nanofiltration (NF) and reverse osmosis process for desalination of MBR effluent. *J. Water Process. Eng.* **2019**, *29*, 100640. Available online: https://doi.org/10.1016/j.jwpe.2018.06.002 (accessed on 20 May 2020). [CrossRef]

20. Lee, J.J.; Kang, J.S. Treatment of reverse osmosis concentrate by electrolysis and MBR process. *Desalination Water Treat.* **2015**, *57*, 1–9. [CrossRef]

21. Asociación Española de Desalación y Reutilización, AEDyR. *Desalination Figures in Spain*; AEDyR: Madrid, Spain, 2019; Available online: http://www.aedyr.com (accessed on 20 February 2020).

22. Urban, J.; Chao, J. Moving Forward on Desalination. Lawrence Berkeley National Laboratory (Berkeley Lab) 2019. Available online: https://newscenter.lbl.gov/2019/07/31/moving-forward-on-desalination/ (accessed on 20 February 2020).

23. De Luis López, M.; Gómez Benítez, M.A. La desalación mediante energía solar como fuente de recursos hídricos. *Colección Informes* **2015**, *111*, 6.

24. Subramani, A.; Jacangelo, J.G. Emerging desalination technologies for water treatment: A critical review. *Water Res.* **2015**, *75*, 164–187. Available online: https://doi.org/10.1016/j.watres.2015.02.032 (accessed on 21 May 2020). [CrossRef]

25. Estevan, A.; García, M. El consumo de energía en la desalación de agua de mar por ósmosis inversa: Situación actual y perspectivas. *Ing. Civ.* **2007**, *148*, 113–121.

26. Jia, X.; Jaromír, J.; Sabev, P.; Wan Alwi, S.R. Analysing the Energy Consumption, GHG Emission, and Cost of Seawater Desalination in China. *Energies* **2019**, *12*, 463. [CrossRef]

27. Water Corporation. *Southern Seawater Desalination Project*; Geon: Perth, Australia, 2011; p. 9. ISBN 1-74043-726-8.

28. Kalogirou, S.A. Seawater desalination using renewable energy sources. *Prog. Energy Combust. Sci.* **2005**, *31*, 242–281. [CrossRef]

29. Charcosset, C. A review of membrane processes and renewable energies for desalination. *Desalination* **2009**, *245*, 214–231. [CrossRef]

30. Eltawil, M.; Zhengming, Z.; Yuan, L. A review of renewable energy technologies integrated with desalination systems. *Renew. Sustain. Energy Rev.* **2009**, *13*, 2245–2262. [CrossRef]

31. Petersen, G.; Fries, S.; Mohn, J.; Müller, A. Wind and solar-powered reverse osmosis desalination units-description of two demonstration projects. *Desalination* **1979**, *31*, 501–509. [CrossRef]

32. Lindermann, J.H. Wind and solar powered seawater desalination applied solutions for the Mediterranean, the Middle East. and the Gulf countries. *Desalination* **2004**, *168*, 73–80. [CrossRef]

33. Ghermandi, A.; Messalem, R. Solar-driven desalination with reverse osmosis: The state of the art. *Desalination Water Treat.* **2009**, *7*, 285–296. [CrossRef]

34. Palenzuela, P.; Zaragoza, G.; Alarcon-Padilla, D.C.; Guillen, E.; Ibarra, M.; Blanco, J. Assessment of different configurations for combined parabolic-trough (PT) solar power and desalination plants in arid regions. *Energy* **2011**, *36*, 4950–4958. [CrossRef]

35. ACWA Power. Taweelah Independent Water Plant (IWP). Available online: https://www.acwapower.com/news/taweelah-iwp-obtains-the-first-ever-sustainable-loan-qualification-for-a-desalination-project/ (accessed on 22 May 2020).

36. ACWA Power. Jubail 3A Independent Water Plant (IWP). Available online: https://www.acwapower.com/en/projects/jubail-3a-iwp/ (accessed on 22 May 2020).

37. Water Corporation. Southern Seawater Desalination Project. Social Impact Management Plan. 2009. Available online: https://pw-cdn.watercorporation.com.au/-/media/WaterCorp/Documents/Our-Water/Desalination/ssdp-social-impact-management-plan.pdf?rev=d58321c9628441909baa0697213a8e33&hash=CCCACB07C0537EDB352A8183369198CB (accessed on 19 June 2020).

38. Mumbida Wind Farm. Available online: https://mumbidawindfarm.com.au/ (accessed on 20 May 2020).

39. Greenough Solar Farm. Available online: https://www.brightenergyinvestments.com.au/greenough-river-solar-farm (accessed on 20 May 2020).

40. Raluy, G.; Serra, L.; Uche, J. Life cycle assessment of MSF, MED and RO desalination technologies. *Energy* **2006**, *31*, 2361–2372. [CrossRef]

41. Mezher, T.; Fath, H.; Abbas, Z.; Khaled, A. Techno-economic assessment and environmental impacts of desalination technologies. *Desalination* **2011**, *266*, 263–273. [CrossRef]

42. Van der Bruggen, B.; Vandecasteele, C. Distillation vs. membrane filtration: Overview of Process. evolutions in Seawater desalination. *Desalination* **2002**, *143*, 207–2018. [CrossRef]

43. Najafi, F.T.; Alsaffar, M.; Schwerer, S.C.; Brown, N. Environmental Impact Cost Analysis of Multi-Stage Flash, Multi-Effect Distillation, Mechanical Vapor Compression, and Reverse Osmosis Medium-Size Desalination Facilities. In Proceedings of the ASEE's 123rd annual Conference & Exposition, New Orleans, LA, USA, 26–29 June 2016.

44. Jones, E.; Qadir, M.; van Vliet, M.T.H.; Kang, S.-M. The state of desalination and brine production: A global outlook. *Sci. Total Environ.* **2019**, *657*, 1343–1356. Available online: https://doi.org/10.1016/j.scitotenv.2018.12.076 (accessed on 21 May 2020). [CrossRef] [PubMed]

45. Ley 21/2013 de Evaluación Ambiental. B.O.E. n° 296. Available online: https://www.boe.es/eli/es/l/2013/12/09/21/con (accessed on 19 June 2020).

46. Martínez de la Vallina, J.J. *Evaluación del Impacto Ambiental de las Desaladoras. Congreso Nacional del Medio Ambiente*; Cumbre del Desarrollo Sostenible: Madrid, Spain, 2008; pp. 5–17.

47. Environmental Protection Authority EPA. Southern Seawater Desalination Project. Report and Recommendations of the EPA. 2008. Available online: http://www.epa.wa.gov.au/sites/default/files/EPA_Report/2797_Rep1302Desal_61008.pdf (accessed on 22 May 2020).

48. Christie, S.; Bonnélye, V. Two-year Feed Back on Operation and Environmental Impact. In Proceedings of the IDA World Congress, Atlantis, Dubai, UAE, 7–12 November 2009; p. 14.

49. Water Corporation. Southern Seawater Desalination Project. 2018 Performance Review Report. Available online: https://www.watercorporation.com.au/Our-water/Desalination/Southern-Seawater-Desalination-Plant (accessed on 19 June 2020).

50. Water Corporation. Perth Seawater Desalination Plant. Compliance Assessment Report Ministerial Statements 655 & 832, 01 July 2017 to 30 June 2018. Available online: https://www.watercorporation.com.au/-/media/files/residential/watersupply/desalination/psdp/psdp-compliance-assessment-report-2017-18.pdf (accessed on 21 May 2020).

51. Water Corporation. Southern Seawater Desalination Plant. Performance and Compliance Report 14 April 2018–13 April 2019. Available online: https://www.watercorporation.com.au/-/media/files/residential/watersupply/desalination/ssdp/ssdp-performance-compliance-report-2018-2019.pdf (accessed on 21 May 2020).

52. Gordon, C.M. Choosing Appropriate Construction Contracting Method. *J. Constr. Eng. Manag. (ASCE)* **1994**, *120*, 196. [CrossRef]

53. Bennett, J.; Pothecary, E.; Robinson, G. *Designing and Building a World-Class Industry*; University of Reading: Reading, UK, 1996; ISBN 9780704911703.

54. Molenaar, K.R. *Public Sector Design-Build: A Model. for Project Selection*; University of Colorado, Civil Engineering: Boulder, CO, USA, 1997.

55. Konchar, M.D.; Sanvido, V.E. Comparison of U.S. Project Delivery Systems. *J. Constr. Eng. Manag. (ASCE)* **1998**, *124*, 435–444. [CrossRef]

56. Ibbs, C.W.; Wong, C.K.; Kwak, Y.H.; Ng, T. Project Delivery Systems and Project Change: Quantitative Analysis. *J. Constr. Eng. Manag. (ASCE)* **2003**, *129*, 328–386. [CrossRef]

57. Gransberg, D.D.; Koch, J.E.; Molenaar, K.R. *Preparing for Design-Build Projects: A Primer for Owners, Engineers, and Contractors*; American Society of Civil Engineers ASCE: Reston, VA, USA, 2006; pp. 13–16.

58. Hale, D.R.; Shrestha, P.P.; Gibson, G.E.; Migliaccio, G.C. Empirical Comparison of Design/Build. and Design/Bid/Build. Project Delivery Methods. *J. Constr. Eng. Manag. (ASCE)* **2009**, *135*, 579–587. [CrossRef]

59. Touran, A.; Gransberg, D.D.; Molenaar, K.R.; Ghavamifar, K.; Mason, D.J.; Fithian, L.A. *A Guidebook for the Evaluation of Project Delivery Methods*; TCRP Report 131; Transportation Research Board, Transit Cooperative Research Program: Washington, DC, USA, 2009.

60. Park, J.; Kwak, Y.H. Design-bid-build (DBB) vs. design-build (db) in the U.S. public transportation projects: The choice and consequences. *Int. J. Proj. Manag.* **2019**, *35*, 280–295. [CrossRef]

61. Sullivan, J.; El Asmar, M.; Chalhoub, J.; Obeid, H. Two decades of performance comparisons for Design-Build., Construction Manager at Risk and Design-Bid-Build.: Quantitative analysis of the state of knowledge about cost, schedule, and project quality. *J. Constr. Eng. Manag. (ASCE)* **2019**, *143*, 04017009. [CrossRef]

62. Jiyong, D.; Wang, N.; Hu, L. Framework for Designing Project Delivery and Contract Strategy in Chinese Construction Industry Based on Value-Added Analysis. *Adv. Civ. Eng.* **2018**, *12*, 1–14.

63. Berenguel-Felices, F.; Lara-Galera, A.; Guirao-Abad, B.; Galindo-Aires, R. Contracting Formulas for Large Engineering Projects. The Case of Desalination Plants. *Sustainability* **2020**, *12*, 219. Available online: https://doi.org/10.3390/su12010219 (accessed on 20 May 2020). [CrossRef]

64. El-Wardani, M.; Messner, J.I.; Horman, M.J. Comparing Procurement Methods for Design-Build Projects. *J. Constr. Eng. Manag. (ASCE)* **2005**, *132*, 230–238. [CrossRef]

65. Ohrn, L.G.; Rogers, T. *Defining Project Delivery Methods for Design, Construction, and Other Construction-Related Services in the United States*; Northern Arizona University: Flagstaff, AZ, USA, 2008; p. 5.

66. Hinze, J. *Construction Contracts 3rd Edition*; McGraw-Hill: Columbus, OH, USA, 2010; ISBN 978-0073397856.

67. Chamarro, I. Contratos internacionales de construcción llave en mano. 2012. Available online: http://www.interempresas.net (accessed on 6 May 2020).

68. Franz, B.; Leicht, R.; Molenaar, K.R.; Messner, J. Impact of team integration and group cohesion on project delivery performance. *J. Constr. Eng. Manag.* **2016**, *143*, 04016088. [CrossRef]

69. Farnsworth, C.B.; Warr, R.O.; Weidman, J.E.; Hutchings, D.M. Effects of CM/GC project delivery on managing process risk in transportation construction. *J. Constr. Eng. Manag.* **2016**, *142*, 04015091. [CrossRef]

70. Hernández Rodríguez, A. Los Contratos Internacionales de Construcción «Llave en Mano». *Cuad. Derecho Transnacional* **2014**, *6*, 161–235.

71. Gad, G.M.; Davis, B.; Shrestha, P.P.; Harder, P. Lessons Learned from Progressive Design-Build. Implementation on Airport Projects. *J. Leg. Aff. Disput. Resolut. Eng. Constr. ASCE* **2019**, *11*, 8. [CrossRef]

72. Gransberg, D.D.; Molenaar, K.R. Critical Comparison of Progressive Design-Build and Construction Manager/General Contractor Project Delivery Methods. *Transp. Res. Record J. Transp. Res. Board Natl. Acad.* **2019**, 261–268. [CrossRef]

MDPI

St. Alban-Anlage 66

4052 Basel

Switzerland

Tel. +41 61 683 77 34

Fax +41 61 302 89 18

www.mdpi.com

Sustainability Editorial Office

E-mail: sustainability@mdpi.com

www.mdpi.com/journal/sustainability